Modern Control Design

With MATLAB and SIMULINK

Ashish Tewari
Indian Institute of Technology, Kanpur, India

JOHN WILEY & SONS, LTD

Copyright © 2002 by John Wiley & Sons Ltd
Baffins Lane, Chichester,
West Sussex, PO19 1UD, England
National   01243 779777
International (+44) 1243 779777
e-mail (for orders and customer service enquiries): cs-books@wiley.co.uk
Visit our Home Page on http://www.wiley.co.uk
or http://www.wiley.com

Other Wiley Editorial Offices

John Wiley & Sons, Inc., 605 Third Avenue,
New York, NY 10158-0012, USA

Wiley-VCH Verlag GmbH, Pappelallee 3,
D-69469 Weinheim, Germany

John Wiley, Australia, Ltd, 33 Park Road, Milton,
Queensland 4064, Australia

John Wiley & Sons (Canada) Ltd, 22 Worcester Road,
Rexdale, Ontario, M9W 1L1, Canada

John Wiley & Sons (Asia) Pte Ltd, 2 Clementi Loop #02-01,
Jin Xing Distripark, Singapore 129809

British Library Cataloguing in Publication Data

A catalogue record for this book is available from the British Library

ISBN 0 471 496790

Typeset in 10/12$\frac{1}{2}$ pt Times by Laserwords Private Limited, Chennai, India
Printed and bound in Great Britain by Biddles Ltd, Guildford and Kings Lynn
This book is printed on acid-free paper responsibly manufactured from sustainable forestry,
in which at least two trees are planted for each one used for paper production.

To the memory of my father,
Dr. Kamaleshwar Sahai Tewari.
To my wife, *Prachi*, and daughter, *Manya*.

Contents

Preface

The motivation for writing this book can be ascribed chiefly to the usual struggle of an average reader to understand and utilize controls concepts, without getting lost in the mathematics. Many textbooks are available on modern control, which do a fine job of presenting the control theory. However, an introductory text on modern control usually stops short of the really useful concepts – such as optimal control and Kalman filters – while an advanced text which covers these topics assumes too much mathematical background of the reader. Furthermore, the examples and exercises contained in many control theory textbooks are too simple to represent modern control applications, because of the computational complexity involved in solving practical problems. This book aims at introducing the reader to the basic concepts and applications of modern control theory in an easy to read manner, while covering in detail what may be normally considered advanced topics, such as multivariable state-space design, solutions to time-varying and nonlinear state-equations, optimal control, Kalman filters, robust control, and digital control. An effort is made to explain the underlying principles behind many controls concepts. The numerical examples and exercises are chosen to represent practical problems in modern control. Perhaps the greatest distinguishing feature of this book is the ready and extensive use of MATLAB (with its *Control System Toolbox*) and SIMULINK®, as practical computational tools to solve problems across the spectrum of modern control. MATLAB/SIMULINK combination has become the single most common – and industry-wide standard – software in the analysis and design of modern control systems. In giving the reader a *hands-on* experience with the MATLAB/SIMULINK and the *Control System Toolbox* as applied to some practical design problems, the book is useful for a practicing engineer, apart from being an introductory text for the beginner.

This book can be used as a textbook in an introductory course on control systems at the third, or fourth year undergraduate level. As stated above, another objective of the book is to make it readable by a practicing engineer *without* a formal controls background. Many modern control applications are *interdisciplinary* in nature, and people from a variety of disciplines are interested in applying control theory to solve practical problems in their own respective fields. Bearing this in mind, the examples and exercises are taken to cover as many different areas as possible, such as aerospace, chemical, electrical and mechanical applications. Continuity in reading is preserved, without frequently referring to an appendix, or other distractions. At the end of each chapter, readers are

® MATLAB, SIMULINK, and *Control System Toolbox* are registered trademarks of the Math Works, Inc.

given a number of exercises, in order to consolidate their grasp of the material presented in the chapter. Answers to selected numerical exercises are provided near the end of the book.

While the main focus of the material presented in the book is on the state-space methods applied to linear, time-invariant control – which forms a majority of modern control applications – the classical frequency domain control design and analysis is not neglected, and large parts of Chapters 2 and 8 cover classical control. Most of the example problems are solved with MATLAB/SIMULINK, using MATLAB *command lines*, and SIMULINK block-diagrams immediately followed by their resulting outputs. The reader can directly reproduce the MATLAB statements and SIMULINK blocks presented in the text to obtain the same results. Also presented are a number of computer programs in the form of *new* MATLAB M-files (i.e. the M-files which are *not* included with MATLAB, or the *Control System Toolbox*) to solve a variety of problems ranging from *step* and *impulse* responses of single-input, single-output systems, to the solution of the *matrix Riccati equation* for the terminal-time weighted, multivariable, optimal control design. This is perhaps the only available controls textbook which gives ready computer programs to solve such a wide range of problems. The reader becomes aware of the power of MATLAB/SIMULINK in going through the examples presented in the book, and gets a good exposure to programming in MATLAB/SIMULINK. The numerical examples presented require MATLAB 6.0, SIMULINK 4.0, and *Control System Toolbox 5.0*. Older versions of this software can also be adapted to run the examples and models presented in the book, with some modifications (refer to the respective *Users' Manuals*).

The numerical examples in the book through MATLAB/SIMULINK and the *Control System Toolbox* have been designed to prevent the use of the software as a *black box*, or by rote. The theoretical background and numerical techniques behind the software commands are explained in the text, so that readers can write their own programs in MATLAB, or another language. Many of the examples contain instructions on programming. It is also explained how many of the important *Control System Toolbox* commands can be replaced by a set of intrinsic MATLAB commands. This is to avoid over-dependence on a particular version of the *Control System Toolbox*, which is frequently updated with new features. After going through the book, readers are better equipped to learn the advanced features of the software for design applications.

Readers are introduced to advanced topics such as H_∞-robust optimal control, structured singular value synthesis, input shaping, rate-weighted optimal control, and nonlinear control in the final chapter of the book. Since the book is intended to be of introductory rather than exhaustive nature, the reader is referred to other articles that cover these advanced topics in detail.

I am grateful to the editorial and production staff at the Wiley college group, Chichester, who diligently worked with many aspects of the book. I would like to specially thank Karen Mossman, Gemma Quilter, Simon Plumtree, Robert Hambrook, Dawn Booth and See Hanson for their encouragement and guidance in the preparation of the manuscript. I found working with Wiley, Chichester, a pleasant experience, and an education into the many aspects of writing and publishing a textbook. I would also like to thank my students and colleagues, who encouraged and inspired me to write this book. I thank all the reviewers for finding the errors in the draft manuscript, and for providing many constructive suggestions. Writing this book would have been impossible without the constant support of my wife, Prachi, and my little daughter, Manya, whose total age in months closely followed the number of chapters as they were being written.

Ashish Tewari

1

Introduction

1.1 What is Control?

When we use the word *control* in everyday life, we are referring to the act of producing a desired result. By this broad definition, control is seen to cover all artificial processes. The temperature inside a refrigerator is controlled by a thermostat. The picture we see on the television is a result of a controlled beam of electrons made to scan the television screen in a selected pattern. A compact-disc player focuses a fine laser beam at the desired spot on the rotating compact-disc in order to produce the desired music. While driving a car, the driver is controlling the speed and direction of the car so as to reach the destination quickly, without hitting anything on the way. The list is endless. Whether the control is automatic (such as in the refrigerator, television or compact-disc player), or caused by a human being (such as the car driver), it is an integral part of our daily existence. However, control is not confined to artificial processes alone. Imagine living in a world where the temperature is unbearably hot (or cold), without the life-supporting oxygen, water or sunlight. We often do not realize how controlled the natural environment we live in is. The composition, temperature and pressure of the earth's atmosphere are kept stable in their livable state by an intricate set of natural processes. The daily variation of temperature caused by the sun controls the metabolism of all living organisms. Even the simplest life form is sustained by unimaginably complex chemical processes. The ultimate control system is the human body, where the controlling mechanism is so complex that even while sleeping, the brain regulates the heartbeat, body temperature and blood-pressure by countless chemical and electrical impulses per second, in a way not quite understood yet. (You have to wonder who designed *that* control system!) Hence, control is everywhere we look, and is crucial for the existence of life itself.

A study of control involves developing a mathematical model for each component of the control system. We have twice used the word *system* without defining it. A system is a set of self-contained processes under study. A *control system* by definition consists of the system to be controlled – called the *plant* – as well as the system which exercises control over the plant, called the *controller*. A controller could be either human, or an artificial device. The controller is said to supply a signal to the plant, called the *input to the plant* (or the *control input*), in order to produce a desired response from the plant, called the *output from the plant*. When referring to an isolated system, the terms *input* and *output* are used to describe the signal that goes into a system, and the signal that comes out of a system, respectively. Let us take the example of the control system consisting of a car and its driver. If we select the car to be the plant, then the driver becomes the controller, who applies an input to the plant in the form of pressing the gas pedal if it is desired to increase the speed of the car. The speed increase can then be the output from the plant. Note that in a control system, what control input can be applied to the plant is determined by the physical processes of the plant (in this case, the car's engine), but the output could be anything that can be directly measured (such as the car's speed or its position). In other words, many different choices of the output can be available at the same time, and the controller can use any number of them, depending upon the application. Say if the driver wants to make sure she is obeying the highway speed limit, she will be focusing on the speedometer. Hence, the speed becomes the plant output. If she wants to stop well before a stop sign, the car's position with respect to the stop sign becomes the plant output. If the driver is overtaking a truck on the highway, both the speed and the position of the car *vis-á-vis* the truck are the plant outputs. Since the plant output is the same as the output of the control system, it is simply called the *output* when referring to the control system as a whole. After understanding the basic terminology of the control system, let us now move on to see what different varieties of control systems there are.

1.2 Open-Loop and Closed-Loop Control Systems

Let us return to the example of the car driver control system. We have encountered the not so rare breed of drivers who generally boast of their driving skills with the following words: "Oh I am so good that I can drive this car with my eyes closed!" Let us imagine we give such a driver an opportunity to live up to that boast (without riding with her, of course) and apply a blindfold. Now ask the driver to accelerate to a particular speed (assuming that she continues driving in a straight line). While driving in this fashion, the driver has absolutely no idea about what her actual speed is. By pressing the gas pedal (control input) she hopes that the car's speed will come up to the desired value, but has no means of verifying the actual increase in speed. Such a control system, in which the control input is applied without the knowledge of the plant output, is called an *open-loop control system*. Figure 1.1 shows a *block-diagram* of an open-loop control system, where the sub-systems (controller and plant) are shown as rectangular blocks, with arrows indicating input and output to each block. By now it must be clear that an open-loop controller is like a rifle shooter who gets only one shot at the target. Hence, open-loop control will be successful only if the controller has a pretty good prior knowledge of the *behavior* of the plant, which can be defined as the relationship between the control input

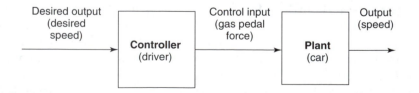

Figure 1.1 An open-loop control system: the controller applies the control input without knowing the plant output

and the plant output. If one knows what output a system will produce when a known input is applied to it, one is said to know the system's behavior.

Mathematically, the relationship between the output of a *linear* plant and the control input (the system's behavior) can be described by a *transfer function* (the concepts of linear systems and transfer functions are explained in Chapter 2). Suppose the driver knows from previous driving experience that, to maintain a speed of 50 kilometers per hour, she needs to apply one kilogram of force on the gas pedal. Then the car's transfer function is said to be 50 km/hr/kg. (This is a very simplified example. *The actual car is not going to have such a simple transfer function.*) Now, if the driver can accurately control the force exerted on the gas pedal, she can be quite confident of achieving her target speed, even though blindfolded. However, as anybody reasonably experienced with driving knows, there are many uncertainties – such as the condition of the road, tyre pressure, the condition of the engine, or even the uncertainty in gas pedal force actually being applied by the driver – which can cause a change in the car's behavior. If the transfer function in the driver's mind was determined on smooth roads, with properly inflated tyres and a well maintained engine, she is going to get a speed of less than 50 km/hr with 1 kg force on the gas pedal if, say, the road she is driving on happens to have rough patches. In addition, if a wind happens to be blowing opposite to the car's direction of motion, a further change in the car's behavior will be produced. Such an unknown and undesirable input to the plant, such as road roughness or the head-wind, is called a *noise*. In the presence of uncertainty about the plant's behavior, or due to a noise (or both), it is clear from the above example that an open-loop control system is unlikely to be successful.

Suppose the driver decides to drive the car like a sane person (i.e. with both eyes wide open). Now she can see her actual speed, as measured by the speedometer. In this situation, the driver can adjust the force she applies to the pedal so as to get the desired speed on the speedometer; it may not be a one shot approach, and some trial and error might be required, causing the speed to initially overshoot or undershoot the desired value. However, after some time (depending on the ability of the driver), the target speed can be achieved (if it is within the capability of the car), irrespective of the condition of the road or the presence of a wind. Note that now the driver – instead of applying a pre-determined control input as in the open-loop case – is adjusting the control input according to the actual observed output. Such a control system in which the control input is a function of the plant's output is called a *closed-loop system*. Since in a closed-loop system the controller is constantly in touch with the actual output, it is likely to succeed in achieving the desired output even in the presence of noise and/or uncertainty in the linear plant's behavior (transfer-function). The mechanism by which the information about the actual output is conveyed to the controller is called *feedback*. On a block-diagram, the path from the plant output to the controller input is called a *feedback-loop*. A block-diagram example of a possible closed-loop system is given in Figure 1.2.

Comparing Figures 1.1 and 1.2, we find a new element in Figure 1.2 denoted by a circle before the controller block, into which two arrows are leading and out of which one arrow is emerging and leading to the controller. This circle is called a *summing junction*, which adds the signals leading into it with the appropriate signs which are indicated adjacent to the respective arrowheads. If a sign is omitted, a positive sign is assumed. The output of

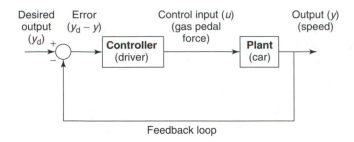

Figure 1.2 Example of a closed-loop control system with feedback; the controller applies a control input based on the plant output

the summing junction is the arithmetic sum of its two (or more) inputs. Using the symbols u (control input), y (output), and y_d (desired output), we can see in Figure 1.2 that the *input* to the *controller* is the error signal $(y_d - y)$. In Figure 1.2, the controller itself is a system which produces an *output* (control input), u, based upon the input it receives in the form of $(y_d - y)$. Hence, the behavior of a *linear* controller could be mathematically described by its transfer-function, which is the relationship between u and $(y_d - y)$. Note that Figure 1.2 shows only a popular kind of closed-loop system. In other closed-loop systems, the input to the controller could be different from the error signal $(y_d - y)$. The controller transfer-function is the main design parameter in the design of a control system and determines how rapidly – and with what *maximum overshoot* (i.e. maximum value of $|y_d - y|$) – the actual output, y, will become equal to the desired output, y_d. We will see later how the controller transfer-function can be obtained, given a set of design requirements. (However, deriving the transfer-function of a human controller is beyond the present science, as mentioned in the previous section.) When the desired output, y_d, is a constant, the resulting controller is called a *regulator*. If the desired output is changing with time, the corresponding control system is called a *tracking system*. In any case, the principal task of a closed-loop controller is to make $(y_d - y) = 0$ as quickly as possible. Figure 1.3 shows a possible plot of the actual output of a closed-loop control system.

Whereas the desired output y_d has been achieved after some time in Figure 1.3, there is a large maximum overshoot which could be unacceptable. A successful closed-loop controller design should achieve both a small maximum overshoot, and a small error magnitude $|y_d - y|$ as quickly as possible. In Chapter 4 we will see that the output of a linear system to an arbitrary input consists of a fluctuating sort of response (called the *transient response*), which begins as soon as the input is applied, and a settled kind of response (called the *steady-state response*) after a long time has elapsed since the input was initially applied. If the linear system is *stable*, the transient response would decay to zero after sometime (*stability* is an important property of a system, and is discussed in Section 2.8), and only the steady-state response would persist for a long time. The transient response of a linear system depends *largely* upon the characteristics and the initial *state* of the system, while the steady-state response depends both upon system's characteristics and the input as a function of time, i.e. $u(t)$. The maximum overshoot is a property of the transient response, but the error magnitude $|y_d - y|$ at large time (or in the limit $t \to \infty$) is a property of the steady-state response of the closed-loop system. In

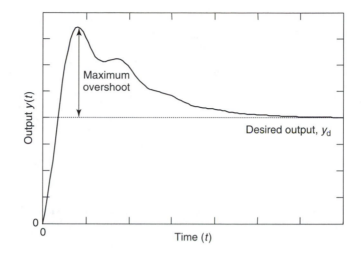

Figure 1.3 Example of a closed-loop control system's response; the desired output is achieved after some time, but there is a large maximum overshoot

Figure 1.3 the steady-state response asymptotically approaches a constant y_d in the limit $t \to \infty$.

Figure 1.3 shows the basic fact that it is impossible to get the desired output *immediately*. The reason why the output of a linear, stable system does not *instantaneously* settle to its steady-state has to do with the inherent physical characteristics of all practical systems that involve either *dissipation* or *storage* of *energy* supplied by the input. Examples of energy storage devices are a spring in a mechanical system, and a capacitor in an electrical system. Examples of energy dissipation processes are mechanical friction, heat transfer, and electrical resistance. Due to a transfer of energy from the applied input to the energy storage or dissipation elements, there is initially a fluctuation of the total energy of the system, which results in the transient response. As the time passes, the energy contribution of storage/dissipative processes in a stable system declines rapidly, and the total energy (hence, the output) of the system tends to the same function of time as that of the applied input. To better understand this behavior of linear, stable systems, consider a bucket with a small hole in its bottom as the system. The input is the flow rate of water supplied to the bucket, which could be a specific function of time, and the output is the *total* flow rate of water coming out of the bucket (from the hole, as well as from the overflowing top). Initially, the bucket takes some time to fill due to the hole (dissipative process) and its internal volume (storage device). However, after the bucket is full, the output largely follows the changing input.

While the most common closed-loop control system is the *feedback control system*, as shown in Figure 1.2, there are other possibilities such as the *feedforward control system*. In a feedforward control system – whose example is shown in Figure 1.4 – in addition to a feedback loop, a feedforward path from the desired output (y_d) to the control input is generally employed to counteract the effect of noise, or to reduce a known undesirable plant behavior. The feedforward controller incorporates some *a priori* knowledge of the plant's behavior, thereby reducing the burden on the feedback controller in controlling

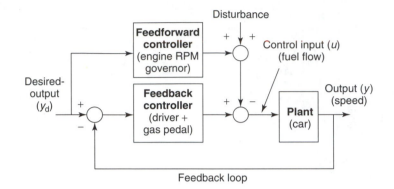

Figure 1.4 A closed-loop control system with a feedforward path; the engine RPM governor takes care of the fuel flow disturbance, leaving the driver free to concentrate on achieving desired speed with gas pedal force

the plant. Note that if the feedback controller is removed from Figure 1.4, the resulting control system becomes open-loop type. Hence, a feedforward control system can be regarded as a *hybrid* of open and closed-loop control systems. In the car driver example, the feedforward controller could be an engine rotational speed governor that keeps the engine's RPM constant in the presence of disturbance (noise) in the fuel flow rate caused by known imperfections in the fuel supply system. This reduces the burden on the driver, who would have been required to apply a rapidly changing gas pedal force to counteract the fuel supply disturbance if there was no feedforward controller. Now the feedback controller consists of the driver and the gas-pedal mechanism, and the control input is the fuel flow into the engine, which is influenced by not only the gas-pedal force, but also by the RPM governor output and the disturbance. It is clear from the present example that many practical control systems can benefit from the feedforward arrangement.

In this section, we have seen that a control system can be classified as either open- or closed-loop, depending upon the physical arrangement of its components. However, there are other ways of classifying control systems, as discussed in the next section.

1.3 Other Classifications of Control Systems

Apart from being open- or closed-loop, a control system can be classified according to the physical nature of the laws obeyed by the system, and the mathematical nature of the governing differential equations. To understand such classifications, we must define the *state* of a system, which is the fundamental concept in modern control. The *state* of a system is any set of physical quantities which need to be specified at a given time in order to completely determine the behavior of the system. This definition is a little confusing, because it introduces another word, *determine*, which needs further explanation given in the following paragraph. We will return to the concept of state in Chapter 3, but here let us only say that the state is all the information we need about a system to tell what the system is doing at any given time. For example, if one is given information about the speed of a car and the positions of other vehicles on the road relative to the car, then one has sufficient information to drive the car safely. Thus, the state of such a system consists of the car's speed and relative positions of other vehicles. However, for the same system one could choose another set of physical quantities to be the system's state, such as velocities of all other vehicles relative to the car, and the position of the car with respect to the road divider. Hence, by definition the state is not a unique set of physical quantities.

A control system is said to be *deterministic* when the set of physical laws governing the system are such that if the state of the system at some time (called the *initial conditions*) and the input are specified, then one can precisely predict the state at a later time. The laws governing a deterministic system are called *deterministic laws*. Since the characteristics of a deterministic system can be found merely by studying its response to initial conditions (transient response), we often study such systems by taking the applied input to be zero. A response to initial conditions when the applied input is zero depicts how the system's state *evolves* from some initial time to that at a later time. Obviously, the *evolution* of only a deterministic system can be determined. Going back to the definition of state, it is clear that the latter is arrived at keeping a deterministic system in mind, but the concept of state can also be used to describe systems that are *not* deterministic. A system that is not deterministic is either *stochastic*, or has *no* laws governing it. A *stochastic* (also called *probabilistic*) system has such governing laws that although the initial conditions (i.e. state of a system at some time) are known in every detail, it is impossible to determine the system's state at a later time. In other words, based upon the *stochastic* governing laws and the initial conditions, one could only determine the probability of a state, rather than the state itself. When we toss a perfect coin, we are dealing with a stochastic law that states that both the possible outcomes of the toss (head or tail) have an equal probability of 50 percent. We should, however, make a distinction between a physically stochastic system, and our *ability* (as humans) to *predict* the behavior of a deterministic system based upon our measurement of the initial conditions and our understanding of the governing laws. Due to an uncertainty in our knowledge of the governing deterministic laws, as well as errors in measuring the initial conditions, we will frequently be unable to predict the state of a deterministic system at a later time. Such a problem of unpredictability is highlighted by a special class of deterministic systems, namely *chaotic* systems. A system is called *chaotic* if even a small change in the initial conditions produces an arbitrarily large change in the system's state at a later time.

An example of chaotic control systems is a *double pendulum* (Figure 1.5). It consists of two masses, m_1 and m_2, joined together and suspended from point O by two rigid massless links of lengths L_1 and L_2 as shown. Here, the state of the system can be defined by the angular displacements of the two links, $\theta_1(t)$ and $\theta_2(t)$, as well as their respective angular velocities, $\theta_1^{(1)}(t)$ and $\theta_2^{(1)}(t)$. (In this book, the notation used for representing a kth order *time derivative* of $f(t)$ is $f^{(k)}(t)$, i.e. $d^k f(t)/dt^k = f^{(k)}(t)$. Thus, $\theta_1^{(1)}(t)$ denotes $d\theta_1(t)/dt$, etc.) Suppose we do not apply an input to the system, and begin observing the system at some time, $t = 0$, at which the initial conditions are, say, $\theta_1(0) = 40°$, $\theta_2(0) = 80°$, $\theta_1^{(1)}(0) = 0°/s$, and $\theta_2^{(1)}(0) = 0°/s$. Then at a later time, say after 100 s, the system's state will be very much different from what it would have been if the initial conditions were, say, $\theta_1(0) = 40.01°$, $\theta_2(0) = 80°$, $\theta_1^{(1)}(0) = 0°/s$, and $\theta_2^{(1)}(0) = 0°/s$. Figure 1.6 shows the time history of the angle $\theta_2(t)$ between 85 s and 100 s

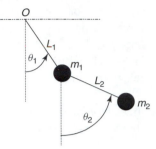

Figure 1.5 A double pendulum is a chaotic system because a small change in its initial conditions produces an arbitrarily large change in the system's state after some time

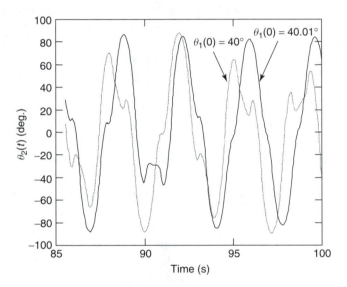

Figure 1.6 Time history between 85 s and 100 s of angle θ_2 of a double pendulum with $m_1 = 1$ kg, $m_2 = 2$ kg, $L_1 = 1$ m, and $L_2 = 2$ m for the two sets of initial conditions $\theta_1(0) = 40°$, $\theta_2(0) = 80°$, $\theta_1^{(1)}(0) = 0°/s$, $\theta_2^{(1)}(0) = 0°/s$ and $\theta_1(0) = 40.01°$, $\theta_2(0) = 80°$, $\theta_1^{(1)}(0) = 0°/s$, $\theta_2^{(1)}(0) = 0°/s$, respectively

for the two sets of initial conditions, for a double pendulum with $m_1 = 1$ kg, $m_2 = 2$ kg, $L_1 = 1$ m, and $L_2 = 2$ m. Note that we know the governing laws of this deterministic system, yet we cannot predict its state after a given time, because there will always be some error in measuring the initial conditions. Chaotic systems are so interesting that they have become the subject of specialization at many physics and engineering departments.

Any unpredictable system can be mistaken to be a stochastic system. Taking the car driver example of Section 1.2, there may exist deterministic laws that govern the road conditions, wind velocity, etc., but our ignorance about them causes us to treat such phenomena as random noise, i.e. stochastic processes. Another situation when a deterministic system may appear to be stochastic is exemplified by the toss of a coin deliberately loaded to fall every time on one particular side (either head or tail). An unwary spectator may believe such a system to be stochastic, when actually it is very much deterministic!

When we analyze and design control systems, we try to express their governing physical laws by differential equations. The mathematical nature of the governing differential equations provides another way of classifying control systems. Here we depart from the realm of physics, and delve into mathematics. Depending upon whether the differential equations used to describe a control system are linear or nonlinear in nature, we can call the system either *linear* or *nonlinear*. Furthermore, a control system whose description requires partial differential equations is called a *distributed parameter system*, whereas a system requiring only ordinary differential equations is called a *lumped parameter system*. A vibrating string, or a membrane is a distributed parameter system, because its properties (mass and stiffness) are distributed in space. A mass suspended by a spring is a lumped parameter system, because its mass and stiffness are concentrated at discrete points in space. (A more common nomenclature of distributed and lumped parameter systems is *continuous* and *discrete* systems, respectively, but we avoid this terminology in this book as it might be confused with *continuous time* and *discrete time* systems.) A particular system can be treated as linear, or nonlinear, distributed, or lumped parameter, depending upon what aspects of its behavior we are interested in. For example, if we want to study only small angular displacements of a simple pendulum, its differential equation of motion can be treated to be linear; but if large angular displacements are to be studied, the same pendulum is treated as a nonlinear system. Similarly, when we are interested in the motion of a car as a whole, its state can be described by only two quantities: the position and the velocity of the car. Hence, it can be treated as a lumped parameter system whose entire mass is concentrated at one point (the center of mass). However, if we want to take into account how the tyres of the car are deforming as it moves along an uneven road, the car becomes a distributed parameter system whose state is described exactly by an infinite set of quantities (such as deformations of all the points on the tyres, and their time derivatives, in addition to the speed and position of the car). Other classifications based upon the mathematical nature of governing differential equations will be discussed in Chapter 2.

Yet another way of classifying control systems is whether their *outputs* are *continuous* or *discontinuous* in time. If one can express the system's state (which is obtained by solving the system's differential equations) as a continuous function of time, the system is called *continuous in time* (or *analog system*). However, a majority of modern control systems produce outputs that 'jump' (or are discontinuous) in time. Such control systems are called *discrete in time* (or *digital systems*). Note that in the limit of very small time steps, a digital system can be approximated as an analog system. In this book, we will make this assumption quite often. If the time steps chosen to sample the discontinuous output are relatively large, then a digital system can have a significantly different behaviour from that of a corresponding analog system. In modern applications, even analog controllers are implemented on a digital processor, which can introduce digital characteristics to the control system. Chapter 8 is devoted to the study of digital systems.

There are other minor classifications of control systems based upon the systems' characteristics, such as *stability, controllability, observability*, etc., which we will take up in subsequent chapters. Frequently, control systems are also classified based upon the number of inputs and outputs of the system, such as *single-input, single-output* system, or *two-input, three-output* system, etc. In *classical* control (an object of Chapter 2) the distinction between *single-input, single-output* (SISO) and *multi-input, multi-output* (MIMO) systems is crucial.

1.4 On the Road to Control System Analysis and Design

When we find an unidentified object on the street, the first thing we may do is prod or poke it with a stick, pick it up and shake it, or even hit it with a hammer and hear the sound it makes, in order to find out something about it. We treat an unknown control system in a similar fashion, i.e. we apply some well known inputs to it and carefully observe how it responds to those inputs. This has been an age old method of analyzing a system. Some of the well known inputs applied to study a system are the *singularity functions*, thus called due to their peculiar nature of being *singular* in the mathematical sense (their time derivative tends to infinity at some time). Two prominent members of this zoo are the *unit step function* and the *unit impulse function*. In Chapter 2, useful computer programs are presented to enable you to find the response to *impulse* and *step* inputs – as well as the response to an *arbitrary* input – of a single-input, single-output control system. Chapter 2 also discusses important properties of a control system, namely, *performance*, *stability*, and *robustness*, and presents the analysis and design of linear control systems using the *classical approach* of *frequency response*, and *transform methods*. Chapter 3 introduces the *state-space* modeling for linear control systems, covering various applications from all walks of engineering. The solution of a linear system's governing equations using the state-space method is discussed in Chapter 4. In this chapter, many new computer programs are presented to help you solve the state-equations for linear or nonlinear systems.

The design of modern control systems using the state-space approach is introduced in Chapter 5, which also discusses two important properties of a plant, namely its *controllability* and *observability*. In this chapter, it is first assumed that all the quantities defining the state of a plant (called *state variables*) are available for exact measurement. However, this assumption is not always practical, since some of the state variables may not be measurable. Hence, we need a procedure for estimating the unmeasurable state variables from the information provided by those variables that we can measure. Later sections of Chapter 5 contains material about how this process of *state estimation* is carried out by an *observer*, and how such an estimation can be incorporated into the control system in the form of a compensator. Chapter 6 introduces the procedure of designing an *optimal control system*, which means a control system meeting all the design requirements in the most efficient manner. Chapter 6 also provides new computer programs for solving important optimal control problems. Chapter 7 introduces the treatment of random signals generated by stochastic systems, and extends the philosophy of state estimation to plants with noise, which is treated as a random signal. Here we also learn how an *optimal* state estimation can be carried out, and how a control system can be made *robust* with respect to *measurement and process noise*. Chapter 8 presents the design and analysis of *digital control systems* (also called *discrete time systems*), and covers many modern digital control applications. Finally, Chapter 9 introduces various advanced topics in modern control, such as *advanced robust control techniques, nonlinear control*, etc. Some of the topics contained in Chapter 9, such as *input shaping control* and *rate-weighted optimal control*, are representative of the latest control techniques.

At the end of each chapter (except Chapter 1), you will find exercises that help you grasp the essential concepts presented in the chapter. These exercises range from analytical to numerical, and are designed to make you think, rather than apply ready-made formulas for their solution. At the end of the book, answers to some numerical exercises are provided to let you check the accuracy of your solutions.

1.5 MATLAB, SIMULINK, and the Control System Toolbox

Modern control design and analysis requires a lot of *linear algebra* (matrix multiplication, inversion, calculation of *eigenvalues* and *eigenvectors*, etc.) which is not very easy to perform manually. Try to remember the last time you attempted to invert a 4×4 matrix by hand! It can be a tedious process for any matrix whose size is greater than 3×3. The repetitive linear algebraic operations required in modern control design and analysis are, however, easily implemented on a computer with the use of standard programming techniques. A useful high-level programming language available for such tasks is the MATLAB®, which not only provides the tools for carrying out the matrix operations, but also contains several other features, such as the time-step integration of linear or nonlinear governing differential equations, which are invaluable in modern control analysis and design. For example, in Figure 1.6 the time-history of a double-pendulum has been obtained by solving the coupled governing nonlinear differential equations using MATLAB. Many of the numerical examples contained in this book have been solved using MATLAB. Although not required for doing the exercises at the end of each chapter, it is recommended that you familiarize yourself with this useful language with the help of Appendix A, which contains information about the commonly used MATLAB operators in modern control applications. Many people, who shied away from modern control courses because of their dread of linear algebra, began taking interest in the subject when MATLAB became handy. Nowadays, personal computer versions of MATLAB are commonly applied to practical problems across the board, including control of aerospace vehicles, magnetically levitated trains, and even stock-market applications. You may find MATLAB available at your university's or organization's computer center. While Appendix A contains useful information about MATLAB which will help you in solving most of the modern control problems, it is recommended that you check with the MATLAB user's guide [1] at your computer center for further details that may be required for advanced applications.

SIMULINK® is a very useful Graphical Users Interface (GUI) tool for modeling control systems, and simulating their time response to specified inputs. It lets you work directly with the block-diagrams (rather than mathematical equations) for designing and analyzing

---

® MATLAB, SIMULINK and Control System Toolbox are registered trademarks of *MathWorks, Inc.*

control systems. For this purpose, numerous linear and nonlinear blocks, input sources, and output devices are available, so that you can easily put together almost any practical control system. Another advantage of using SIMULINK is that it works seamlessly with MATLAB, and can draw upon the vast programming features and function library of MATLAB. A SIMULINK block-diagram can be converted into a MATLAB program (called *M-file*). In other words, a SIMULINK block-diagram does all the programming for you, so that you are free to worry about other practical aspects of a control system's design and implementation. With advanced features (such as the *Real Time Workshop* for C-code generation, and specialized block-sets) one can also use SIMULINK for practical implementation of control systems [2]. We will be using SIMULINK as a design and analysis tool, especially in simulating the response of a control system designed with MATLAB.

For solving many problems in control, you will find the *Control System Toolbox*® [3] for MATLAB very useful. It contains a set of MATLAB M-files of numerical procedures that are commonly used to design and analyze modern control systems. The Control System Toolbox is available at a small extra cost when you purchase MATLAB, and is likely to be installed at your computer center if it has MATLAB. Many solved examples presented in this book require the Control System Toolbox. In the solved examples, effort has been made to ensure that the application of MATLAB is clear and direct. This is done by directly presenting the MATLAB line commands – and some MATLAB *M-files* – followed by the numerical values resulting after executing those commands. Since the commands are presented *exactly* as they would appear in a MATLAB *workspace*, the reader can easily reproduce all the computations presented in the book. Again, take some time to familiarize yourself with MATLAB, SIMULINK and the Control System Toolbox by reading Appendix A.

References

1. *MATLAB*® *6.0 – User's Guide*, The Math Works Inc., Natick, MA, USA, 2000.
2. *SIMULINK*® *4.0 – User's Guide*, The Math Works Inc., Natick, MA, USA, 2000.
3. *Control System Toolbox 5.0 for Use with MATLAB*® *– User's Guide*, The Math Works Inc., Natick, MA, USA, 2000.

2

Linear Systems and Classical Control

2.1 How Valid is the Assumption of Linearity?

It was mentioned in Chapter 1 that we need differential equations to describe the behavior of a system, and that the mathematical nature of the governing differential equations is another way of classifying control systems. In a large class of engineering applications, the governing differential equations can be assumed to be linear. The concept of linearity is one of the most important assumptions often employed in studying control systems. However, the following questions naturally arise: what is this assumption and how valid is it anyway? To answer these questions, let us consider lumped parameter systems for simplicity, even though all the arguments presented below are equally applicable to distributed systems. (Recall that *lumped parameter* systems are those systems whose behavior can be described by *ordinary* differential equations.) Furthermore, we shall confine our attention (until Section 2.13) to single-input, single-output (SISO) systems. For a general lumped parameter, SISO system (Figure 2.1) with input $u(t)$ and output $y(t)$, the governing ordinary differential equation can be written as

$$y^{(n)}(t) = f(y^{(n-1)}(t), y^{(n-2)}(t), \ldots, y^{(1)}(t), y(t), u^{(m)}(t), u^{(m-1)}(t), \ldots, u^{(1)}(t), u(t), t) \tag{2.1}$$

where $y^{(k)}$ denotes the kth derivative of $y(t)$ with respect to time, t, e.g. $y^{(n)} = d^n y / dt^n$, $y^{(n-1)} = d^{n-1} y / dt^{n-1}$, and $u^{(k)}$ denotes the kth time derivative of $u(t)$. This notation for derivatives of a function will be used throughout the book. In Eq. (2.1), $f()$ denotes a function of all the time derivatives of $y(t)$ of order $(n-1)$ and less, as well as the time derivatives of $u(t)$ of order m and less, and time, t. For most systems $m \leq n$, and such systems are said to be *proper*.

Since n is the order of the highest time derivative of $y(t)$ in Eq. (2.1), the system is said to be *of order n*. To determine the output $y(t)$, Eq. (2.1) must be somehow integrated in time, with $u(t)$ known and for specific initial conditions $y(0), y^{(1)}(0), y^{(2)}(0), \ldots, y^{(n-1)}(0)$. Suppose we are capable of solving Eq. (2.1), given any time varying input, $u(t)$, and the initial conditions. For simplicity, let us assume that the initial conditions are zero, and we apply an input, $u(t)$, which is a *linear combination* of two different inputs, $u_1(t)$, and $u_2(t)$, given by

$$u(t) = c_1 u_1(t) + c_2 u_2(t) \tag{2.2}$$

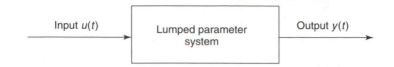

Figure 2.1 A general lumped parameter system with input, $u(t)$, and output, $y(t)$

where c_1 and c_2 are constants. If the resulting output, $y(t)$, can be written as

$$y(t) = c_1 y_1(t) + c_2 y_2(t) \tag{2.3}$$

where $y_1(t)$ is the output when $u_1(t)$ is the input, and $y_2(t)$ is the output when $u_2(t)$ is the input, then the system is said to be *linear*; otherwise it is called *nonlinear*. In short, a linear system is said to obey the *superposition principle*, which states that the output of a linear system to an input consisting of linear combination of two different inputs (Eq. (2.2)) can be obtained by linearly superposing the outputs to the respective inputs (Eq. (2.3)). (The superposition principle is also applicable for *non-zero* initial conditions, if the initial conditions on $y(t)$ and its time derivatives are linear combinations of the initial conditions on $y_1(t)$ and $y_2(t)$, and their corresponding time derivatives, with the constants c_1 and c_2.) Since linearity is a mathematical property of the governing differential equations, it is possible to say merely by inspecting the differential equation whether a system is linear. If the function $f()$ in Eq. (2.1) contains no powers (other than one) of $y(t)$ and its derivatives, or the mixed products of $y(t)$, its derivatives, and $u(t)$ and its derivatives, or transcendental functions of $y(t)$ and $u(t)$, then the system will obey the superposition principle, and its linear differential equation can be written as

$$a_n y^{(n)}(t) + a_{n-1} y^{(n-1)}(t) + \cdots + a_1 y^{(1)}(t) + a_0 y(t)$$
$$= b_m u^{(m)}(t) + b_{m-1} u^{(m-1)}(t) + \cdots + b_1 u^{(1)}(t) + b_0 u(t) \tag{2.4}$$

Note that even though the coefficients $a_0, a_1, \ldots, a_n$ and $b_0, b_1, \ldots, b_m$ (called the *parameters* of a system) in Eq. (2.4) may be *varying* with time, the system given by Eq. (2.4) is still linear. A system with time-varying parameters is called a *time-varying* system, while a system whose parameters are constant with time is called *time-invariant* system. In the present chapter, we will be dealing only with linear, time-invariant systems. It is possible to express Eq. (2.4) as a *set* of lower order differential equations, whose individual orders add up to n. Hence, the order of a system is the *sum* of orders of all the differential equations needed to describe its behavior.

Example 2.1

For an electrical network shown in Figure 2.2, the governing differential equations are the following:

$$v_1^{(1)}(t) = -(v_1(t)/C_1)(1/R_1 + 1/R_3) + v_2(t)/(C_1 R_3) + e(t)/(R_1 C_1) \tag{2.5a}$$

$$v_2^{(1)}(t) = v_1(t)/(C_2 R_3) - (v_2(t)/C_2)(1/R_2 + 1/R_3) + e(t)/(R_2 C_2) \tag{2.5b}$$

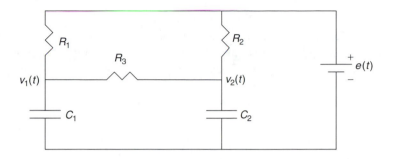

Figure 2.2 Electrical network for Example 2.1

where $v_1(t)$ and $v_2(t)$ are the voltages of the two capacitors, C_1 and C_2, $e(t)$ is the applied voltage, and R_1, R_2, and R_3 are the three resistances as shown.

On inspection of Eq. (2.5), we can see that the system is described by two *first order*, ordinary differential equations. Therefore, the system is of *second order*. Upon the substitution of Eq. (2.5b) into Eq. (2.5a), and by eliminating v_2, we get the following second order differential equation:

$$C_1 R_3 v_1^{(2)}(t) + [R_3/R_1 + 1 + (C_1/C_2)(R_3/R_2 + 1)]v_1^{(1)}(t)$$
$$+ [(1/R_2 + 1/R_3)(R_3/R_1 + 1) - 1/R_3]v_1(t)$$
$$= (R_3/R_1)(1/R_1 + 1/R_3)e(t)/C_2 + (R_3/R_1)e^{(1)}(t) \qquad (2.6)$$

Assuming $y(t) = v_1(t)$ and $u(t) = e(t)$, and comparing Eq. (2.6) with Eq. (2.4), we can see that there are no higher powers, transcendental functions, or mixed products of the output, input, and their time derivatives. Hence, *the system is linear*.

Suppose we *do not* have an input, $u(t)$, applied to the system in Figure 2.1. Such a system is called an *unforced system*. Substituting $u(t) = u^{(1)}(t) = u^{(2)}(t) = \cdots = u^{(m)}(t) = 0$ into Eq. (2.1) we can obtain the following governing differential equation for the unforced system:

$$y^{(n)}(t) = f(y^{(n-1)}(t), y^{(n-2)}(t), \ldots, y^{(1)}(t), y(t), 0, 0, \ldots, 0, 0, t) \qquad (2.7)$$

In general, the solution, $y(t)$, to Eq. (2.7) for a given set of initial conditions is a function of time. However, there may also exist special solutions to Eq. (2.7) which are constant. Such constant solutions for an unforced system are called its *equilibrium points*, because the system continues to be at rest when it is already at such points. A large majority of control systems are designed for keeping a plant at one of its equilibrium points, such as the cruise-control system of a car and the autopilot of an airplane or missile, which keep the vehicle moving at a constant velocity. When a control system is designed for maintaining the plant at an equilibrium point, then only small deviations from the equilibrium point need to be considered for evaluating the performance of such a control system. Under such circumstances, the time behavior of the plant and the resulting control system can generally be assumed to be governed by linear differential equations, even though the governing differential equations of the plant and the control system may be nonlinear. The following examples demonstrate how a nonlinear system can be linearized near its equilibrium points. Also included is an example which illustrates that such a linearization may not always be possible.

Example 2.2

Consider a simple pendulum (Figure 2.3) consisting of a point mass, m, suspended from hinge at point O by a rigid massless link of length L. The equation of motion of the simple pendulum in the absence of an externally applied torque about point O in terms of the angular displacement, $\theta(t)$, can be written as

$$L\theta^{(2)}(t) + g.\sin(\theta(t)) = 0 \qquad (2.8)$$

This governing equation indicates a second-order system. Due to the presence of $\sin(\theta)$ – a transcendental function of θ – Eq. (2.8) is nonlinear. From our everyday experience with a simple pendulum, it is clear that it can be brought to rest at only *two* positions, namely $\theta = 0$ and $\theta = \pi$ rad. (180°). Therefore, these two positions are the equilibrium points of the system given by Eq. (2.8). Let us examine the behavior of the system near each of these equilibrium points.

Since the only nonlinear term in Eq. (2.8) is $\sin(\theta)$, if we can show that $\sin(\theta)$ can be approximated by a linear term, then Eq. (2.8) can be linearized. Expanding $\sin(\theta)$ about the equilibrium point $\theta = 0$, we get the following *Taylor's series* expansion:

$$\sin(\theta) = \theta - \theta^3/3! + \theta^5/5! - \theta^7/7! + \cdots \qquad (2.9)$$

If we assume that motion of the pendulum about $\theta = 0$ consists of small angular displacements (say $\theta < 10°$), then $\sin(\theta) \approx \theta$, and Eq. (2.8) becomes

$$L\theta^{(2)}(t) + g\theta(t) = 0 \qquad (2.10)$$

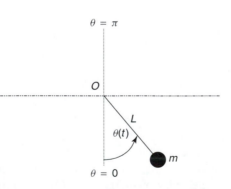

Figure 2.3   A simple pendulum of length L and mass m

Similarly, expanding $\sin(\theta)$ about the *other* equilibrium point, $\theta = \pi$, by assuming small angular displacement, ϕ, such that $\theta = \pi - \phi$, and noting that $\sin(\theta) = -\sin(\phi) \approx -\phi$, we can write Eq. (2.8) as

$$L\phi^{(2)}(t) - g\phi(t) = 0 \qquad (2.11)$$

We can see that both Eqs. (2.10) and (2.11) are linear. Hence, the nonlinear system given by Eq. (2.8) has been linearized about both of its equilibrium points. Second order linear ordinary differential equations (especially the *homogeneous* ones like Eqs. (2.10) and (2.11)) can be be solved analytically. It is well known (and you may verify) that the solution to Eq. (2.10) is of the form $\theta(t) = A.\sin(t(g/L)^{1/2} + B.\cos(t(g/L)^{1/2})$, where the constants A and B are determined from the initial conditions, $\theta(0)$ and $\theta^{(1)}(0)$. This solution implies that $\theta(t)$ oscillates about the equilibrium point $\theta = 0$. However, the solution to Eq. (2.11) is of the form $\phi(t) = C.\exp(t(g/L)^{1/2})$, where C is a constant, which indicates an exponentially increasing $\phi(t)$ if $\phi(0) \neq 0$. (This nature of the equilibrium point at $\theta = \pi$ can be experimentally verified by anybody trying to stand on one's head for any length of time!) The comparison of the solutions to the linearized governing equations close to the equilibrium points (Figure 2.4) brings us to an important property of an equilibrium point, called *stability*.

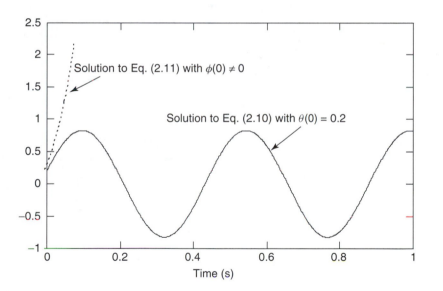

Figure 2.4 Solutions to the governing differential equation linearized about the two equilibrium points ($\theta = 0$ and $\theta = \pi$)

Stability is defined as the ability of a system to approach one of its equilibrium points once displaced from it. We will discuss stability in detail later. Here, suffice it to say that the pendulum is stable about the equilibrium point $\theta = 0$, but unstable about the equilibrium point $\theta = \pi$. While Example 2.2 showed how a nonlinear system can be linearized close to its equilibrium points, the following example illustrates how a nonlinear system's description can be transformed into a linear system description through a clever change of coordinates.

Example 2.3

Consider a satellite of mass m in an orbit about a planet of mass M (Figure 2.5). The distance of the satellite from the center of the planet is denoted $r(t)$, while its orientation with respect to the planet's equatorial plane is indicated by the angle $\theta(t)$, as shown. Assuming there are no gravitational anomalies that cause a departure from Newton's inverse-square law of gravitation, the governing equation of motion of the satellite can be written as

$$r^{(2)}(t) - h^2/r(t)^3 + k^2/r(t)^2 = 0 \qquad (2.12)$$

where h is the constant angular momentum, given by

$$h = r(t)^2\theta^{(1)}(t) = \text{constant} \qquad (2.13)$$

and $k = GM$, with G being the universal gravitational constant.

Equation (2.12) represents a nonlinear, second order system. However, since we are usually interested in the *path* (or the shape of the orbit) of the satellite, given by $r(\theta)$, rather than its distance from the planet's center as a function of time, r(t), we can transform Eq. (2.12) to the following linear differential equation by using the co-ordinate transformation $u(\theta) = 1/r(\theta)$:

$$u^{(2)}(\theta) + u(\theta) - k^2/h^2 = 0 \qquad (2.14)$$

Being a linear, *second* order ordinary differential equation (similar to Eq. (2.10)), Eq. (2.14) is easily solved for $u(\theta)$, and the solution transformed back to $r(\theta)$

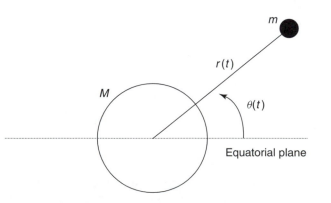

Figure 2.5   A satellite of mass m in orbit around a planet of mass M at a distance $r(t)$ from the planet's center, and azimuth angle $\theta(t)$ from the equatorial plane

given by

$$r(\theta) = (h^2/k^2)/[1 + A(h^2/k^2)\cos(\theta - B)] \qquad (2.15)$$

where the constants A and B are determined from $r(\theta)$ and $r^{(1)}(\theta)$ specified at given values of θ. Such specifications are called *boundary conditions*, because they refer to points in space, as opposed to *initial conditions* when quantities at given instants of time are specified. Equation (2.15) can represent a circle, an ellipse, a parabola, or a hyperbola, depending upon the magnitude of $A(h^2/k^2)$ (called the *eccentricity* of the orbit).

Note that we could also have linearized Eq. (2.12) about one of its equilibrium points, as we did in Example 2.2. One such equilibrium point is given by $r(t) =$ constant, which represents a circular orbit. Many practical orbit control applications consist of minimizing deviations from a given circular orbit using rocket thrusters to provide *radial acceleration* (i.e. acceleration along the line joining the satellite and the planet) as an input, $u(t)$, which is based upon the measured deviation from the circular path fed back to an onboard controller, as shown in Figure 2.6. In such a case, the governing differential equation is no longer homogeneous as Eq. (2.12), but has a *non-homogeneous* forcing term on the right-hand side given by

$$r^{(2)}(t) - h^2/r(t)^3 + k^2/r(t)^2 = u(t) \qquad (2.16)$$

Since the deviations from a given circular orbit are usually small, Eq. (2.16) can be suitably linearized about the equilibrium point $r(t) = C$. (This linearization is left as an exercise for you at the end of the chapter.)

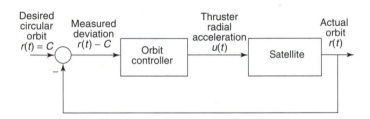

Figure 2.6 On orbit feedback control system for maintaining a circular orbit of a satellite around a planet

Examples 2.2 and 2.3 illustrated how a nonlinear system can be linearized for practical control applications. However, as pointed out earlier, it is not always possible to do so. If a nonlinear system has to be moved from one equilibrium point to another (such as changing the speed or altitude of a cruising airplane), the assumption of linearity that is possible in the close neighborhood of each equilibrium point disappears as we cross the nonlinear region between the equilibrium points. Also, if the motion of a nonlinear system consists of large deviations from an equilibrium point, again the concept of linearity is not valid. Lastly, the characteristics of a nonlinear system may be such that it does not have any equilibrium point about which it can be linearized. The following missile guidance example illustrates such a nonlinear system.

Example 2.4

Radar or laser-guided missiles used in modern warfare employ a special guidance scheme which aims at flying the missile along a radar or laser beam that is illuminating a moving target. The guidance strategy is such that a correcting command signal (input) is provided to the missile if its flight path deviates from the moving beam. For simplicity, let us assume that both the missile and the target are moving in the same plane (Figure 2.7). Although the distance from the beam source to the target, $R_T(t)$, is not known, it is assumed that the angles made by the missile and the target with respect to the beam source, $\theta_M(t)$ and $\theta_T(t)$, are available for precise measurement. In addition, the distance of the missile from the beam source, $R_M(t)$, is also known at each instant.

A *guidance law* provides the following normal acceleration command signal, $a_c(t)$, to the missile

$$a_c(t) = K R_M(t)[\theta_T(t) - \theta_M(t)] \tag{2.17}$$

As the missile is usually faster than the target, if the angular deviation $[\theta_T(t) - \theta_M(t)]$ is made small enough, the missile will intercept the target. The feedback guidance scheme of Eq. (2.17) is called *beam-rider guidance*, and is shown in Figure 2.8.

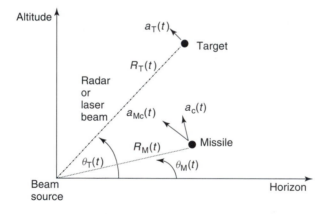

Figure 2.7 Beam guided missile follows a beam that continuously illuminates a moving target located at distance $R_T(t)$ from the beam source

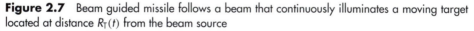
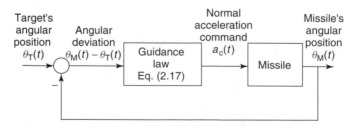

Figure 2.8 Beam-rider closed-loop guidance for a missile

The beam-rider guidance can be significantly improved in performance if we can measure the angular velocity, $\theta_T^{(1)}(t)$, and the angular acceleration, $\theta_T^{(2)}(t)$, of the target. Then the beam's normal acceleration can be determined from the following equation:

$$a_T = R_T(t)\theta_T^{(2)}(t) + 2R_T^{(1)}(t)\theta_T^{(1)}(t) \tag{2.18}$$

In such a case, along with $a_c(t)$ given by Eq. (2.17), an additional command signal (input) can be provided to the missile in the form of missile's acceleration perpendicular to the beam, $a_{Mc}(t)$, given by

$$a_{Mc}(t) = R_M(t)\theta_M^{(2)}(t) + 2R_M^{(1)}(t)\theta_M^{(1)}(t) \tag{2.19}$$

Since the final objective is to make the missile intercept the target, it must be ensured that $\theta_M^{(1)}(t) = \theta_T^{(1)}(t)$ and $\theta_M^{(2)}(t) = \theta_T^{(2)}(t)$, even though $[\theta_T(t) - \theta_M(t)]$ may not be *exactly* zero. (To understand this philosophy, remember how we catch up with a friend's car so that we can chat with her. We accelerate (or decelerate) until our velocity (and acceleration) become identical with our friend's car, then we can talk with her; although the two cars are abreast, *they are not exactly in the same position*.) Hence, the following command signal for missile's normal acceleration perpendicular to the beam must be provided:

$$a_{Mc}(t) = R_M(t)\theta_T^{(2)}(t) + 2R_M^{(1)}(t)\theta_T^{(1)}(t) \tag{2.20}$$

The guidance law given by Eq. (2.20) is called *command line-of-sight guidance*, and its implementation along with the beam-rider guidance is shown in the block diagram of Figure 2.9. It can be seen in Figure 2.9 that while $\theta_T(t)$ is being fed back, the angular velocity and acceleration of the target, $\theta_T^{(1)}(t)$, and $\theta_T^{(2)}(t)$, respectively, are being fed forward to the controller. Hence, similar to the control system of Figure 1.4, additional information about the target is being provided by a feedforward loop to improve the closed-loop performance of the missile guidance system.

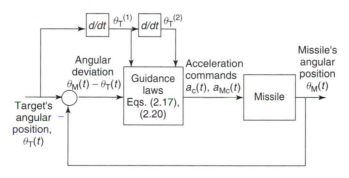

Figure 2.9 Beam-rider and command line-of-sight guidance for a missile

Note that both Eq. (2.17) and Eq. (2.20) are nonlinear in nature, and generally cannot be linearized about an equilibrium point. This example shows that the concept of linearity is not always valid. For more information on missile guidance strategies, you may refer to the excellent book by Zarchan [1].

2.2 Singularity Functions

It was mentioned briefly in Chapter 1 that some peculiar, well known input functions are generally applied to test the behavior of an unknown system. A set of such test functions is called *singularity functions*. The singularity functions are important because they can be used as building blocks to construct any arbitrary input function and, by the superposition principle (Eq. (2.3)), the response of a linear system to any arbitrary input can be easily obtained as the linear superposition of responses to singularity functions. The two distinct singularity functions commonly used for determining an unknown system's behavior are the *unit impulse* and *unit step* functions. A common property of these functions is that they are continuous in time, *except at a given time*. Another interesting fact about the singularity functions is that they can be derived from each other by differentiation or integration in time.

The unit impulse function (also called the *Dirac delta function*), $\delta(t-a)$, is seen in Figure 2.10 to be a very large spike occurring for a very small duration, applied at time $t = a$, such that the total area under the curve (shaded region) is unity. A unit impulse function can be multiplied by a constant to give a general impulse function (whose area under the curve is not unity). From this description, we recognize an impulse function to be the force one feels when hit by a car – and in all other kinds of impacts.

The height of the rectangular pulse in Figure 2.10 is $1/\varepsilon$, whereas its width is ε seconds, ε being a very small number. In the limit $\varepsilon \to 0$, the unit impulse function tends to infinity (i.e. $\delta(t-a) \to \infty$). The unit impulse function shown in Figure 2.10 is an idealization of the actual impulse whose shape is not rectangular, because it takes some time to reach the maximum value, unlike the unit impulse function (which becomes very large instantaneously). Mathematically, the unit impulse function can be described by the following equations:

$$\delta(t-a) = 0, \quad \text{for } t \neq a \tag{2.21}$$

$$\int_{-\infty}^{\infty} \delta(t-a)\,dt = 1 \tag{2.22}$$

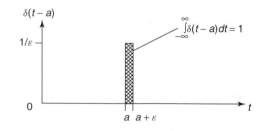

Figure 2.10 The unit impulse function; a pulse of infinitesimal duration (ε) and very large magnitude ($1/\varepsilon$) such that its total area is unity

Note that $\delta(t - a)$ is discontinuous at $t = a$. Furthermore, since the unit impulse function is non-zero only in the period $a \leq t \leq a + \varepsilon$, we can also express Eqs. (2.21) and (2.22) by

$$\int_a^{a+\varepsilon} \delta(t - a)\, dt = 1 \tag{2.23}$$

However, when utilizing the unit impulse function for control applications, Eq. (2.22) is much more useful. In fact, if $\delta(t - a)$ appears inside an integral with infinite integration limits, then such an integral is very easily carried out with the use of Eqs. (2.21) and (2.22). For example, if $f(t)$ is a continuous function, then the well known *Mean Value Theorem* of integral calculus can be applied to show that

$$\int_{T_1}^{T_2} f(t)\delta(t - a)\, dt = f(a) \int_{T_1}^{T_2} \delta(t - a)\, dt = f(a) \tag{2.24}$$

where $T_1 < a < T_2$. Equation (2.24) indicates an important property of the unit impulse function called the *sampling property*, which allows the time integral of any continuous function $f(t)$ weighted by $\delta(t - a)$ to be simply equal to the function $f(t)$ evaluated at $t = a$, provided the limits of integration bracket the time $t = a$.

The unit step function, $u_s(t - a)$, is shown in Figure 2.11 to be a jump of unit magnitude at time $t = a$. It is aptly named, because it resembles a step of a staircase. Like the unit impulse function, the unit step function is also a mathematical idealization, because it is impossible to apply a non-zero input instantaneously. Mathematically, the unit step function can be defined as follows:

$$u_s(t - a) = \begin{bmatrix} 0 & \text{for } t < a \\ 1 & \text{for } t > a \end{bmatrix} \tag{2.25}$$

It is clear that $u_s(t - a)$ is discontinuous at $t = a$, and its time derivative at $t = a$ is infinite. Recalling from Figure 2.10 that in the limit $\varepsilon \to 0$, the unit impulse function tends to infinity (i.e. $\delta(t - a) \to \infty$), we can express the unit impulse function, $\delta(t - a)$, as the time derivative of the unit step function, $u_s(t - a)$, at time $t = a$. Also, since the time derivative of $u_s(t - a)$ is zero at all times, except at $t = a$ (where it is infinite), we can write

$$\delta(t - a) = du_s(t - a)/dt \tag{2.26}$$

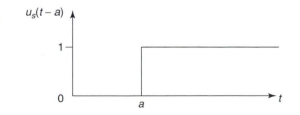

Figure 2.11 The unit step function, $u_s(t - a)$; a jump of unit magnitude at time $t = a$

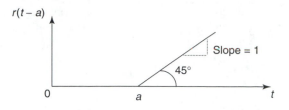

Figure 2.12 The unit ramp function; a ramp of unit slope applied at time $t = a$

Or, conversely, the unit step function is the time integral of the unit impulse function, given by

$$u_s(t - a) = \int_{-\infty}^{t} \delta(\tau - a)\,d\tau \tag{2.27}$$

A useful function related to the unit step function is the *unit ramp* function, $r(t - a)$, which is seen in Figure 2.12 to be a ramp of unit slope applied at time $t = a$. It is like an upslope of $45°$ angle you suddenly encounter while driving down a perfectly flat highway at $t = a$. Mathematically, $r(t - a)$ is given by

$$r(t - a) = \begin{bmatrix} 0 & \text{for } t < a \\ (t - a) & \text{for } t > a \end{bmatrix} \tag{2.28}$$

Note that $r(t - a)$ is continuous everywhere, but its slope is discontinuous at $t = a$. Comparing Eq. (2.28) with Eq. (2.25), it is clear that

$$r(t - a) = (t - a)u_s(t - a) \tag{2.29}$$

or

$$r(t - a) = \int_{-\infty}^{t} u_s(\tau - a)\,d\tau \tag{2.30}$$

Thus, the unit ramp function is the *time integral* of the unit step function, or conversely, the unit step function is the time derivative of the unit ramp function, given by

$$u_s(t - a) = dr(t - a)/dt \tag{2.31}$$

The basic singularity functions (unit impulse and step), and their relatives (unit ramp function) can be used to synthesize more complicated functions, as illustrated by the following examples.

Example 2.5

The *rectangular pulse function*, $f(t)$, shown in Figure 2.13, can be expressed by subtracting one step function from another as

$$f(t) = f_o[u_s(t + T/2) - u_s(t - T/2)] \tag{2.32}$$

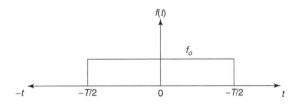

Figure 2.13   The rectangular pulse function of magnitude f_o

Example 2.6

The *decaying exponential function,* $f(t)$ (Figure 2.14) is zero before $t = 0$, and decays exponentially from a magnitude of f_o at $t = 0$. It can be expressed by multiplying the unit step function with f_o and a decaying exponential term, given by

$$f(t) = f_o \mathrm{e}^{-t/\tau} u_s(t) \tag{2.33}$$

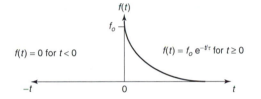

Figure 2.14   The decaying exponential function of magnitude f_o

Example 2.7

The *sawtooth pulse function,* $f(t)$, shown in Figure 2.15, can be expressed in terms of the unit step and unit ramp functions as follows:

$$f(t) = (f_o/T)[r(t) - r(t - T)] - f_o u_s(t - T) \tag{2.34}$$

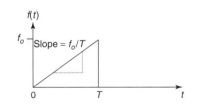

Figure 2.15   The sawtooth pulse of height f_o and width T

After going through Examples 2.5–2.7, and with a little practice, you can decide merely by looking at a given function how to synthesize it using the singularity functions. The unit impulse function has a special place among the singularity functions, because it can be

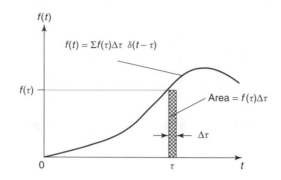

Figure 2.16 Any arbitrary function, $f(t)$, can be represented by summing up unit impulse functions, $\delta(t - \tau)$ applied at $t = \tau$ and multiplied by the area $f(\tau)\Delta\tau$ for all values of τ from $-\infty$ to t

used to describe *any* arbitrary shaped function as a sum of suitably scaled unit impulses, $\delta(t - a)$, applied at appropriate time, $t = a$. This fact is illustrated in Figure 2.16, where the function $f(t)$ is represented by

$$f(t) = \sum_{\tau=-\infty}^{\infty} f(\tau)\Delta\tau\delta(t - \tau) \tag{2.35}$$

or, in the limit $\Delta\tau \to 0$,

$$f(t) = \int_{-\infty}^{\infty} f(\tau)\delta(t - \tau)\,d\tau \tag{2.36}$$

   Equation (2.36) is one of the most important equations of modern control theory, because it lets us evaluate the response of a linear system to any arbitrary input, $f(t)$, by the use of the *superposition principle*. We will see how this is done when we discuss the response to singularity functions in Section 2.5. While the singularity functions and their relatives are useful as test inputs for studying the behavior of control systems, we can also apply some well known *continuous time* functions as inputs to a control system. Examples of continuous time test functions are the *harmonic* functions $\sin(\omega t)$ and $\cos(\omega t)$, where ω is a frequency, called the *excitation frequency*. As an alternative to singularity inputs (which are often difficult to apply in practical cases), measuring the output of a linear system to harmonic inputs gives essential information about the system's behavior, which can be used to construct a model of the system that will be useful in designing a control system. We shall study next how such a model can be obtained.

2.3 Frequency Response

Frequency response is related to the *steady-state response* of a system when a *harmonic function* is applied as the input. Recall from Section 1.2 that *steady-state response* is the linear system's output after the *transient response* has decayed to zero. Of course, the requirement that the transient response should have decayed to zero after some time calls for the linear system to be *stable*. (An *unstable* system will have a transient response shooting to infinite magnitudes, irrespective of what input is applied.) The steady-state response of a linear system is generally of the same *shape* as that of the applied input, e.g. a step input applied to a linear, stable system yields a steady-state output which is also a step function. Similarly, the steady-state response of a linear, stable system to a harmonic input is also harmonic. Studying a linear system's characteristics based upon the steady-state response to *harmonic* inputs constitutes a range of *classical control* methods called the *frequency response methods*. Such methods formed the backbone of the *classical control theory* developed between 1900–60, because the modern *state-space* methods (to be discussed in Chapter 3) were unavailable then to give the response of a linear system to any arbitrary input directly in the *time domain* (i.e. as a function of time). Modern control techniques still employ frequency response methods to shed light on some important characteristics of an unknown control system, such as the *robustness* of multi-variable (i.e. multi-input, multi-output) systems. For these reasons, we will discuss frequency response methods here.

A simple choice of the harmonic input, $u(t)$, can be

$$u(t) = u_o \cos(\omega t) \quad \text{or} \quad u(t) = u_o \sin(\omega t) \tag{2.37}$$

where u_o is the constant *amplitude* and ω is the *frequency* of excitation (sometimes called the *driving frequency*). If we choose to write the input (and output) of a linear system as *complex functions*, the governing differential equation can be replaced by *complex algebraic equations*. This is an advantage, because complex algebra is easier to deal with than differential equations. Furthermore, there is a vast factory of analytical machinery for dealing with complex functions, as we will sample later in this chapter. For these powerful reasons, let us express the harmonic input in the *complex space* as

$$u(t) = u_o e^{i\omega t} \tag{2.38}$$

where $i = \sqrt{-1}$ (a purely imaginary quantity), and

$$e^{i\omega t} = \cos(\omega t) + i \sin(\omega t) \tag{2.39}$$

Equation (2.39) is a *complex* representation in which $\cos(\omega t)$ is called the *real part* of $e^{i\omega t}$ and $\sin(\omega t)$ is called the *imaginary part* of $e^{i\omega t}$ (because it is multiplied by the imaginary number i). The complex space representation of the harmonic input given by Eq. (2.38) is shown in Figure 2.17. The two axes of the complex plane are called the *real*

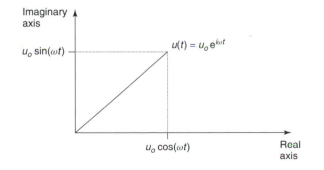

Figure 2.17 Complex space representation of a harmonic input, $u(t)$

and *imaginary axis*, respectively, as shown. Hence, complex space representation of a harmonic function is a device of representing both the possibilities of a simple harmonic input, namely $u_o \cos(\omega t)$ and $u_o \sin(\omega t)$, respectively, in one expression. By obtaining a steady-state response to the complex input given by Eq. (2.38), we will be obtaining simultaneously the steady-state responses of a linear, stable system to $u_o \cos(\omega t)$ and $u_o \sin(\omega t)$.

When you studied solution to ordinary differential equations, you learnt that their solution consists of two parts – the *complimentary solution* (or the solution to the *unforced* differential equation (Eq. (2.7)), and a *particular solution* which depends upon the input. While the transient response of a linear, stable system is *largely* described by the complimentary solution, the steady-state response is the same as the particular solution at large times. The particular solution is of the *same form* as the input, and must by itself satisfy the differential equation. Hence, you can verify that the steady-state responses to $u(t) = u_o \cos(\omega t)$ and $u(t) = u_o \sin(\omega t)$, are given by $y_{ss}(t) = y_o \cos(\omega t)$ and $y_{ss}(t) = y_o \sin(\omega t)$, respectively (where y_o is the amplitude of the resulting harmonic, steady-state output, $y_{ss}(t)$) by plugging the corresponding expressions of $u(t)$ and $y_{ss}(t)$ into Eq. (2.4), which represents a general linear system. You will see that the equation is satisfied in each case. In the complex space, we can write the steady-state response to harmonic input as follows:

$$y_{ss}(t) = y_o(i\omega)e^{i\omega t} \qquad (2.40)$$

Here, the steady-state response amplitude, y_o, is a *complex* function of the frequency of excitation, ω. We will shortly see the implications of a complex response amplitude. Consider a linear, lumped parameter, control system governed by Eq. (2.4) which can be re-written as follows

$$D_1\{y_{ss}(t)\} = D_2\{u(t)\} \qquad (2.41)$$

where $D_1\{\cdot\}$ and $D_2\{\cdot\}$ are *differential operators* (i.e. they *operate* on the steady-state output, $y_{ss}(t)$, and the input, $u(t)$, respectively, by *differentiating* them), given by

$$D_1\{\cdot\} = a_n d^n/dt^n + a_{n-1}d^{n-1}/dt^{n-1} + \cdots + a_1 d/dt + a_0 \qquad (2.42)$$

and

$$D_2\{\cdot\} = b_m d^m/dt^m + b_{m-1}d^{m-1}/dt^{m-1} + \cdots + b_1 d/dt + b_0 \qquad (2.43)$$

Then noting that

$$D_1(e^{i\omega t}) = [(i\omega)^n a_n d^n/dt^n + (i\omega)^{n-1}a_{n-1}d^{n-1}/dt^{n-1} + \cdots + (i\omega)a_1 d/dt + a_0]e^{i\omega t}$$
$$(2.44)$$

and

$$D_2(e^{i\omega t}) = [(i\omega)^m b_m d^m/dt^m + (i\omega)^{m-1}b_{m-1}d^{m-1}/dt^{m-1} + \cdots + (i\omega)b_1 d/dt + b_0]e^{i\omega t}$$
$$(2.45)$$

we can write, using Eq. (2.41),

$$y_o(i\omega) = G(i\omega)u_o \qquad (2.46)$$

where $G(i\omega)$ is called the *frequency response* of the system, and is given by

$$G(i\omega) = [(i\omega)^m b_m + (i\omega)^{m-1} b_{m-1} + \cdots + (i\omega)b_1 + b_o]/[(i\omega)^n a_n$$

$$+ (i\omega)^{n-1} a_{n-1} + \cdots + (i\omega)a_1 + a_o] \qquad (2.47)$$

Needless to say, the frequency response $G(i\omega)$ is also a complex quantity, consisting of both real and imaginary parts. Equations (2.46) and (2.47) describe how the steady-state output of a linear system is related to its input through the frequency response, $G(i\omega)$. Instead of the real and imaginary parts, an alternative description of a complex quantity is in terms of its *magnitude* and the *phase*, which can be thought of as a vector's *length* and *direction*, respectively. Representation of a complex quantity as a vector in the complex space is called a *phasor*. The length of the phasor in the complex space is called its *magnitude*, while the angle made by the phasor with the real axis is called its *phase*. The magnitude of a phasor represents the amplitude of a harmonic function, while the phase determines the value of the function at $t = 0$. The phasor description of the steady-state output amplitude is given by

$$y_o(i\omega) = |y_o(i\omega)|e^{i\alpha(\omega)} \qquad (2.48)$$

where $|y_o(i\omega)|$ is the *magnitude* and $\alpha(\omega)$ is the *phase* of $y_o(i\omega)$. It is easy to see that

$$|y_o(i\omega)| = [\text{real }\{y_o(i\omega)\}^2 + \text{imag }\{y_o(i\omega)\}^2]^{1/2};$$

$$\alpha(\omega) = \tan^{-1}[\text{imag }\{y_o(i\omega)\}/\text{real }\{y_o(i\omega)\}] \qquad (2.49)$$

where real$\{\cdot\}$ and imag$\{\cdot\}$ denote the real and imaginary parts of a complex number. We can also express the frequency response, $G(i\omega)$, in terms of its magnitude, $|G(i\omega)|$, and phase, $\phi(\omega)$, as follows:

$$G(i\omega) = |G(i\omega)|e^{i\phi(\omega)} \qquad (2.50)$$

Substituting Eqs. (2.48) and (2.50) into Eq. (2.46), it is clear that $|y_o(i\omega)| = |G(i\omega)|u_o$ and $\alpha(\omega) = \phi(\omega)$. Hence, the steady-state response of a linear system excited by a harmonic input of amplitude u_o and *zero* phase ($u_o = u_o e^{i0}$) is given through Eq. (2.40) by

$$y_{ss}(t) = y_o(i\omega)e^{i\omega t} = |G(i\omega)|u_o e^{i\phi(\omega)}e^{i\omega t} = |G(i\omega)|u_o e^{i[\omega t + \phi(\omega)]} \qquad (2.51)$$

Thus, the steady-state response to a zero phase harmonic input acquires its phase from the frequency response, which is purely a characteristic of the linear system. You can easily show that if the harmonic input has a *non-zero* phase, then the phase of the steady-state response is the *sum* of the input phase and the phase of the frequency response, $\phi(\omega)$. The phasor representation of the steady-state response amplitude is depicted in Figure 2.18.

From Eq. (2.51), it is clear that the steady-state response is governed by the amplitude of the harmonic input, u_o, and magnitude and phase of the frequency response, $G(i\omega)$, which represent the characteristics of the system, and are functions of the frequency of excitation. If we excite the system at various frequencies, and measure the magnitude and phase of the steady-state response, we could obtain $G(i\omega)$ using Eq. (2.51), and consequently, crucial information about the system's characteristics (such as the coefficients a_k

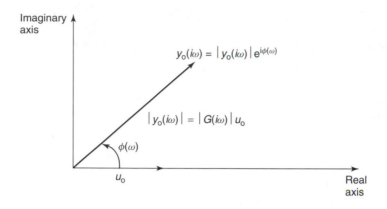

Figure 2.18 Phasor representations of a harmonic input, $u(t)$, with zero phase and amplitude u_0, and steady-state response amplitude, $y_0(i\omega)$, of a linear system with frequency response, $G(i\omega)$

and b_k, in Eq. (2.47)). In general, we would require $G(i\omega)$ at as many frequencies as are the number of unknowns, a_k and b_k, in Eq. (2.47). Conversely, if we know a system's parameters, we can study some of its properties, such as *stability* and *robustness*, using frequency response plots (as discussed later in this chapter). Therefore, plots of magnitude and phase of $G(i\omega)$ with frequency, ω, serve as important tools in the analysis and design of control systems. Alternatively, we could derive the same information as obtained from the magnitude and phase plots of $G(i\omega)$ from the *path* traced by the *tip* of the frequency response phasor in the complex space as the frequency of excitation is varied. Such a plot of $G(i\omega)$ in the complex space is called a *polar* plot (since it represents $G(i\omega)$ in terms of the *polar coordinates*, $|G(i\omega)|$ and $\phi(\omega)$). Polar plots have an advantage over the frequency plots of magnitude and phase in that both magnitude and phase can be seen in *one* (rather than two) plots. Referring to Figure 2.18, it is easily seen that a phase $\phi(\omega) = 0°$ corresponds to the real part of $G(i\omega)$, while the phase $\phi(\omega) = 90°$ corresponds to the imaginary part of $G(i\omega)$. When talking about stability and robustness properties, we will refer again to the polar plot.

Since the range of frequencies required to study a linear system is usually very large, it is often useful to plot the magnitude, $|G(i\omega)|$, and phase, $\phi(\omega)$, with respect to the frequency, ω, on a *logarithmic scale* of frequency, called *Bode plots*. In Bode plots, the magnitude is usually converted to *gain* in *decibels* (dB) by taking the logarithm of $|G(i\omega)|$ to the base 10, and multiplying the result with 20 as follows:

$$\text{Gain} = 20 \log_{10} |G(i\omega)| \tag{2.52}$$

As we will see later in this chapter, important information about a linear, single-input, single-output system's behavior (such as *stability* and *robustness*) can be obtained from the Bode plots, which serve as a cornerstone of classical control design techniques. *Factoring* the polynomials in $G(i\omega)$ (Eq. (2.47)) just produces *addition* of terms in $\log_{10} |G(i\omega)|$, which enables us to construct Bode plots by log-paper and pencil. Despite this, Bode plots are cumbersome to construct by hand. With the availability of personal computers and software with mathematical functions and graphics capability – such as MATLAB – Bode plots can be plotted quite easily. In MATLAB, all you have to do is specify a set of frequencies, ω, at which the gain and phase plots are desired, and use the intrinsic functions *abs* and *angle* which calculate the magnitude and phase (in radians), respectively, of a complex number. If you have the MATLAB's *Control System Toolbox* (CST), the task of obtaining a Bode plot becomes even simpler through the use of the command *bode* as follows:

```
>>G=tf(num,den); bode(G,w) <enter> %a Bode plot will appear on the screen
```

Here >> is the MATLAB prompt, <enter> denotes the pressing of the 'enter' (or 'return') key, and the % sign indicates that everything to its right is a comment. In the *bode* command, *w* is the specified frequency vector consisting of *equally spaced* frequency values at which the gain and phase are desired, *G* is the name given to the frequency response of the linear, time-invariant system created using the CST *LTI object* function *tf* which requires *num* and *den* as the vectors containing the coefficients of *numerator* and *denominator polynomials*, respectively, of $G(i\omega)$ in (Eq. (2.47)) in *decreasing* powers of *s*. These coefficients should be be specified as follows, before using the *tf* and *bode* commands:

```
>>num=[bm bm−1 ... b0]; den=[an an−1 ... a0]; <enter>
```

By using the MATLAB command *logspace*, the *w* vector can also be pre-specified as follows:

```
>>w=logspace(-2,3); <enter> %w consists of  equally spaced frequencies in the
    range 0.01-1000 rad/s.
```

(Using a semicolon after a MATLAB command suppresses the print-out of the result on the screen.)

Obviously, *w* must be specified *before* you use the *bode* command. If you don't specify *w*, MATLAB will automatically generate an appropriate *w* vector, and create the plot.

Instead of plotting the Bode plot, you may like to store the magnitude (*mag*), $|G(i\omega)|$, and the *phase*, $\phi(\omega)$, at given set of frequencies, *w*, for further processing by using the following MATLAB command:

```
>>[mag,phase,w]=bode(num,den,w); <enter>
```

For more information about Bode plots, do the following:

```
>>help bode <enter>
```

The same procedure can be used to get help on any other MATLAB command. The example given below will illustrate what Bode plots look like. Before we do that, let us try to understand in physical terms what a frequency response (given by the Bode plot) is.

Musical notes produced by a guitar are related to its frequency response. The guitar player makes each string vibrate at a particular frequency, and the notes produced by the various strings are the measure of whether the guitar is being played well or not. Each string of the guitar is capable of being excited at many frequencies, depending upon where the string is struck, and where it is held. Just like the guitar, any system can be *excited* at a set of frequencies. When we use the word *excited*, it is quite in the literal sense, because it denotes the condition (called *resonance*) when the magnitude of the frequency response, $|G(i\omega)|$, becomes very large, or infinite. The frequencies at which a system can be excited are called its *natural (or resonant) frequencies*. High pitched voice of many a diva has shattered the opera-house window panes while accidently singing at one of the natural frequencies of the window! If a system contains energy dissipative processes (called *damping*), the frequency response magnitude at natural frequencies is large, but *finite*. An *undamped* system, however, has *infinite* response at each natural frequency. A natural frequency is indicated by a *peak* in the gain plot, or as the frequency where the phase changes by 180°. A practical limitation of Bode plots is that they show only an *interpolation* of the gain and phase through selected frequency points. The frequencies where $|G(i\omega)|$ becomes *zero* or *infinite* are excluded from the gain plot (since logarithm of zero is undefined, and an infinite gain cannot be shown on any scale). Instead, only frequency points located *close* to the zero magnitude frequency and the infinite gain frequencies of the system can be used in the gain plot. Thus, the Bode gain plot for a guitar will consist of several *peaks*, corresponding to the natural frequencies of the notes being struck. One could determine from the peaks the *approximate* values of the natural frequencies.

Example 2.8

Consider the electrical network shown in Figure 2.19 consisting of three resistances, R_1, R_2, and R_3, a capacitor, C, and an inductor, L, connected to a voltage source, $e(t)$, and a switch, S. When the switch, S, is closed at time $t = 0$, the current passing through the resistance R_1 is $i_1(t)$, and that passing through the inductor, L, is $i_2(t)$. The input to the system is the applied voltage, $e(t)$, and the output is the current, $i_2(t)$.

The two governing equations of the network are

$$e(t) = R_1 i_1(t) + R_3[i_1(t) - i_2(t)] \tag{2.53}$$

$$0 = R_2 i_2(t) + R_3[i_2(t) - i_1(t)] + L i_2^{(1)}(t) + (1/C) \int_0^t i_2(\tau)\, d\tau \tag{2.54}$$

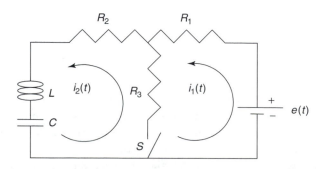

Figure 2.19 Electrical network for Example 2.8

Differentiating Eq. (2.54) and eliminating $i_1(t)$, we can write

$$Li_2^{(2)}(t) + [(R_1R_3 + R_1R_2 + R_2R_3)/(R_1 + R_3)]i_2^{(1)}(t)$$

$$+ (1/C)i_2(t) = [R_3/(R_1 + R_3)]e^{(1)}(t) \qquad (2.55)$$

Comparing Eq. (2.55) with Eq. (2.4) we find that the system is linear and of second order, with $y(t) = i_2(t)$, $u(t) = e(t)$, $a_0 = 1/C$, $a_1 = (R_1R_3 + R_1R_2 + R_2R_3)/(R_1 + R_3)$, $b_0 = 0$, and $b_1 = R_3/(R_1 + R_3)$. Hence, from Eq. (2.47), the frequency response of the system is given by

$$G(i\omega) = (i\omega)[R_3/(R_1 + R_3)]/[(i\omega)^2 L + (i\omega)(R_1R_3$$

$$+ R_1R_2 + R_2R_3)/(R_1 + R_3) + 1/C] \qquad (2.56)$$

For $R_1 = R_3 = 10$ ohms, $R_2 = 25$ ohms, $L = 1$ henry, and $C = 10^{-6}$ farad, the frequency response is the following:

$$G(i\omega) = 0.5(i\omega)/[(i\omega)^2 + 30(i\omega) + 10^6] \qquad (2.57)$$

Bode gain and phase plots of frequency response given by Eq. (2.57) can be plotted in Figure 2.20 using the following MATLAB commands:

```
>>w=logspace(-1,4); <enter>
```

(This command produces equally spaced frequency points on logarithmic scale from 0.1 to 10 000 rad/s, and stores them in the vector w.)

```
>>G=i*w*0.5./(-w.*w+30*i*w+1e6); <enter>
```

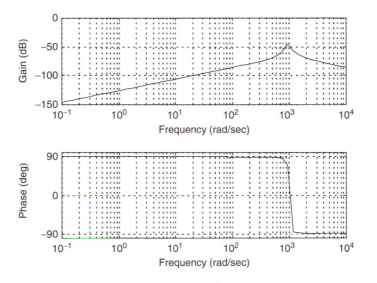

Figure 2.20 Bode plot for the electrical network in Example 2.8; a peak in the gain plot and the corresponding phase change of $180°$ denotes the natural frequency of the system

(This command calculates the value of $G(i\omega)$ by Eq. (2.57) at each of the specified frequency points in w, and stores them in the vector G. Note the MATLAB operations .* and ./ which allow element by element multiplication and division, respectively, of two arrays (see Appendix B).)

```
>>gain=20*log10(abs(G)); phase=180*angle(G)/pi; <enter>
```

(This command calculates the gain and phase of $G(i\omega)$ at each frequency point in w using the MATLAB intrinsic functions *abs, angle,* and *log10*, and stores them in the vectors *gain* and *phase*, respectively. We are assuming, however, that G does not become zero or infinite at any of the frequencies contained in w.)

```
>>subplot(211), semilogx(w,gain), grid, subplot(212), semilogx(w,phase),
   grid <enter>
```

(This command produces gain and phase Bode plots as two (unlabeled) subplots, as shown in Figure 2.20. Labels for the axes can be added using the MATLAB commands *xlabel* and *ylabel*.)

The Bode plots shown in Figure 2.20 are obtained much more easily through the Control System Toolbox (CST) command *bode* as follows:

```
>>num=[0.5 0]; den=[1 30 1e6]; g=tf(num,den), bode(g,w) <enter>
```

Note the peak in the gain plot of Figure 2.20 at the frequency, $\omega = 1000$ rad/s. At the same frequency the phase changes by $180°$. Hence, $\omega = 1000$ rad/s is the system's natural frequency. To verify whether this is the *exact* natural frequency, we can rationalize the denominator in Eq. (2.57) (i.e. make it a real number by multiplying both numerator and denominator by a suitable complex factor – in this case $(-\omega^2 + 10^6) - 30i\omega$ and express the magnitude and phase as follows:

$$|G(i\omega)| = [225\omega^4 + 0.25\omega^2(-\omega^2 + 10^6)^2]^{1/2}/[(-\omega^2 + 10^6)^2 + 900\omega^2];$$

$$\phi(\omega) = \tan^{-1}(-\omega^2 + 10^6)/(30\omega) \tag{2.58}$$

From Eq. (2.58), it is clear that $|G(i\omega)|$ has a *maximum* value (0.0167 or -35.547 dB) – and $\phi(\omega)$ jumps by $180°$ – at $\omega = 1000$ rad/s. Hence, the natural frequency is *exactly* 1000 rad/s. Figure 2.20 also shows that the gain at $\omega = 0.1$ rad/s is -150 dB, which corresponds to $|G(0.1i)| = 10^{-7.5} = 3.1623 \times 10^{-8}$, a small number. Equation (2.58) indicates that $|G(0)| = 0$. Hence, $\omega = 0.1$ rad/s approximates quite well the zero-frequency gain (called the *DC gain*) of the system. The frequency response is used to define a linear system's property called *bandwidth* defined as the range of frequencies from zero *up to* the frequency, ω_b, where $|G(i\omega_b)| = 0.707|G(0)|$. Examining the numerator of $|G(i\omega)|$ in Eq. (2.58), we see that $|G(i\omega)|$ vanishes at $\omega = 0$ and $\omega = 1999\,100$ rad/s (the numerator roots can be obtained using the MATLAB intrinsic function *roots*). Since $|G(0)| = 0$, the present system's bandwidth is $\omega_b = 1999\,100$ rad/s (which lies beyond the frequency range of Figure 2.20). Since the degree of the denominator polynomial of $G(i\omega)$ in Eq. (2.47) is *greater* than that of the numerator polynomial, it follows that $|G(i\omega)| \to 0$ as $\omega \to \infty$. Linear systems with $G(i\omega)$ having a higher degree denominator polynomial (than the numerator polynomial) in Eq. (2.47) are called *strictly proper* systems. Equation (2.58) also shows that $\phi(\omega) \to 90°$ as $\omega \to 0$, and $\phi(\omega) \to -90°$ as $\omega \to \infty$. For a general system, $\phi(\omega) \to -k90°$ as $\omega \to \infty$, where k is the number by which the degree of the denominator polynomial of $G(i\omega)$ *exceeds* that of the numerator polynomial (in the present example, $k = 1$).

Let us now draw a polar plot of $G(i\omega)$ as follows (note that we need more frequency points close to the natural frequency for a smooth polar plot, because of the 180° phase jump at the natural frequency):

```
>>w=[logspace(-1,2.5) 350:2:1500 logspace(3.18,5)]; <enter>
```

(This command creates a frequency vector, w, with more frequency points close to 1000 rad/s.)

```
>>G=i*w*0.5./(-w.*w+30*i*w+1e6); <enter>
```

```
>>polar(angle(G), abs(G)); <enter>
```

(This command for generating a polar plot requires phase angles in *radians*, but the plot *shows* the phase in *degrees*.)

The resulting polar plot is shown in Figure 2.21. The plot is in polar coordinates, $|G(i\omega)|$ and $\phi(\omega)$, with circles of constant radius, $|G(i\omega)|$, and *radial* lines of constant $\phi(\omega)$ overlaid on the plot. Conventionally, polar plots show either *all positive*, or *all negative* phase angles. In the present plot, the negative phase angles have been shown as positive angles using the transformation $\phi \to (\phi + 360°)$, which is acceptable since both sine and cosine functions are *invariant* under this transformation for $\phi < 0$ (e.g. $\phi = -90°$ is the same as $\phi = 270°$). Note that the 0° and

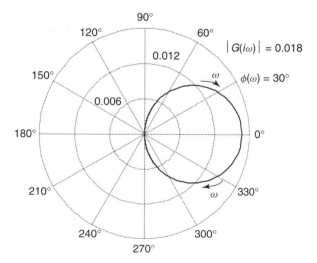

Figure 2.21 Polar plot of the frequency response, $G(i\omega)$, of the electrical system of Example 2.8

90° radial lines represent the real and imaginary parts, respectively, of $G(i\omega)$. The polar curve is seen in Figure 2.21 to be a circle of radius 0.00835 centered at the point 0.00835 on the real axis. The direction of increasing ω is shown by arrows on the polar curve. The shape and direction (with increasing ω) of a polar plot gives valuable insight about a linear system's stability, which will be seen in Section 2.10.

2.4 Laplace Transform and the Transfer Function

In the previous section we had confined our attention to the *steady-state* response of a linear system to harmonic inputs. Here we would like to consider the *total* response (both transient and steady-state) of a linear, single-input, single-output system when the applied input is some *arbitrary* function of time. We saw how the representation of a harmonic input by a complex function *transformed* the governing differential equations into a complex algebraic expression for the frequency response. For a general input, a similar complex expression can be obtained by applying the *Laplace transformation* (denoted by $\mathcal{L}$) to the input, $u(t)$, defined as

$$U(s) = \mathcal{L}u(t) = \int_0^\infty e^{-st} u(t)\,dt \qquad (2.59)$$

where s denotes the *Laplace variable* (a complex number), and $U(s)$ is called the *Laplace transform* of $u(t)$. The Laplace transform of a function $u(t)$ is defined only if the infinite integral in Eq. (2.59) exists, and converges to a functional form, $U(s)$. However, if $U(s)$ exists, then it is *unique*. The convergence of the Laplace integral depends solely upon the shape of the function, $u(t)$. It can be shown rigorously that the Laplace integral converges only if $u(t)$ is *piecewise continuous* (i.e. any time interval, however large, can be broken up into a finite number of sub-intervals over each of which $u(t)$ is continuous, and at the ends of each sub-interval, $u(t)$ is finite) and *bounded by an exponential* (i.e. there exists a constant a such that $e^{-at}|u(t)|$ is *bounded* at all times). The term *bounded* implies that a function's value lies between two *finite* limits. Most of the commonly used input functions are Laplace transformable. For example, if $u(t)$, is *discontinuous* (i.e. it has a jump) at $t = 0$, such as $u(t) = \delta(t)$ or $u(t) = u_s(t)$, we can obtain its Laplace transform. In such a case, the lower limit of integration in Eq. (2.59) is understood to be *just before* $t = 0$, i.e. just prior to the discontinuity in $u(t)$. Some important properties of the Laplace transform are stated below, and you may verify each of them using the definition given by Eq. (2.59):

(a) Linearity:
   If a is a constant (or independent of s and t) and $\mathcal{L}f(t) = F(s)$, then

$$\mathcal{L}\{af(t)\} = a\mathcal{L}f(t) = aF(s) \qquad (2.60)$$

Also, if $\mathcal{L}f_1(t) = F_1(s)$ and $\mathcal{L}f_2(t) = F_2(s)$, then

$$\mathcal{L}\{f_1(t) + f_2(t)\} = F_1(s) + F_2(s) \qquad (2.61)$$

(b) Complex differentiation:

If $\mathcal{L}f(t) = F(s)$, then

$$\mathcal{L}\{tf(t)\} = -dF(s)/ds \qquad (2.62)$$

(c) Complex integration:

If $\mathcal{L}f(t) = F(s)$, and if $\lim_{t\to 0} f(t)/t$ exists as $t = 0$ is approached from the *positive side*, then

$$\mathcal{L}\{f(t)/t\} = \int_s^\infty F(s)\,ds \qquad (2.63)$$

(d) Translation in time:

If $\mathcal{L}f(t) = F(s)$, and a is a positive, real number such that $f(t - a) = 0$ for $0 < t < a$, then

$$\mathcal{L}f(t - a) = e^{-as} F(s) \qquad (2.64)$$

(e) Translation in Laplace domain:

If $\mathcal{L}f(t) = F(s)$, and a is a complex number, then

$$\mathcal{L}\{e^{at} f(t)\} = F(s - a) \qquad (2.65)$$

(f) Real differentiation:

If $\mathcal{L}f(t) = F(s)$, and if $f^{(1)}(t)$ is Laplace transformable, then

$$\mathcal{L}f^{(1)}(t) = s F(s) - f(0^+) \qquad (2.66)$$

where $f(0^+)$ denotes the value of $f(t)$ in the limit $t \to 0$, approaching $t = 0$ from the *positive side*. If we apply the real differentiation property successively to the higher order time derivatives of $f(t)$ (assuming they are Laplace transformable), we can write the Laplace transform of the kth derivative, $f^{(k)}(t)$, as follows:

$$\mathcal{L}f^{(k)}(t) = s^k F(s) - s^{k-1} f(0^+) - s^{k-2} f^{(1)}(0^+) - \cdots - f^{(k-1)}(0^+) \qquad (2.67)$$

(g) Real integration:

If $\mathcal{L}f(t) = F(s)$, and the indefinite integral $\int f(t)\,dt$ is Laplace transformable, then

$$\mathcal{L}\left\{\int f(t)\,dt\right\} = F(s)/s + (1/s)\int_{-\infty}^0 f(t)\,dt \qquad (2.68)$$

Note that the integral term on the right-hand side of Eq. (2.68) is zero if $f(t) = 0$ for $t < 0$.

(h) Initial value theorem:

If $\mathcal{L}f(t) = F(s)$, $f^{(1)}(t)$ is Laplace transformable, and $\lim_{s\to\infty} s F(s)$ exists, then

$$f(0^+) = \lim_{s\to\infty} s F(s) \qquad (2.69)$$

(i) Final value theorem:

If $\mathcal{L}f(t) = F(s)$, $f^{(1)}(t)$ is Laplace transformable, and $\lim_{t\to\infty} f(t) = f(\infty)$ exists, then

$$f(\infty) = \lim_{s\to 0} s F(s) \qquad (2.70)$$

Since we are usually dealing with positive values of time, we will replace 0^+ by 0 in all relevant applications of the Laplace transform. It is easy to see that if the input, $u(t)$, and its time derivatives are Laplace transformable, then the differential equation (Eq. (2.4)) of a linear, *time-invariant* system is Laplace transformable, which implies that the output, $y(t)$, is also Laplace transformable, whose Laplace transform is $Y(s)$. For simplicity, we assume that all initial conditions for the input, $u(t)$, and its derivatives and the output, $y(t)$, and its derivatives are zeros. Then, using Eq. (2.67) we can transform the governing equation of the system (Eq. (2.4)) to the Laplace domain as follows:

$$(s^n a_n + s^{n-1} a_{n-1} + \cdots + s a_1 + a_o) Y(s) = (s^m b_m + s^{m-1} b_{m-1} + \cdots + s b_1 + b_o) U(s)$$
(2.71)

Equation (2.71) brings us to one of the most important concepts in control theory, namely the *transfer function*, $G(s)$, which is defined as the ratio of the Laplace transform of the output, $Y(s)$, and that of the input, $U(s)$, given by

$$G(s) = Y(s)/U(s)$$
(2.72)

Substituting Eq. (2.71) into (2.72), we obtain the following expression for the transfer function of a linear, single-input, single-output system:

$$G(s) = (s^m b_m + s^{m-1} b_{m-1} + \cdots + s b_1 + b_o)/(s^n a_n + s^{n-1} a_{n-1} + \cdots + s a_1 + a_o)$$
(2.73)

As we saw in Chapter 1, the transfer function, $G(s)$, represents how an input, $U(s)$, is *transferred* to the output, $Y(s)$, or, in other words, the *relationship* between the input and output, when the initial conditions are *zero*. The transfer function representation of a system is widely used in block diagrams, such as Figure 2.22, and is very useful for even such systems for which the governing differential equations are not available. For such unknown systems, the transfer function is like a *black-box* defining the system's characteristics.

By applying known inputs (such as the singularity functions or harmonic signals) and measuring the output, one can determine an unknown system's transfer function experimentally. To do so, we have to see what are the relationships between the transfer function and the responses to singularity functions, and between the transfer function and the frequency response. The latter relationship is easily obtained by comparing Eq. (2.73) defining the transfer function, $G(s)$, with Eq. (2.47), which defines the frequency response, $G(i\omega)$. We see that the two quantities can be obtained from one another by using the relationship $s = i\omega$ (that is the reason why we knowingly used the same symbol, $G(\cdot)$, for both transfer function and the frequency response). A special transform, called the *Fourier transform*, can be defined by substituting $s = i\omega$ in the definition of the Laplace transform (Eq. (2.59). Fourier transform is widely used as a method of calculating the

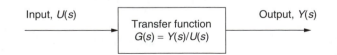

Figure 2.22 Transfer function representation of a single-input, single-output system

response of linear systems to arbitrary inputs by transforming an arbitrary input, $u(t)$, to its *frequency domain* counterpart, $U(i\omega)$ as follows:

$$U(i\omega) = \int_0^\infty e^{-i\omega t} u(t)\, dt \qquad (2.74)$$

(The lower limit of integration in Eq. (2.74) is replaced by $-\infty$ if $u(t) \neq 0$ for $t < 0$.) Then, from Eq. (2.72), we can determine the resulting output (assuming zero initial conditions) in the frequency domain as $Y(i\omega) = G(i\omega)U(i\omega)$ (where $G(i\omega)$ is the pre-determined frequency response), and apply the *inverse Fourier transform* to obtain the output in the time-domain as follows:

$$y(t) = 1/(2\pi) \int_{-\infty}^\infty e^{i\omega t} Y(i\omega)\, d\omega \qquad (2.75)$$

Note that in Eqs. (2.74) and (2.75), the Fourier transforms of the input and the output, $U(i\omega)$ and $Y(i\omega)$, do not have any physical significance, and in this respect they are similar to the Laplace transforms, $U(s)$ and $Y(s)$. However, the frequency response, $G(i\omega)$, is related to the steady-state response to harmonic input (as seen in Section 2.3), and can be experimentally measured. The transfer function, $G(s)$, however, is a useful mathematical abstraction, and cannot be experimentally measured in the Laplace domain. The Laplace variable, s, is a complex quantity, $s = \sigma \pm i\omega$, whose real part, σ, denotes whether the amplitude of the input (or output) is increasing or decreasing with time. We can grasp this fact by applying the *inverse Laplace transform*, $\mathcal{L}^{-1}$ (i.e. going from the Laplace domain to the time domain) to Eq. (2.59)

$$y(t) = \mathcal{L}^{-1} Y(s) = 1/(2\pi i) \int_{\sigma - i\infty}^{\sigma + i\infty} Y(s) e^{st}\, ds \qquad (2.76)$$

where the integral is performed along an *infinitely* long line, parallel to the imaginary axis with a constant real part, σ (Figure 2.23). Note that inverse Laplace transform is possible, because $Y(s)$ (if it exists) is unique.

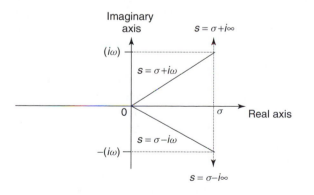

Figure 2.23 The Laplace domain

From Eq. (2.76) we can see that a general output, $y(t)$, will consist of such terms as $y_0 e^{st}$ (where y_0 is a *constant*), which can be expressed as $y_0 e^{\sigma t} e^{\pm i \omega t}$. The latter term indicates a periodically changing quantity of frequency, ω, whose amplitude is a function of time given by $y_0 e^{\sigma t}$. When dealing with non-harmonic inputs and outputs, the use of the Laplace transform and the transfer function, $G(s)$, is more rewarding than working with the Fourier transform and the frequency response, $G(i\omega)$, because the resulting algebraic expressions are much simpler through the use of s rather than $(i\omega)$. However, use of $G(s)$ involves interpreting system characteristics from complex (rather than purely imaginary) numbers.

The roots of the numerator and denominator polynomials of the transfer function, $G(s)$, given by Eq. (2.73) represent the characteristics of the linear, time-invariant system. The *denominator* polynomial of the transfer function, $G(s)$, equated to zero is called the *characteristic equation* of the system, given by

$$s^n a_n + s^{n-1} a_{n-1} + \cdots + s a_1 + a_0 = 0 \qquad (2.77)$$

The roots of the characteristic equation are called the *poles* of the system. The roots of the *numerator* polynomial of $G(s)$ equated to zero are called the *zeros* of the transfer function, given by

$$s^m b_m + s^{m-1} b_{m-1} + \cdots + s b_1 + b_0 = 0 \qquad (2.78)$$

In terms of its poles and zeros, a transfer function can be represented as a ratio of *factorized* numerator and denominator polynomials, given by the following *rational expression*:

$$G(s) = K(s - z_1)(s - z_2) \ldots (s - z_m)/[(s - p_1)(s - p_2) \ldots (s - p_n)]$$

$$= K \prod_{i=1}^{m}(s - z_i) / \prod_{j=1}^{n}(s - p_j) \qquad (2.79)$$

where K is a constant (sometimes referred to as the *gain*), $z_i (i = 1, 2, \ldots, m)$ and $p_j (j = 1, 2, \ldots, n)$ are the zeros and poles of the system, respectively, and Π is a short-hand notation denoting a *product* of many terms (in the same manner as Σ denotes a *summation* of many terms). Equation (2.79) is also called *zero-pole-gain* description of a linear, time-invariant system, which can be modeled by the MATLAB Control System Toolbox's (CST) *LTI* object, *zpk*. As in Eq. (2.1), we repeat that for most linear, time-invariant systems $m \leq n$. Such systems are said to be *proper*. If $m < n$, the system is said to be *strictly proper*. Also, note that some zeros, z_i, and poles, p_j, may be *repeated* (i.e. two or more poles (or zeros) having identical values). Such a pole (or zero) is said to be *multiple*, and its *degree of multiplicity* is defined as the number of times it occurs. Finally, it may happen for some systems that a pole has the same value as a zero (i.e. $p_j = z_i$ for some pair (i,j)). Then the transfer function representation of Eq. (2.79) will not contain those poles and zeros, because they have canceled each other out. Pole-zero cancelations have a great impact on a system's *controllabilty* or *observability* (which will be studied in Chapter 5).

Example 2.9

Revisiting the electrical network of Example 2.8, we can write the system's transfer function as

$$G(s) = 0.5s/(s^2 + 30s + 10^6) \tag{2.80}$$

which indicates a zero at the origin ($z_1 = 0$), and the two complex poles given by the solution of the following quadratic characteristic equation:

$$s^2 + 30s + 10^6 = 0 \tag{2.81}$$

To get a better insight into the characteristics of a system, we can express *each* quadratic factor (such as that on the left-hand side of Eq. (2.81)) of the denominator polynomial as $s^2 + 2\varsigma\omega_n s + \omega_n^2$, where ω_n is a *natural frequency* of the system (see Section 2.3), and ς is called the *damping ratio*. The damping ratio, ς, governs how rapidly the *magnitude* of the response of an *unforced* system decays with time. For a mechanical or electrical system, *damping* is the property which converts a part of the unforced system's *energy* to heat, thereby causing the system's energy – and consequently the output – to dissipate with time. Examples of damping are resistances in electrical circuits and friction in mechanical systems. From the discussion following Eq. (2.76), it can be seen that ς is closely related to the real part, σ, of a complex root of the characteristic equation (pole) given by $s = \sigma \pm i\omega$. The roots of the characteristic equation (or, in other words, the poles of the system) expressed as

$$s^2 + 2\varsigma\omega_n s + \omega_n^2 = 0 \tag{2.82}$$

are

$$s = p_1 = -\varsigma\omega_n - i\omega_n(\varsigma^2 - 1)^{1/2} \tag{2.83}$$

and

$$s = p_2 = -\varsigma\omega_n + i\omega_n(\varsigma^2 - 1)^{1/2} \tag{2.84}$$

Note that the real part of each pole is $\sigma = -\varsigma\omega_n$, while the imaginary parts are $\pm\omega = \pm\omega_n(\varsigma^2 - 1)^{1/2}$. For the present example, the poles are found by solving Eq. (2.81) to be $p_{1,2} = -15 \pm 999.9i$, which implies that the natural frequency and damping-ratio are, $\omega_n = 1000$ rad/s and $\varsigma = 0.015$, respectively. These numbers could also have been obtained by comparing Eq. (2.81) and Eq. (2.82). The natural frequency agrees with our calculation in Example 2.8, which was also observed as a peak in the Bode gain plot of Figure 2.20. The positive damping-ratio (or the negative real part of the complex poles) indicates that the amplitude of the response to any input will decay with time due to the presence of terms such as $y_0 e^{\sigma t} e^{\pm i\omega t}$ in the expression for the output, $y(t)$.

One can see the dependence of the response, $y(t)$, on the damping-ratio, ς, in Figure 2.24, which is a plot of a typical initial response of an unforced second order system. $\varsigma = 1$ is the limiting case, called *critical damping*, because it denotes the boundary between *oscillatory* and *exponentially decaying* response. For $0 < \varsigma < 1$,

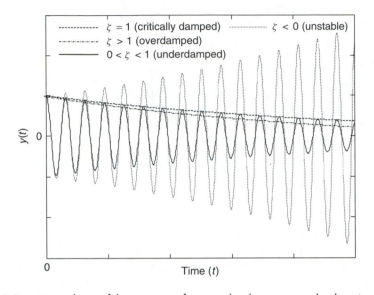

Figure 2.24 Dependence of the response of a second order system on the damping-ratio, ς

the response is oscillatory with amplitude decreasing with time (called the *under-damped case*), while for $\varsigma > 1$, the response decays exponentially (called the *over-damped case*). Clearly, the larger the value of the damping-ratio, ς, the faster the response decays to zero. The case for which $\varsigma < 0$ denotes a response with *exponentially increasing amplitude*. A response, $y(t)$, whose limit as $t \to \infty$, either does not exist or is *infinite*, is called an *unbounded* response. Clearly, $\varsigma < 0$ case has an unbounded response. As soon as we see a linear system producing an unbounded response to a *bounded* input (i.e. an input whose finite limit exists as $t \to \infty$) and *finite* initial conditions, we call the system *unstable*. A further discussion of *stability* follows a little later.

Locations of poles and zeros in the Laplace domain determine the characteristics of a linear, time-invariant system. Some indication of the locations of a poles and zeros can be obtained from the frequency response, $G(i\omega)$. Let us go back to Figure 2.20, showing the Bode plots of the electrical system of Examples 2.8 and 2.9. Due to the presence of a zero at the origin (see Eq. (2.80)), there is a phase of $90°$ and a *non-zero* (dB) gain at $\omega = 0$. The presence of a complex conjugate pair of poles is indicated by a peak in the gain plot and a phase change of $180°$. The difference between the number of zeros and poles in a system affects the phase and the *slope* of the Bode gain plot with frequency (in units of dB per *decade* of frequency), when the frequency is very large (i.e. in the limit $\omega \to \infty$). From Eq. (2.79), we can say the following about gain-slope and phase in the high-frequency limit:

$$\lim_{\omega \to \infty} d\{20 \log_{10} |G(i\omega)|\}/d\omega \approx 20(m - n) \text{ dB/decade}$$

$$\lim_{\omega \to \infty} \phi(\omega) \approx \begin{bmatrix} (m - n)90° & \text{if } K > 0 \\ (m - n)90° - 180° & \text{if } K < 0 \end{bmatrix} \qquad (2.85)$$

Note that the expressions in Eq. (2.85) are only *approximate*. For example, the transfer function in Eq. (2.80) has $K = 0.5$, $m = 1$, and $n = 2$, which implies that the gain-slope and phase in the limit $\omega \to \infty$ should be -20 dB/decade and $-90°$, respectively. These values are very good estimates (the phase is exactly $-90°$) of the frequency response plotted in Figure 2.20.

Example 2.10

Consider a linear model describing the longitudinal dynamics of an aircraft (Figure 2.25). *Three* different output variables (in the Laplace domain) are of interest when the aircraft is displaced from the equilibrium point (defined by a constant *angle of attack*, α_0, a constant longitudinal *velocity*, v_0, and a constant *pitch-angle*, θ_0): the *change* in airspeed, $v(s)$, the *change* in the angle of attack, $\alpha(s)$, and the change in pitch angle, $\theta(s)$. The input variable in the Laplace domain is the *elevator angle*, $\delta(s)$. The three transfer functions separately defining the relationship between the input, $\delta(s)$, and the three respective outputs, $v(s)$, $\alpha(s)$, and $\theta(s)$, are as follows:

$$v(s)/\delta(s) = -0.0005(s - 70)(s + 0.5)/[(s^2 + 0.005s + 0.006)(s^2 + s + 1.4)] \tag{2.86}$$

$$\alpha(s)/\delta(s) = -0.02(s + 80)(s^2 + 0.0065s + 0.006)/$$
$$[(s^2 + 0.005s + 0.006)(s^2 + s + 1.4)] \tag{2.87}$$

$$\theta(s)/\delta(s) = -1.4(s + 0.02)(s + 0.4)/[(s^2 + 0.005s + 0.006)(s^2 + s + 1.4)] \tag{2.88}$$

It should be noted that all three transfer functions have the *same* denominator polynomial, $(s^2 + 0.005s + 0.006)(s^2 + s + 1.4)$. Since we know that the denominator polynomial equated to zero denotes the characteristic equation of the system, we can write the characteristic equation for the aircraft's longitudinal dynamics as

$$(s^2 + 0.005s + 0.006)(s^2 + s + 1.4) = 0 \tag{2.89}$$

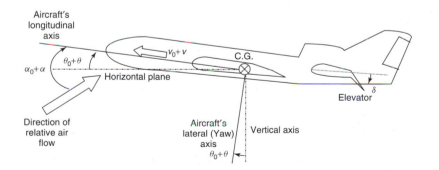

Figure 2.25 Longitudinal dynamics of an airplane, with outputs α, θ, and v denoting small changes in angle of attack, pitch angle, and velocity component along longitudinal axis, respectively, and input, elevator deflection, δ. The equilibrium condition is denoted by $\alpha = \theta = v = \delta = 0$

Equation (2.89) indicates that the systems complex poles are given by *two quadratic factors* $(s^2 + 0.005s + 0.006)$ and $(s^2 + s + 1.4)$. Comparing the result with that of Example 2.9, where the quadratic factor in the characteristic polynomial was expressed as $s^2 + 2\varsigma\omega_n s + \omega_n^2$, we can see that here we should expect *two* values of the natural frequency, ω_n, and the damping-ratio, ς, i.e. one set of values for each of the *two* quadratic factors. These values are the following:

(a) $\varsigma = 0.4226$; $\omega_n = 1.1832$ rad/s   (*short-period mode*)

(b) $\varsigma = 0.0323$; $\omega_n = 0.0775$ rad/s   (*long-period*, or *phugoid mode*)

Using MATLAB's Control System Toolbox (CST) command *damp*, the damping-ratio and natural frequency associated with each quadratic factor in the characteristic equation can be easily obtained as follows:

```
>>a=[1 0.005 0.006]; damp(a) % first quadratic factor <enter>
```

Eigenvalue	Damping	Freq. (rad/sec)
-0.0025+0.0774i	0.0323	0.0775
-0.0025-0.0774i	0.0323	0.0775

```
>>b=[1 1 1.4]; damp(b) % second quadratic factor <enter>
```

Eigenvalue	Damping	Freq. (rad/sec)
-0.5000+1.0724i	0.4226	1.1832
-0.5000-1.0724i	0.4226	1.1832

Note that the CST command *damp* also lists the *eigenvalues*, which are nothing but the roots of the characteristic polynomial (same as the *poles* of the system). We will discuss the *eigenvalues* in Chapter 3. (Alternatively, we could have used the intrinsic MATLAB function *roots* to get the pole locations as the roots of each quadratic factor.) As expected, the poles for each quadratic factor in the characteristic equation are complex conjugates. Instead of calculating the roots of each quadratic factor separately, we can multiply the two quadratic factors of Eq. (2.89) using the intrinsic MATLAB command *conv*, and then directly compute the roots of the characteristic polynomial as follows:

```
>>damp(conv(a,b))% roots of the characteristic polynomial <enter>
```

Eigenvalue	Damping	Freq. (rad/sec)
-0.0025+0.0774i	0.0323	0.0775
-0.0025-0.0774i	0.0323	0.0775
-0.5000+1.0724i	0.4226	1.1832
-0.5000-1.0724i	0.4226	1.1832

The pair of natural frequencies and damping-ratios denote two *natural modes* of the system, i.e. the two ways in which one can excite the system. The first mode is highly damped, with a larger natural frequency (1.1832 rad/s), and is called the *short-period mode* (because the *time-period* of the oscillation, $T = 2\pi/\omega_n$ is smaller for this mode). The second characteristic mode is very lightly damped with a smaller natural frequency (0.0775 rad/s) – hence, a longer time-period – and is called the *long-period* (or *phugoid*) *mode*. While an arbitrary input will excite a response containing both of these modes, it is sometimes instructive to study the two modes separately. There are special elevator inputs, $\delta(s)$, which largely excite either one or the other mode at a time. (You may refer to Blakelock [3] for details of longitudinal dynamics and control of aircraft and missiles.)

We now examine the Bode plots of each of the *three* transfer functions, $v(s)/\delta(s)$, $\alpha(s)/\delta(s)$, and $\theta(s)/\delta(s)$, respectively, to see how much is each output variable influenced by each of the two characteristic modes. Figures 2.26, 2.27, and 2.28 show the gain and phase Bode plots for the three transfer functions in the limit $s = i\omega$ (they are the frequency responses of the concerned output variable). Using Control Systems Toolbox (CST), these plots are directly obtained by the command *bode*, after constructing each transfer function using the LTI object *tf*. Bode plot of transfer function $v(s)/\delta(s)$ (Figure 2.26) is generated using the following MATLAB statements:

```
>>a=[1 -70]; b=[1 0.5]; num=-0.0005*conv(a,b) <enter>

num =
 -0.0005  0.0348  0.0175

>>a=[1 0.005 0.006]; b=[1 1 1.4]; den=conv(a,b) <enter>

den =
   1.0000  1.0050  1.4110  0.0130  0.0084
```

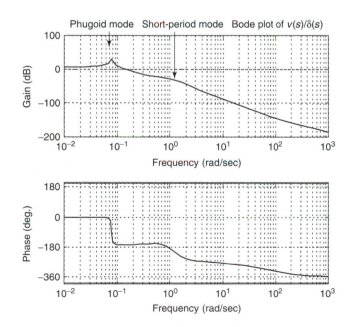

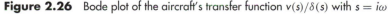

Figure 2.26   Bode plot of the aircraft's transfer function $v(s)/\delta(s)$ with $s = i\omega$

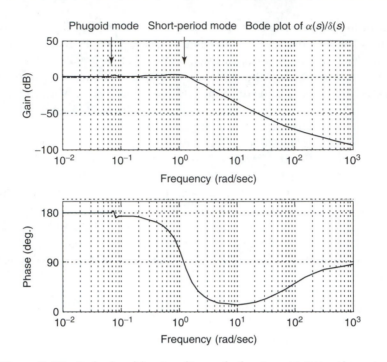

Figure 2.27   Bode plot of the aircraft's transfer function $\alpha(s)/\delta(s)$ with $s = i\omega$

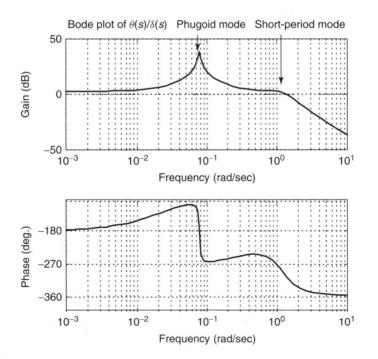

Figure 2.28   Bode plot of the aircraft's transfer function $\theta(s)/\delta(s)$ with $s = i\omega$

```
>>vdelta = tf(num,den) <enter>

Transfer function:
  -0.0005 s^2+0.03475 s+0.0175
 -------------------------------------------
s^4+1.005 s^3+1.411 s^2+0.013 s+0.0084

>>bode(vdelta) % Figure 2.26 <enter>
```

The Bode plot of transfer function $\alpha(s)/\delta(s)$ (Figure 2.27) is generated using the following MATLAB statements:

```
>>a=[1 80];b=[1  0.0065  0.006]; num=-0.02*conv(a,b) <enter>

num =
 -0.0200  -1.6001  -0.0105  -0.0096

>>alphadelta = tf(num,den) <enter>

Transfer function:
  -0.02 s^3-1.6 s^2-0.01052 s-0.0096
 -------------------------------------------
s^4+1.005 s^3+1.411 s^2+0.013 s+0.0084

>>bode(alphadelta) % Figure 2.27 <enter>
```

(Note that the denominator polynomial, *den*, of $\alpha(s)/\delta(s)$ is same as that of $v(s)/\delta(s)$, and does not have to be re-calculated.)

Finally, the Bode plot of transfer function $\theta(s)/\delta(s)$ (Figure 2.28) is generated using the following MATLAB statements:

```
>>a=[1  0.02]; b=[1  0.4]; num=-1.4*conv(a,b) <enter>

num =
  -1.4000  -0.5880  -0.0112

>>thetadelta = tf(num,den) <enter>

Transfer function:
  -1.4 s^2-0.588 s-0.0112
 -------------------------------------------
s^4+1.005 s^3+1.411 s^2+0.013 s+0.0084

>>bode(thetadelta) % Figure 2.28 <enter>
```

From the Bode plots (Figures 2.26–2.28), we can note the natural frequencies of the phugoid and the short period modes, respectively, as either the *peaks* or *changes of slope* (called *breaks*) in the respective *gain* plots. The peaks due to complex poles sometimes disappear due to the presence of zeros in the vicinity of the poles. As expected, the natural frequencies agree with the values already calculated from the characteristic polynomial, because all the three transfer functions have the same characteristic (denominator) polynomial. Figure 2.26 shows that the magnitude (gain) of $v(i\omega)/\delta(i\omega)$ at the short period natural frequency is very small, which indicates that the short period mode oscillation is characterized by very small changes in forward velocity, $v(i\omega)$, which can be neglected (i.e. $v(i\omega) \approx 0$) to obtain a *short period approximation*. As expected, near each natural frequency the phase changes by $180°$, except for the phugoid mode in $\alpha(i\omega)/\delta(i\omega)$ (Figure 2.27). The latter strange behavior of the phugoid mode is due to the fact that in the transfer function $\alpha(i\omega)/\delta(i\omega)$, one of the numerator quadratics (i.e. a pair of complex zeros) *almost cancels out* the quadratic corresponding to the phugoid mode in the denominator polynomial (i.e. a pair of complex poles), indicating that there is *essentially* no change in the angle-of-attack, $\alpha(i\omega)$, in the phugoid mode. Also, the magnitude (gain) of $\alpha(i\omega)/\delta(i\omega)$ at the phugoid natural frequency is seen to be very small in Figure 2.27 as compared to the gain at the same frequency in Figures 2.26 and 2.28. The fact that the phugoid oscillation does not involve an appreciable change in the angle-of-attack, $\alpha(i\omega)$, forms the basis of the *phugoid approximation* in which $\alpha(i\omega) \approx 0$. However, Figure 2.28 shows that considerable magnitude (gain) of $\theta(i\omega)/\delta(i\omega)$ exists at *both* short period and the phugoid natural frequencies. Hence, both modes essentially consist of oscillations in the pitch angle, $\theta(i\omega)$. The present example shows how one can obtain an insight into a system's behavior just by analyzing the frequency response of its transfer function(s).

Note from Figures 2.26–2.28 that the gains of all three transfer functions decay rapidly with frequency at high frequencies. Such a decay in the gain at high frequencies is a desirable feature, called *roll-off*, and provides *attenuation* of high frequency *noise* arising due to *unmodeled* dynamics in the system. We will define *sensitivity* (or *robustness*) of a system to transfer function variations later in this chapter, and formally study the effects of *noise* in Chapter 7. Using Eq. (2.85), we can estimate the high-frequency gain-slope and phase of the three transfer functions given by Eqs. (2.86)–(2.88). For $v(s)/\delta(s)$, $K < 0$, $m = 2$, and $n = 4$, which implies a gain-slope (or roll-off) of -40 dB/decade and a phase of $-360°$ (or $0°$) in the limit $\omega \to \infty$, which are confirmed in Figure 2.26. For $\alpha(s)/\delta(s)$, $K < 0$, $m = 3$, and $n = 4$, which implies a roll-off of -20 dB/decade and a phase of $-270°$ (or $90°$) in the limit $\omega \to \infty$, which are evident in Figure 2.27. Finally, for $\theta(s)/\delta(s)$, $K < 0$, $m = 2$, and $n = 4$, which implies a gain-slope (or roll-off) of -40 dB/decade and a phase of $-360°$ (or $0°$) in the limit $\omega \to \infty$, which are also seen in Figure 2.28.

The transfer function $v(s)/\delta(s)$ has a peculiarity which is absent in the other two transfer functions – namely, a *zero* at $s = 70$. A system with transfer function having *poles* or *zeros* in the *right-half* s-plane is called a *non-minimum phase* system, while a system with all the poles and zeros in the left-half s-plane, or on the imaginary axis is called a *minimum phase* system. We will see below that systems which have *poles* in the right-half s-plane are *unstable*. Hence, *stable* non-minimum phase systems have only zeros in the right-half s-plane, such as the system denoted by $v(s)/\delta(s)$. Stable non-minimum phase systems have a markedly different phase in the limit $\omega \to \infty$ (we may have to add or subtract $360°$ to find non-minimum phase from Eq. (2.85)), when compared to a *corresponding* minimum phase system (i.e. a similar system with no zeros in the right-half s-plane). This usually results in an unacceptable transient response. A non-minimum phase system with only one right-half plane zero (such as $v(s)/\delta(s)$) results in a transient response which is of *opposite sign* when compared to the input. Popular examples of such systems are aircraft or missiles controlled by forces applied *aft* of the center of mass. For this reason, a right-half plane zero in an aircraft (or missile) transfer function is called 'tail-wags-the-dog zero'. Control of non-minimum phase systems requires special attention.

Before we can apply the transfer function approach to a general system, we must know how to derive Laplace transform (and inverse Laplace transform) of some frequently encountered functions. This information is tabulated in Table 2.1, using the definitions and properties of the Laplace transform (Eqs. (2.59)–(2.70)). Note that Table 2.1 gives the Laplace transform of some commonly encountered functions, $f(t)$, which are defined for $t \geq 0$. At $t = 0$, $f(t)$ can have a discontinuity, such as $f(t) = u_s(t)$ or $f(t) = \delta(t)$. It is interesting to see in Table 2.1 that the Laplace transform of the unit impulse function, $\delta(t)$, is unity, while that of the unit step function, $u_s(t)$, is $1/s$. Since $du_s(t)/dt = \delta(t)$, the Laplace transforms of these two singularity functions agree with the properties given by Eqs. (2.66) and (2.68).

Table 2.1   Laplace transforms of some common functions

S. No.	$f(t)$ $(t \geq 0)$	$F(s) = \mathcal{L}f(t) = \int_0^\infty e^{-st} f(t)\, dt$
1	e^{-at}	$1/(s+a)$
2	$e^{-at} f(t)$	$F(s+a)$
3	t^n	$n!/s^{n+1}$
4	Unit Step Function, $u_s(t)$	$1/s$
5	$\sin(\omega t)$	$\omega/(s^2 + \omega^2)$
6	$\cos(\omega t)$	$s/(s^2 + \omega^2)$
7	$f^{(k)}(t)$	$s^k F(s) - s^{k-1} f(0) - s^{k-2} f^{(1)}(0) - \cdots - f^{(k-1)}(0)$
8	$\int_{-\infty}^t f(t)\, dt$	$F(s)/s + (1/s) \int_{-\infty}^0 f(t)\, dt$
9	Unit Impulse Function, $\delta(t)$	1

Example 2.11

Consider a system with the following transfer function:

$$G(s) = (s+3)/[(s+1)(s+2)] \qquad (2.90)$$

The second order system (denominator polynomial is of degree 2) has a zero, $z_1 = -3$, and two poles, $p_1 = -1$ and $p_2 = -2$. Let us assume that the system has the input, $u(t)$, and initial conditions as follows:

$$u(t) = 0, \ y(0) = y_0, \ y^{(1)}(0) = 0 \tag{2.91}$$

Since $G(s) = Y(s)/U(s)$ when the *initial conditions are zero* (which is not the case here), we cannot directly use the transfer function to determine the system's response, $y(t)$, for $t > 0$. Let us first derive the system's governing differential equation by applying inverse Laplace transform to the transfer function (with zero initial conditions, because that is how a transfer function is defined) as follows:

$$(s + 1)(s + 2)Y(s) = (s + 3)U(s) \tag{2.92}$$

or

$$s^2 Y(s) + 3s Y(s) + 2Y(s) = sU(s) + 3U(s) \tag{2.93}$$

and

$$\mathcal{L}^{-1}[s^2 Y(s) + 3s Y(s) + 2Y(s)] = \mathcal{L}^{-1}[sU(s) + 3U(s)] \tag{2.94}$$

which, using the real differentiation property (Eq. (2.67)) with zero initial conditions for both input, $u(t)$, and output, $y(t)$, yields the following differential equation:

$$y^{(2)}(t) + 3y^{(1)}(t) + 2y(t) = u^{(1)}(t) + 3u(t) \tag{2.95}$$

Now, we can apply the Laplace transform to this governing differential equation using real differentiation property *with the input and the initial conditions* given by Eq. (2.91) as

$$\mathcal{L}[y^{(2)}(t) + 3y^{(1)}(t) + 2y(t)] = \mathcal{L}[u^{(1)}(t) + 3u(t)] \tag{2.96}$$

or

$$s^2 Y(s) - sy_0 + 3s Y(s) - 3y_0 + 2Y(s) = 0 \tag{2.97}$$

and it follows that

$$Y(s) = (s + 3)y_0/[(s + 1)(s + 2)] \tag{2.98}$$

We can express $Y(s)$ as

$$Y(s) = y_0[2/(s + 1) - 1/(s + 2)] \tag{2.99}$$

Equation (2.99) is called the *partial fraction expansion* of Eq. (2.98), where the contribution of each pole is expressed separately as a fraction and added up. In Eq. (2.99) the two numerator coefficients, 2 and -1, corresponding to the two fractions are called the *residues*.

The output, $y(t)$, of the system can then be obtained by applying inverse Laplace transform to Eq. (2.99) for $t \geq 0$ as

$$y(t) = y_0\{\mathcal{L}^{-1}[2/(s + 1)] + \mathcal{L}^{-1}[-1/(s + 2)]\} \tag{2.100}$$

or, using the *translation in Laplace domain* property given by Eq. 2.65, we can write the output finally as

$$y(t) = 2y_0 e^{-t} - y_0 e^{-2t}; \quad (t \geq 0) \tag{2.101}$$

In Example 2.11 we have seen how we can evaluate a single-input, single-output system's response if we know its transfer function, applied input and initial conditions, by using a partial fraction expansion of $Y(s)$. For a system with complex poles (such as Example 2.8), finding partial fraction expansion can be very difficult. Fortunately, the MATLAB intrinsic command *residue* makes finding partial fraction expansion a simple affair. All one has to do is to specify the numerator and denominator polynomials of the *rational function* in s – such as Eq. (2.98) – for which a partial fraction expansion is desired. For example, if the rational function is $N(s)/D(s)$, then the coefficients of the polynomials $N(s)$ and $D(s)$ *in decreasing powers of* s are specified in two vectors, say, n and d. Then the *residue* command is used as follows to give the terms of the partial fraction expansion:

```
>>[k,p,c] = residue(n,d) <enter>
```

where p is a vector containing the *poles* of $N(s)/D(s)$, k is a vector containing the corresponding *residues*, and c is the *direct constant*. In terms of the elements of p and k, the partial fraction expansion is given by

$$N(s)/D(s) = c + k_1/(s - p_1) + \cdots + k_n/(s - p_n) \tag{2.102}$$

where all the poles, p_j, are *distinct* (i.e. they appear *only once* – as in Example 2.11). If a pole, say p_m, is *repeated q times*, then the partial fraction expansion obtained from the *residue* command is given by

$$N(s)/D(s) = c + k_1/(s - p_1) + \cdots + k_m/(s - p_m) + k_{m+1}/(s - p_m)^2$$
$$+ k_{m+2}/(s - p_m)^3 + \cdots + k_{m+q-1}/(s - p_m)^q + \cdots + k_n/(s - p_n) \tag{2.103}$$

Now we are well equipped to talk about a linear system's response to singularity functions.

2.5 Response to Singularity Functions

In the previous two sections, we saw how frequency response, Laplace transform, and transfer function can be used to evaluate a linear system's characteristics, and its response to initial conditions (Example 2.11). Here we will apply a similar approach to find out a linear system's response to singularity functions, and extend the method for the case of arbitrary inputs. We had ended Section 2.2 with a remark on the special place held by the unit impulse function in control theory. To understand why this is so, let us define *impulse response*, $g(t)$, as the response of a system to a unit impulse, $\delta(t)$, applied as input at time $t = 0$. Furthermore, it is assumed that the system is at rest at $t = 0$, i.e. all initial conditions (in terms of the output, $y(t)$, and its time derivatives) are zero. We know from Table 2.1 that the Laplace transform of $\delta(t)$ is unity. Also, from the definition of the transfer function for a single-input, single-output system (Eq. (2.72)) $Y(s) = G(s)U(s)$. Since, in this case, $y(t) = g(t)$ and Laplace transform of the input, $U(s)=1$, it implies that the following must be true:

$$g(t) = \mathcal{L}^{-1}Y(s) = \mathcal{L}^{-1}[G(s)U(s)] = \mathcal{L}^{-1}G(s) \qquad (2.104)$$

Equation (2.104) denotes a very important property of the impulse response, namely that the impulse response of a linear, time-invariant system with *zero initial conditions* is equal to the *inverse Laplace transform* of the system's *transfer function*. Hence, the symbol $g(t)$ for the impulse response! One can thus obtain $G(s)$ from $g(t)$ by applying the Laplace transform, or $g(t)$ from $G(s)$ by applying the inverse Laplace transform. Since the transfer function contains information about a linear system's characteristics, we can now understand why impulse response (and the unit impulse function) deserve a special place in control theory. In a manner similar to the impulse response, we can define the *step response*, $s(t)$, as a linear, time-invariant system's response to unit step input, $u_s(t)$, applied at time $t = 0$ with *zero initial conditions*. Again, using Table 2.1, we note that Laplace transform of the unit step function is given by $U(s) = 1/s$, and the step response can be expressed as

$$s(t) = \mathcal{L}^{-1}[G(s)U(s)] = \mathcal{L}^{-1}[G(s)/s] \qquad (2.105)$$

which shows that the step response is also intimately related with a system's transfer function, and hence with its characteristics.

Example 2.12

Let us revisit the second order system consisting of the electrical network of Examples 2.8 and 2.9, and evaluate the impulse response, $g(t)$, and step response, $s(t)$, for this system with zero initial conditions. Equation (2.80), gives the system's transfer function as $G(s) = 0.5s/(s^2 + 30s + 10^6)$. Using the partial fractions expansion of $G(s)$, we can write

$$G(s) = k_1/(s - p_1) + k_2/(s - p_2) \qquad (2.106)$$

where p_1 and p_2 are the two *poles* (roots of the denominator polynomial of $G(s)$), and the *residues* k_1 and k_2 are evaluated as follows:

$$k_1 = [(s - p_1)G(s)]|_{s=p1} = 0.5p_1/(p_1 - p_2) \qquad (2.107)$$

$$k_2 = [(s - p_2)G(s)]|_{s=p2} = 0.5p_2/(p_2 - p_1) \qquad (2.108)$$

Again, from Example 2.9 we know that $p_1 = -15 - 999.9i$ and $p_2 = -15 + 999.9i$. Then from Eqs. (2.107) and (2.108), $k_1 = 0.25 - 0.00375i$ and $k_2 = 0.25 + 0.00375i$. The residues can be verified by using the MATLAB intrinsic command *residue* as follows:

```
>>N=[0.5 0]; D=[1 30 1e6]; [k,p,c]=residue(N,D) <enter>
```

```
k =
   0.25000000000000+0.00375042194620i
   0.25000000000000-0.00375042194620i

p =
   1.0e+002*
  -0.15000000000000+9.99887493671163i
  -0.15000000000000-9.99887493671163i

c =
   [ ]
```

Taking the inverse Laplace transform of Eq. (2.106) with the use of Table 2.1, we get the following expression for the impulse response, $g(t)$:

$$g(t) = k_1 \exp(p_1 t) + k_2 \exp(p_2 t); \quad (t \geq 0) \tag{2.109}$$

Using the fact that p_1 and p_2 are *complex conjugates* (and k_1 and k_2 are also complex conjugates), we can simplify Eq. (2.109) to give

$$g(t) = e^{-15t}[0.5\cos(999.9t) - 0.0075\sin(999.9t)]; \quad (t \geq 0) \tag{2.110}$$

Note that the impulse response, $g(t)$, given by Eq. (2.110) has an amplitude which decreases *exponentially* with time due to the term e^{-15t}. This is a characteristic of a underdamped, *stable* system, as seen in Figure 2.24.

Since the poles can also be represented in terms of their natural frequency, ω_n, and damping-ratio, ς, as $p_{2,1} = -\varsigma\omega_n \pm i\omega_n(\varsigma^2 - 1)^{1/2}$, (see Eqs. (2.83) and (2.84)) we can also write

$$g(t) = y_0 \exp(-\varsigma\omega_n t) \sin[\omega_n t\ (1 - \varsigma^2)^{1/2} + \theta]/(1 - \varsigma^2)^{1/2}; \quad (t \geq 0) \tag{2.111}$$

where $\theta = \cos^{-1}(\varsigma)$ and

$$y_0 = 2k_1[(1 - \varsigma^2)^{1/2}]/e^{i(\theta - \pi/2)} = 2k_2[(1 - \varsigma^2)^{1/2}]/e^{-i(\theta - \pi/2)} \tag{2.112}$$

You can verify Eqs. (2.111) and (2.112) using complex algebra, but don't worry if you don't feel like doing so, because Eq. (2.109) can be directly obtained using MATLAB to get the impulse response, $g(t)$, as follows:

```
>>p1=-15-999.9i; p2=conj(p1); k1=0.25-0.00375i; k2=conj(k1); <enter>

>>t=0:0.001:1; g=k1*exp(p1*t)+k2*exp(p2*t); plot(t,g),
   xlabel('Time (s)'), ylabel('g(t)') <enter>
```

Here *conj*( ) is the intrinsic MATLAB operator that calculates the complex conjugate, *exp*( ) is the exponential function, and *plot(x,y)* is the MATLAB command for plotting the vector x against the vector y. Note that both t and g are vectors of the same size. Also, note the ease by which complex vector calculations have been made using MATLAB. The same computation in a low-level language – such as

Fortran, Basic, or C – would require many lines of programming. For more information on the usage of a MATLAB command type *help < name of command> <enter>* at the MATLAB prompt.

We know from Example 2.9 that for the example electrical network, the natural frequency and damping-ratio are, $\omega_n = 1000$ rad/s and $\varsigma = 0.015$, respectively. Substituting these numerical values in Eq. (2.112), we get $y_0 = 0.5$ amperes.

We can also evaluate the step response, $s(t)$, of the electrical network by taking the inverse Laplace transform of $G(s)/s$ as follows:

$$s(t) = \mathcal{L}^{-1}[G(s)/s] = \mathcal{L}^{-1}[0.5/(s^2 + 30s + 10^6)]$$
$$= \mathcal{L}^{-1}[2.5 \times 10^{-4}i/(s - p_1) - 2.5 \times 10^{-4}i/(s - p_2)] \qquad (2.113)$$

or

$$s(t) = 2.5 \times 10^{-4}i[\exp(p_1 t) - \exp(p_2 t)]$$
$$= 5 \times 10^{-4}e^{-15t}\sin(999.9t); \quad (t \geq 0) \qquad (2.114)$$

Note that $s(t)$ given by Eq. (2.114) also indicates a stable system due to the oscillatory step response with a decaying amplitude of $5 \times 10^{-4}e^{-15t}$.

The calculation of step and impulse responses can be generalized by using the partial fraction expansion of $G(s)/s$ and $G(s)$, respectively. Taking the inverse Laplace transform of Eq. (2.104), we can express the impulse response for a unit impulse input applied at $t = 0$ as follows:

$$g(t) = c\delta(t) + k_1 \exp(p_1 t) + \cdots + k_m \exp(p_m t) + k_{m+1}t \exp(p_m t)$$
$$+ k_{m+2}t^2 \exp(p_m t)/2 + \cdots + k_{m+q-1}t^{q-1} \exp(p_m t)/(q - 1)!$$
$$+ \cdots + k_n \exp(p_n t) \quad (t \geq 0) \qquad (2.115)$$

Note that in deriving Eq. (2.115), we have used the *translation in Laplace domain* property of the Laplace transform (Eq. (2.65)), and have written the inverse Laplace transform of $1/(s - p)^k$ as $t^{k-1}e^{pt}/(k - 1)!$. If the impulse is applied at $t = t_0$, the time t in Eq. (2.115) should be replaced by $(t - t_0)$, since the Laplace transform of $g(t - t_0)$ is $G(s)\exp(-st_0)$. If the system is *strictly proper* (i.e. the degree of the numerator polynomial of the transfer function, $G(s)$, is *less than* that of the denominator polynomial), then the direct constant, c, in the partial fraction expansion of $G(s)$ is zero, and the impulse response does not go to infinity at $t = 0$ (Eq. (2.115)). Hence, for strictly proper transfer functions, we can write a computer program using MATLAB to evaluate the impulse response using Eq. (2.115) with $c = 0$. Such a program is the M-file named *impresp.m*, which is listed in Table 2.2, and can be called as follows:

```
>>[g,t] = impresp(num,den,t0,dt,tf) <enter>
```

Table 2.2 Listing of the M-file *impresp.m*, which calculates the impulse response of a strictly proper, single-input, single-output system

---

impresp.m

```
function [y,t]=impresp(num,den,t0,dt,tf);
%Program for calculation of impulse response of strictly proper SISO
   systems
%num = numerator polynomial coefficients of transfer function
%den = denominator polynomial coefficients of transfer function
%(Coefficients of 'num' and 'den' are specified as a row vector, in
%decreasing powers of 's')
%t0 = time at which unit impulse input is applied
%dt = time-step (should be smaller than 1/(largest natural freq.))
%tf = final time for impulse response calculation
%y = impulse response; t= vector of time points
%copyright(c)2000 by Ashish Tewari
%
%Find a partial fraction expansion of num/(den):-
[r,p,k]=residue(num,den);
%Calculate the time points for impulse response:-
t=t0:dt:tf;
%Find the multiplicity of each pole, p(j):-
for j=1:size(p)
n=1;
          for i=1:size(p)
                   if p(j)==p(i)
                            if(i~=j)
                            n=n+1;
                            end
                   end
          end
mult(:,j)=n;
end
%Calculate the impulse response by inverse Laplace transform of
%partial-fraction expansion:-
y=zeros(size(t));
j=1;
while j<=size(p,1)
          for i=1:mult(:,j)
          y=y+r(j+i-1)*((t-t0).^(i-1)).*exp(p(j)*(t-t0))/factorial(i-1);
          end
          j=j+i;
end
end
```

---

where *num* and *den* are row vectors containing numerator and denominator polynomial coefficients, respectively, of the transfer function, $G(s)$, in decreasing powers of s, $t0$ is the time at which the unit impulse input is applied, dt is the *time-step* size, tf is the *final time* for the response, g is the returned impulse response, and t is the returned vector of time points at which $g(t)$ is calculated. Instead of having to do inverse Laplace transformation by hand, we can easily use *impresp* to quickly get the impulse response of a strictly proper

Table 2.3 Listing of the M-file *stepresp.m*, which calculates the step response of a proper, single-input, single-output system

stepresp.m

```
function [y,t]=stepresp(num,den,t0,dt,tf);
%Program for calculation of step response of proper SISO systems
%num = numerator polynomial coefficients of transfer function
%den = denominator polynomial coefficients of transfer function
%(Coefficients of 'num' and 'den' are specified as a row vector, in
%decreasing powers of 's')
%t0 = time at which unit step input is applied
%dt = time-step (should be smaller than 1/(largest natural freq.))
%tf = final time for step response calculation
%y = step response; t= vector of time points
%copyright(c)2000 by Ashish Tewari
%
%Find a partial fraction expansion of num/(den.s):-
[r,p,k]=residue(num,conv(den,[1 0]));
%Calculate the time points for step response:-
t=t0:dt:tf;
%Find the multiplicity of each pole, p(j):-
for j=1:size(p)
n=1;
        for i=1:size(p)
                if p(j)==p(i)
                        if(i~=j)
                        n=n+1;
                        end
                end
        end
mult(:,j)=n;
end
%Calculate the step response by inverse Laplace transform of
%partial-fraction expansion:-
y=zeros(size(t));
j=1;
while j<=size(p,1)
        for i=1:mult(:,j)
        y=y+r(j+i-1)*((t-t0).^(i-1)).*exp(p(j)*(t-t0))/factorial(i-1);
        end
        j=j+i;
end
end
```

plant. The M-file *impresp.m* uses only the intrinsic MATLAB functions, and is useful for those who do not have Control System Toolbox (CST). Usage of CST command *impulse* yields the same result. (We postpone the discussion of the CST command *impulse* until Chapter 4, as it uses a *state-space* model of the system to calculate impulse response.) Note the programming steps required in *impresp* to identify the multiplicity of each pole of $G(s)$. For increased accuracy, the time-step, *dt*, should be as small as possible, and, in any case, should *not exceed* the reciprocal of the *largest natural frequency* of the system.

In a manner similar to the impulse response, the step response calculation can be generalized by taking the inverse Laplace transform of the partial fraction expansion of

$G(s)/s$ as follows:

$$s(t) = k_1 + k_2 \exp(p_2 t) + \cdots + k_m \exp(p_m t) + k_{m+1} t \exp(p_m t)$$
$$+ k_{m+2} t^2 \exp(p_m t)/2 + \cdots + k_{m+q-1} t^{q-1} \exp(p_m t)/(q-1)!$$
$$+ \cdots + k_n \exp(p_n t)(t \geq 0) \tag{2.116}$$

where k_1 is the residue corresponding to the pole at $s = 0$, i.e. $p_1 = 0$, and $k_2 \ldots k_n$ are the residues corresponding to the poles of $G(s)$, $p_1 \ldots p_n$, in the partial fraction expansion of $G(s)/s$. Note that if $G(s)$ is a *proper* transfer function (i.e. numerator polynomial is of *lesser* or *equal degree* than the denominator polynomial), then $G(s)/s$ is *strictly proper*, and the direct term, c, is zero in the partial fraction expansion of $G(s)/s$. Thus, we can evaluate the step response of a proper system using Eq. (2.116), which should be modified for a unit step input applied at $t = t_0$ by replacing t in Eq. (2.116) by $(t - t_0)$. A MATLAB program called *stepresp.m*, which evaluates the step response of proper system by Eq. (2.116), is listed in Table 2.3, and can be used as follows:

```
>>[s,t] = stepresp(num,den,t0,dt,tf) <enter>
```

where *num* and *den* are row vectors containing numerator and denominator polynomial coefficients, respectively, of the transfer function, $G(s)$, in decreasing powers of s, *t0* is the time at which the unit step input is applied, *dt* is the *time-step* size, *tf* is the *final time* for the response, *s* is the returned step response, and *t* is a vector containing time points at which $s(t)$ is calculated. The M-file *stepresp.m* uses only intrinsic MATLAB functions, and is useful in case you do not have access to Control System Toolbox (CST). The CST command *step* is a quick way of calculating the step response. The GUI tool associated with the command *step* also lets you get the values of $s(t)$ and t at any point on the step response curve by merely clicking at that point.

Example 2.13

Let us compute and plot the step and impulse responses of the aircraft transfer function, $\theta(s)/\delta(s)$, of Example 2.10, given by Eq. (2.88). We must begin with the specification of the transfer function as follows:

```
>>a=[1 0.02]; b=[1 0.4]; num=-1.4*conv(a,b) <enter>
num =
  -1.4000   -0.5880   -0.0112

>>a=[1 0.005 0.006]; b=[1 1 1.4]; den=conv(a,b) <enter>
den =
   1.0000  1.0050  1.4110  0.0130  0.0084
```

Note that the transfer function is *strictly proper* (the numerator polynomial is of *second degree*, while the denominator polynomial is of *fourth degree*). Hence, we can use *impresp.m* to compute the impulse response, $g(t)$, and plot the result as follows:

```
>>[g,t] = impresp(num,den,0,0.5,250); plot(t,g) <enter>
```

Note that the time-step for calculating the impulse response is *smaller* than inverse of the *largest natural frequency* of the system (1.18 rad/s). Similarly, the step response, $s(t)$, is computed using *stepresp* and plotted as follows:

```
>>[s,t] = stepresp(num,den,0,0.5,250); plot(t,s) <enter>
```

The resulting plots of impulse and step responses are shown in Figure 2.29. Note that the plot of the impulse response clearly shows an initial, well damped oscillation of high-frequency (short-period mode) and a *lightly damped*, long-period oscillation (phugoid mode). This behavior meets our expectation from the natural frequencies and damping-ratios of the two modes calculated in Example 2.10 (recall that the natural frequency and damping of the short-period mode are more than 10 times those of the phugoid mode). It is also clear that the impulse response excites both the modes, while the step response is dominated by the phugoid mode. The time taken by the phugoid mode to decay to zero is an indicator of the *sluggishness* of the longitudinal dynamics of the airplane. The non-minimum phase character of the transfer function due to the zero at $s = 70$ is evident in the large initial undershoot in the impulse response.

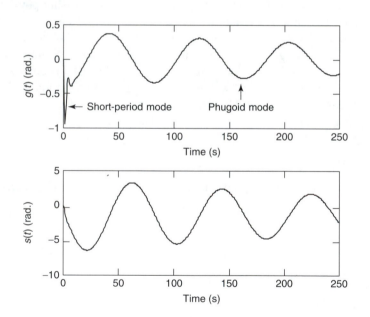

Figure 2.29   Impulse response, $g(t)$, and step response, $s(t)$, for the aircraft transfer function, $\theta(s)/\delta(s)$ (Example 2.13)

2.6 Response to Arbitrary Inputs

After learning how to find step and impulse responses of a given linear system, the next logical step is to find the response to an arbitrary input, $u(t)$, which is applied to a linear system with zero initial conditions. At the end of Section 2.2 (Figure 2.16), we saw how an arbitrary function can be expressed as a sum (or integral) of the suitably scaled impulse functions, $\delta(t - a)$, as given by Eq. (2.36). Since we know how to find the response of a linear, time-invariant system to a unit impulse input for zero initial conditions, we can apply the superposition principle to get the linear system's response, $y(t)$, to an arbitrary input, $u(t)$, which is expressed as

$$u(t) = \sum_{\tau=-\infty}^{\infty} u(\tau)\Delta\tau\delta(t - \tau) \tag{2.117}$$

and the response to the impulse input $u(\tau)\Delta\tau\delta(t - \tau)$ applied at $t = \tau$ is given by

$$\Delta y(t, \tau) = u(\tau)\Delta\tau g(t - \tau) \tag{2.118}$$

where $g(t - \tau)$ is the impulse response to unit impulse, $\delta(t - \tau)$, applied at $t = \tau$. Then by the superposition principle (Eq. (2.36)), the response to a linear combination of impulses (given by Eq. (2.35)) is nothing else but a linear combination of individual impulse responses, $g(t - \tau)$, each multiplied by the corresponding factor, $u(\tau)\Delta\tau$. Hence, we can write the response, $y(t)$, as

$$y(t) = \sum_{\tau=-\infty}^{\infty} u(\tau)\Delta\tau g(t - \tau) \tag{2.119}$$

or, in the limit $\Delta\tau \to 0$,

$$y(t) = \int_{-\infty}^{\infty} u(\tau)g(t - \tau)\,d\tau \tag{2.120}$$

Equation (2.120) is one of the most important equations in control theory, since it lets us determine a linear system's response to an arbitrary input. The integral on the right-hand side of Eq. (2.120) is called the *superposition integral* (or *convolution integral*). Note that we can apply a change of integration variable, and show that the convolution integral is *symmetric* with respect to $u(t)$ and $g(t)$ as follows:

$$y(t) = \int_{-\infty}^{\infty} u(\tau)g(t - \tau)\,d\tau = \int_{-\infty}^{\infty} u(t - \tau)g(\tau)\,d\tau \tag{2.121}$$

Most commonly, the input, $u(t)$, is non-zero only for $t > 0$. Also, since $g(t - \tau) = 0$ for $\tau > t$, we can change the upper limit of the convolution integral to t. Hence, the convolution integral becomes the following:

$$y(t) = \int_{0}^{t} u(\tau)g(t - \tau)\,d\tau = \int_{0}^{t} u(t - \tau)g(\tau)\,d\tau \tag{2.122}$$

We could have obtained Eq. (2.122) alternatively by applying inverse Laplace transform to $Y(s) = G(s)U(s)$ (since $G(s)$ is the Laplace transform of $g(t)$).

Example 2.14

We can use the convolution integral to obtain the step response, $s(t)$, of the second order system of Example 2.12 as follows:

$$s(t) = \int_0^t u_s(t - \tau)g(\tau)\,d\tau \qquad (2.123)$$

where $u_s(t)$ is the unit step function and the impulse response, $g(t)$, is given by Eq. (2.110). Substituting Eq. (2.110) into Eq. (2.123), and noting that $u_s(t - \tau) = 1$ for $\tau < t$ and $u_s(t - \tau) = 0$ for $\tau > t$, we get the following expression for the step response:

$$s(t) = \int_0^t e^{-15\tau}[0.5\cos(999.9\tau) - 0.0075\sin(999.9\tau)]\,d\tau \qquad (2.124)$$

Carrying out the integration in Eq. (2.124) by parts, we get the same expression for $s(t)$ as we obtained in Eq. (2.114), i.e. $s(t) = 5 \times 10^{-4}\,e^{-15t}\sin(999.9t)$ for $t \geq 0$.

The use of convolution integral of Eq. (2.120) can get difficult to evaluate by hand if the input $u(t)$ is a more complicated function than the unit step function. Hence, for a really arbitrary input, it is advisable to use a *numerical approximation* of the convolution integral as a *summation* over a large number of finite time-steps, $\Delta\tau$. Such an

Table 2.4 Listing of the M-file *response.m*, which calculates the response of a strictly proper, single-input, single-output system to an arbitrary input

---

response.m

```
function y=response(num,den,t,u);
%Program for calculation of the response of strictly proper SISO systems
%to arbitrary input by the convolution integral.
%num = numerator polynomial coefficients of transfer function
%den = denominator polynomial coefficients of transfer function
%(Coefficients of 'num' and 'den' are specified as a row vector, in
%decreasing powers of 's')
%t = row vector of time points (specified by the user)
%u = vector of input values at the time points contained in t.
%y = calculated response
%copyright(c)2000 by Ashish Tewari
%
%Calculate the time-step:-
dt=t(2)-t(1);
m=size(t,2)
tf=t(m);
%Calculate the convolution integral:-
y=zeros(size(t));
G=y;
[g,T]=impresp(num,den,t(1),dt,tf);
for i=1:m
y=y+dt*u(i)*[G(1:i-1) g(1:m-i+1)];
end
```

approximation of an integral by a summation is called *quadrature*, and there are many numerical techniques available of varying efficiency for carrying out quadrature. The simplest numerical integration (or quadrature) is the assumption that the integrand is *constant* in each time interval, $\Delta\tau$, which is the same as Eq. (2.119). Hence, we can use Eq. (2.119) with a *finite* number of time steps, $\Delta\tau$, to evaluate the convolution integral, and the response to an arbitrary input, $u(t)$. Such a numerical evaluation of the convolution integral is performed by the M-file, called *response.m*, listed in Table 2.4. Note that *response* calls *impresp.m* internally to evaluate the impulse response, $g(t - \tau)$, for each time interval, $\Delta\tau$. The M-file *response.m* is called as follows:

```
>>y = response(num,den,t,u) <enter>
```

where *num, den* are the numerator and denominator polynomial coefficients of the transfer function, $G(s)$, (same as in *impresp* and *stepresp*), t is the row vector containing time points at which the response, y, is desired, and u is a vector containing values of the applied arbitrary input at the time points contained in t.

Example 2.15

Let us determine the pitch response, $\theta(t)$, of the aircraft transfer function, $\theta(s)/\delta(s)$, (Examples 2.10, 2.13) if the applied input is $\delta(t) = 0.01t.\sin(1.3t)$, beginning at $t = 0$. We first specify the time vector, t, and input vector, u, as follows:

```
>>t=0:0.1:20; u=0.01*t.*sin(1.3*t); <enter>
```

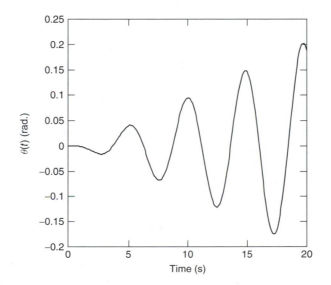

Figure 2.30   Response of the pitch angle, $\theta(t)$, for the aircraft when the elevator deflection is $\delta(t) = 0.01t.\sin(1.3t)$ radians

Then the response is calculated using the M-file *response.m* as follows:

```
>>y = response(num,den,t,u); <enter>
```

where *num* and *den* were specified in Example 2.13. The calculated response is plotted in Figure 2.30. Note the ease with which *response.m* calculates the response to a complicated input function. A more general method of evaluating response of even multi-input, multi-output systems based on the *state-space* approach will be given in Chapter 4. MATLAB (CST) functions use the *state-space* approach for calculating step, impulse, and arbitrary input responses; hence, discussion of these functions will be postponed until Chapter 4.

There are three properties which determine whether a control system is good or bad, namely its *performance, stability*, and *robustness*. We briefly discussed the implications of each of these in Chapter 1 using the car-driver example. Now we are well equipped to define each of these three properties precisely. Let us first consider the performance of a control system.

2.7 Performance

Performance is all about *how successfully* a control system meets its desired objectives. Figure 1.3 showed an example of a closed-loop system's performance in terms of the *maximum overshoot* of the actual output, $y(t)$, from the desired *constant* output, y_d. More generally, the desired output may be a specific function of time, $y_d(t)$. In such a case, the difference between the actual output and the desired output, called *error*, $e(t) = y_d(t) - y(t)$, is an important measure of the control system's performance. If the error, $e(t)$, becomes zero very rapidly, the control system is said to perform very well. However, the error of certain control systems may not exactly reach zero for even very large times. For such systems, another performance parameter is considered important, namely the *steady-state error*, e_{ss}, defined as the value of the error, $e(t)$, in the limit $t \to \infty$. The smaller the *magnitude* of the steady-state error, $|e_{ss}|$, the better a control system is said to perform. There are some performance parameters that indicate the *speed* of a control system's response, such as the *rise time*, T_r, defined as the time taken by the output, $y(t)$, to *first reach* within a specified band, $\pm\varepsilon$, of the *steady-state value*, $y(\infty)$, the *peak time*, T_p, defined as the time taken to reach the first peak (or *maximum overshoot*), and the *settling time*, T_s, defined as the time taken until the output, $y(t)$, finally settles (or comes closer) to within a specified band, $\pm\varepsilon$, of its *steady-state value*, $y(\infty)$.

The performance parameters are usually defined for the step response, $s(t)$, of a system, which implies a *constant* value of the desired output, y_d. Figure 2.31 shows a typical step response for a control system, indicating the various performance parameters. Obviously, we assume that the control system reaches a steady-state value in the limit $t \to \infty$. Not all systems have this property, called *asymptotic stability*, which we will discuss in the next section.

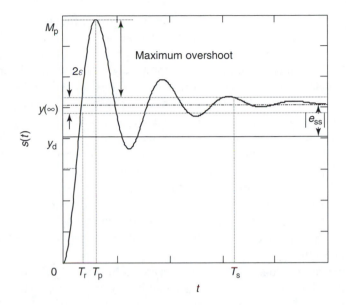

Figure 2.31 Step response, $s(t)$, defining a control system's performance

Note that while the swiftness, or alacrity, with which a system responds to a given input is described by the rise time, T_r, the peak time, T_p, and the settling time, T_s, the departure of the output from its steady-state value is measured by the maximum overshoot, or the *first peak value*, M_p, and the accuracy of the control system in reaching a final desired value, y_d, is indicated by the steady-state error, e_{ss}. For a second order system (such as the one considered in Example 2.12) with damping-ratio, ς, and natural frequency, ω_n, we can find simple expressions for many of the performance parameters *when a unit step input is applied*, given by

$$T_s = 4/(\varsigma \omega_n) \tag{2.125}$$

$$T_p = \pi/[\omega_n(1 - \varsigma^2)^{1/2}] \tag{2.126}$$

and

$$M_p = 1 + \exp\{-\varsigma\pi/(1 - \varsigma^2)^{1/2}\} \tag{2.127}$$

Equations (2.125)–(2.127) can be obtained from the step response, $s(t)$, of a second-order system, using Eq. (2.116). Note that when a successful second order control system reaches its steady-state value *asymptotically* in the limit $t \rightarrow \infty$, then for large times, it behaves in a manner quite similar to a first order control system with a pole at $s = -\varsigma\omega_n$, whose output can be expressed as $y(t) = y(\infty)[1 - \exp\{-\varsigma\omega_n t\}]$. Then the settling time can be determined as the time taken when the $y(t)$ settles to within 2 percent of $y(\infty)$, or $0.02 = \exp\{-\varsigma\omega_n T_s\}$, which gives $T_s = -\log(0.02)/(\varsigma\omega_n)$, or $T_s = 4/(\varsigma\omega_n)$, which is the same as Eq. (2.125). In other words, the output, $y(t)$, reaches within 2 percent of the steady-state value, $y(\infty)$ (i.e. $\varepsilon = 0.02$) after four leaps of the *time-constant*, $1/(\varsigma\omega_n)$. Equation (2.127) can be obtained by using the fact that at the maximum overshoot, or the first peak value, M_p, the slope of the step response, $ds(t)/dt$, is zero.

Note that the performance parameters are intimately related to the damping-ratio, ς, and natural frequency, ω_n, of a second order control system. If $\varsigma \geq 1$, a second order system behaves like a *first order* system, with an exponentially decaying step response (see Figure 2.24). Also, you can see from Eqs. (2.125)–(2.127) that the performance parameters determining the swiftness of response (such as the peak time, T_p) and those determining the deviation of the response from the desired steady-state value (such as peak value, M_p) are *contradictory*. In other words, if we try to *increase* the swiftness of the response by suitably adjusting a control system's characteristics (which are given by ς and ω_n for a second order system), we will have to accept *larger* overshoots from the steady-state value, $y(\infty)$, and *vice versa*. How the control system characteristics are modified to achieve a desired set of performance parameters is an essential part of the control system design.

The performance of a control system is determined by the locations of its poles in the Laplace domain. Generally, the poles of a control system may be such that there are a few poles very close to the imaginary axis, and some poles far away from the imaginary axis. As may be clear from examining expressions for step or impulse response, such as Eqs. (2.115) and (2.116), a control system's response is largely dictated by those poles that are the *closest* to the imaginary axis, i.e. the poles that have the *smallest* real part magnitudes. Such poles that dominate the control system's performance are called the *dominant poles*. Many times, it is possible to identify a single pole, or a pair of poles, as the dominant poles. In such cases, a fair idea of the control system's performance can be obtained from the damping and natural frequency of the dominant poles, by using Eqs. (2.125)–(2.127).

The *steady-state error*, e_{ss}, to an *arbitrary input* is an important measure of control system performance. Consider a general single-input, single-output closed-loop system shown in Figure 2.32, where $G(s)$ and $H(s)$ are the transfer-functions of the plant and the *controller* (also called *compensator*), respectively. Such a closed-loop control system is said to have the controller, $H(s)$, in *cascade* (or *series*) with the plant, $G(s)$. (Another closed-loop configuration is also possible in which $H(s)$ is placed in the *feedback path* of (or in *parallel* with) $G(s)$.) The controller applies an input, $U(s)$, to the plant based upon the error, $E(s) = Y_d(s) - Y(s)$. We saw in Chapter 1 an example of how a controller performs the task of controlling a plant in a closed-loop system by ensuring that the plant output, $y(t)$, becomes as close as possible to the desired output, $y_d(t)$, as quickly

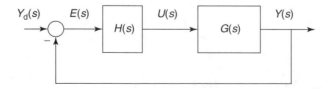

Figure 2.32 A single-input, single-output feedback control system with controller transfer function, $H(s)$, and plant transfer function, $G(s)$

as possible. In any case, a successful control system must bring $y(t)$ very close to $y_d(t)$ when time t becomes very large, i.e. *in the limit that time tends to infinity* ($t \to \infty$). Such a system is called a *tracking system*, because its output, $y(t)$, continuously *tracks* a changing desired output, $y_d(t)$. Examples of tracking systems are a telescope tracking a comet, an antenna tracking a satellite, a missile tracking an aircraft, a rifle-shooter tracking a pigeon, etc. The *error* ($e(t) = y_d(t) - y(t)$) which persists in the limit $t \to \infty$ is called the *steady-state error*, e_{ss}. Obviously, the closed-loop system should first be *able* to reach a *steady-state* (i.e. its response, $y(t)$, must be *finite* and *constant* in the limit $t \to \infty$) before its steady-state error can be defined (it is like saying *you should first be able to stand, before I can measure your height*). An *unstable* system cannot reach a steady-state; therefore, there is no point in talking about steady-state error of unstable systems. We will discuss later what are the precise requirements for stability, but at present let us confine our discussion to *stable closed-loop systems*, which we tentatively define here as those systems in which a *bounded* $y_d(t)$ leads to a *bounded* $y(t)$, for all values of t.

   Going back to Figure 2.32, we can see that the Laplace transforms of the output, $y(t)$, desired output, $y_d(t)$, input, $u(t)$, and error, $e(t)$, are given by $Y(s)$, $Y_d(s)$, $U(s)$, and $E(s)$, respectively. Then the steady-state error is expressed as

$$e_{ss} = \lim_{t \to \infty} \mathrm{e}(t) = \lim_{t \to \infty} \mathcal{L}^{-1}(E(s)) \tag{2.128}$$

However, we can avoid evaluating the inverse Laplace transform of $E(s)$ to calculate the steady-state error if we can utilize an important property of the Laplace transform, namely the *final value theorem* given by Eq. (2.70), which yields the following result:

$$e_{ss} = \lim_{t \to \infty} e(t) = \lim_{s \to 0} s E(s) \tag{2.129}$$

Of course, Eq. (2.129) requires that the limit of $e(t)$ when $t \to \infty$ (or of $sE(s)$ when $s \to 0$) must exist. Looking at the block-diagram of Figure 2.32, we can express $E(s)$ as follows:

$$E(s) = Y_d(s) - Y(s) = Y_d(s) - G(s)U(s) = Y_d(s) - G(s)H(s)E(s) \tag{2.130}$$

Thus, we can write

$$E(s) = Y_d(s)/[1 + G(s)H(s)] \tag{2.131}$$

On substituting Eq. (2.131) into Eq. (2.129), we get

$$e_{ss} = \lim_{s \to 0} s E(s) = \lim_{s \to 0} s Y_d(s)/[1 + G(s)H(s)] \tag{2.132}$$

Equation (2.132) implies that the steady-state error, e_{ss}, depends not only upon the two transfer functions, $G(s)$ and $H(s)$, but also on the desired output, $Y_d(s)$.

Example 2.16

Consider the closed-loop system of Figure 2.32 with $G(s) = (2s^2 + 5s + 1)/(s^2 + 2s + 3)$ and $H(s) = K$, where K is a constant. Let us determine the steady-state error of this system if the desired output, $y_\mathrm{d}(t)$ is (a) a unit step function, $u_s(t)$, and (b) a unit ramp function, $r(t) = t \cdot u_s(t)$. If $y_\mathrm{d}(t) = u_s(t)$ then $Y_\mathrm{d}(s) = 1/s$ (see Table 2.1). Hence, the steady-state error is given by Eq. (2.132) as

$$e_\mathrm{ss} = \lim_{s \to 0} sY_\mathrm{d}(s)/[1 + G(s)H(s)] = \lim_{s \to 0} s(1/s)/[1 + KG(s)]$$

$$= 1/[1 + K \lim_{s \to 0} G(s)] \tag{2.133}$$

where $\lim_{s \to 0} G(s)$ is called the *DC gain of* $G(s)$, because it is a property of the system in the limit $s \to 0$, or frequency of oscillation, $\omega \to 0$, in the frequency response $G(i\omega)$ – something like the *direct current* which is the limiting case of *alternating current* in the limit $\omega \to 0$. Here $\lim_{s \to 0} G(s) = 1/3$. Therefore, the steady-state error to unit step function is

$$e_\mathrm{ss} = 1/(1 + K/3) = 3/(3 + K) \tag{2.134}$$

The CST of MATLAB provides a useful command called *dcgain* for calculating the DC gain of a transfer function, which is used as follows:

```
>>sys= tf(num,den); dcgain(sys) <enter>
```

where *sys* is the name of the system's transfer function calculated using the LTI object *sys*, and *num* and *den* are the numerator and denominator polynomial coefficients (in decreasing powers of *s*), respectively, of the system's transfer function. This command is quite useful when the transfer function is too complicated to be easily manipulated by hand.

Note that the DC gain of the closed-loop transfer function in Figure 2.32, $G(s)H(s)/[1 + G(s)H(s)]$, is also the steady-state value of the output, $y(t)$, when the desired output is a unit step function. Hence, the steady-state error to a step desired output is nothing but $e_\mathrm{ss} = 1 - $ DC gain of the *closed-loop* transfer function.

The steady-state error given by Eq. (2.134) can be decreased by making the controller gain, K, large. However, for any finite value of K, we will be left with a non-zero steady-state error. Also, there is a physical limit upto which K (and the resulting input, $U(s)$) can be increased; the larger the value of K, the greater will be the control input, $U(s)$, which increases the *cost* of controlling the system. Hence, this closed-loop system is not very attractive for tracking a desired output which changes by a step.

If $y_\mathrm{d}(t) = r(t)$, then noting that $r(t)$ is the time-integral of $u_\mathrm{s}(t)$, we can get the Laplace transform, $Y_\mathrm{d}(s)$ from the *real integration* property (Eq. (2.68)) as follows:

$$Y_\mathrm{d}(s) = \mathcal{L}(r(t)) = \mathcal{L}\left(\int_{-\infty}^{t} u_\mathrm{s}(t)\,dt \right) = 1/s^2 \tag{2.135}$$

Hence, the steady-state error is given by

$$e_{ss} = \lim_{s \to 0} sY_d(s)/[1 + G(s)H(s)] = \lim_{s \to 0} s(1/s^2)/[1 + KG(s)]$$
$$= 1/[\lim_{s \to 0} s + sKG(s)] = \infty \qquad (2.136)$$

Thus, the steady-state error of the present closed-loop system is *infinite* when the desired output is a ramp function, which is clearly unacceptable. An example of tracking systems whose desired output is a ramp function is an *antenna* which is required to track an object moving at a *constant velocity*. This calls for the antenna to move at a *constant angular velocity*, c. Then the desired output of the antenna is $y_d(t) = c \cdot r(t)$.

Let us see what can be done to reduce the steady-state error of system in Example 2.16 when the desired output is either a unit step or a ramp function.

Example 2.17

In a control system, we can change the controller transfer function, $H(s)$, to meet the desired objectives. This process is called *control system design*. From Example 2.16, it is clear that $H(s) = K$ is a *bad design* for a closed-loop tracking system when the desired output is changing like a step, or like a ramp. If we can make the steady-state error to a ramp function *finite* by somehow changing the system, the steady-state error to a step function will automatically become *zero* (this fact is obvious from Eqs. (2.133) and (2.136)). Let us see what kind of controller transfer function, $H(s)$, will make the steady-state error to a ramp function finite (or possibly zero). For $Y_d(s) = 1/s^2$, the steady-state error is

$$e_{ss} = \lim_{s \to 0} s(1/s^2)/[1 + G(s)H(s)]$$
$$= \lim_{s \to 0}(s^2 + 2s + 3)/s[s^2 + 2s + 3 + (2s^2 + 5s + 1)H(s)]$$
$$= 3/[\lim_{s \to 0} sH(s)] \qquad (2.137)$$

If we choose $H(s) = K/s$, then Eq. (2.137) implies that $e_{ss} = 3/K$, which is a finite quantity. If $H(s) = K/s^2$ then $e_{ss} = 0$ from Eq. (2.137). For both the choices of $H(s)$, the steady-state error is zero when $y_d(t) = u_s(t)$. The choice $H(s) = K/s^2$ thus makes the steady-state error zero for both step and ramp functions.

Note that for the closed-loop system of Figure 2.32, the closed-loop transfer function, $Y(s)/Y_d(s)$, can be derived using Eq. (2.131) as follows:

$$Y(s) = G(s)H(s)E(s) = G(s)H(s)Y_d(s)/[1 + G(s)H(s)] \qquad (2.138)$$

or

$$Y(s)/Y_d(s) = G(s)H(s)/[1 + G(s)H(s)] \qquad (2.139)$$

For single-input, single-output systems such as that shown in Figure 2.32, we can calculate the closed-loop response, $y(t)$, to a specified function, $y_d(t)$, applying the inverse Laplace transform to Eq. (2.139).

Example 2.18

Let us calculate the response of the closed-loop system in Example 2.17 to a unit ramp function, $y_d(t) = r(t)$ and zero initial conditions, when (a) $H(s) = 1/s$, and (b) $H(s) = 1/s^2$. Using Eq. (2.139) for $H(s) = 1/s$ we can write

$$Y(s)/Y_d(s) = [(2s^2 + 5s + 1)/(s^2 + 2s + 3)](1/s)/$$
$$[1 + \{(2s^2 + 5s + 1)/(s^2 + 2s + 3)\}(1/s)]$$
$$= (2s^2 + 5s + 1)/(s^3 + 4s^2 + 8s + 1) \tag{2.140}$$

or

$$Y(s) = (2s^2 + 5s + 1)Y_d(s)/(s^3 + 4s^2 + 8s + 1)$$
$$= (2s^2 + 5s + 1)(1/s^2)/(s^3 + 4s^2 + 8s + 1) \tag{2.141}$$

Equation (2.141) can be expressed in a partial fraction expansion as follows:

$$Y(s) = k_1/(s - p_1) + k_2/(s - p_2) + k_3/(s - p_3) + k_4/s + k_5/s^2 \tag{2.142}$$

where p_1, p_2, and p_3 are the poles of the closed-loop transfer function, $(2s^2 + 5s + 1)/(s^3 + 4s^2 + 8s + 1)$, and k_1, k_2, and k_3 are the corresponding residues. k_4 and k_5 are the residues due to the ramp function, $Y_d(s) = 1/s^2$. We know that the poles of the closed-loop transfer function are *distinct* (i.e. not repeated) because we used the Control System Toolbox (CST) command *damp* to get the poles as follows:

```
>>damp([1 4 8 1]) <enter>

Eigenvalue        Damping   Freq. (rad/sec)
-0.1336           1.0000    0.1336
-1.9332+1.9355i   0.7067    2.7356
-1.9332-1.9355i   0.7067    2.7356
```

Note that the closed-loop system has a real pole, -0.1336, and a pair of complex conjugate poles, $-1.9332 + 1.9355i$, and $-1.9332 - 1.9355i$. The residues of the partial fraction expansion (Eq. (2.142)) can be calculated using MATLAB intrinsic command *residue* as follows:

```
>>num=[2 5 1]; den=conv([1 0 0],[1 4 8 1]); <enter>

>>[k,p,c]=residue(num,den) <enter>
```

```
k  =
 -0.0265-0.1301i
 -0.0265+0.1301i
 -2.9470
  3.0000
       0


p  =
 -1.9332+1.9355i
 -1.9332-1.9355i
 -0.1336
       0
       0


c  =
   [ ]
```

The roots of the denominator polynomial of Eq. (2.141) are contained in the vector p, while the vector k contains the corresponding residues of Eq. (2.142). The direct term c is a *null vector*, because the numerator polynomial is of a degree *smaller* than the denominator polynomial in Eq. (2.141). Taking the inverse Laplace transform of Eq. (2.142), we can express $y(t)$ as follows:

$$y(t) = k_1 \exp(p_1 t) + k_2 \exp(p_2 t) + k_3 \exp(p_3 t) + k_4 + k_5 t; \quad (t \geq 0) \quad (2.143)$$

The error $e(t) = y_d(t) - y(t)$, where $y_d(t) = r(t)$, and $y(t)$ is given by Eq. (2.143) is plotted in Figure 2.33 using MATLAB as follows (we could also have

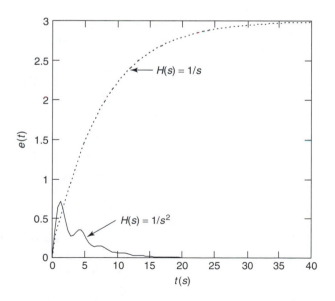

Figure 2.33 Error, $e(t) = y_d(t) - y(t)$, for the closed-loop systems of Example 2.18 when $y_d(t) = r(t)$

obtained $y(t)$ directly by using the M-file *response.m* listed in Table 2.4 by specifying a ramp input):

```
>>t=0:0.4:40; y=k(1)*exp(p(1)*t)+k(2)*exp(p(2)*t)+k(3)*exp(p(3)*t)
  +k(4)+k(5)*t; e=t-y <enter>
>>plot(t,e)
```

Similarly, when $H(s) = 1/s^2$, the closed-loop transfer function is

$$Y(s)/Y_d(s) = (2s^2 + 5s + 1)/(s^4 + 2s^3 + 5s^2 + 5s + 1) \tag{2.144}$$

and the inverse Laplace transform applied to $Y(s)$ with $Y_d(s) = 1/s^2$ yields (you may
verify using MATLAB)

$$y(t) = k_1 \exp(p_1 t) + k_2 \exp(p_2 t) + k_3 \exp(p_3 t) + k_4 \exp(p_4 t) + k_5 + k_6 t; \ (t \geq 0) \tag{2.145}$$

where $p_1 = -0.3686 + 1.9158i$, $p_2 = -0.3686 - 1.9158i$, $p_3 = -1.0$, $p_4 = -0.2627$,
$k_1 = -0.1352 - 0.1252i$, $k_2 = -0.1352 + 0.1252i$, $k_3 = -0.6667$, $k_4 = 0.9371$, $k_5 = 0$,
$k_6 = 0$. The error, $e(t) = y_d(t) - y(t)$, for $H(s) = 1/s^2$ is also plotted in Figure 2.33.
Note that $e_{ss} = 3$ for $H(s) = 1/s$, and $e_{ss} = 0$ for $H(s) = 1/s^2$, as expected from
Example 2.17.

We have seen in Examples 2.16–2.18 that for a plant transfer function, $G(s)$, of a
particular form, the controller transfer function, $H(s)$, must have either one or two poles
at the origin ($s = 0$) in order to reduce the closed-loop error due to ramp function.
Precisely how many poles $H(s)$ should have to reduce the steady-state error of a closed-
loop system to a particular desired output, $y_d(t)$, depends upon the plant transfer function,
$G(s)$, and $y_d(t)$. When $y_d(t)$ is a ramp function, Eq. (2.132) implies that

$$e_{ss} = \lim_{s \to 0} s(1/s^2)/[1 + G(s)H(s)] = 1/[\lim_{s \to 0}(s + sG(s)H(s))]$$

$$= 1/[\lim_{s \to 0} sG(s)H(s)] \tag{2.146}$$

Clearly, if we want zero steady-state error when desired output is a ramp function,
then Eq. (2.144) requires that $\lim_{s \to 0} sG(s)H(s) = \infty$, which is possible only if the
transfer function $G(s)H(s)$ has *two or more* poles at the origin, $s = 0$. Since $G(s)$ in
Examples 2.17 had *no poles* at the origin, we had to choose $H(s)$ with *two* poles at the
origin (i.e. $H(s) = 1/s^2$) to make $e_{ss} = 0$ when $y_d(t) = r(t)$. Classical control assigns a
type to a closed-loop system of Figure 2.32 according to *how many* poles the transfer
function, $G(s)H(s)$, has at the origin. Thus, a *type 1* system has exactly one pole of
$G(s)H(s)$ at origin, a *type 2* system has exactly *two* poles of $G(s)H(s)$ at the origin,
and so on. The transfer function, $G(s)H(s)$, is called the *open-loop transfer function*
of the system in Figure 2.32, because $Y(s)/Y_d(s) = G(s)H(s)$ if the feedback loop is
broken (or *opened*). We know from the real integration property of the Laplace transform
(Eq. (2.68)) that a pole at the origin results from a time-integration. Hence, it is said in

Table 2.5 Steady-state error according to system type for selected desired outputs

Desired Output, $y_d(t)$	Type 0 Steady-State Error	Type 1 Steady-State Error	Type 2 Steady-State Error
Unit step, $u_s(t)$	$1/[1 + \lim_{s \to 0} G(s)H(s)]$	0	0
Unit ramp, $t \cdot u_s(t)$	∞	$1/\lim_{s \to 0} sG(s)H(s)$	0
Unit parabola, $t^2 \cdot u_s(t)/2$	∞	∞	$1/\lim_{s \to 0} s^2 G(s)H(s)$

classical control parlance that the system type is equal to the *number of pure integrations in the open-loop transfer function*, $G(s)H(s)$. The system of Example 2.16 with $H(s) = K$ is of type 0, while the system with the same plant, $G(s)$, in Example 2.17 becomes of type 1 with $H(s) = 1/s$, and of type 2 with $H(s) = 1/s^2$.

Based on our experience with Examples 2.16–2.18, we can tabulate (and you may verify) the steady-state errors of *stable* closed-loop systems, according to their type and the desired output in Table 2.5.

In Table 2.5 we have introduced a new animal called the *unit parabolic function*, given by $t^2 \cdot u_s(t)/2$ (also $t \cdot r(t)/2$). An example of $y_d(t)$ as a parabolic function is when it is desired to track an object moving with a *constant acceleration* (recall that $y_d(t) = r(t)$ represented an object moving with a *constant velocity*). From Table 2.5, it is evident that to track an object moving with constant acceleration, the closed-loop system must be at least of type 2.

2.8 Stability

As stated previously, one of the most important qualities of a control system is its *stability*. In Example 2.2 we saw that an inverted pendulum is *unstable* about the equilibrium point $\theta = \pi$. In addition, while discussing the transfer function in Section 2.4, we saw that a second order system whose poles have *negative real parts* (or positive damping-ratio, $\varsigma > 0$) exhibits step and impulse responses with exponentially decaying amplitudes, and we called such a system *stable*. While discussing steady-state error, we required that a closed-loop system must be stable before its steady-state error can be calculated, and defined stability tentatively as the property which results in a *bounded* output if the applied input is *bounded*. From Examples 2.2 and 2.12, we have a rough idea about stability, i.e. the tendency of a system (either linear, or nonlinear) to *regain* its equilibrium point once displaced from it. While nonlinear systems can have more than one equilibrium points, their stability must be examined about each equilibrium point. Hence, for nonlinear systems stability is a property of the equilibrium point. The pendulum in Example 2.2 has two equilibrium points, one of which is unstable while the other is stable. We can now define stability (and instability) more precisely for *linear* systems. For simplicity, we will focus on the *initial response* of a system (i.e. response to initial conditions when the applied input is *zero*), and by looking at it, try to determine whether a linear system is stable. In this manner, we avoid having to classify the stability of a system according to the nature of the applied input, since stability is an intrinsic property of the linear system, independent of the input. There are the following three categories under which all *linear* control systems fall in terms of stability:

1. If the real parts of all the poles (roots of the denominator polynomial of the transfer function) are *negative*, then the initial response to finite initial conditions tends to a *finite* steady-state value in the limit $t \to \infty$. Such linear systems are said to be *asymptotically stable*. The aircraft of Example 2.10 is asymptotically stable, because all four poles have negative real parts.

2. If any pole of the linear system has a *positive* real part, then its initial response to finite initial conditions will be infinite in magnitude in the limit $t \to \infty$. Such a system is said to be *unstable*.

3. If all the poles of a system have real parts *less than or equal to zero*, and all the poles which have *zero* real parts are *simple*, i.e. they are not *repeated* (or *multiple*) poles (recall the discussion following Eq. (2.79)), the initial response of the system to finite initial conditions will keep on oscillating with a *finite* amplitude in the limit $t \to \infty$. Such a system is said to be *stable but not asymptotically stable* (because the response does not tend to an infinite magnitude, but also *does not* approach a constant steady-state value in the limit $t \to \infty$). However, if the poles having *zero* real part are *repeated* (i.e. they are *multiple poles* with the same imaginary part) the initial response of the system to finite initial conditions tends to infinity in the limit $t \to \infty$, and such a system is said to be *unstable*. In physical systems, complex poles occur in conjugate pairs (see Examples 2.10, 2.12). Thus, the only physical possibility of two (or more) repeated poles having zero real parts is that all such poles should be at the *origin* (i.e. their imaginary parts should also be zero).

We can summarize the stability criteria 1–3 by saying that if either the real part of any one pole is positive, or any one repeated pole has zero real part then the linear system is unstable. Otherwise, it is stable. A stable linear system having all poles with negative real parts is asymptotically stable. Using MATLAB you can easily obtain a location of the poles (and zeros) of a system in the Laplace domain with either the intrinsic command *roots(num)* and *roots(den)*, or the Control System Toolbox (CST) commands *pole(sys)*, *zero(sys)*, or *pzmap(sys)*, where *sys* is the transfer function LTI object of the system. From such a plot, it can be seen whether the system is asymptotically stable, stable (but not asymptotically stable), or unstable, from the above stated stability criteria. Another MATLAB (CST) command available for determining the poles of a system is *damp(den)* (see Example 2.10 for use of *damp*). Since the poles of a transfer function can be directly computed using MATLAB, one does not have to perform such mental calisthenics as the *Routh–Hurwitz* stability criteria (D'Azzo and Houpis [2]), which is a method for predicting the *number* of poles in the left and right half planes from the *coefficients* of the denominator polynomial by laboriously constructing a *Routh array* (see D'Azzo and Houpis [2] for details on Routh–Hurwitz stability criteria). Tabular methods such as Routh–Hurwitz were indispensible before the availability of digital computers and software such as MATLAB, which can directly solve the characteristic equation and give the location of the poles, rather than their number in the left and right half planes.

We shall further discuss stability from the viewpoint of *state-space* methods (as opposed to the classical frequency and Laplace domain methods) in Chapter 3. Before the advent of state-space methods, it was customary to employ *graphical methods* in the Laplace or the frequency domain for determining whether a closed-loop system was stable. The Bode plot is a graphical method that we have already seen. However, we are yet to discuss Bode plots from the stability viewpoint. Other graphical methods are the *root-locus*, the *Nyquist* plot, and the *Nichols* plot. Design using graphical methods is an instructive process for single-input, single-output, linear, time-invariant systems. Also, sometimes the graphical methods help in visualizing the results of a *multi-variable state-space* design (Chapters 5–7), and in that sense they have an important place in the modern control system design.

2.9 Root-Locus Method

To understand the root-locus method of stability analysis, reconsider the single-input, single-output feedback control system of Figure 2.32. The closed-loop transfer function, $Y(s)/Y_d(s)$, given by Eq. (2.139) is said to have a *return difference function*, $1 + G(s)H(s)$, which is the same as the *characteristic polynomial* of the *closed-loop system*. The *return difference* is a property of the feedback loop, and comes from the fact that if $Y_d(s) = 0$, then $Y(s) = G(s)U(s)$ and $U(s) = -H(s)Y(s)$; combining these two equations we can write $Y(s) = -G(s)H(s)Y(s)$, i.e. $Y(s)$ has *returned* to itself, or $[1 + G(s)H(s)]Y(s) = 0$. (The function $G(s)H(s)$ is called the *return ratio*.) The root-locus method determine stability simply by investigating whether the return difference becomes zero for any value of s in the *right* half of the Laplace domain (i.e. values of s with positive real parts). If $1 + G(s)H(s) = 0$ for some s in the right-half plane, it implies that there must be a closed-loop pole with *positive real part*, thus (according to stability criteria 2) the closed-loop system must be *unstable*. By drawing a *locus* of the each of the *roots* of the *return difference function*, $1 + G(s)H(s)$, as a *design parameter* is varied, we can find those values of the design parameter for which the system is stable (i.e. for which the loci do not enter the right half s-plane). How far away the locus closest to the imaginary axis is from crossing over into the right half s-plane also indicates how far away the system is from being unstable – in other words, the *stability margin* of the closed-loop system. A pole on the imaginary axis indicates *zero stability margin*, which we called the case of stable but not asymptotically stable system in stability criteria 3. By the same criteria, more than one pole at the *origin* indicates instability. If we see the loci of one (or more) poles crossing into the right-half s-plane, or the loci of two (or more) poles *simultaneously* approaching the origin, we should realize that the closed-loop system is heading towards instability.

Constructing a root-locus plot by hand is difficult, and standard classical controls textbooks, such as D'Azzo and Houpis [2], contain information about doing so. However, by using either the intrinsic MATLAB command *roots* repeatedly for each value of the design parameter, or the Control System Toolbox (CST) command *rlocus* the root-locus plot is very easily constructed. In the CST command *rlocus(sys)*, *sys* is the LTI object of the open-loop transfer function, $G(s)H(s)$. The *rlocus* program utilizes the MATLAB *roots* command repeatedly to plot the roots of $1 + KG(s)H(s)$ as a *design parameter*, K, is varied automatically. The design parameter, K, is thus a *scaling factor* for the controller transfer function, $H(s)$, and the root-locus shows what happens to the closed-loop poles as the controller transfer function is *scaled* to $KH(s)$. The user can also specify a vector k containing all the values of K for which the roots are to be computed by entering *rlocus(sys,k)* at the MATLAB prompt.

Example 2.19

For the closed-loop system of Figure 2.32 with plant, $G(s) = (2s^2 + 5s + 1)/(s^2 + 2s + 3)$, and controller, $H(s) = 1$, the root-locus is plotted in Figure 2.34 using the MATLAB command *rlocus* when the design parameter, K, is varied from -0.4 to 0.4 as follows:

```
>>num=[2  5  1]; den=[1  2  3]; k=-0.4:0.02:0.4; GH=tf(num,den);
  p= rlocus(GH,k); <enter>

>>plot(p(1,:),'o'); hold on; plot(p(2,:),'x') <enter>
```

It is clear from Figure 2.34 that both the poles of the closed-loop system have real part *zero* when $K = -0.4$, and then move towards the left as K is increased.

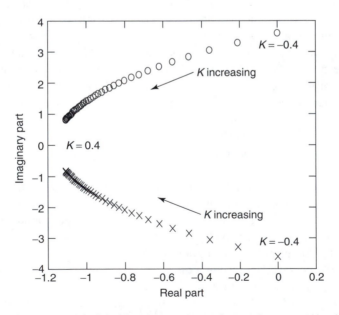

Figure 2.34 Root-locus plot of the closed-loop system of Example 2.19, as the design parameter, K, is varied from -0.4 to 0.4

For $K = -0.4$, the two loci are on the verge of crossing into the right half plane. Hence the closed-loop system is *stable* for $K > -0.4$ and *unstable* for $K < -0.4$. The value of K for which both the poles are on the imaginary axis (i.e. real part $= 0$) can be calculated by hand for this second order system by finding the characteristic equation as follows:

$$1 + KG(s)H(s) = 1 + K(2s^2 + 5s + 1)/(s^2 + 2s + 3) = 0 \qquad (2.147)$$

or

$$(1 + 2K)s^2 + (2 + 5K)s + (3 + K) = 0 \qquad (2.148)$$

Note that Eq. (2.148) is the second order characteristic equation, which can be expressed in terms of the natural frequency, ω_n, and damping-ratio, ζ, as Eq. (2.82). The poles with zero real part correspond to a damping-ratio, $\zeta = 0$. Comparing Eq. (2.131) with Eq. (2.82) and putting $\zeta = 0$, we get $(2 + 5K) = 0$, or $K = -0.4$, for which $\omega_n = [(3 + K)/(1 + 2K)]^{1/2} = 3.606$ rad/s, and the corresponding pole locations are $p_{1,2} = \pm i\omega_n = \pm3.606i$, which are the same as the pole locations in Figure 2.34 for $K = -0.4$.

We can find the values of the design parameter K for *specific pole locations* along the root-loci using the CST command *rlocfind* as follows:

```
>>[k,poles]=rlocfind(sys,p) <enter>
```

where p is the specified vector containing the pole locations for which values of K (returned in the vector k) are desired. You can get the value of gain, K, the pole location, associated damping, natural frequency, and maximum overshoot of step response by clicking at any point on the root-locus generated by the command *rlocus(sys)*. Similarly, the command *rlocfind(sys)* lets you move a cross-hair along the root-loci with the mouse to a desired pole location, and then returns the corresponding value of K when you click the left mouse button. To do this you must first plot a root-locus, and then use *rlocfind* on the same plot as follows:

```
>>rlocus(sys); hold on; rlocfind(sys) <enter>
```

The CST command *rlocus* is of limited utility in the stability analysis of closed-loop system, because it merely scales the controller transfer function, $H(s)$, by a factor K. For the root-locus plot of more general systems, we should use the MATLAB command *roots*, as shown in Example 2.20.

Example 2.20

Consider a closed-loop system shown in Figure 2.35. Note that this system is different from the one shown in Figure 2.32, because the controller, $H(s)$, is in the *feedback path* whereas the controller was placed in the *forward path*, in series (or

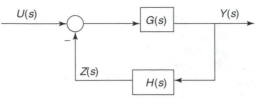

Figure 2.35   A closed-loop system with plant, $G(s)$, and feedback controller, $H(s)$

cascade) with the plant, $G(s)$, in Figure 2.32. You can show that both the systems have the *same* closed-loop transfer function, $Y(s)/Y_d(s)$, if $Z(s) = H(s)Y_d(s)$. The closed-loop transfer function, $Y(s)/U(s)$, for the system in Figure 2.35 is derived in the following steps:

$$Y(s) = G(s)[U(s) - Z(s)] \qquad (2.149)$$

where

$$Z(s) = H(s)Y(s) \qquad (2.150)$$

Substituting Eq. (2.150) into Eq. (2.149), we get the following expression for $Y(s)$:

$$Y(s) = G(s)U(s) - G(s)H(s)Y(s) \qquad (2.151)$$

or

$$Y(s)/U(s) = G(s)/[1 + G(s)H(s)] \qquad (2.152)$$

Note that the characteristic equation is still given by $1 + G(s)H(s) = 0$. For the present example, let us take $G(s) = (2s^2 + 5s + 1)/(s^2 - 2s + 3)$ and $H(s) = 1/(Ks + 1)$. This choice of $H(s)$ allows introducing a pole of $G(s)H(s)$ at a location that can be varied by changing the design parameter, K. (Introduction of poles and zeros in the open-loop transfer function, $G(s)H(s)$, through a suitable $H(s)$ to achieved desired closed-loop performance is called *compensation*, and will be studied in Section 2.12.) Let us plot the root-locus of the closed-loop system as the controller design parameter, K, is varied. Then the closed-loop characteristic equation is given by

$$Ks^3 + (3 - 2K)s^2 + (3 + 3K)s + 4 = 0 \qquad (2.153)$$

Since $H(s)$ is not merely scaled by K (which was the case in Example 2.19), we cannot use the MATLAB CST command *rlocus* to get the root-locus. Instead, the intrinsic MATLAB command *roots* is used repeatedly to plot the root-locus as K is varied from 1.0 to 1.3 as follows:

```
>>i=1; for k=1:0.02:1.3; r=roots([k 3-2*k 3+3*k 4]); R(i,:)=r'; i=i+1;
    end; plot(R,'x') <enter>
```

Note that the matrix R is used to *store* the roots in its *rows*. The resulting plot is shown in Figure 2.36. The complex conjugate roots are seen to cross into the right half plane for $K = 1.15$. Hence, the system is *stable* for $1 < K < 1.15$ and *unstable* for $K > 1.15$.

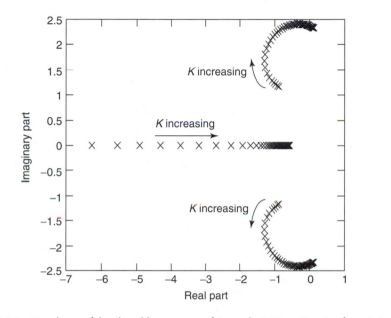

Figure 2.36   Root-locus of the closed-loop system of Example 2.20 as K varies from 0.3 to 1.3

2.10 Nyquist Stability Criterion

The *Nyquist* criterion determines a closed-loop system's stability using the *complex-variable* theory. It employs a polar plot (see Section 2.4) of the *open-loop frequency response*, $G(i\omega)H(i\omega)$, as ω increases from $-\infty$ to ∞. Such a polar plot is called a *Nyquist* plot. In the s-plane, the imaginary axis denotes an increase of ω from $-\infty$ to ∞. Hence, the Nyquist plot of $G(i\omega)H(i\omega)$, as ω increases from $-\infty$ to ∞ is said to be a *mapping* in the $G(s)H(s)$ *plane* (i.e. the plane defined by real and imaginary parts of $G(s)H(s)$) of all the points on the *imaginary axis* in the s-plane. The direction of the Nyquist plot indicates the direction of increasing ω. A polar plot is usually restricted to ω between 0 and ∞. However, note that a polar plot of $G(i\omega)H(i\omega)$ drawn for *negative frequencies* is the *complex conjugate* of the polar plot of $G(i\omega)H(i\omega)$ drawn for *positive frequencies*. In other words, the Nyquist plot is *symmetrical* about the real axis. Hence, the practical technique of plotting Nyquist plot is to first make the polar plot for ω increasing from 0 to ∞, and then draw a *mirror image* of the polar plot (about the real axis) to represent the other half of the Nyquist plot (i.e. the frequency range $-\infty < \omega < 0$). The direction of the *mirror image* polar plot would be clear from the $\omega = 0$ point location on the *original* polar plot, where the two plots should necessarily meet. The two mirror images should also meet at $\omega \to \infty$ and $\omega \to -\infty$, respectively. Hence, $\omega \to \pm\infty$ is a *single* point on the Nyquist plot, and the positive and negative frequency branches of a Nyquist plot form a *closed contour*. It is quite possible that the two polar plots for positive and negative frequencies of some functions may overlap. Figure 2.21 showed the polar plot for positive frequencies. You may verify that a polar plot of the same frequency response for negative frequencies overlaps the curve shown in Figure 2.21, but has an *opposite direction* for increasing ω.

Since the Nyquist plot of $G(s)H(s)$ is a *closed* contour for $s = i\omega$ when $-\infty < \omega < \infty$, whose direction is indicated by *increasing* ω, the only possibility of the positive and negative frequency branches meeting at $\omega \to \pm\infty$ is that the Laplace variable, s, must traverse an *infinite* semi-circle in the *right-half* s-plane. In other words, the region enclosed by the Nyquist plot in the $G(s)H(s)$ plane is a *mapping* of the entire right half s-plane. This fact is depicted in Figure 2.37. The direction of the curves in Figure 2.37 indicate the direction in which the frequency, ω is increasing. The point $G(s)H(s) = -1$ has a special significance in the Nyquist plot, since it denotes the closed-loop characteristic equation, $1 + G(s)H(s) = 0$.

The application of Nyquist stability criteria is restricted to linear, time-invariant control systems with a proper open-loop transfer function, $G(s)H(s)$. Since $G(s)H(s)$ of a linear, time-invariant control system is a *rational function* of s, a point in the s-plane corresponds to *only one* point in the $G(s)H(s)$ plane (such a mapping is called *one-to-one* mapping). Since $G(s)H(s)$ is proper, it implies that $\lim_{s \to \infty} G(s)H(s)$ must be either *zero*, or a non-zero *constant*. A one-to-one mapping from s-plane to the $G(s)H(s)$-plane in which

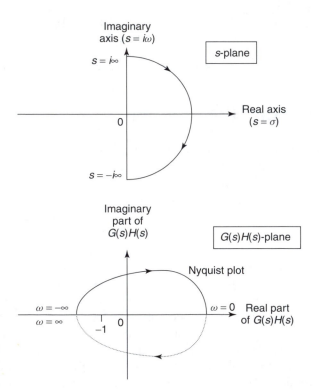

Figure 2.37 The mapping of right-half s-plane into a region enclosed by the Nyquist plot in the $G(s)H(s)$ plane

the limit $\lim_{s\to\infty} G(s)H(s)$ exists and is finite is called a *conformal mapping*. The Nyquist stability criterion is based on a fundamental principle of complex algebra, called *Cauchy's theorem*, which states that if a closed contour with an *anti-clockwise* direction in the s-plane, which *encloses* P poles and Z zeros of $1 + G(s)H(s)$, and which *does not* pass *through* any poles or zeros of $1 + G(s)H(s)$, is *conformally* mapped into a closed contour (i.e. Nyquist plot) in the $G(s)H(s)$ plane, then the latter will encircle the point $G(s)H(s) = -1$ *exactly* N times in an *anti-clockwise* direction, where $N = P - Z$. An anti-clockwise encirclement of $G(s)H(s) = -1$ is considered *positive*, while a clockwise encirclement is considered *negative*. Hence, N could be either *positive*, *negative*, or *zero*. The proof of Cauchy's theorem is beyond the scope of this book, but can be found in D'Azzo and Houpis [2], or in a textbook on complex variables. Applying Cauchy's theorem to the contour enclosing the entire right-half s-plane (shown in Figure 2.37), we find that for closed-loop stability we must have *no zeros* of the closed-loop characteristic polynomial, $1 + G(s)H(s)$, in the right-half s-plane (i.e. $Z = 0$ must hold), and thus $N = P$, which implies that for close-loop stability we must have *exactly* as many *anti-clockwise* encirclements of the point $G(s)H(s) = -1$ as the number poles of $G(s)H(s)$ in the right-half s-plane. (The Nyquist plot shown in Figure 2.37 contains one anti-clockwise encirclement of -1, i.e. $N = 1$. The system shown in Figure 2.37 would be stable if there is exactly one pole of $G(s)H(s)$ in the right-half s-plane.)

Note that Cauchy's theorem does not allow the presence of poles of $G(s)H(s)$ anywhere on the imaginary axis of s-plane. If $G(s)H(s)$ has poles on the imaginary axis, the closed contour in the s-plane (Figure 2.37) should be modified such that it passes *just around* the poles on the imaginary axis. Hence, the closed contour should have loops of infinitesimal radius passing around the poles of $G(s)H(s)$ on the imaginary axis, and the Nyquist plot would be a conformal mapping of such a contour in the $G(s)H(s)$ plane. Studying the effect of each detour around imaginary axis poles on the Nyquist plot is necessary, and could be a tedious process by hand [2].

You can make the Nyquist plot using the polar plot of $G(i\omega)H(i\omega)$ drawn for *positive frequencies* and its complex conjugate for negative frequencies, using the intrinsic MATLAB command *polar* (see Example 2.8). The abscissa of the polar plot must be modified to locate the point $G(s)H(s) = -1$ on the negative real axis (i.e. the radial line corresponding to $\phi = 180°$ in Figure 2.21). (Actually, the polar plot for negative frequencies is not required, since the number of encirclements can be determined from the shape of the polar plot of $G(i\omega)H(i\omega)$ drawn for positive frequencies near the point -1, and the fact that the negative frequency plot is the mirror image of the positive frequency plot.) An easier way of making the Nyquist plot is by the MATLAB Control System Toolbox (CST) command *nyquist(sys)*, where *sys* denotes the LTI object of the open-loop transfer function, $G(s)H(s)$.

Example 2.21

The Nyquist plot of the open-loop transfer function, $G(s)H(s)$ of the system in Figure 2.32 with $G(s) = (2s^2 + 5s + 1)/(s^2 - 2s + 3)$, and $H(s) = 1$ is obtained in Figure 2.38, using the following MATLAB command:

```
>>num=[2 5 1]; den=[1 -2 3]; GH=tf(num,den); nyquist(GH) <enter>
```

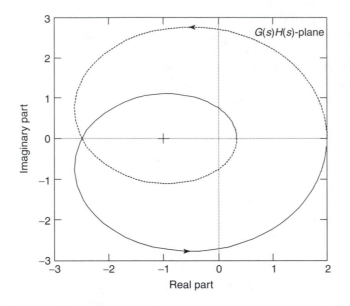

Figure 2.38   Nyquist plot of the closed-loop system in Example 2.21

It is clear from the Nyquist plot in Figure 2.38 that the point -1 is *encircled in the anticlockwise direction exactly twice*. Hence, $N = 2$. The open-loop transfer function $G(s)H(s)$ has *two* poles with positive real parts (i.e. $P = 2$), which is seen by using the MATLAB CST command *damp* as follows:

```
>>damp(den) <enter>

Eigenvalue       Damping   Freq. (rad/sec)
  1.0000+1.4142i  -0.5774   1.7321
  1.0000-1.4142i  -0.5774   1.7321
```

Since $N = P = 2$, Cauchy's theorem dictates that the number of zeros of $1 + G(s)H(s)$ in the right-half plane is $Z = P - N = 0$, which implies that the *closed-loop* transfer function has no poles in the right half plane. Hence by the Nyquist stability criterion, the closed-loop system is *stable*. You can verify that the closed-loop characteristic equation is $1 + G(s)H(s) = 0$, or $3s^2 + 3s + 4 = 0$, resulting in the closed-loop poles $-0.5 \pm 1.0408i$.

A major limitation of the Nyquist stability criterion is that it does not give any indication about the presence of *multiple zeros* of $1 + G(s)H(s)$ at $s = 0$, which cause closed-loop instability due to stability criterion 3 of Section 2.8. Another limitation of the Nyquist stability criterion is that it cannot be applied in cases where the Nyquist plot of $G(s)H(s)$ *passes through* the point $G(s)H(s) = -1$, because in that case the number of encirclements of -1 point are *indeterminate*.

2.11 Robustness

Robustness of a control system is related to its *sensitivity* to *unmodeled dynamics*, i.e. part of the behavior of an actual control system which is *not* included in the mathematical model of the control system (governing differential equations, transfer function, etc.). Since we have to deal with actual control systems in which it is impossible (or difficult) to mathematically model *all* physical processes, we are always left with the question: will the control system based upon a mathematical model really work? In Chapter 1, we saw that *disturbances (or noise)* such as road roughness, tyre condition, wind velocity, etc., cannot be mathematically modeled when discussing a control system for the car driver example. If a control system meets its performance and stability objectives in the presence of all kinds of expected *noises* (whose mathematical models are *uncertain*), then the control system is said to be *robust*. Hence, robustness is a desirable property that dictates whether a control system is immune to uncertainties in its mathematical model. More specifically, robustness can be subdivided into *stability robustness* and *performance robustness*, depending upon whether we are looking at the robustness of the *stability* of a system (determined by the location of the system's poles), or that of its *performance objectives* (such as peak overshoot, settling time, etc.).

We intuitively felt in Chapter 1 that a closed-loop system is *more* robust than an open-loop system. We are now in a position to mathematically compare the *sensitivities* of open and closed-loop systems, representatives of which are shown in Figure 2.39. For simplicity, it is assumed that for both open and closed loop systems, the controller transfer function is a constant, given by K, while the plant transfer function is $G(s)$. The transfer function of the open-loop control-system, $F_o(s)$, is given by

$$F_o(s) = Y(s)/Y_d(s) = KG(s) \tag{2.154}$$

while that of the closed-loop system ($F_c(s)$) is given by Eq. (2.124)

$$F_c(s) = Y(s)/Y_d(s) = KG(s)/[1 + KG(s)] \tag{2.155}$$

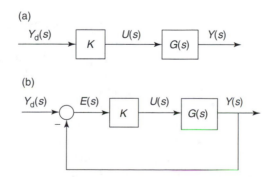

Figure 2.39 Single-input, single-output open-loop (a) and closed-loop (b) control systems with controller transfer function K and plant transfer function $G(s)$

Since the controller transfer function, K, is derived based on some mathematical model of $G(s)$, we can determine each control system's robustness by calculating the *sensitivity* of the overall transfer function, $Y(s)/Y_d(s)$, to variation in K. Mathematically, the sensitivity of either open-loop, or closed-loop system to variation in controller gain, K, can be expressed as

$$S(s) = [Y_d(s)/Y(s)]\partial[Y(s)/Y_d(s)]/\partial K \qquad (2.156)$$

where $\partial[Y(s)/Y_d(s)]/\partial K$ denotes the change in the transfer function due to a change in K. Then the sensitivities of the open and closed-loop systems to variation in K are given by

$$S_o(s) = [1/F_o(s)]\partial F_o(s)/\partial K = 1/K \qquad (2.157)$$

and

$$S_c(s) = [1/F_c(s)]\partial F_c(s)/\partial K = 1/[K(1 + KG(s))] \qquad (2.158)$$

The ratio of the open-loop sensitivity to the closed-loop sensitivity $S_o(s)/S_c(s)$ is thus

$$S_o(s)/S_c(s) = [1 + KG(s)] \qquad (2.159)$$

which is nothing else but our well known acquaintance, the *return difference function* (or the closed-loop characteristic polynomial)! The *magnitude* of the return difference, $|1 + KG(s)|$, is *greater than* 1, which confirms that an open-loop system is *more sensitive* (or *less robust*) when compared to the closed-loop system to variations in controller gain. The *greater* the value of the return difference, the *larger* is the robustness of a closed-loop system (Eq. (2.158)). Hence, one can measure a closed-loop system's robustness by determining the return difference function in the frequency domain, $s = i\omega$, for a range of frequencies, ω.

We need not confine ourselves to closed-loop systems with a constant controller transfer function when talking about robustness; let us consider a cascade closed-loop system (Figure 2.32) with a controller transfer function, $H(s)$. Either the Bode plot of $G(i\omega)H(i\omega)$, or the Nyquist plot of $G(s)H(s)$ can be utilized to convey information about a system's *stability robustness* (i.e. *how robust* is the *stability* of the system with respect to *variations* in the system's model). The Nyquist plot is more intuitive for analyzing stability robustness. From Nyquist stability theorem, we know that the closed-loop system's stability is determined from the encirclements by the locus of $G(s)H(s)$ of the point -1 in the $G(s)H(s)$ plane. Therefore, a measure of stability robustness can be *how far away* the locus of $G(s)H(s)$ is to the point -1, which indicates how far the system is from being unstable. The farther away the $G(s)H(s)$ locus is from -1, the greater is its stability robustness, which is also called the *margin of stability*. The *closest distance* of $G(s)H(s)$ from -1 can be defined in the complex $G(s)H(s)$ plane by two quantities, called the *gain margin* and the *phase margin*, which are illustrated in Figure 2.40 depicting a typical Nyquist diagram for a stable closed-loop system. A circle of unit radius is overlaid on the Nyquist plot. The closest distance of the Nyquist plot of $G(s)H(s)$ to -1 is indicated by two points A and B. Point A denotes the intersection of $G(s)H(s)$ with the negative real axis *nearest to the point* -1, while point B denotes the intersection of $G(s)H(s)$ with the *unit circle*. Point A is situated at a distance α from

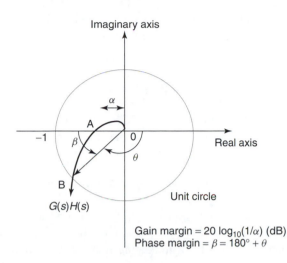

Figure 2.40 The Nyquist plot of $G(s)H(s)$ showing the gain and phase margins

the origin, while point B is located on the unit circle at an *angle θ from the positive real axis*. It is clear that the gain of $G(s)H(s)$ at point B is unity, while its phase at the same point is θ.

The *gain margin* is defined as the *factor* by which the gain of $G(s)H(s)$ can be *increased* before the locus of $G(s)H(s)$ hits the point -1. From Figure 2.40, it is evident that the gain margin is equal to $1/\alpha$, or in dB it is given by

$$\text{Gain Margin in dB} = 20 \ \log_{10}(1/\alpha) \qquad (2.160)$$

A *negative* gain margin indicates that the system is *unstable*. The *phase margin* is defined as the difference between the phase of $G(s)H(s)$ at point B, θ, and the phase of the *negative real axis*, $-180°$ (on which the point -1 is located). Thus, phase margin is given by the angle β in Figure 2.40:

$$\text{Phase Margin} = \beta = \theta - (-180°) = \theta + 180° \qquad (2.161)$$

The gain margin indicates how far away the *gain* of $G(s)H(s)$ is from 1 (i.e. the gain of the point -1) when its phase is $-180°$ (point A). Similarly, the phase margin indicates how far away the *phase* of $G(s)H(s)$ is from $-180°$ (i.e. the phase of the point -1) when the gain of $G(s)H(s)$ is 1 (point B).

Since the Bode plot is a plot of gain and phase in frequency domain (i.e. when $s = i\omega$), we can use the Bode plot of $G(s)H(s)$ to determine the gain and phase margins. In the Bode gain plot, the unit circle of the Nyquist plot (Figure 2.40) translates into the line for or *zero* dB gain (i.e. *unit* magnitude, $|G(i\omega)H(i\omega)|$), while the negative real axis of the Nyquist plot transforms into the line for $-180°$ phase in the Bode phase plot. Therefore, the gain margin is simply the gain of $G(i\omega)H(i\omega)$ when its phase crosses the $-180°$ phase line, and the phase margin is the difference between the phase of $G(i\omega)H(i\omega)$ and $-180°$ when the gain of $G(i\omega)H(i\omega)$ crosses the 0 dB line. The frequency for which the phase of $G(i\omega)H(i\omega)$ crosses the $-180°$ line is called the *phase crossover frequency*, ω_P, while the frequency at which the gain of $G(i\omega)H(i\omega)$ crosses the 0 dB line is called the *gain crossover frequency*, ω_G.

Example 2.22

Consider the closed-loop system of Figure 2.32, with $G(s) = (2s^2 + 5s + 1)/(s^2 + 2s + 3)$ and $H(s) = 1/s^2$. Figure 2.41 shows the Bode plot of $G(i\omega)H(i\omega)$ which is obtained by the following MATLAB CST command:

```
>>num=[2 5 1]; den=conv([1 2 3],[1 0 0]); G=tf(num,den); bode(G) <enter>
```

Figure 2.41 shows that the phase crosses the $-180°$ line at the phase crossover frequency, $\omega_P = 3.606$ rad/s. The gain present at this frequency is -14.32 dB. Hence, the gain margin is 14.32 dB, indicating that the gain of $G(s)H(s)$ can be increased by 14.32 dB before its Nyquist locus hits the point -1. In Figure 2.41, we can see that the 0 dB gain line is crossed at gain crossover frequency, $\omega_G = 1.691$ rad/s, for which the corresponding phase angle is $-148.47°$. Therefore, the phase margin is $-148.47° + 180° = 31.53°$.

The numerical values of gain margin, phase margin, gain crossover frequency, and phase crossover frequency, can be directly obtained using the MATLAB CST command *margin(sys)*, or *margin(mag,phase,w)*, where *mag,phase,w* are the magnitude, phase, and frequency vectors obtained using the command *[mag,phase,w] = bode(sys,w)*. For a system having a frequency response, $G(i\omega)H(i\omega)$, which is changing rapidly with frequency (such as in the present

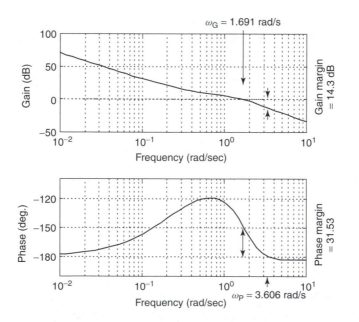

Figure 2.41　Bode plot of $G(s)H(s)$ for the closed-loop system of Example 2.22

example), the command *margin(sys)* may yield inaccurate results. Therefore, it is advisable to use the command *margin(mag,phase,w)*, with the magnitude and phase generated at a large number of frequencies in the range where the frequency response is changing rapidly. This is done by using the command $w = logspace(a, b, n)$, which generates n equally space frequencies between 10^a and 10^b rad/s, and stores them in vector w. Then the command $[mag,phase,w] = bode(sys,w)$ will give the desired magnitude and phase vectors from which the gain and phase margins can be calculated. This procedure is illustrated for the present example by the following commands:

```
>>num=[2 5 1]; den=conv([1 2 3],[1 0 0]); w=logspace(-2,1,1000);
  G=tf(num,den);

>>[mag,phase,w]=bode(G); <enter>

>>margin(mag,phase,w) <enter>
```

A Bode plot results, with computed gain and phase margins and the corresponding crossover frequencies indicated at the top of the figure (Figure 2.42).

Another way of obtaining gain and phase margins is from the *Nichols* plot, which is a plot of the gain of the open-loop frequency response, $G(i\omega)H(i\omega)$, against its phase. The gain and phase margins can be directly read from the resulting plot in which the point -1 of the Nyquist plot corresponds to the point 0 dB, $-180°$.

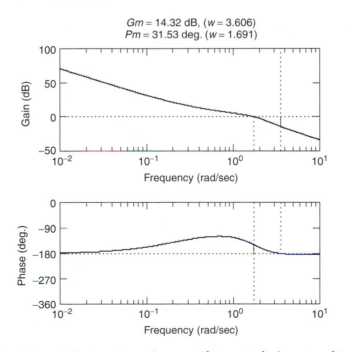

Figure 2.42   Gain and phase margins and crossover frequencies for the system of Example 2.22 obtained using the MATLAB Control System Toolbox command *margin*

In addition to the open-loop gain and phase, the Nichols plot shows *contours* of constant *closed-loop magnitude and phase*, $|Y(i\omega)/Y_d(i\omega)|$, and $\phi(\omega)$, given by

$$Y(i\omega)/Y_d(i\omega) = G(i\omega)H(i\omega)/[1 + G(i\omega)H(i\omega)]$$

$$= |Y(i\omega)/Y_d(i\omega)| \, e^{i\phi(\omega)} \tag{2.162}$$

It can be shown that the closed-loop gain and phase are related to the open-loop gain and phase $|G(i\omega)H(i\omega)|$ and θ, respectively, by the following equations:

$$|Y(i\omega)/Y_d(i\omega)| = 1/[1 + 1/|G(i\omega)H(i\omega|^2$$

$$+ 2\cos(\theta)/|G(i\omega)H(i\omega|]^{1/2} \tag{2.163}$$

$$\phi(\omega) = -\tan^{-1}\{\sin(\theta)/[\cos(\theta) + |G(i\omega)H(i\omega|]\} \tag{2.164}$$

where $G(i\omega)H(i\omega) = |G(i\omega)H(i\omega|e^{i\theta}$. Seeing the formidable nature of Eqs. (2.163) and (2.164), plotting the contours of constant closed-loop gain and phase by hand appears impossible. However, MATLAB again comes to our rescue by providing the Control System Toolbox (CST) command *nichols(sys)* for plotting the open-loop gain and phase, and the command *ngrid* for plotting the contours of the closed-loop gain and phase. Here *sys* denotes the LTI object of the open-loop transfer function, $G(s)H(s)$. Figure 2.43 shows the Nichols plot of the system in Example 2.22 obtained using the following MATLAB commands:

```
>>num=[2 5 1]; den=conv([1 0 0],[1 2 3]); G=tf(num,den); ngrid('new');
  nichols(G) <enter>
```

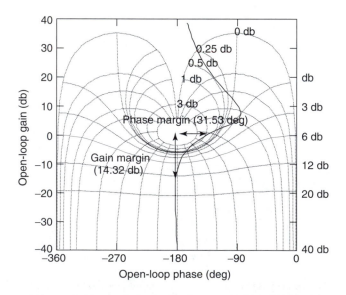

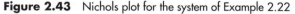

Figure 2.43　Nichols plot for the system of Example 2.22

The gain and phase margins are directly read from Figure 2.43 as shown. However, the crossover frequencies cannot be obtained from the Nichols plot. Apart from giving the gain and phase margins, the Nichols plot can be used to design and analyze the closed-loop frequency response of the system. The intersection of the $G(i\omega)H(i\omega)$ Nichols plot with the closed-loop gain contours give the closed-loop gain frequency response, $|Y(i\omega)/Y_d(i\omega)|$. The frequencies at the intersection points can be obtained from a Bode gain plot of $G(i\omega)H(i\omega)$. However, such a procedure is fairly complicated, and it is easier to get the closed-loop frequency response directly from the Bode plot of $Y(i\omega)/Y_d(i\omega)$.

We have seen how the stability robustness of a closed-loop control system is defined by the gain and phase margins determined from the open-loop transfer function, $G(s)H(s)$. We have considered variations in the overall transfer function, $Y(s)/Y_d(s)$, for defining robustness, and shown that closed-loop systems are more robust to such variations than open-loop systems. Variations in the overall transfer function are called *process noise*. However, closed-loop systems are susceptible to another form of noise, which is absent in open-loop systems, and which arises due to *errors* in *measuring* (and feeding back) the output, $Y(s)$. Such a noise is called *measurement noise*. Invariably, the measurement noise is caused by imperfections in the sensors used to measure the output, and usually occurs at higher frequencies than the natural frequencies of the closed-loop system. To reduce the sensitivity of a closed-loop system to measurement noise (or to make the system robust with respect to measurement noise), the frequency response of the closed-loop system must have *smaller gains at higher frequencies*. This requirement results in the magnitude Bode plot of the closed-loop system *decaying rapidly* (or *rolling-off*) with frequency, at high frequencies. Hence, a controller transfer function must be selected to provide not only good gain and phase margins, which indicate the closed-loop system's robustness to *process noise*, but also a large decay (or *roll-off*) of the gain at high frequencies, indicating robustness due to the measurement noise.

While designing a closed-loop system, one is not only interested in stability in the presence of modeling uncertainties (such as process and measurement noise), but also in maintaining a desired level of *performance*, i.e. one would like to achieve *performance robustness* as well. It can be appreciated that for achieving performance robustness, we should first achieve stability robustness (i.e. there is no point in talking about the performance of an unstable control system). Criteria for achieving stability and performance robustness of *multivariable systems* are more generally expressed with the use of modern *state-space* methods (rather than the classical single-input, single-output, frequency domain procedures), as discussed in Chapter 7.

2.12 Closed-Loop Compensation Techniques for Single-Input, Single-Output Systems

We have seen in the previous sections how the steady-state error, stability, and gain and phase margins can be affected by changing the controller transfer function, $H(s)$, in a typical single-input, single-output closed-loop system of Figure 2.32. Since we can use $H(s)$ to *compensate* for the poor characteristics of a plant, $G(s)$, such a controller, $H(s)$, is called a *compensator*, and the procedure of selecting a controller, $H(s)$, in order to remove the *deficiencies* of the plant, $G(s)$, (such as improving stability, performance, and robustness), is called *closed-loop compensation. Classical* control techniques of designing single-input, single-output control systems based upon their frequency response characteristics largely rely upon *closed-loop compensation.* We will consider some commonly employed closed-loop compensation techniques. Closed-loop compensation is generally of two types: cascade (or series) compensation in which $H(s)$ is placed in series with the plant (Figure 2.32), and feedback (or parallel) compensation in which $H(s)$ is placed in the feedback path of $G(s)$ (Figure 2.35). Cascade and feedback compensation represent two alternatives for achieving the same closed-loop characteristics. Where we insert the compensator in the control system depends largely upon the physical aspects of implementation. We study only cascade compensation techniques here, which can be easily extended to feedback compensation.

2.12.1 Proportional-integral-derivative compensation

A popularly used compensator with the transfer function, $H(s) = (K_D s^2 + K_P s + K_I)/s$, where K_D, K_P, and K_I are constants, is called the *proportional plus integral plus derivative* (PID) *compensator*, because its transfer function can be expressed as $H(s) = K_P + K_D s + K_I/s$, signifying that the output of the controller, $U(s)$, is a sum of its input, $E(s)$, multiplied by *constant* K_P, the *integral* of the input times K_I, and the *derivative* of the input times K_D, i.e. $U(s) = K_P E(s) + K_I E(s)/s + K_D s E(s)$ (recall that $E(s)/s$ is the Laplace transform of the *integral* of the function, $e(t)$, and $sE(s)$ is the Laplace transform of the derivative of $e(t)$ if the initial conditions are zero). PID compensators are very common in industrial applications due to their good *robustness* over a wide frequency range. The presence of the integral term, K_I/s, in $H(s)$ increases the type of the closed-loop system due to the pole at the origin, thereby reducing the steady-state error. The derivative term, $K_D s$, and the proportional term, K_P, can be used to place two *zeros* of $H(s)$ at suitable locations, to change the phase characteristics of the closed-loop system. Let us consider an example of PID compensation.

Example 2.23

Consider a hard-disk read/write head positioning system with the following transfer function:

$$Y(s)/U(s) = G(s) = 700/(s^2 + 15s + 100\,000) \qquad (2.165)$$

The output, $Y(s)$, is the angular position of the head in radians, while the input, $U(s)$, is the current in milli-Amperes (mA) supplied to the head positioning solenoid. The poles, natural frequencies, and damping-ratios for this second order plant are as follows:

```
>>num=700; den = [1 15 1e5]; damp(den) <enter>
```

```
Eigenvalue                     Damping      Freq. (rad/sec)
-7.5000e+000+3.1614e+002i   2.3717e-002   3.1623e+002
-7.5000e+000-3.1614e+002i   2.3717e-002   3.1623e+002
```

Note that the damping is very small, and hence the head positioning system, will oscillate a lot before coming to a steady-state. Since the plant is of type 0, the steady-state error due to a step input will be non-zero. To make the steady-state error zero for a step input, we initially choose a PID compensator with $K_D = 0$, $K_P = 1$, and $K_I = 1100$, which makes the controller transfer function the following:

$$H(s) = (s + 1100)/s \qquad (2.166)$$

Since $H(s)$ has no derivative term, it is essentially a *proportional-plus-integral* (PI) compensator, with a zero at $s = -1100$ and a pole at $s = 0$. Connecting $H(s)$ and $G(s)$ in series, and then closing the feedback loop as shown in Figure 2.32, we get the following closed-loop transfer function:

$$Y(s)/Y_d(s) = (700s + 7\,70\,000)/(s^3 + 15s^2 + 1\,00\,700s + 7\,70\,000) \qquad (2.167)$$

The Bode plots of the plant, $G(s)$, the compensator, $H(s)$, and the closed-loop system are obtained as follows, and are shown in Figure 2.44:

```
>>G=tf(num,den); w=logspace(-1,4); [mag1,phase1,w]=bode(G,w); %
  Bode plot of the plant <enter>
```

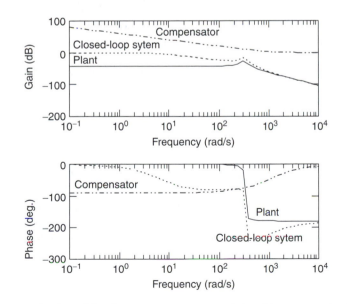

Figure 2.44   Bode plots of the plant, PI compensator, and closed-loop system for the hard-disk read/write head

```
>>n1=[1 1100]; d1=[1 0];H=tf(n1,d1); [mag2,phase2,w]=bode(H,w); %
  Bode plot of the compensator <enter>

>>nCL=[700 770 000]; dCL=[1 15 100 700 770 000]; GCL=tf(nCL,dCL);
  <enter>

>>[mag3,phase3,w]=bode(GCL,w); % Bode plot of the closed-loop system
  <enter>
```

Note from Figure 2.44 that the PI compensator *increases the gain*, while *decreasing the phase* at low frequencies, and leaves the *high frequency gain* unaffected. The *DC gain* of the closed-loop system is brought to 0 dB, indicating that the step response will have a *zero* steady-state error. There is a *reduction* in the *gain margin* of the closed-loop system due to the increased gain at low frequencies, when compared to that of the plant, indicating a loss of robustness. The plant's natural frequency of 316.23 rad/s is visible as a peak in the gain plot. The closed-loop poles are the following:

```
>>damp(dCL) <enter>
```

Eigenvalue	Damping	Freq. (rad/sec)
-3.6746e+000+3.1722e+002i	1.1583e-002	3.1724e+002
-3.6746e+000-3.1722e+002i	1.1583e-002	3.1724e+002
-7.6507e+000	1.0000e+000	7.6507e+000

Note that the closed-loop damping near plant's natural frequency is *slightly reduced*, while another pole is placed at $s = -7.6507$. How does this closed-loop pole configuration affect the step response of the system? This question is best answered by comparing the plant's step response with that of the closed-loop system. The step responses are calculated using the M-file *stepresp.m* of Table 2.3 as follows:

```
>>[s1,t] = stepresp(num,den,0,0.01,1); % step response of the plant
  <enter>

>>[s2,t] = stepresp(nCL,dCL,0,0.01,1); % step response of the
  closed-loop system <enter>
```

The two step responses are compared in Figure 2.45. Note that while the closed-loop steady-state error is brought to zero, the settling time and the large number of oscillations of the plant are unaffected.

Furthermore, let us compare the gain and phase margins of the closed-loop system with those of the plant, to see how the stability robustness is affected. The gain and phase margins of the plant, with the respective crossover frequencies, are calculated using the command *margin* as follows:

```
>>[gm,pm,wg,wp]=margin(G) <enter>

gm =
  Inf
```

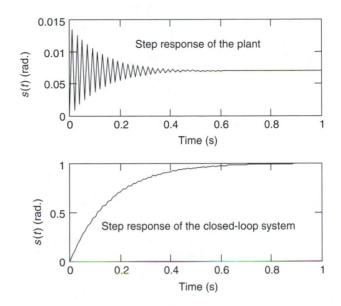

Figure 2.45 Step responses of the plant and the PI compensated closed-loop system for the hard-disk read/write head

```
pm =
  Inf
wg =
  NaN
wp =
  NaN
```

where *inf* denotes ∞, and *NaN* stands for 'not a number', i.e. an *undefined* quantity. The plant thus has *infinite* gain and phase margins, and the corresponding crossover frequencies are thus undefined. The margins for the closed-loop system are the following:

```
>>[gmCL,pmCL,wgCL,wpCL] = margin(GCL) <enter>

gmCL =
  0.9750
pmCL =
  -3.8267
wgCL =
  318.4062
wpCL =
  318.6843
```

Both gain and phase margins of the closed-loop system have been drastically reduced to only 0.975 (−0.22 dB) and −3.8267°, respectively, with the gain and phase crossover frequencies quite close together at 318.7 rad/s and 318.4 rad/s, respectively. Clearly, the closed-loop system is quite less robust than the plant. In summary, the PI compensator given by Eq. (2.166) not only results in a highly oscillatory response, but also a significant loss in robustness.

Let us now improve the PID compensation by *selecting* the values of K_D, K_P, and K_I, that lead to a *desired* closed-loop response. It is desirable to have a well damped closed-loop response, to reduce the number of oscillations, as well as the settling time. Hence, let us select a damping-ratio of $\zeta = 0.707$ for the closed-loop system, without changing the plant's natural frequency, $\omega_n = 316.23$ rad/s. The general transfer function of the forward path, $G(s)H(s)$, can then be written as follows:

$$G(s)H(s) = 700(K_D s^2 + K_P s + K_I)/[s(s^2 + 15s + 100\,000)] \qquad (2.168)$$

Note that the closed-loop system's type is increased from 0 to 1, due to the presence of the pole at $s = 0$ in $G(s)H(s)$. Hence, the steady-state error to step input will be zero, irrespective of the values of K_D, K_P, and K_I. The closed-loop transfer function is $G(s)H(s)/[1 + G(s)H(s)]$, which can be expressed as $N(s)/D(s)$, where $N(s)$ and $D(s)$ are the following numerator and denominator polynomials, respectively:

$$N(s) = 700(K_D s^2 + K_P s + K_I); \; D(s) = s^3 + (15 + 700K_D)s^2$$

$$+ (100\,000 + 700K_P)s + 700K_I \qquad (2.169)$$

Note that the closed-loop system is of third order. We can write $D(s)$ as a quadratic factor in s multiplied by a first order polynomial in s, as follows:

$$D(s) = (s^2 + 2\zeta\omega_n s + \omega_n^2)(s + p) \qquad (2.170)$$

Note from Eq. (2.116) that the step (or impulse) response of a system is *dominated* by the poles with the *smallest real part magnitudes*. If we select the *two* poles of the closed-loop system resulting from the roots of $(s^2 + 2\zeta\omega_n s + \omega_n^2)$ to be *closer* to the imaginary axis than the *third pole* $s = -p$, the closed-loop response would be *dominated* by the quadratic factor in $D(s)$, with $(s + p)$ influencing the closed-loop response by a lesser extent. In such a case, the roots of $(s^2 + 2\zeta\omega_n s + \omega_n^2)$ are called the *dominant poles* of the closed-loop system. Since we have already selected the closed-loop damping-ratio as $\zeta = 0.707$, and natural frequency as, $\omega_n = 316.23$ rad/s, we can choose the pole $s = -p$ to be *further away* from the imaginary axis in the left-half plane by having $p > \zeta\omega_n$ (i.e. $p > 223.57$). Let us take $p = 300$. Substituting the values of ζ, ω_n, and p into Eq. (2.170) and comparing with $D(s)$ given in Eq. (2.169), we get the following values of the PID compensator constants that would result in the desired closed-loop dynamics: $K_I = 42857.14$, $K_D = 1.0459$, $K_P = 191.64$, and the compensator's transfer function is given by

$$H(s) = (1.0459s^2 + 191.64s + 42\,857.14)/s \qquad (2.171)$$

With this PID compensator, we get the following closed-loop transfer function:

$$Y(s)/Y_d(s)$$

$$= (732.15s^2 + 134\,140s + 3 \times 10^7)/[(s + 300)(s^2 + 447.15s + 1\,00\,000)] \qquad (2.172)$$

Let us check the closed-loop step response as follows:

```
>>nCL = [732.15 134140 3e7]; dCL = conv([1 300],[1 447.15 1e5]); <enter>

>>[s,t] = stepresp(nCL,dCL,0,0.0005,0.03); plot(t,s) <enter>
```

The resulting closed-loop step response is plotted in Figure 2.46. Note the settling time of 0.025 seconds (as compared to 0.7 seconds for the plant), and a *complete lack* of high frequency oscillations in the step response, with a zero steady-state error. The performance is thus greatly improved by the PID compensation.

Finally, let us determine the gain and phase margins of the PID compensated closed-loop system as follows:

```
>>[gmCL,pmCL,wgCL,wpCL]=margin(GCL) <enter>

gmCL =
  Inf
pmCL =
  144.8891
wgCL =
  NaN
wpCL =
  583.6209
```

Note that the closed-loop gain margin is infinite, and the phase margin is 144.9° occurring at gain crossover frequency of 583.6 rad/s. Although the phase margin is no longer infinite as for the plant, it is quite large and adequate. The large

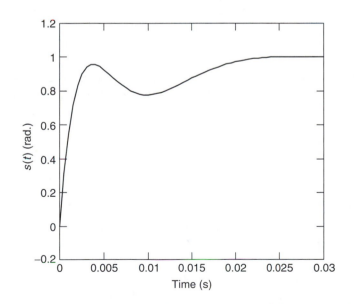

Figure 2.46 Closed-loop step response of the PID compensated hard-disk read/write head positioning system

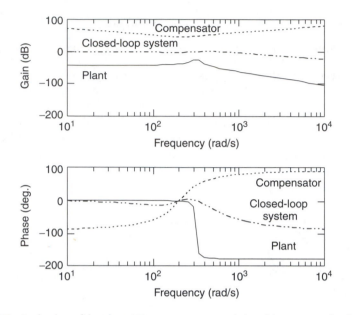

Figure 2.47 Bode plots of the plant, PID compensator, and closed-loop system for the hard-disk read/write head

value of the gain crossover frequency indicates a very fast response, which is evident in Figure 2.46. In short, the PID compensated closed-loop system has a good combination of performance and stability robustness. The Bode plots of the PID compensator and the closed-loop system are compared with those of the plant in Figure 2.47. Note that the PID compensator has a *decreasing* gain at low frequencies, and an *increasing* gain at high frequencies. Also, the PID compensator provides a *phase-lag* at frequencies below the plant's natural frequency, and a *phase-lead* at higher frequencies. The resulting closed-loop gain and phase plots are much *flatter* (compared to the plant). However, due to an increased closed-loop gain at high frequencies, there is an *increased* sensitivity (and decreased robustness) with respect to the high frequency *measurement noise*, which is undesirable.

The process of finding suitable PID constants K_D, K_P, and K_I (as illustrated very simply in Example 2.23) is called PID *tuning*. Often, PID tuning in trying to achieve desirable closed-loop characteristics is an iterative procedure A PID (or PI) compensator contains an *integrator* (i.e. a pole $s = 0$) which requires special implementation techniques. Mechanically, a pure integration is possible using a *rate-integrating gyroscope* – a commonly used (but expensive) device in aircraft, missile, and spacecraft control systems. Figure 2.48(a) depicts the schematic diagram of a *single degree of freedom* rate gyroscope, with *gimbal angle* (i.e. rotation of the wheel assembly about the x-axis), θ, as the output and the *angular velocity of the case* about the z-axis, Ω, as the input. If we neglect the stiffness and inertia of the gimbal and wheel about the x-axis, we can write the following transfer

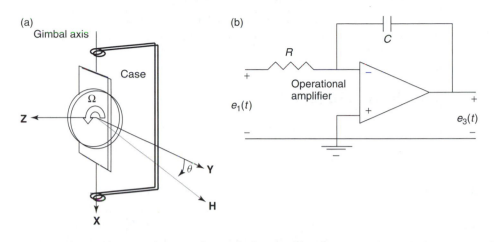

Figure 2.48 (a) Schematic diagram of a single degree of freedom, rate integrating gyroscope with output, $\theta(s)$ and input, $\Omega(s)$. **H** is the angular momentum vector with a constant magnitude, H; (b) Active circuit of an operational amplifier connected as an integrator with input, $e_1(t)$ and output, $e_3(t)$

function for the rate-integrating gyroscope:

$$\theta(s)/\Omega(s) = H/(cs) \tag{2.173}$$

where H is the constant magnitude of the *angular momentum vector*, **H**, of the wheel about the y-axis, and c is the *viscous damping coefficient* of the gimbal (i.e. rotary mechanism about the x-axis) (see Figure 2.48(a)). Equation (2.173) indicates that the gimbal angle output is the time integral of the input angular velocity of the case.

Modern control systems are relatively inexpensively implemented using either *passive* (unpowered) or *active* (powered) electronic circuits. Figure 2.48(b) shows an active circuit which uses an *operational amplifier* to represent a pure integrator. An operational amplifier has two input voltages, $e_1(t)$ and $e_2(t)$, and an output voltage, $e_3(t)$. In the circuit shown in Figure 2.48(b), $e_2(t) = 0$, and the equation governing the circuit is the following:

$$e_3(t) = -1/(RC) \int e_1(t)\, dt \tag{2.174}$$

which implies that the output voltage, $e_3(t)$, is the *time integral* of the input voltage, $e_1(t)$, multiplied by the constant $-1/(RC)$. Operational amplifiers can also be used to represent summing junctions, and other useful devices.

Control systems based upon operational amplifiers (and other active circuits), are generally more expensive and sensitive to noise than those based on passive circuits. A simpler alternative to operational amplifier is the *approximation* of the pure integrator by a *passive* circuit, called *lag circuit*. The chief difficulty in implementing PID compensators is the *ideal differentiator* (i.e. the term $K_d s$ in the expression for $H(s)$). An ideal differentiator is difficult to set up, and leads to the amplification of any noise present in the input signal. The noise amplification may interfere with the working of the entire control system. For these reasons, a pure differentiator is never used practically, but only an *approximate* differentiation is implemented using a *passive* circuit (called a *lead circuit*). Hence, the ideal PID compensator is practically implemented by a combination of lag and lead circuits (called a *lead-lag* compensator), which represent *approximate* integration and differentiation.

2.12.2 Lag, lead, and lead-lag compensation

Consider a compensator with the following transfer function:

$$H(s) = (s + \omega_o)/(s + \omega_o/\alpha) \qquad (2.175)$$

where α is a real constant and ω_o is a constant frequency. If $\alpha > 1$, the compensator is called a *lag compensator*, because it always has a *negative* phase angle, i.e. a *phase-lag*. In the limit $\alpha \to \infty$, the lag compensator approaches an ideal PI compensator. Hence, a lag compensator can be used to approximate a PI compensator in practical implementations. A *lag compensator* is useful in reducing the steady-state error of type 0 plants, and decreasing the gain of the closed-loop system at high frequencies (which is desirable for reducing the sensitivity to the measurement noise). However, lag compensation *slows down* the closed-loop transient response (i.e. increases the settling time). Lag compensation is relatively simple to use, because the *passive circuit* through which it can be implemented is quite inexpensive.

In Eq. (2.175), if $\alpha < 1$, the resulting compensator is called a *lead compensator*, because it always has a *positive* phase angle, i.e. a *phase-lead*. A *lead compensator* is useful for increasing the *speed* of the closed-loop response (i.e. decreasing the *settling time*), and increasing the *phase margin* of the closed-loop system, which also results in smaller overshoots in the transient response. Lead compensation usually requires amplification of error signals, which results in an expensive electrical circuit for implementation. Also, lead compensation increases the gain at high frequencies, which is undesirable due to increased sensitivity to measurement noise. A lead compensator given by the transfer function of Eq. (2.175) would *decrease* the DC gain of the type 0 open-loop transfer function, G(s)H(s), which is undesirable as it would *increase* the steady-state error due to a step input. So that the DC gain of the type 0 open-loop transfer function, $G(s)H(s)$, is *unchanged* with a lead compensator, the lead compensator transfer function is usually multiplied by the factor $1/\alpha$, resulting in

$$H(s) = (s + \omega_o)/(\alpha s + \omega_o) \qquad (2.176)$$

Conversely, if we do not wish to *increase* the DC gain of $G(s)H(s)$ (which may be infinite due to poles of $G(s)$ at $s = 0$) with *lag compensation*, we would choose a lag compensator with the transfer function given by Eq. (2.176). Phase lag (or lead) compensation is traditionally employed in a variety of control applications.

To combine the desirable properties of the lead and lag compensators, sometimes it is better to use a compensator which has *both* a phase-lag at low frequencies, *and* a phase-lead at high frequencies. Such a compensator is called a *lead-lag compensator*, and has the transfer function

$$H(s) = [(s + \omega_1)/(s + \omega_1/\alpha)][(s + \omega_2)/(s/\alpha + \omega_2)] \qquad (2.177)$$

where $\alpha > 1$, and ω_1 and ω_2 are constant frequencies. Note that the transfer function suggests that the lead-lag compensator consists of a lag compensator in series with a lead compensator. Also, note that the lag part of the transfer function is designed to increase the DC gain of open-loop system, $G(s)H(s)$. The frequencies, ω_1 and ω_2, and the constant, α, must be selected to achieve desired closed-loop characteristics. It is easy to see that in the limit $\alpha \to \infty$, the lead-lag compensator approaches an ideal PID compensator.

Figure 2.49 shows the passive circuits used to implement the lag, lead, and lead-lag compensators. In each circuit, $e_1(t)$ and $e_2(t)$ are the input and output voltages, respectively. The transfer function of the *lag circuit* (Figure 2.49(a)) is expressed as

$$H(s) = E_2(s)/E_1(s) = (1 + R_2Cs)/[1 + (R_1 + R_2)Cs] \qquad (2.178)$$

Comparing Eqs. (2.178) and (2.176), it is clear that $\omega_o = 1/(R_2C)$ and $\alpha = (R_1 + R_2)/R_2$. The *lead* circuit in Figure 2.49(b) has the transfer function

$$H(s) = E_2(s)/E_1(s) = R_2(1 + R_1Cs)/[(R_1 + R_2) + R_1R_2Cs] \qquad (2.179)$$

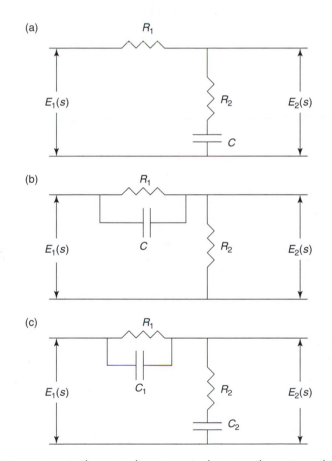

Figure 2.49 (a) Lag circuit with input voltage $E_1(s)$ and output voltage $E_2(s)$; (b) Lead circuit with input voltage $E_1(s)$ and output voltage $E_2(s)$; (c) Lead-lag circuit with input voltage $E_1(s)$ and output voltage $E_2(s)$

which implies that $\omega_o = 1/(R_1C)$ and $\alpha = R_2/(R_1 + R_2)$. Finally, the transfer function of the *lead-lag* circuit shown in Figure 2.49(c) is the following:

$$H(s) = E_2(s)/E_1(s) = [(1 + R_1C_1s)(1 + R_2C_2s)]/[(1 + R_1C_1s)(1 + R_2C_2s) + R_1C_2s] \quad (2.180)$$

Comparing Eqs. (2.180) and (2.177), you may verify that $\omega_1 = 1/(R_1C_1)$, $\omega_2 = 1/(R_2C_2)$, and α is obtained by factoring the denominator (i.e. solving a quadratic equation) in Eq. (2.180).

Example 2.24

Consider the *roll dynamics* of a fighter aircraft with the following transfer function:

$$\phi(s)/\delta(s) = G(s) = 1000/[s(s+5)] \quad (2.181)$$

The output, $Y(s) = \phi(s)$, is the *roll-angle*, while the input, $U(s) = \delta(s)$, is the *aileron deflection angle*. The maneuverability of an aircraft depends upon the time taken to achieve a *desired roll-angle*, (or *bank-angle*) $Y_d(s) = \phi_d(s)$. It is required to design a closed-loop control system to achieve a *unit step* desired roll-angle in about three seconds, with a maximum overshoot less than 2 percent. Since the plant has a pole at $s = 0$, it is not *asymptotically stable* and the steady-state condition would *not* be achieved by plant alone, thereby requiring a feedback compensation that would provide asymptotic stability. Since the plant is of type 1 due to the pole at the origin, a simple feedback system with $H(s) = K$ (where K is a constant) will achieve a zero steady-state error to a step desired output. With such a compensator, the closed-loop characteristic equation would be $s^2 + 5s + 1000K = 0$, and the closed-loop poles would be $s_{1,2} = -2.5 \pm (1000K^2 - 6.25)^{1/2}i$, with a damping-ratio, $\zeta = 0.0791/\sqrt{K}$, and natural frequency, $\omega_n = 31.6228\sqrt{K}$. If we want to achieve a settling time, T_s, of 3 seconds, then from Eq. (2.125) we have $T_s = 4/(\zeta\omega_n)$, or $\zeta\omega_n = 4/3$. The closed-loop system with $H(s) = K$ can give us $\zeta\omega_n = 0.0791 \times 31.6228 = 2.5$, or $T_s = 4/2.5 = 1.6$ seconds, which is a *smaller* settling time than required. The maximum overshoot requirement of 2 percent of steady-state response, according to Eq. (2.127) results in the condition $M_P = 1.02 = 1 + \exp[-\zeta\pi/(1 - \zeta^2)^{1/2}]$, or $\zeta = 0.7797$, which requires $\sqrt{K} = 0.0791/0.7797 = 0.1014$, or $K = 0.0103$. Hence, we should have $K \leq 0.0103$ in order to satisfy *both* settling time and maximum overshoot requirements. Let us choose $K = 0.01$, which gives us a closed-loop settling time, $T_s = 1.6$ seconds, and maximum overshoot, $M_P = 1.0172$, or the maximum percentage overshoot of 1.72 percent. Thus, the performance requirements are met quite successfully with $H(s) = 0.01$.

Let us now consider a *lag-compensator* for this task. Since $G(s)$ is of type 1, we do not need to increase the DC gain of $G(s)H(s)$ to reduce the steady-state error to a step input. Hence, we would like to use a lag compensator transfer function of the form $H(s) = (s + \omega_o)/(\alpha s + \omega_o)$, with $\alpha > 1$, which leaves the DC gain of $G(s)H(s)$ unchanged. A candidate lag-compensator transfer function which achieves the desired performance is $H(s) = (s + 0.001)/(100s + 0.001)$, i.e. $\omega_0 = 0.001$ rad/s and $\alpha = 100$. The closed-loop transfer function is thus the following:

$$\phi(s)/\phi_d(s) = 1000(s + 0.001)/(100s^3 + 500s^2 + 1000s + 1) \qquad (2.182)$$

The natural frequencies and damping-ratios of the closed-loop system are the following:

```
>>nCL=[0 0 1000 1]; dCL=[100 500 1000 1]; damp(dCL) <enter>

Eigenvalue          Damping   Freq. (rad/sec)
 -0.0010            1.0000    0.0010
 -2.4995+1.9359i    0.7906    3.1615
 -2.4995-1.9359i    0.7906    3.1615
```

Note that there is a closed-loop pole at $s = -0.001$. Since there is a closed-loop zero at the same location, after *canceling* the pole with the zero at $s = -0.001$, the closed-loop transfer function can be written as:

$$\phi(s)/\phi_d(s) = 10/(s^2 + 5s + 10) \qquad (2.183)$$

which is the *same* transfer function as that obtained with $H(s) = 0.01$! This is an interesting result, and shows that the lag compensator has resulted in a *third order* closed-loop system, which essentially behaves as a *second order* system due to a pole-zero cancelation. The performance objectives are met by such a lag compensator, as shown above for $H(s) = 0.01$.

Figure 2.50 compares the Bode plots of the plant, the lag compensator, and the closed-loop system. The lag compensator provides a negative phase and a reduction

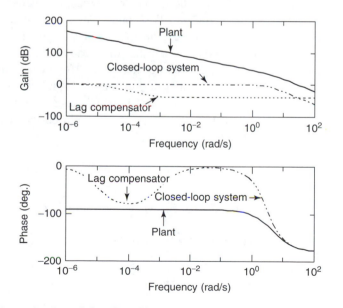

Figure 2.50 Bode plots of the plant, lag compensator, and closed-loop system for the roll control of a fighter aircraft

in gain in the frequencies below $\alpha\omega_0$ (i.e. 0.1 rad/s). Since the closed-loop gain never crosses the 0 dB line, and the closed-loop phase tends asymptotically to $-180°$, the gain and phase margins are *infinite* for the closed-loop system. Also, note that there is no change in the *slope* of gain reduction with frequency (i.e. *roll-off*) at high frequencies of the closed-loop system, when compared to that of the plant, which implies that there is no reduction in the *robustness* with respect to high-frequency *measurement noise*.

Example 2.25

Consider a large chemical plant with the following transfer function:

$$Y(s)/U(s) = G(s) = 0.025/[(s^2 + 0.3s + 0.01)(s + 0.33)] \qquad (2.184)$$

where the output, $Y(s)$, is the temperature, and the input, $U(s)$, is the mass flow-rate of Xylene gas. The natural frequencies of the plant are calculated as follows:

```
>>num=0.025; den=conv([1 0.33],[1 0.3 0.01]); damp(den) <enter>
```

```
Eigenvalue      Damping      Freq. (rad/sec)
 -0.0382        1.0000       0.0382
 -0.2618        1.0000       0.2618
 -0.3300        1.0000       0.3300
```

The plant's response is dominated by the pole, $s = -0.0382$, which is very much closer to the imaginary axis than the other two poles. Hence, the plant has a settling time of approximately $T_s = 4/0.0382 = 105$ seconds. The steady-state error of the plant is $e(\infty) = 1 - \lim_{s \to 0} sY(s) = -6.5758$ for a unit step input. It is required to design a closed-loop control system such that the steady-state error is brought down to less than ±0.15, with a maximum overshoot of 10 percent, and a settling time less than 20 seconds to a unit step desired output, $Y_d(s)$. Consider a *lead compensator* with $\omega_0 = 0.15$ rad/s and $\alpha = 0.01$, which results in a compensator transfer function, $H(s) = (s + 0.15)/(0.01s + 0.15)$. The closed-loop transfer function, $G(s)H(s)/[1 + G(s)H(s)]$, is thus

$$Y(s)/Y_d(s) = 0.025(s + 0.15)/(0.01s^4 + 0.1563s^3 + 0.09559s^2 + 0.041383s + 0.004245)$$

$$(2.185)$$

The closed-loop poles and natural frequencies are the following:

```
>>nCL=0.025*[1 0.15]; dCL=[0.01 0.1563 0.09559 0.041383 0.004245]; damp
  (dCL) <enter>
```

```
Eigenvalue           Damping       Freq. (rad/sec)
 -0.1358             1.0000        0.1358
 -0.2414+0.3873i     0.5289        0.4563
 -0.2414-0.3873i     0.5289        0.4563
 -15.0115            1.0000        15.0115
```

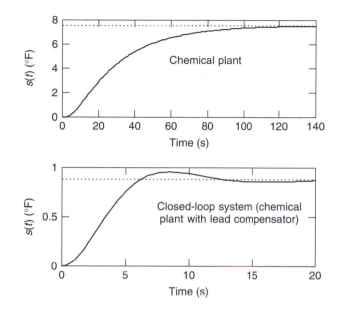

Figure 2.51 Step responses of the chemical plant and the closed-loop system with the lead compensator

Since there is no clearly dominant pole (or conjugate pair of poles), the settling time of the closed-loop system is determined by the first three poles, which indicates a dominant third order system. The step response of the closed-loop system is compared with that of the plant in Figure 2.51, obtained using the M-file *stepresp.m*. Note from Figure 2.51 that the closed-loop settling time is about 19 seconds, while the maximum overshoot is about 8 percent. The closed-loop steady-state error is calculated to be $e(\infty) = 1 - \lim_{s\to 0} sY(s) = 0.1166$. Hence, the performance objectives have been met.

The robustness properties of the closed-loop chemical plant control system are indicated by a Bode plot of the closed-loop transfer function, which is shown along with the Bode plots of the plant and the lead compensator in Figure 2.52. Note that the lead compensator provides a phase lead and a gain increase in the frequency range $\alpha\omega_0 \leq \omega \leq \omega_0/\alpha$ (i.e. $0.0015 \leq \omega \leq 1500$ rad/s). This results in a *speeding-up* of the closed-loop response, which is evident in a reduced settling time and an increased phase crossover frequency. Using the CST command *margin*, the gain and phase margins of the plant are calculated to be 8.349 dB and 33.6°, respectively, with gain and phase crossover frequencies of 0.1957 rad/s and 0.3302 rad/s, respectively. The closed-loop system has, however, a gain margin of 32.79 dB and an *infinite* phase margin, with a phase crossover frequency of 2.691 rad/s. Hence, the closed-loop system has a greater robustness to transfer function variations, and a faster response than the plant. However, due to an increased closed-loop gain at high frequencies, the *roll-off* at high frequencies is reduced for the closed-loop system, when compared to the plant. This implies an increased sensitivity of the closed-loop system to the high-frequency measurement noise.

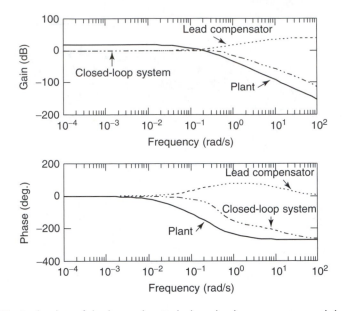

Figure 2.52 Bode plots of the large chemical plant, lead compensator, and the closed-loop system

From Examples 2.24 and 2.25, it is clear that while a lag compensator reduces the closed-loop steady-state error by providing a phase-lag at low frequencies, a lead compensator speeds-up the closed-loop response by providing a phase-lead at high frequencies. Let us see an application of the lead-lag compensation, which combines the desirable features of lag and lead compensators.

Example 2.26

In Example 2.25, the use of lead compensation resulted in a faster closed-loop response of a chemical plant. However, the closed-loop steady-state error with lead compensation was non-zero. Let us try to reduce the steady-state error, while speeding up the response, with the help of lead-lag compensation. In Example 2.25, a lead compensator with transfer function, $H(s) = (s + 0.15)/(0.01s + 0.15)$, produced a closed-loop steady-state error of 0.1166 and a settling time of 19 seconds. Let us choose the same transfer function for the lead part of the lag-lead compensator, i.e. $\omega_2 = 0.15$ rad/s and $1/\alpha = 0.01$ (i.e. $\alpha = 100$). The lag part of the lead-lag compensator would reduce the steady-state error by increasing the DC gain of the open-loop transfer function, $G(s)H(s)$. Let us therefore modify our design requirements to a closed-loop steady-state error of less than 0.002 and a settling time less than 20 seconds, while accepting a 10 percent maximum overshoot in the step response. To meet the new design requirements, a lag compensation frequency of $\omega_1 = 0.13$ rad/s is selected, resulting in the following lead-lag compensator transfer function:

$$H(s) = [(s + 0.15)(s + 0.13)]/[(0.01s + 0.15)(s + 0.0013)] \qquad (2.186)$$

which gives us the following closed-loop transfer function:

$Y(s)/Y_d(s)$

$$= (0.025s + 0.007)/(0.01s^4 + 0.15631s^3 + 0.095793s^2 + 0.041507s + 0.0075163)$$

$$(2.187)$$

The closed-loop poles, damping, and natural frequencies are as follows:

```
>>nCL=[0.025 0.007];dCL=[1e-2 1.5631e-1 9.5793e-2 4.1507e-2 7.5163e-3];
  damp(dCL) <enter>
```

```
Eigenvalue                    Damping          Freq. (rad/sec)
-1.3483e-001+4.8777e-002i     9.4036e-001      1.4339e-001
-1.3483e-001-4.8777e-002i     9.4036e-001      1.4339e-001
-1.7514e-001+3.5707e-001i     4.4037e-001      3.9770e-001
-1.7514e-001-3.5707e-001i     4.4037e-001      3.9770e-001
-1.5011e+001                  1.0000e+000      1.5011e+001
```

The first four poles, roughly located the same distance from the imaginary axis, are the dominant poles of the closed-loop system. The DC gain of the closed-loop system indicates the steady-state value of the step response, and is calculated using the MATLAB (CST) command *dcgain* as follows:

```
>>GCL=tf(nCL,dCL); dcgain(GCL) <enter>=

ans =

   0.9987
```

Hence, the closed-loop steady-state error is brought down to $1 - 0.9987 = 0.0013$, which is acceptable. The closed-loop step response is plotted in Figure 2.53 for three different values of the lag frequency, ω_1. For $\omega_1 = 0.13$ rad/s (used in the above calculations), the closed-loop response has a settling time of about 16 seconds, and a maximum overshoot of about 30 percent, which is unacceptable. Figure 2.53 shows that the maximum percentage overshoot can be reduced by decreasing ω_1. For $\omega_1 = 0.05$ rad/s, the maximum percentage overshoot is about 10 percent, and the settling time is about 13 seconds. However, decreasing ω_1 below 0.05 rad/s results in an *increase* in the settling time, as is evident from the step response for $\omega_2 = 0.03$ rad/s which has a maximum overshoot of about 5 percent, but a settling time of about 22 seconds. The steady-state error is unaffected by changing ω_1. We select $\omega_1 = 0.05$ rad/s, which gives a fast response and an acceptable maximum overshoot.

The Bode plots of the plant, the lead-lag compensator with $\omega_2 = 0.15$ rad/s, $\omega_1 = 0.05$ rad/s, $\alpha = 100$, and the resulting closed-loop system are shown in Figure 2.54.

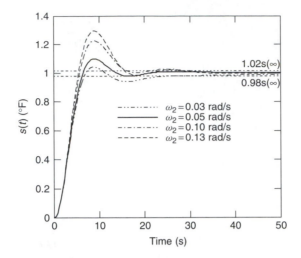

Figure 2.53 Closed-loop step response of the chemical plant with a lead-lag compensator for various values of the lag frequency, ω_2

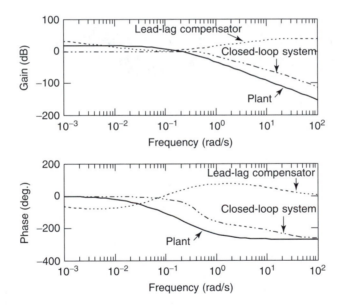

Figure 2.54 Bode plots of the large chemical plant, the lead-lag compensator, and the closed-loop system

Comparing Figure 2.54 with Figure 2.52, the phase-lag and decreasing compensator gain below 0.1 rad/s, and phase-lag and increasing compensator gain above 0.1 rad/s are evident for the lead-lag compensator. The high-frequency closed-loop behavior is largely unaffected, when compared to the lead compensation in Figure 2.52. The closed-loop gain and phase margins are computed by the command *margin* to be 38.9 dB and 104.5°, respectively, with gain and phase crossover frequencies of

0.3656 rad/s and 2.5456 rad/s, respectively. Comparing these values with those of Example 2.25 with lead compensation, we find that there is a slight increase in the gain margin, while the phase margin has been reduced from infinite to 104.5°. Furthermore, there is a slight slowing down of the closed-loop response due to the added lag compensation, which is evident in a slight reduction in the phase crossover frequency. The slowing down of the closed-loop response is also clear in the overshoot and settling time of the step response. The lead compensated system of Example 2.25 had an 8 percent overshoot and settling time of 19 seconds. To achieve the same overshoot, the lead-lag compensated system would require a settling time slightly larger than 20 seconds, which is obvious from Figure 2.53. Example 2.26 illustrates the iterative process of determining the three unknowns (α, ω_1, ω_2) of a lead-lag compensator to meet a conflicting set of performance requirements. Since a lead-lag compensator is a practical (approximate) implementation of the ideal PID compensator, we have witnessed an example of PID tuning in which the three unknown constants K_P, K_D, and K_I are determined in a similar manner.

The Control System Toolbox (CST) offers an interactive *Graphical Users Interface* (GUI) tool, called the *SISO Design Tool*, to carry out the steps illustrated in this section for the design of closed-loop compensators in a user-friendly manner. To use this design feature, go to MATLAB *launch pad*, click on the *Control System Toolbox*, and select the icon for *SISO Design Tool* from the resulting menu. Another interactive GUI tool available in the CST menu is the *LTI Viewer*, which lets you directly view all the important graphs of a linear, time-invariant system (step and impulse responses, Bode, Nyquist, root-locus, Nichols plots, and pole-zero map, etc.), as well as the performance characteristics of the system (peak response, rise time, settling time, and steady-state), along with several graphical operations. As an exercise to familiarize yourself with these GUI tools, repeat Examples 2.23–2.26 with the *SISO Design Tool*, and use *LTI Viewer* to check the closed-loop system's characteristics.

2.13 Multivariable Systems

So far we have considered only single-input, single-output systems (an exception was the aircraft in Example 2.10, which had three outputs, and one input). A practical modern control system may simultaneously contain several inputs and several outputs. Such systems are called *multivariable systems*. For multivariable systems, we must re-define the relationship between the Laplace transform of the input, $\mathbf{U}(s)$, and the Laplace transform of the output, $\mathbf{Y}(s)$, both of which are now *vectors*, consisting of more than one signal. Each element of $\mathbf{U}(s)$ is the Laplace transform of the *corresponding element* of the input vector, $\mathbf{u}(t)$. Similarly, each element of $\mathbf{Y}(s)$ is the Laplace transform of the *corresponding element* of the output vector, $\mathbf{y}(t)$. The input and output vectors are related in the Laplace domain by a *transfer matrix* (instead of the transfer function for the single-input, single-output systems), $\mathbf{G}(s)$, which is defined as follows:

$$\mathbf{Y}(s) = \mathbf{G}(s)\mathbf{U}(s) \tag{2.188}$$

Note that we are using *bold letters* to denote *vectors* and *matrices*. This notation will be followed throughout the book. Refer to Appendix B, or Gantmacher [4] and Kreyszig [5] for vectors, matrices, and their properties.

In Eq. (2.188) if the system consists of p outputs and r inputs, then the *sizes* of $\mathbf{Y}(s)$, $\mathbf{U}(s)$, and $\mathbf{G}(s)$ are $(p \times 1)$, $(r \times 1)$, and $(p \times r)$, respectively. (Clearly, the single-input, single-output system is a special case of the multivariable systems where $p = 1$ and $r = 1$.) Since we are dealing with vectors and matrices here, we have to be careful how we *multiply* them such that the sizes of the multiplied quantities are *compatible* (e.g. we can not write Eq. (2.188) as $\mathbf{Y}(s) = \mathbf{U}(s)\mathbf{G}(s)$!).

Example 2.27

Let us derive the transfer matrix of the multivariable system of Example 2.10. Recall that the system (longitudinal dynamics of an aircraft) consists of a *single* input whose Laplace transform is $\delta(s)$, and *three* outputs with Laplace transforms $v(s)$, $\alpha(s)$, and $\theta(s)$. Thus, we can write

$$\mathbf{U}(s) = \delta(s); \quad \mathbf{Y}(s) = \begin{bmatrix} v(s) \\ \alpha(s) \\ \theta(s) \end{bmatrix} \tag{2.189}$$

The transfer matrix between $\mathbf{G}(s)$ can be obtained using the relationships between the input and the individual elements of $\mathbf{Y}(s)$ given by Eqs. (2.86)–(2.88) as follows:

$$\mathbf{Y}(s) = \begin{bmatrix} v(s) \\ \alpha(s) \\ \theta(s) \end{bmatrix} = \begin{bmatrix} v(s)/\delta(s) \\ \alpha(s)/\delta(s) \\ \theta(s)/\delta(s) \end{bmatrix} \mathbf{U}(s) \tag{2.190}$$

or

$\mathbf{Y}(s)$

$$= \begin{bmatrix} -0.0005(s-70)(s+0.5)/[(s^2+0.005s+0.006)(s^2+s+1.4)] \\ -0.02(s+80)(s^2+0.0065s+0.006)/[(s^2+0.005s+0.006)(s^2+s+1.4)] \\ -1.4(s+0.02)(s+0.4)/[(s^2+0.005s+0.006)(s^2+s+1.4)] \end{bmatrix} \mathbf{U}(s)$$

$$\tag{2.191}$$

which, from Eq. (2.188), gives

$\mathbf{G}(s)$

$$= \begin{bmatrix} -0.0005(s-70)(s+0.5)/[(s^2+0.005s+0.006)(s^2+s+1.4)] \\ -0.02(s+80)(s^2+0.0065s+0.006)/[(s^2+0.005s+0.006)(s^2+s+1.4)] \\ -1.4(s+0.02)(s+0.4)/[(s^2+0.005s+0.006)(s^2+s+1.4)] \end{bmatrix}$$

$$\tag{2.192}$$

Note that the transfer matrix, $\mathbf{G}(s)$, is of size (3×1), denoting a single-input, three-output system.

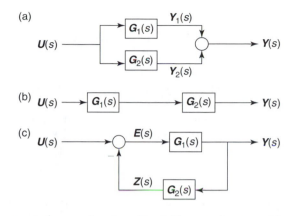

Figure 2.55 Three types of connections in multivariable control systems: (a) parallel, (b) series, and (c) negative feedback

We must be able to determine a system's transfer matrix from the block diagram of a multivariable system, which may include transfer matrices of the various sub-systems. As in single-input, single-output (SISO) systems, there are three distinct ways in which the multivariable sub-systems can be connected to each other, as shown in Figure 2.55. Figure 2.55(a) depicts two *sub-systems* of transfer matrices $\mathbf{G}_1(s)$ and $\mathbf{G}_2(s)$, respectively, connected in *parallel*. Here, the outputs of the two sub-systems, $\mathbf{Y}_1(s)$ and $\mathbf{Y}_2(s)$ (which clearly must be vectors of the *same size*) are added-up at the *summing junction* to give the system's output, $\mathbf{Y}(s)$, by

$$\mathbf{Y}(s) = \mathbf{Y}_1(s) + \mathbf{Y}_2(s) = \mathbf{G}_1(s)\mathbf{U}(s) + \mathbf{G}_2(s)\mathbf{U}(s) = [\mathbf{G}_1(s) + \mathbf{G}_2(s)]\mathbf{U}(s) \quad (2.193)$$

Hence, the parallel system's transfer matrix is the sum of the transfer matrices of its two (or more) components, i.e. $\mathbf{G}(s) = \mathbf{G}_1(s) + \mathbf{G}_2(s)$. The system of Figure 2.55(b) is in *series*, and it can be easily seen (left as an exercise for you) that the transfer matrix of such a system is given by the *product* of the transfer matrices of the two (or more) components, i.e. $\mathbf{G}(s) = \mathbf{G}_2(s)\mathbf{G}_1(s)$. In series combinations, one has to be careful in the *sequence of multiplication* (i.e. $\mathbf{G}_2(s)\mathbf{G}_1(s)$, and *not* $\mathbf{G}_1(s)\mathbf{G}_2(s)$, etc.), because the matrix multiplication is not always *conformable* (i.e. for a matrix product $\mathbf{G}(s) = \mathbf{G}_2(s)\mathbf{G}_1(s)$ to be defined, the number of *columns* in $\mathbf{G}_1(s)$ must be equal to the number of *rows* in $\mathbf{G}_2(s)$) (see Appendix B). The third kind of connection, shown in Figure 2.55(c) is the *feedback* arrangement. The two sub-systems of this combination are the *forward trans-mission*, given by the transfer matrix $\mathbf{G}_1(s)$, and the *feedback transmission*, whose transfer matrix is $\mathbf{G}_2(s)$. If the *feedback signal*, $\mathbf{Z}(s)$, is subtracted from the input, $\mathbf{U}(s)$ – as in Figure 2.55(c) – then the system is said to be of a *negative feedback* kind, and if $\mathbf{Z}(s)$ is added to $\mathbf{U}(s)$ at the summing junction, the system is called a *positive feedback system*. Again, as in SISO systems, we usually encounter the negative feedback arrangement in multivariable control systems, whose objective is to drive the inverse Laplace transform of the *error vector*, $\mathbf{E}(s) = \mathbf{U}(s) - \mathbf{Z}(s)$, to zero as $t \to \infty$. For a negative feedback system, we can write

$$\mathbf{Y}(s) = \mathbf{G}_1(s)\mathbf{E}(s) = \mathbf{G}_1(s)[\mathbf{U}(s) - \mathbf{G}_2(s)\mathbf{Y}(s)] \quad (2.194)$$

or

$$[\mathbf{I} + \mathbf{G}_1(s)\mathbf{G}_2(s)]\mathbf{Y}(s) = \mathbf{G}_1(s)\mathbf{U}(s) \tag{2.195}$$

and it follows that

$$\mathbf{Y}(s) = \mathbf{G}(s)\mathbf{U}(s) = [\mathbf{I} + \mathbf{G}_1(s)\mathbf{G}_2(s)]^{-1}\mathbf{G}_1(s)\mathbf{U}(s) \tag{2.196}$$

where $[\ ]^{-1}$ denotes the *inverse* of a (square) matrix, and $\mathbf{I}$ is the *identity matrix* of the same size as the product $\mathbf{G}_1(s)\mathbf{G}_2(s)$ (Appendix B). Hence, from Eq. (2.196), the transfer matrix of a negative feedback system is given by $\mathbf{G}(s) = [\mathbf{I} + \mathbf{G}_1(s)\mathbf{G}_2(s)]^{-1}\mathbf{G}_1(s)$. (Note that for a SISO system, the transfer matrix (function) becomes $G(s) = G_1(s)/[1 + G_1(s)G_2(s)]$.) Again, we must be careful with the sequence of matrix multiplication involved in expressions such as Eq. (2.196). Not all multivariable closed-loop systems are of the kind shown in Figure 2.55(c).

Trying to find the transfer matrices of multivariable systems from their block diagrams can be a daunting task. The Control System Toolbox (CST) helps in constructing the transfer function from its individual components with model building commands, such as *series, parallel, feedback*, and *append*. For example, the multivariable system of Figure 2.55(a) can be constructed using the command *parallel* as follows:

```
>>sys = parallel (G1,G2) <enter>
```

where $G1$, $G2$ are the LTI objects (either transfer matrices, or *state-space* models (see Chapter 3)) of the two parallel sub-systems, $\mathbf{G}_1$ and $\mathbf{G}_2$, respectively. Alternatively, we could type

```
>>sys = G1+G2 <enter>
```

to get the transfer matrix (or *state-space* model) of the overall system. Similarly, we can use the *series* command for the system of Figure 2.55(b) as follows:

```
>>sys = series(G1,G2) <enter>
```

or type

```
>>sys = G2*G1 <enter>
```

For multivariable systems in which only *some* of the outputs of G1 are the inputs to G2, both *parallel* and *series* commands can be used with *two additional input arguments* such as

```
>>sys = series(G1,G2,OUTPUTS1,INPUTS2) <enter>
```

which specifies that the outputs of G1 denoted by the *index vector* OUTPUTS1 are connected to the inputs of G2 denoted by INPUTS2. (The vectors OUTPUTS1 and INPUTS2 contain indices into the outputs and inputs of G1 and G2, respectively.) For the feedback multivariable system the CST command *feedback* gives the closed-loop transfer matrix as follows:

```
>>sys = feedback(G1,G2,FEEDIN,FEEDOUT,sign) <enter>
```

The vector FEEDIN contains indices into the input vector of G1, and specifies which inputs are involved in the feedback loop. Similarly, FEEDOUT specifies which outputs of G1 are used for feedback. If the sign is omitted, a negative feedback system is assumed, while sign=1 results in a positive feedback system. If FEEDIN and FFEDOUT are omitted from the feedback command, it is assumed that all the outputs of G1 are the inputs to G2 (as shown in Figure 2.55(c)).

Example 2.28

Consider the multivariable control system of Figure 2.56. The system's block-diagram indicates that only the *second* and *third* outputs of $\mathbf{G}_1(s)$ are inputs to $\mathbf{G}_2(s)$, while the *second* output of $\mathbf{G}_2(s)$ is an input to $\mathbf{G}_3(s)$. It is also clear that $\mathbf{G}_1(s)$ and $\mathbf{G}_2(s)$ are in series, while $\mathbf{G}_3(s)$ is in a negative feedback arrangement with $\mathbf{G}_1(s)$ and $\mathbf{G}_2(s)$. The closed-loop system's transfer matrix from the input vector, $\mathbf{U}(s) = U_1(s)$, and the output vector, $\mathbf{Y}(s) = [Y_2(s)Y_3(s)]^{\mathrm{T}}$ can be determined using the CST commands as follows:

```
>>G1=tf({1; -2; 1},{[1 1]; [1 0]; [1 0 5]}) % transfer matrix,G1 <enter>

Transfer function from input to output...

       1
#1:  ----
      s+1

      -2
#2:  --
       s
```

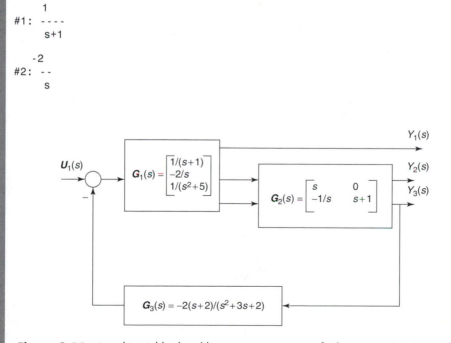

Figure 2.56 A multivariable closed-loop system consisting of sub-systems, $\mathbf{G}_1$, $\mathbf{G}_2$, and $\mathbf{G}_3$

```
         1
#3:  -----
       s^2+5

>>G2=tf({[1 0] 0;-1 [1 1]},{1 1;[1 0] 1}) % transfer matrix, G2 <enter>


Transfer function from input 1 to output...

#1: s

       -1
#2:  --
       s

Transfer function from input 2 to output...

#1: 0

#2: s+1

>>G3=tf(-2*[1 2],[1 3 2]) % transfer matrix, G3 <enter>

Transfer function:

  -2s-4
 ----------
  s^2+3s+2
```

(Note the ease with which the transfer matrices are constructed using the LTI object,
tf. The coefficients of the numerator and denominator polynomials are specified using
{·}, {·}, with the transfer functions on different rows separated by;.)

```
>>G2G1=series(G1,G2,[2 3],[1 2]) %transfer matrix of the series
   connection <enter>

Transfer function from input to output...

      -2s
#1:  ----
       s

    s^3+3s^2+10
#2:  ---------
     s^4+5s^2

>>GCL=feedback(G2G1, G3, 1, 2) % closed-loop transfer matrix from U1(s)
  to [Y2(s); Y3(s)] <enter>

Transfer function from input to output...

    -2s^6-6s^5-14s^4-30s^3-20s^2+7.43e-015s
#1: ----------------------------------------
      s^6+3s^5+5s^4+5s^3-2s^2-20s-40
```

```
        s^5+6s^4+11s^3+16s^2+30s+20
#2:  - - - - - - - - - - - - - - - - - - - - - - - - - - - - - - -
        s^6+3s^5+5s^4+5s^3-2s^2-20s-40
```

Thus the closed-loop transfer matrix indicates a sixth order, single-input, two-output system. The transfer function of the forward path from $\mathbf{U}(s) = U_1(s)$ to $Y_1(s)$ is easily seen from Figure 2.56 to be $1/(s + 1)$. Note that the denominator polynomials of the two transfer functions constituting *GCL* are the same, which indicates that the closed-loop system's poles can be obtained by factoring the common denominator polynomial. The closed-loop zeros are the zeros of transfer functions constituting *GCL*, which are easily identified from Figure 2.56 as $z_1 = -1$ and $z_2 = -2$. Let us determine the closed-loop zeros and poles using CST as follows:

```
>>zero(GCL) <enter>

ans =
 -1.0000
 -2.0000
>>pole(GCL) <enter>
ans =
 -2.0000
 -1.3584+1.3634i
 -1.3584-1.3634i
  0.1132+1.9000i
  0.1132-1.9000i
  1.4904
```

Due to the presence of three poles with positive real parts, the closed-loop system is *unstable*.

Example 2.28 indicates that, to determine the performance, stability, and robustness of a multivariable, linear, time-invariant control system, we should be first able to express its transfer matrix $\mathbf{G}(s)$, as a matrix of rational transfer functions, $G_{ij}(s)$, given by

$$\mathbf{G}(s) = \begin{bmatrix} G_{11}(s) & G_{12}(s) & \dots & G_{1m}(s) \\ \dots\dots\dots\dots\dots\dots\dots\dots \\ G_{n1}(s) & G_{n2}(s) & \dots & G_{nm}(s) \end{bmatrix} \tag{2.197}$$

where n is the number of outputs and m is the number of inputs. Example 2.28 also illustrates that we can extend the logic of *poles* and *zeros* of a single-input, single-output system to multivariable systems by defining the roots of the denominator and numerator polynomials of all the elemental transfer functions, $G_{ij}(s)$, as the system's poles and zeros, respectively. Such a definition also suggests that the *characteristic polynomial* of the multivariable system be defined as a single common denominator polynomial, and the roots of the characteristic polynomial are the poles of the system. Then we can apply the same stability criteria as for the single-input, single output systems by looking at the locations of the system's poles. Similarly, the determination of responses to the singularity functions (or to an arbitrary input) can be carried out from $\mathbf{G}(s)$ in the same manner as we did for the single-input, single output systems, if we consider one input at a time. The robustness of a multivariable system can be determined using frequency response methods (see Chapter 7), provided we define scalar measures for multivariable systems analogous to gain and phase margins.

For analyzing the robustness of a general closed-loop multivariable system, consider the multivariable counterpart of the single-input, single-output closed-loop system of Figure 2.32, in which we take $\mathbf{G}(s)$ and $\mathbf{H}(s)$ to denote the plant and controller transfer matrices, respectively. Note that for the system of Figure 2.32, the closed-loop transfer matrix, $\mathbf{G}_c(s)$, defined as $\mathbf{Y}(s) = \mathbf{G}_c(s)\mathbf{Y}_d(s)$, is given by

$$\mathbf{G}_c(s) = [\mathbf{I} + \mathbf{G}(s)\mathbf{H}(s)]^{-1}\mathbf{G}(s)\mathbf{H}(s) \qquad (2.198)$$

where $\mathbf{I}$ is identity matrix, $\mathbf{G}(s)$ is the plant's transfer matrix, and $\mathbf{H}(s)$ is the transfer matrix of the controller. In a manner similar to the single-loop feedback systems, we can define a *return difference matrix*. However, there are *two* different *return difference matrices* for the multivariable systems: the *return difference matrix at the plant's output*, $[\mathbf{I} + \mathbf{G}(s)\mathbf{H}(s)]$, and the *return difference matrix at the plant's input*, $[\mathbf{I} + \mathbf{H}(s)\mathbf{G}(s)]$. These return difference matrices are the *coefficient* matrices of the output, $\mathbf{Y}(s)$, and the input, $\mathbf{U}(s)$, respectively, when the expressions relating the desired output, $\mathbf{Y}_d(s)$, with $\mathbf{Y}(s)$ and $\mathbf{U}(s)$, respectively, are written as $[\mathbf{I} + \mathbf{G}(s)\mathbf{H}(s)]\mathbf{Y}(s) = \mathbf{G}(s)\mathbf{H}(s)\mathbf{Y}_d(s)$ and $[\mathbf{I} + \mathbf{H}(s)\mathbf{G}(s)]\mathbf{U}(s) = \mathbf{H}(s)\mathbf{Y}_d(s)$. Continuing the analogy with the single-input, single-output systems, the matrices $\mathbf{G}(s)\mathbf{H}(s)$ and $\mathbf{H}(s)\mathbf{G}(s)$ are called the *return ratio matrices* at the plant *output* and *input*, respectively. For studying robustness of the closed-loop system, the return difference (and return ratio) matrices are considered important, as we shall discuss in Chapter 7.

Before we can define quantities analogous to gain and phase margins of single-loop systems, we have to assign a *scalar* measure to the return difference matrix at the output, $[\mathbf{I} + \mathbf{G}(s)\mathbf{H}(s)]$. Such a scalar measure assigned to a matrix is its *determinant*. Since the determinant of the return difference matrix at the output is set to zero (i.e. $|\mathbf{I} + \mathbf{G}(s)\mathbf{H}(s)| = 0$) is the characteristic equation which determines the closed-loop system's poles, we obtain stability robustness information by studying Bode or Nyquist plot of the determinant of the return difference matrix at the output, $|\mathbf{I} + \mathbf{G}(s)\mathbf{H}(s)|$.

Since the return difference matrix at the output appears as an inverse in the closed-loop transfer function, we expect $[\mathbf{I} + \mathbf{G}(s)\mathbf{H}(s)]$ to be *singular* (i.e. a matrix whose inverse is not defined because its determinant is zero) at the poles of the system. If the closed-loop system is asymptotically stable, the poles must lie in the left-half s-plane. The stability robustness in this case is, therefore, a measure of how much the controller transfer matrix, $\mathbf{H}(s)$, is allowed to vary before the poles cross into the right half s-plane.

Example 2.29

Consider a two-input, two-output chemical process plant with transfer matrix $\mathbf{G}(s)$ (Figure 2.52) controlled by three flow control servo-valves, each of which has a constant transfer function (or gain), K. The outputs, $Y_1(s)$, and $Y_2(s)$, are fed back to the individual servo-control valves as shown. The plant's transfer matrix, $\mathbf{G}(s)$, is expressed as

$$\mathbf{G}(s) = \begin{bmatrix} 3/s & -200/[s(s+6)(s+30)] \\ 0.05/s & -250/[s(s+6)(s+30)] \end{bmatrix} \qquad (2.199)$$

The closed-loop system of Figure 2.57 can be reduced to that of Figure 2.55(c), with the *plant* transfer matrix, $\mathbf{G}_1(s) = \mathbf{G}(s)$, given by Eq. (2.199), and the *controller* transfer matrix, $\mathbf{G}_2(s) = \mathbf{H}(s)$, given by

$$\mathbf{H}(s) = \begin{bmatrix} K & 0 \\ 0 & K \end{bmatrix} = K\mathbf{I} \qquad (2.200)$$

For a given value of controller design parameter, K, the resulting *closed-loop* transfer matrix, $\mathbf{G}_c(s) = [\mathbf{I} + \mathbf{G}_1(s)\mathbf{G}_2(s)]^{-1}\mathbf{G}_1(s) = [\mathbf{I} + \mathbf{G}(s)\mathbf{H}(s)]^{-1}\mathbf{G}(s)$, can

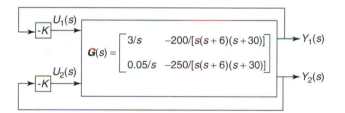

Figure 2.57 A chemical plant with transfer matrix $\mathbf{G}(s)$, controlled by two feedback servo-valves, each of transfer function, K

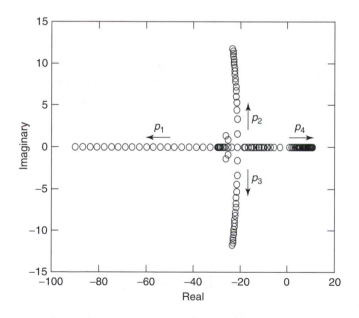

Figure 2.58 Root-locus of the multivariable chemical control system as the controller gain, K, varies from 1 to 10. The arrows indicate how each of the four poles p_1, p_2, p_3, and p_4 move as K is increased

be calculated using MATLAB (CST) with the use of the *feedback* command. Alternatively, the closed-loop poles of the system are obtained by setting the determinant of the return difference matrix at the output to zero, i.e. $|\mathbf{I} + \mathbf{G}(s)\mathbf{H}(s)| = 0$, which results in the following characteristic equation:

$$|\mathbf{I} + \mathbf{G}(s)\mathbf{H}(s)| = s^4 + (36 + 3K)s^3 + (180 + 108K)s^2$$
$$+ 290Ks - 740K^2 = 0 \qquad (2.201)$$

Figure 2.58 is a plot of the closed-loop poles (i.e. roots of Eq. (2.201)) p_1, p_2, p_3, and p_4 as the controller gain K varies from 1 to 10. It is noted that for all values of K, one of the poles (p_4) is always in the right-half s-plane, indicating instability. Hence, the system is unstable in the range $1 < K < 10$. Although for this two-loop system we can plot the root-locus by using the determinant $|\mathbf{I} + \mathbf{G}(s)\mathbf{H}(s)|$, such a procedure quickly becomes unwieldy as the number of inputs and outputs is increased.

Example 2.29 showed how $|\mathbf{I} + \mathbf{G}(s)\mathbf{H}(s)| = 0$ (i.e. the return difference matrix at the output, $[\mathbf{I} + \mathbf{G}(s)\mathbf{H}(s)]$, being singular) can be used to find out whether a *multivariable* system is stable. However, the determinant of a matrix is not a very good measure of *how close the matrix is to being singular*. Thus a Bode or Nyquist plot of $|\mathbf{I} + \mathbf{G}(s)\mathbf{H}(s)|$ in the frequency domain is *not* going to give us a very *accurate* information about the stability robustness (such as gain and phase margins), and we need another *scalar* measure which is a better indicator of how close the return difference matrix is to being singular. One such measure is the set of *singular values* of the return difference matrix in the frequency domain ($s = i\omega$), which we will study in Chapter 7.

Examples 2.28 and 2.29 illustrate the increased difficulty of estimating stability of a multivariable system using the classical frequency domain methods. While MATLAB allows that multivariable systems can be represented by their transfer matrices, an evaluation of their performance would require calculation of step (or impulse) responses of individual elements of the transfer matrix – a task which becomes unwieldy as the number of inputs and outputs are increased. A better mathematical procedure of handling arbitrary input responses of multivariable systems is to employ the methods of matrix algebra in a more tractable manner through the introduction of a *state-space* model of the system. The chief advantage of the state-space methodology is that time responses of a multivariable system are obtained directly by solving a set of first-order differential equations (called *state equations*) which are used to represent the governing differential equations. The direct calculation of a system's response to an arbitrary input enables the design and analysis of multivariable systems in the time (rather than frequency, or Laplace) domain. The state-space approach treats a single-input, single-output system in exactly the same way as it does a multivariable system, irrespective of the number of inputs and outputs involved. The remainder of this book will discuss the application of the state-space methods in control system analysis and design. However, as we will see in Chapter 7, analyzing robustness of multivariable systems requires a return to frequency domain analysis through complex function theory.

Exercises

2.1. Consider an inverted pendulum of length L and mass m on a moving cart of mass, M (Figure 2.59). It is assumed that the cart moves on a frictionless, flat plane. The force, $f(t)$, applied to the cart is the input, while the output is the angular deflection, $\theta(t)$, of the pendulum from the vertical. The displacement of the cart from a fixed point is given by $x(t)$. The governing differential equations of the system are as follows:

$$(M + m)x^{(2)}(t) + mL\theta^{(2)}(t)\cos(\theta(t)) - mL[\theta^{(1)}(t)]^2 \sin(\theta(t)) = f(t) \quad (2.202)$$

$$mx^{(2)}(t)\cos(\theta(t)) + mL\theta^{(2)}(t) - mg\sin(\theta(t)) = 0 \quad (2.203)$$

(a) What is the order of the system?

(b) Eliminate $x(t)$ from Eqs. (2.202) and (2.203) to get a differential equation in terms of the output, $\theta(t)$, and input, $f(t)$.

(c) Linearize the system about the equilibrium point, $\theta(t) = 0$.

(d) Find the transfer function of the linearized system, $\Theta(s)/F(s)$, where $\Theta(s) = \mathcal{L}\theta(t)$ and $F(s) = \mathcal{L}f(t)$.

(e) What are the poles of the linearized system? Is the linearized system stable?

2.2. Certain unforced physical systems obey the following governing differential equation, called *van der Pol* equation:

$$x^{(2)}(t) + a[x(t)^2 - 1]x^{(1)}(t) + bx(t) = 0 \quad (2.204)$$

where $x(t)$ is the output variable, and a and b are positive constants. Can you linearize such a system about an equilibrium point?

2.3. The governing equations of an electrical network (Figure 2.60(a)) are as follows:

$$i_1(t) = f(v_1(t)) \quad (2.205)$$

$$v_1^{(1)}(t) = [v_2(t) - v_1(t)]/(R_2C_1) - i_1(t)/C_1 \quad (2.206)$$

$$v_2^{(1)}(t) = [v_1(t) - v_2(t)]/(R_2C_2) - i_2(t)/C_2 \quad (2.207)$$

$$i_2^{(1)}(t) = -v_2(t)/L \quad (2.208)$$

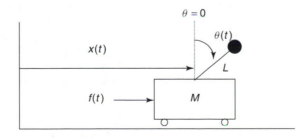

Figure 2.59 Inverted pendulum on a moving cart

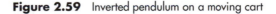

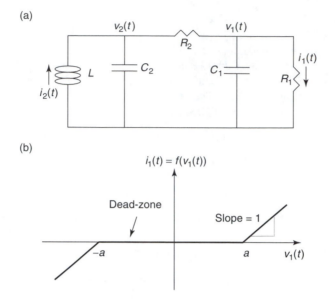

Figure 2.60 (a) Electrical network for Exercise 2.3; (b) Nonlinear current, $i_1(t)$, with a dead-zone in the voltage band $-a < v_1(t) < a$

where $i_1(t)$ is the current flowing through a nonlinear resistance, R_1, $i_2(t)$ is the current flowing through the inductor, L, and $v_1(t)$ and $v_2(t)$ are the voltages across the capacitors, C_1 and C_2, respectively. The function $f(v_1(t))$ is shown in Figure 2.60(b). Such a nonlinear function is said to have a *dead-zone* in the voltage band, $-a < v_1(t) < a$.

(a) Find the equilibrium points of the system, and linearize the system about each of them.

(b) Derive the transfer function, $G(s)$, between the current, $i_1(t)$, as the output and the current, $i_2(t)$, as the input, of the system linearized about each equilibrium point.

(c) Investigate the stability of the system about each equilibrium point.

(d) What is the step response of the linearized system when the current $i_1(t)$ is in the dead-zone?

2.4. Linearize the governing differential equation of a satellite in an orbit around a planet (Eq. 2.16) about the equilibrium point denoted by the circular orbit, $r(t) = C$. What are the natural frequency and damping-ratio of the linearized orbital dynamics? Is the orbital system stable about the equilibrium point? Find the response of the deviation of the satellite from the equilibrium point denoted by the circular orbit, if the input radial acceleration is 1000 m/s² applied as a step input at $t = 0$, i.e. $u(t) = 1000u_s(t)$. What are the maximum overshoot, settling time, and steady-state value of the response? (Assume that the satellite is in a circular orbit of radius, $C = 3.0 \times 10^8$m, around Mars, with an angular momentum, $h = 3.0 \times 10^{12}$ m²/s. For Mars, $k^2 = 4.27 \times 10^{13}$ m³/s².)

2.5. Repeat Exercise 2.4 for a satellite in orbit around Jupiter, with $C = 4.0 \times 10^9$ m, $h = 4.8 \times 10^{13}$ m²/s, and $k^2 = 1.27 \times 10^{17}$ m³/s².

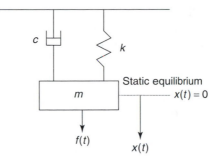

Figure 2.61 Spring-mass-damper system with input, $f(t)$, and output, $x(t)$

2.6. Consider the spring-mass-damper system shown in Figure 2.61, with mass, m, spring stiffness, k, and damping coefficient, c. The deflection, $x(t)$, of the mass is measured from its *static equilibrium position* (given by $x(t) = 0$), and is the system's output. A force, $f(t)$, applied to the mass is the input. The governing differential equation of the system is

$$mx^{(2)}(t) + cx^{(1)}(t) + kx(t) = f(t) \qquad (2.209)$$

(a) Derive an expression for the system's frequency response.

(b) What are the expressions for the natural frequency, ω_n, and damping-ratio, ς, of the system?

(c) Assuming zero initial conditions, derive the step response and impulse response of the system. Plot the step response for $\varsigma = 0.1$ and $\varsigma = 2.0$. Calculate the maximum percentage overshoot, peak-time and settling-time in both the values of ς.

(d) If $c = 0$, is the system stable?

2.7. Calculate the impulse response of the following transfer functions:

(a) $(s + 1)/[(s + 0.01)(s + 0.3)]$

(b) $(s^2 + 3s + 1)/(s^3 + 2s^2 + 7s + 10)$

(c) $10s/(s^2 + 2s - 1)$

2.8. Plot the step response of the following transfer functions, and calculate the maximum overshoot, settling time, and the steady-state output:

(a) $s/(s + 5)$

(b) $(s^2 + 1)/(s^2 + 3s + 2)$

(c) $10(s^3 + 2s^2 + 4s + 1)/(s^4 + s^3 + 10s^2 + 5s + 10)$

2.9. Calculate and plot the response of a system with transfer function, $G(s) = 100(s^2 + 2)/(s^3 + s^2 + 5s + 1)$ to the input, $u(t) = 10e^{-t}\sin(5t)$, assuming zero initial conditions. What is the magnitude of the maximum overshoot? What is the steady-state value of the output?

2.10. For the linearized satellite orbital dynamics of Exercise 2.4, plot the deviation from the circular orbit if the input is given by $u(t) = 5000[\sin(0.1t) + \cos(0.2t)]$, assuming zero initial conditions. What is the magnitude of the maximum deviation from the circular orbit?

2.11. Plot the response of the spring-mass-damper system of Exercise 2.5, if $m = 1$ kg, $k = 10$ N/m, $c = 0.1$ Ns/m, and $f(t) = 10e^{-2t}\cos(10t)$, assuming zero initial conditions.

2.12. The transfer function of a control system is given by

$$Y(s)/U(s) = G(s) = 20(s^3 + 4s^2 + 5s + 2)/(s^5 + 15s^4 + 30s^3) \qquad (2.210)$$

(a) Investigate the stability of the system.

(b) Plot the response to a unit ramp function, $r(t)$, for zero initial conditions.

(c) Draw a Bode plot of the system.

2.13. For a second-order linear system, prove that the step response goes to infinity in the limit $t \to \infty$ if the system has *repeated* poles with *real part* equal to *zero*.

2.14. Consider the control system shown in Figure 2.32, where the plant's transfer function is $G(s) = 1/[(s + 0.1)(s + 0.2)]$. It is intended to use a PID compensator for controlling this plant.

(a) What are the values of the PID compensator constants, K_P, K_I, and K_D, for achieving a zero steady-state error with a closed-loop pole at $s = -1$, and two complex conjugate poles with damping ratio, $\zeta = 0.707$, and natural frequency, $\omega_n = 1$ rad/s?

(b) Derive the closed-loop transfer function, $Y(s)/Y_d(s)$, with the compensator in part (a), and compute the closed-loop system step response. What is the maximum percentage overshoot, settling time, and steady-state error of the closed-loop step response?

(c) Plot the locus of the closed-loop poles as K_I varies from 0 to 10, with K_D and K_P remaining constant at the values calculated in part (a). What are the values of K_I for which the closed-loop system is stable?

(d) Draw a Bode plot of the closed-loop system of part (a), and determine the gain and phase margins, and the respective crossover frequencies.

2.15. For the control system shown in Figure 2.32, it is desired to track an object moving with a *constant acceleration* in m/s² given by $y_d(t) = 5t^2 \cdot u_s(t)$, by moving an antenna whose transfer function is $G(s) = 20/[s(s + 100)]$. Find a controller transfer function, $H(s)$, such that the steady-state tracking error is less than 0.001 m/s², with a settling time of about 0.1 seconds. Plot the closed-loop error response as a function of time. Determine the gain and phase margins, and the respective crossover frequencies of the closed-loop system. What are the maximum overshoot, settling time, and steady-state error of the closed-loop system if the desired output is a unit step function?

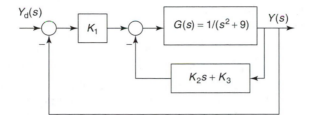

Figure 2.62 Two loop control system for Exercise 2.16

2.16. (a) Derive the transfer function, $Y(s)/Y_d(s)$, of the closed-loop system shown in Figure 2.62.

(b) What is the type of this system?

(c) Find the values of K_1, K_2, and K_3 such that the closed-loop transfer function becomes

$$Y(s)/Y_d(s) = (s + 10)/(s + 2)(s^2 + 4s + 5) \qquad (2.211)$$

(d) Determine the step response of the system in part (c).

(e) Using the convolution integral, determine the response of the system in part (c) to a rectangular pulse (Figure 2.13) of magnitude, $f_o = 1$ and duration, $T = 10$ s.

(f) Determine the gain and phase margins, and the respective crossover frequencies of the system in part (c).

2.17. For the control system in Exercise 2.16, determine the gain and phase margins if $K_1 = 10$, $K_2 = 25$, and $K_3 = 150$.

2.18. For the control system shown in Figure 2.63:

(a) Derive the closed-loop transfer function, $Y(s)/Y_d(s)$. If this control system is to be expressed as a *negative feedback* connection of two systems, $G(s)$ and $H(s)$, as shown in Figure 2.35, identify the *plant* transfer function, $G(s)$, and the *controller* transfer function, $H(s)$.

(b) What is the type of this system?

(c) Is the plant, $G(s)$, stable?

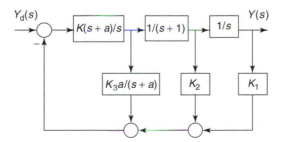

Figure 2.63 Control system for Exercise 2.18

(d) Plot the locus of closed-loop poles if $K = 10$, $a = 2$, $K_2 = 25$, $K_3 = 150$, and K_1 is varied from 1 to 500.

(e) What is the value of K_1 for which the closed-loop system has a dominant pair of complex poles with natural frequency, $\omega_n = 1.83$ rad/s and damping-ratio, $\varsigma = 0.753$?

(f) Plot the step response of the closed-loop system for the value of K_1 desired in part (e). What is the resulting maximum percentage overshoot, settling-time and the steady-state error?

(g) Draw the Bode plots of the plant, $G(s)$, the controller, $H(s)$, and the closed-loop system, and find the gain and phase margins of the control system with K_1 of part (e).

2.19. For the electrical system of Example 2.8, consider a feedback control system of Figure 2.32 using a constant controller transfer function $H(s) = K$.

(a) Draw a root-locus of the closed-loop poles as K is varied. What is the range of K for which the closed-loop system is stable?

(b) Find the value of K for which the closed-loop system has a damping ratio, $\zeta = 0.707$. Plot the step response of the closed-loop system, and find the maximum overshoot, settling time, and steady-state error.

2.20. Re-design the controller, $H(s)$, in Exercise 2.19 to make the steady-state error zero, and the damping ratio, $\zeta = 0.707$. Plot the step response of the closed-loop system, and find the maximum overshoot and settling time.

2.21. For the closed-loop system shown in Figure 2.32, the plant's transfer function is $G(s) = 150\,000/(s^3 + 110s^2 + 1000s)$. It is desired to use a *lag compensator* to control this plant.

(a) Draw a Bode plot of the plant, $G(s)$, and find the gain and phase margins of the plant.

(b) Draw a Bode plot of the *appropriate* lag compensator, $H(s)$, for $\omega_o = 1$ rad/s and $\alpha = 10$. What is the frequency at which the phase of $H(s)$ is the minimum? What is the change in the gain of $H(s)$ between 0 and 10 rad/s?

(c) Compare the steady-state error, maximum overshoot, and settling time of the plant with those of the closed-loop system with the controller parameters of part (b), if the desired output is a unit step function.

(d) Draw a Bode plot of the closed-loop system with the lag compensator of part (b). Find the gain and phase margins of the closed-loop system.

(e) Repeat parts (b)–(d) with $\omega_o = 1$ rad/s and $\alpha = 1.1$.

2.22. For the closed-loop system shown in Figure 2.32, the plant's transfer function is $G(s) = 1000/(s^2 + 2s + 5)$. It is desired to use a *lead compensator* to control this plant.

(a) Draw a Bode plot of the plant, $G(s)$, and find the gain and phase margins of the plant.

(b) Draw a Bode plot of the lead compensator, $H(s)$, for $\omega_o = 40$ rad/s and $\alpha = 0.01$. What is the frequency at which the phase of $H(s)$ becomes a maximum? What is the total change in gain over the entire frequency range?

(c) Compare the steady-state error, maximum overshoot, and settling time of the plant with those of the closed-loop system if the desired output is a unit step function.

(d) Draw a Bode plot of the closed-loop system with the lead compensator of part (b). Find the gain and phase margins of the closed-loop system.

(e) Repeat parts (b)–(d) with $\omega_o = 40$ rad/s and $\alpha = 0.45$.

2.23. A plant's transfer function is $G(s) = 10^6/(s^2 + 0.99s - 0.01)$. It is desired to use a *lead-lag compensator* in the control system of Figure 2.32 to control this plant.

(a) Draw a Bode plot of the plant, $G(s)$, and find the gain and phase margins of the plant.

(b) Draw a Bode plot of the lead-lag compensator, $H(s)$, for $\omega_1 = 0.1$ rad/s, $\omega_2 = 1000$ rad/s and $\alpha = 5$. What are the frequencies corresponding to minimum and maximum phase of $H(s)$?

(c) Compare the steady-state error, maximum overshoot, and settling time of the plant with those of the closed-loop system if the desired output is a unit step function.

(d) Draw a Bode plot of the closed-loop system with the lead-lag compensator of part (b). Find the gain and phase margins of the closed-loop system.

(e) Keeping the values of ω_1 and ω_2 the same as in part (b), what is the value of α for which the maximum overshoot of the closed-loop step response is 30 percent? What is the corresponding settling time? Calculate the new closed-loop gain and phase margins for this value of α.

2.24. Suppose the closed-loop system shown in Figure 2.32 has plant's transfer function, $G(s) = (s + 1)/(s^2 + 2s - 3)$ and controller's transfer function, $H(s) = K/s$. Such a controller is called an *integral compensator*, since it is obtained from the PID compensator by setting $K_I = K$, and $K_P = K_D = 0$.

(a) Is the plant alone stable?

(b) Derive the closed-loop transfer function, $Y(s)/Y_d(s)$.

(c) From a root-locus determine the range of variation of K for which the closed-loop system stable.

(d) Derive the step response of the closed-loop system for $K = 50$. What is the maximum percentage overshoot, peak-time, settling-time and the steady-state error of the system?

(e) What is the steady-state error of the closed-loop system with $K = 50$ if $y_d(t) = t \cdot u_s(t)$?

(f) Determine the gain and phase margins, and the respective crossover frequencies of the system for $K = 50$.

(g) Repeat (d)–(f) for $K = 15$.

2.25. For the aircraft in Example 2.10, it is desired to increase the damping in *phugoid mode* by using a *lead compensator*, with the plant's transfer function is $G(s) = v(s)/\delta(s)$.

(a) Find the gain and phase margins for the plant, $G(s) = v(s)/\delta(s)$.

(b) Find the values of the lead compensator parameters, ω_o and α, such that the closed-loop system has a gain margin greater than 10 dB and a phase margin greater than 130°.

(c) Compare the settling time and maximum overshoot of the plant with those of the closed-loop system with the lead compensator designed in part (b), if the desired output is a unit step function, $u_s(t)$.

2.26. For the aircraft in Example 2.10, it is desired to reduce the *steady-state error* in the angle of attack, $\alpha(s)$, (which is largely influenced by the *short-period mode*) by using a *lag compensator* as a controller in the closed-loop configuration of Figure 2.32, where the plant's transfer function is $G(s) = \alpha(s)/\delta(s)$. From Figure 2.27, it is clear that the *phugoid mode* does not appreciably affect, $\alpha(s)/\delta(s)$, implying that the denominator quadratic corresponding to the phugoid mode, $(s^2 + 0.005s + 0.006)$, gets *approximately* canceled by the numerator quadratic $(s^2 + 0.0065s + 0.006)$ in the expression for $\alpha(s)/\delta(s)$ given by Eq. (2.87). Thus, we can write the *approximate* plant transfer function, $\alpha(s)/\delta(s)$, as follows:

$$\alpha(s)/\delta(s) \approx -0.02(s + 80)/(s^2 + s + 1.4) \qquad (2.212)$$

(a) Compare the Bode plot of the approximate $\alpha(s)/\delta(s)$ given by Eq. (2.212) with that shown in Figure 2.27 for the exact $\alpha(s)/\delta(s)$.

(b) Compare the gain and phase margins for the approximate and exact plant transfer function, $G(s) = \alpha(s)/\delta(s)$.

(c) Find the values of the appropriate lag compensator parameters, ω_o and α, such that the closed-loop system has a gain margin greater than 11 dB and a phase margin greater than 130°.

(d) Compare the settling time and maximum overshoot of the plant with those of the closed-loop system with the lag compensator designed in part (b), if the desired output is a unit step function, $u_s(t)$.

2.27. In an aircraft, the *actuator* of a *control surface* – such as the *elevator* in Example 2.10 – takes some (*non-zero*) *time* to achieve the *desired* control-surface deflection angle, $\delta(t)$. The simplest model for such an actuator is given by a *first-order* transfer function, $\delta(s)/\delta_d(s) = 1/(Ts + 1)$, where T is a time-constant, and $\delta_d(s)$ is the desired deflection angle.

(a) If $T = 0.02$ second, find the step response of $\delta(s)/\delta_d(s) = 1/(Ts + 1)$ (i.e. $\delta(t)$ when $\delta_d(t) = u_s(t)$). What is the settling time and the steady-state error?

(b) Make a Bode plot of the actuator transfer function for $T = 0.02$ second. What are the DC gain, and gain and phase margins of the actuator?

(c) For the actuator of parts (a) and (b), plot the actual control-surface deflection, $\delta(t)$, when the desired deflection, $\delta_d(t)$, is a *rectangular pulse* of width 0.5 second and height 0.05 radian, applied at $t = 0$. (Such a pulse *elevator* input is normally applied by the pilot to excite the *phugoid* mode.)

(d) To account for the presence of the actuator, the longitudinal transfer functions of Example 2.10, are multiplied by the elevator actuator transfer function, $\delta(s)/\delta_{\mathrm{d}}(s) = 1/(Ts+1)$, resulting in the transfer functions, $v(s)/\delta_{\mathrm{d}}(s)$, $\alpha(s)/\delta_{\mathrm{d}}(s)$, and $\theta(s)/\delta_{\mathrm{d}}(s)$. If $T = 0.02$ second, compare the Bode plots of $v(s)/\delta_{\mathrm{d}}(s)$, $\alpha(s)/\delta_{\mathrm{d}}(s)$, and $\theta(s)/\delta_{\mathrm{d}}(s)$, with those of $v(s)/\delta(s)$, $\alpha(s)/\delta(s)$, and $\theta(s)/\delta(s)$ (shown in Figures 2.26–2.28), respectively. Is there any difference in the corresponding gain and phase margins?

(e) Plot the step response of $\theta(s)/\delta_{\mathrm{d}}(s)$, and compare it with that of $\theta(s)/\delta(s)$ shown in Figure 2.29. Is there a difference in the two step responses?

(f) Repeat the design of a lead compensator for controlling the phugoid mode carried out in Exercise 2.25 with a plant transfer function, $G(s) = v(s)/\delta_{\mathrm{d}}(s)$, instead of $G(s) = v(s)/\delta(s)$.

2.28. For controlling the deviation of a satellite from a circular orbit (Exercise 2.4), it is desired to use a *derivative* compensator, $H(s) = Ks$, in the *feedback* arrangement of Figure 2.35, such that the plant's transfer function is $G(s)$, and the feedback controller transfer function is $H(s)$. It is required that the closed-loop response to a *step input* of magnitude 1000 m/s^2 (i.e. $u(t) = 1000u_s(t)$) should have a maximum overshoot of 2 m and a settling time less than 13 seconds. Find the value of K which achieves this, and plot the closed-loop step response. What are the maximum overshoot and the settling time? Plot the *output*, $z(t)$, of the compensator, $H(s)$. What is the maximum magnitude of the controller output, $z(t)$? Draw the Bode plots of the plant, the compensator, and the closed-loop system, and compare the gain and phase margins of the plant with those of the closed-loop system.

2.29. Find the poles and zeros, and analyze the stability of multivariable systems with the following transfer matrices:

(a) $\mathbf{G}(s) = [(s+1)/(s+3)s/(s^2-7s+1)]$

(b) $\mathbf{G}(s) = \begin{bmatrix} 1/(s+1) & 0 \\ -1/(s+2) & 2(s+1)/(s^2-1) \end{bmatrix}$

(c) $\mathbf{G}(s) = \begin{bmatrix} (s+4)/(s^2+3s+2) & s/(s+2) \\ 1/(s+1) & 1/s \\ 1/s & 1/(s+4) \end{bmatrix}$

2.30. Consider a multivariable system of Figure 2.32, with plant transfer matrix, $\mathbf{G}(s)$, and controller transfer matrix, $\mathbf{H}(s)$, given by

$$\mathbf{G}(s) = (1/s)\begin{bmatrix} 10 & 9 \\ 9 & 8 \end{bmatrix}; \quad \mathbf{H}(s) = \begin{bmatrix} K & 1 \\ 1 & K \end{bmatrix} \tag{2.213}$$

(a) Derive the characteristic equation of the system.

(b) Plot the root-locus as K varies from 0.01 to 10. For what range of variation of K is the system stable?

References

1. Zarchan, P. *Tactical and Strategic Missile Guidance*, 2nd Edition, Vol. 157. Progress in Aeronautics and Astronautics, American Institute of Aeronautics and Astronautics, Washington, DC, 1994.
2. D'Azzo, J.J. and Houpis, C. *Feedback Control System Analysis and Synthesis*, 2nd Edition. McGraw-Hill, 1966.
3. Blakelock, J.H. *Automatic Control of Aircraft and Missiles*, 2nd Edition. Wiley, 1991.
4. Gantmacher, F.R. *Theory of Matrices*, Vol. 1. Chelsea, New York, 1959.
5. Kreyszig, E. *Advanced Engineering Mathematics*. Wiley, New York, 1972.

3

State-Space Representation

3.1 The State-Space: Why Do I Need It?

In Chapter 1, we defined the *state* of a system as any set of quantities which must be specified at a given time in order to completely determine the behavior of the system. The quantities constituting the state are called the *state variables*, and the hypothetical space spanned by the state variables is called the *state-space*. In a manner of speaking, we put the cart before the horse – we went ahead and defined the state before really understanding what it was. In the good old car-driver example, we said that the state variables could be the car's speed and the positions of all other vehicles on the road. We also said that the state variables are not unique; we might as well have taken the velocities of all other vehicles relative to the car, and the position of the car with respect to the road divider to be the state variables of the car-driver system. Let us try to understand what the state of a system really means by considering the example of a simple pendulum.

Example 3.1

Recall from Example 2.2 that the governing differential equation for the motion of a simple pendulum on which no external input is applied (Figure 2.3) is given by Eq. (2.8). If we apply a torque, $M(t)$, about the hinge, O, as an input to the pendulum, the governing differential equation can be written as

$$L\theta^{(2)}(t) + g\sin(\theta(t)) = M(t)/(\text{mL}) \qquad (3.1)$$

where $\theta^{(2)}(t)$ represents the second order time derivative of $\theta(t)$, as per our notation (i.e. $d^2\theta(t)/dt^2 = \theta^{(2)}(t)$). Let the output of the system be the angle, $\theta(t)$, of the pendulum. We would like to determine the state of this system. To begin, we must know how many quantities (i.e. state variables) need to be specified to completely determine the motion of the pendulum. Going back to Chapter 2, we know that for a system of order n, we have to specify precisely n *initial conditions* to solve the governing differential equation. Hence, it must follow that the *state* of an nth order system should consist of *precisely n* state variables, which must be specified at some time (e.g. $t = 0$) as initial conditions in order to completely determine the solution to the governing differential equation. Here we are dealing with a *second order* system – which implies that the state must consist of *two* state variables. Let

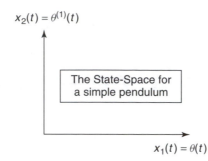

$$x_2(t) = \theta^{(1)}(t)$$

The State-Space for
a simple pendulum

$$x_1(t) = \theta(t)$$

Figure 3.1   The two-dimensional state-space for a simple pendulum (Example 3.1)

us call these state variables $x_1(t)$ and $x_2(t)$, and arbitrarily choose them to be the following:

$$x_1(t) = \theta(t) \qquad\qquad (3.2)$$

$$x_2(t) = \theta^{(1)}(t) \qquad\qquad (3.3)$$

The *state-space* is thus *two-dimensional* for the simple pendulum whose axes are $x_1(t)$ and $x_2(t)$ (Figure 3.1).

It is now required that we express the governing differential equation (Eq. (3.1)) in terms of the state variables defined by Eqs. (3.2) and (3.3). Substituting Eqs. (3.2) and (3.3) into Eq. (3.1), we get the following first order differential equation:

$$x_2^{(1)}(t) = M(t)/(mL^2) - (g/L)\sin(x_1(t)) \qquad\qquad (3.4)$$

Have we *transformed* a second order differential equation (Eq. (3.1)) into a first order differential equation (Eq. (3.4)) by using the state-variables? *Not really*, because there is a *another* first order differential equation that we have forgotten about – the one obtained by substituting Eq. (3.2) into Eq. (3.3), and written as

$$x_1^{(1)}(t) = x_2(t) \qquad\qquad (3.5)$$

Equations (3.4) and (3.5) are *two first order* differential equations, called the *state equations*, into which the governing equation (Eq. (3.1)) has been transformed. The *order* of the system, which is its important characteristic, remains *unchanged* when we express it in terms of the state variables. In addition to the state equations (Eqs. (3.4) and (3.5)), we need an *output equation* which defines the relationship between the output, $\theta(t)$, and the state variables $x_1(t)$ and $x_2(t)$. Equation (3.2) simply gives the output equation as

$$\theta(t) = x_1(t) \qquad\qquad (3.6)$$

The state equations, Eqs. (3.4) and (3.5), along with the output equation, Eq. (3.6), are called the *state-space representation* of the system.

Instead of choosing the state variables as $\theta(t)$ and $\theta^{(1)}(t)$, we could have selected a different set of state variables, such as

$$x_1(t) = L\theta(t) \tag{3.7}$$

and

$$x_2(t) = L^2\theta^{(1)}(t) \tag{3.8}$$

which would result in the following state equations:

$$x_1^{(1)}(t) = x_2(t)/L \tag{3.9}$$

$$x_2^{(1)}(t) = M(t)/m - gL\sin(x_1(t)/L) \tag{3.10}$$

and the output equation would be given by

$$\theta(t) = x_1(t)/L \tag{3.11}$$

Although the state-space representation given by Eqs. (3.9)–(3.11) is different from that given by Eqs. (3.4)–(3.6), both descriptions are for the same system. Hence, we expect that the solution of *either* set of equations would yield the *same* essential characteristics of the system, such as performance, stability, and robustness. Hence, the state-space representation of a system is not *unique*, and all *legitimate* state-space representations should give the same system characteristics. What do we mean by a *legitimate* state-space representation? While we have freedom to choose our state variables, we have to ensure that we have chosen the *minimum* number of state variables that are required to describe the system. In other words, we should not have *too many* or *too few* state variables. One way of ensuring this is by taking *precisely* n state variables, where n is the order of the system. If we are deriving state-space representation from the system's governing differential equation (such as in Example 3.1), the number of state-variables is easily determined by the order of the differential equation. However, if we are deriving the state-space representation from a transfer function (or transfer matrix), some poles may be *canceled* by the zeros, thereby yielding an *erroneous* order of the system which is *less* than the correct order.

Example 3.2

Consider a system with input, $u(t)$, and output, $y(t)$, described by the following differential equation:

$$y^{(2)}(t) + (b - a)y^{(1)}(t) - ab\, y(t) = u^{(1)}(t) - au(t) \tag{3.12}$$

where a and b are positive constants. The transfer function, $Y(s)/U(s)$, of this system can be obtained by taking the Laplace transform of Eq. (3.12) with *zero initial conditions*, and written as follows:

$$Y(s)/U(s) = (s - a)/[s^2 + (b - a)s - ab] = (s - a)/[(s - a)(s + b)] \tag{3.13}$$

In Eq. (3.13), if we cannot resist the temptation to cancel the pole at $s = a$ with the zero at $s = a$, we will be left with the following transfer function:

$$Y(s)/U(s) = 1/(s + b) \tag{3.14}$$

which yields the following *incorrect* differential equation for the system:

$$y^{(1)}(t) + by(t) = u(t) \tag{3.15}$$

Since the pole cancelled at $s = a$ has a *positive real part*, the actual system given by the transfer function of Eq. (3.13) is *unstable*, while that given by Eq. (3.14) is *stable*. Needless to say, basing a state-space representation on the transfer function given by Eq. (3.14) will be incorrect. This example illustrates one of the hazards associated with the transfer function description of a system, which can be avoided if we directly obtain state-space representation from the governing differential equation.

Another cause of illegitimacy in a state-space representation is when two (or more) state variables are *linearly dependent*. For example, if $x_1(t) = \theta(t)$ is a state variable, then $x_2(t) = L\theta(t)$ *cannot* be another state variable in the *same state-space representation*, because that would make $x_1(t)$ and $x_2(t)$ linearly dependent. You can demonstrate that with such a choice of state variables in Example 3.1, the state equations will *not* be *two* first order differential equations. In general, for a system of order n, if $x_1(t), x_2(t), \ldots, x_{n-1}(t)$ are state variables, then $x_n(t)$ is *not* a legitimate state variable if it can be expressed as a linear combination of the other state variables given by

$$x_n(t) = c_1 x_1(t) + c_2 x_2(t) + \cdots + c_{n-1} x_{n-1}(t) \tag{3.16}$$

where $c_1, c_2, \ldots, c_{n-1}$ are constants. Thus, while we have an unlimited choice in selecting state variables for a given system, we should ensure that their number is *equal* to the order of the system, and also that *each* state variable is *linearly independent* of the other state variables in a state-space representation.

In Chapter 2, we saw how *single-input, single-output, linear systems* can be designed and analyzed using the *classical* methods of frequency response and transfer function. The transfer function – or frequency response – representations of linear systems were indispensable before the wide availability of fast digital computers, necessitating the use of tables (such as the *Routh* table [1]) and graphical methods, such as *Bode, Nyquist, root-locus*, and *Nichols* plots for the analysis and design of control systems. As we saw in Chapter 2, the classical methods require a lot of complex variable analysis, such as interpretation of gain and phase plots and complex mapping, which becomes complicated for multivariable systems. Obtaining information about a multivariable system's time-response to an arbitrary input using classical methods is a difficult and indirect process, requiring inverse Laplace transformation. Clearly, design and analysis of modern control systems which are usually multivariable (such as Example 2.10) will be very difficult using the classical methods of Chapter 2.

In contrast to classical methods, the state-space methods work directly with the governing differential equations of the system in the time-domain. Representing the governing differential equations by first order state equations makes it possible to directly solve the state equations in time, using standard numerical methods and efficient algorithms on today's fast digital computers. Since the state equations are always of first order irrespective of the system's order or the number of inputs and outputs, the greatest advantage of state-space methods is that they *do not formally distinguish* between single-input, single-output systems and multivariable systems, allowing efficient design and analysis of multivariable systems with the same ease as for single variable systems. Furthermore, using state-space methods it is possible to directly design and analyze *nonlinear* systems (such as Example 3.1), which is utterly impossible using classical methods. When dealing with linear systems, state-space methods result in repetitive *linear algebraic* manipulations (such as matrix multiplication, inversion, solution of a linear matrix equation, etc.), which are easily programmed on a digital computer. This saves a lot of drudgery that is common when working with inverse Laplace transforms of transfer matrices. With the use of a high-level programming language, such as MATLAB, the linear algebraic manipulations for state-space methods are a breeze. Let us find a state-space representation for a multivariable nonlinear system.

Example 3.3

Consider an inverted pendulum on a moving cart (see Exercise 2.1), for which the governing differential equations are the following:

$$(M + m)x^{(2)}(t) + mL\theta^{(2)}(t)\cos(\theta(t)) - mL[\theta^{(1)}(t)]^2\sin(\theta(t)) = f(t) \quad (3.17)$$

$$mx^{(2)}(t)\cos(\theta(t)) + mL\theta^{(2)}(t) - mg\sin(\theta(t)) = 0 \quad (3.18)$$

where m and L are the mass and length, respectively, of the inverted pendulum, M is the mass of the cart, $\theta(t)$ is the angular position of the pendulum from the vertical, $x(t)$ is the horizontal displacement of the cart, $f(t)$ is the applied force on the cart in the same direction as $x(t)$ (see Figure 2.59), and g is the acceleration due to gravity. Assuming $f(t)$ to be the input to the system, and $x(t)$ and $\theta(t)$ to be the two outputs, let us derive a state-space representation of the system.

The system is described by *two* second order differential equations; hence, the order of the system is *four*. Thus, we need precisely four linearly independent state-variables to describe the system. When dealing with a physical system, it is often desirable to select *physical quantities* as state variables. Let us take the state variables to be the angular position of the pendulum, $\theta(t)$, the cart displacement, $x(t)$, the angular velocity of the pendulum, $\theta^{(1)}(t)$, and the cart's velocity, $x^{(1)}(t)$. We can arbitrarily number the state variables as follows:

$$x_1(t) = \theta(t) \quad (3.19)$$
$$x_2(t) = x(t) \quad (3.20)$$
$$x_3(t) = \theta^{(1)}(t) \quad (3.21)$$
$$x_4(t) = x^{(1)}(t) \quad (3.22)$$

From Eqs. (3.19) and (3.21), we get our first state-equation as follows:

$$x_1^{(1)}(t) = x_3(t) \tag{3.23}$$

while the second state-equation follows from Eqs. (3.20) and (3.22) as

$$x_2^{(1)}(t) = x_4(t) \tag{3.24}$$

The two remaining state-equations are derived by substituting Eqs. (3.19)–(3.22) into Eq. (3.17) and Eq. (3.18), respectively, yielding

$$x_3^{(1)}(t) = g \sin(x_1(t))/L - x_4^{(1)}(t) \cos(x_1(t))/L \tag{3.25}$$

$$x_4^{(1)}(t) = [mL/(M + m)][x_3^{(1)}(t)]^2 \sin(x_1(t))$$
$$- [mL/(M + m)]x_3^{(1)}(t) \cos(x_1(t)) + f(t)/(M + m) \tag{3.26}$$

The two output equations are given by

$$\theta(t) = x_1(t) \tag{3.27}$$

$$x(t) = x_2(t) \tag{3.28}$$

Note that due to the nonlinear nature of the system, we cannot express the last two state-equations (Eqs. (3.25), (3.26)) in a form such that each equation contains the *time derivative* of only *one* state variable. Such a form is called an *explicit form* of the state-equations. If the motion of the pendulum is *small* about the *equilibrium point*, $\theta = 0$, we can *linearize* Eqs. (3.25) and (3.26) by assuming $\cos(\theta(t)) = \cos(x_1(t)) = 1$, $\sin(\theta(t)) = \sin(x_1(t)) = x_1(t)$, and $[\theta^{(1)}(t)]^2 \sin(\theta(t)) = [x_3^{(1)}(t)]^2 \sin(x_1(t)) = 0$. The corresponding *linearized* state equations can then be written in explicit form as follows:

$$x_3^{(1)}(t) = [(M + m)g/(ML)]x_1(t) - f(t)/(ML) \tag{3.29}$$

$$x_4^{(1)}(t) = -(mg/M)x_1(t) + f(t)/M \tag{3.30}$$

The *linearized* state-equations of the system, Eqs. (3.23), (3.24), (3.29), and (3.30), can be expressed in the following *matrix* form, where all coefficients are collected together by suitable *coefficient matrices*:

$$\begin{bmatrix} x_1^{(1)}(t) \\ x_2^{(1)}(t) \\ x_3^{(1)}(t) \\ x_4^{(1)}(t) \end{bmatrix} = \begin{bmatrix} 0 & 0 & 1 & 0 \\ 0 & 0 & 0 & 1 \\ (M+m)g/(ML) & 0 & 0 & 0 \\ -mg/M & 0 & 0 & 0 \end{bmatrix} \begin{bmatrix} x_1(t) \\ x_2(t) \\ x_3(t) \\ x_4(t) \end{bmatrix} + \begin{bmatrix} 0 \\ 0 \\ -1/(ML) \\ 1/M \end{bmatrix} f(t)$$

$$\tag{3.31}$$

with the output matrix equation given by

$$\begin{bmatrix} \theta(t) \\ x(t) \end{bmatrix} = \begin{bmatrix} 1 & 0 & 0 & 0 \\ 0 & 1 & 0 & 0 \end{bmatrix} \begin{bmatrix} x_1(t) \\ x_2(t) \\ x_3(t) \\ x_4(t) \end{bmatrix} + \begin{bmatrix} 0 \\ 0 \end{bmatrix} f(t) \tag{3.32}$$

Note that the state-space representation of the linearized system consists of linear state-equations, Eq. (3.31), and a linear output equation, Eq. (3.32).

Taking a cue from Example 3.3, we can write the state-equations of a general linear system of order n, with m inputs and p outputs, in the following matrix form:

$$\mathbf{x}^{(1)}(t) = \mathbf{A}\mathbf{x}(t) + \mathbf{B}\mathbf{u}(t) \tag{3.33}$$

and the general output equation is

$$\mathbf{y}(t) = \mathbf{C}\mathbf{x}(t) + \mathbf{D}\mathbf{u}(t) \tag{3.34}$$

where $\mathbf{x}(t) = [x_1(t); x_2(t); \ldots; x_n(t)]^T$ is the *state vector* consisting of n state variables as its elements, $\mathbf{x}^{(1)}(t) = [x_1^{(1)}(t); x_2^{(1)}(t); \ldots; x_n^{(1)}(t)]^T$ is the *time derivative of the state vector*, $\mathbf{u}(t) = [u_1(t); u_2(t); \ldots; u_r(t)]^T$ is the *input vector* consisting of r inputs as its elements, $\mathbf{y}(t) = [y_1(t); y_2(t); \ldots; y_p(t)]^T$ is the *output vector* consisting of p outputs as its elements, and $\mathbf{A}, \mathbf{B}, \mathbf{C}, \mathbf{D}$ are the *coefficient matrices*. Note that the row dimension (i.e. the number of rows) of the state vector is equal to the order of the system, n, while those of the input and output vectors are r and p, respectively. Correspondingly, for the matrix multiplications in Eqs. (3.33) and (3.34) to be defined, the sizes of the coefficient matrices, $\mathbf{A}, \mathbf{B}, \mathbf{C}, \mathbf{D}$, should be $(n \times n)$, $(n \times r)$, $(p \times n)$, and $(p \times r)$, respectively. The coefficient matrices in Example 3.3 were all *constant*, i.e. they were not varying with time. Such a state-space representation in which all the coefficient matrices are constants is said to be *time-invariant*. In general, there are linear systems with coefficient matrices that are functions of time. Such state-space representations are said to be linear, but *time-varying*. Let us take another example of a *linear, time-invariant* state-space representation, which is a little more difficult to derive than the state-space representation of Example 3.3.

Example 3.4

Re-consider the electrical network presented in Example 2.8 whose governing differential equations are as follows:

$$R_1 i_1(t) + R_3[i_1(t) - i_2(t)] = e(t) \tag{3.35}$$

$$L i_2^{(2)}(t) + [(R_1 R_3 + R_1 R_2 + R_2 R_3)/(R_1 + R_3)]i_2^{(1)}(t) + (1/C)i_2(t)$$
$$= [R_3/(R_1 + R_3)]e^{(1)}(t) \tag{3.36}$$

If the input is the applied voltage, $e(t)$, and the output, $y_1(t)$, is the current in the resistor R_3 (given by $i_1(t) - i_2(t)$) when the switch S is closed (see Figure 2.19), we have to find a state-space representation of the system. Looking at Eq. (3.36), we find that the time derivative of the input appears on the right-hand side. For a linear, time-invariant state-space form of Eqs. (3.33) and (3.34), the state variables must be selected in such a way that the time derivative of the input, $e^{(1)}(t)$, vanishes from the state and output equations. One possible choice of state variables which accomplishes this is the following:

$$x_1(t) = i_2(t) \tag{3.37}$$

$$x_2(t) = i_2^{(1)}(t) - R_3 e(t)/[L(R_1 + R_3)] \tag{3.38}$$

Then the first state-equation is obtained by substituting Eq. (3.37) into Eq. (3.38), and expressed as

$$x_1^{(1)}(t) = x_2(t) + R_3 e(t)/[L(R_1 + R_3)] \tag{3.39}$$

Substitution of Eqs. (3.37) and (3.38) into Eq. (3.36) yields the second state-equation, given by

$$x_2^{(1)}(t) = -[(R_1 R_3 + R_1 R_2 + R_2 R_3)/[L(R_1 + R_3)]$$
$$\times [x_2(t) + R_3 e(t)/L(R_1 + R_3)] - x_1(t)/(LC) \tag{3.40}$$

The output equation is given by using Eq. (3.35) as follows:

$$y_1(t) = i_1(t) - i_2(t) = [e(t) + R_3 x_1(t)]/(R_1 + R_3) - x_1(t)$$
$$= -R_1 x_1(t)/(R_1 + R_3) + e(t)/(R_1 + R_3) \tag{3.41}$$

In the matrix notation, Eqs. (3.39)–(3.41) are expressed as

$$\begin{bmatrix} x_1^{(1)}(t) \\ x_2^{(1)}(t) \end{bmatrix} = \begin{bmatrix} 0 & 1 \\ -1/(LC) & -[(R_1 R_3 + R_1 R_2 + R_2 R_3)/[L(R_1 + R_3)] \end{bmatrix} \begin{bmatrix} x_1(t) \\ x_2(t) \end{bmatrix}$$
$$+ \begin{bmatrix} R_3/[L(R_1 + R_3)] \\ -[R_3(R_1 R_3 + R_1 R_2 + R_2 R_3)/[L(R_1 + R_3)]^2 \end{bmatrix} e(t) \tag{3.42}$$

$$y_1(t) = \begin{bmatrix} -R_1/(R_1 + R_3) & 0 \end{bmatrix} \begin{bmatrix} x_1(t) \\ x_1(t) \end{bmatrix} + 1/(R_1 + R_3)e(t) \tag{3.43}$$

Comparing Eqs. (3.42) and (3.43) with Eqs. (3.33) and (3.34), we can find the constant coefficient matrices, **A**, **B**, **C**, **D**, of the system, with the input vector, $\mathbf{u}(t) = e(t)$, and the output vector, $\mathbf{y}(t) = y_1(t)$.

If we compare Examples 3.3 and 3.4, it is *harder* to select the state variables in Example 3.4 due to the presence of the time derivative of the input in the governing differential equation. A general linear (or nonlinear) system may have several higher-order time derivatives of the input in its governing differential equations (such as Eq. (2.4)). To simplify the selection of state variables in such cases, it is often useful to first draw a *schematic diagram* of the governing differential equations. The schematic diagram is drawn using elements similar to those used in the *block diagram* of a system. These elements are the *summing-junction* (which adds two or more signals with appropriate

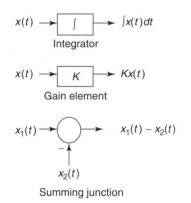

Figure 3.2 The state-space schematic diagram elements

signs), and the two transfer elements, namely the *gain element* (which multiplies a signal by a constant), and the *integrator* (which integrates a signal). The *arrows* are used to indicate the direction in which the signals are flowing into these elements. Figure 3.2 shows what the schematic diagram elements look like.

Let us use the schematic diagram approach to find another state-space representation for the system in Example 3.4.

Example 3.5

The system of Example 3.4 has two governing equations, Eqs. (3.35) and (3.36). While Eq. (3.35) is an *algebraic* equation (i.e. a *zero* order differential equation), Eq. (3.36) is a second order differential equation. Let us express Eq. (3.36) in terms of a *dummy variable* (so called because it is neither a state variable, an input, nor output) $z(t)$, such that

$$z^{(2)}(t) + [(R_1 R_3 + R_1 R_2 + R_2 R_3)/[L(R_1 + R_3)]z^{(1)}(t) + 1/(LC)z(t) = e(t) \tag{3.44}$$

where

$$i_2(t) = R_3/[L(R_1 + R_3)]z^{(1)}(t) \tag{3.45}$$

We have split Eq. (3.36) into Eqs. (3.44) and (3.45) because we want to eliminate the time derivative of the input, $e^{(1)}(t)$, from the state-equations. You may verify that substituting Eq. (3.45) into Eq. (3.44) yields the original differential equation, Eq. (3.36). The schematic diagram of Eqs. (3.44) and (3.45) is drawn in Figure 3.3. Furthermore, Figure 3.3 uses Eq. (3.35) to represent the output, $y_1(t) = i_1(t) - i_2(t)$. Note the similarity between a block diagram, such as Figure 2.1, and a schematic diagram. Both have the inputs coming in from the left, and the outputs going out at the right. The difference between a block diagram and a schematic diagram is that, while the former usually represents the input-output relationship as a transfer function (or transfer matrix) in the Laplace domain, the latter represents

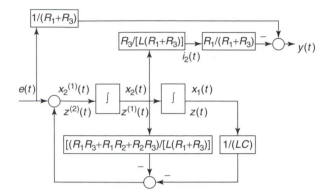

Figure 3.3 Schematic diagram for the electrical system of Example 3.4

the same relationship as a set of differential equations in time. Note that the number of integrators in a schematic diagram is equal to the order of the system.

A state-space representation of the system can be obtained from Figure 3.3 by choosing the *outputs of the integrators* as state variables, as shown in Figure 3.3. Then using the fact that the output from the *second integrator* from the left, $x_1(t)$, is the time integral of its input, $x_2(t)$, the first state-equation is given by

$$x_1^{(1)}(t) = x_2(t) \tag{3.46}$$

The second state-equation is obtained seeing what is happening at the *first summing junction* from the left. The *output* of that summing junction is the *input* to the *first integrator* from left, $x_2^{(1)}(t)$, and the two signals being added at the summing junction are $e(t)$ and $-x_1(t)/(LC) - [(R_1R_3 + R_1R_2 + R_2R_3)/[L(R_1 + R_3)]x_2(t)$. Therefore, the second state-equation is given by

$$x_2^{(1)}(t) = -x_1(t)/(LC) - [(R_1R_3 + R_1R_2 + R_2R_3)/[L(R_1 + R_3)]x_2(t) + e(t) \tag{3.47}$$

The output equation is obtained by expressing the output, $y_1(t) = i_1(t) - i_2(t)$, in terms of the state variables. Before relating the output to the state variables, we should express each state variable in terms of the physical quantities, $i_1(t)$, $i_2(t)$, and $e(t)$. We see from Figure 3.3 that $x_2(t) = z^{(1)}(t)$; thus, from Eq. (3.45), it follows that

$$x_2(t) = L(R_1 + R_3)i_2(t)/R_3 \tag{3.48}$$

Then, substitution of Eq. (3.48) into Eq. (3.47) yields

$$x_1(t) = LC[e(t) - i_2(t)(R_1R_3 + R_1R_2 + R_2R_3)/R_3 - L(R_1 + R_3)i_2^{(1)}(t)/R_3] \tag{3.49}$$

Using the algebraic relationship among $i_1(t)$, $i_2(t)$ and $e(t)$ by Eq. (3.35), we can write the output equation as follows:

$$y_1(t) = -R_1 R_3/[L(R_1 + R_3)^2]x_2(t) + e(t)/(R_1 + R_3) \qquad (3.50)$$

In matrix form, the state-space representation is given by

$$\begin{bmatrix} x_1^{(1)}(t) \\ x_2^{(1)}(t) \end{bmatrix} = \begin{bmatrix} 0 & 1 \\ -1/(LC & -(R_1 R_3 + R_1 R_2 + R_2 R_3)/[L(R_1 + R_3)] \end{bmatrix}$$
$$\times \begin{bmatrix} x_1(t) \\ x_2(t) \end{bmatrix} + \begin{bmatrix} 0 \\ 1 \end{bmatrix} e(t) \qquad (3.51)$$

$$y_1(t) = [0 - R_1 R_3/\{L(R_1 + R_3)^2\}] \begin{bmatrix} x_1(t) \\ x_2(t) \end{bmatrix} + 1/(R_1 + R_3)e(t) \qquad (3.52)$$

Note the difference in the state-space representations of the same system given by Eqs. (3.42), (3.43) and by Eqs. (3.51), (3.52). Also, note that *another* state-space representation could have been obtained by numbering the state variables in Figure 3.3 starting from the *left* rather than from the *right*, which we did in Example 3.5.

Example 3.6

Let us find a state-space representation using the schematic diagram for the system with input, $u(t)$, and output, $y(t)$, described by the following differential equation:

$$y^{(n)}(t) + a_{n-1}y^{(n-1)}(t) + \cdots + a_1 y^{(1)}(t) + a_0 y(t)$$
$$= b_n u^{(n)}(t) + b_{n-1}u^{(n-1)}(t) + \cdots + b_1 u^{(1)}(t) + b_0 u(t) \qquad (3.53)$$

Since the right-hand side of Eq. (3.53) contains time derivatives of the input, we should introduce a dummy variable, $z(t)$, in a manner similar to Example 3.5, such that

$$z^{(n)}(t) + a_{n-1}z^{(n-1)}(t) + \cdots + a_1 z^{(1)}(t) + a_0 z(t) = u(t) \qquad (3.54)$$

and

$$b_n z^{(n)}(t) + b_{n-1}z^{(n-1)}(t) + \cdots + b_1 z^{(1)}(t) + b_0 z(t) = y(t) \qquad (3.55)$$

Figure 3.4 shows the schematic diagram of Eqs. (3.54) and (3.55). Note that Figure 3.4 has n integrators arranged in a series.

As in Example 3.5, let us choose the state variables to be the integrator outputs, and number them beginning from the right of Figure 3.4. Then the state-equations are as follows:

$$x_1^{(1)}(t) = x_2(t) \qquad (3.56a)$$

$$x_2^{(1)}(t) = x_3(t) \qquad (3.56b)$$

$$\vdots$$

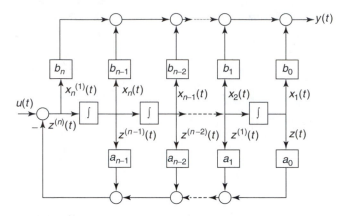

Figure 3.4 Schematic diagram for the controller companion form of the system in Example 3.6

$$x_{n-1}^{(1)}(t) = x_n(t) \tag{3.56c}$$

$$x_n^{(1)}(t) = -[a_{n-1}x_n(t) + a_{n-2}x_{n-1}(t) + \cdots + a_0 x_1(t) - u(t)] \tag{3.56d}$$

The output equation is obtained by substituting the definitions of the state variables, namely, $x_1(t) = z(t)$, $x_2(t) = z^{(1)}(t)$, ..., $x_n(t) = z^{(n-1)}(t)$ into Eq. (3.55), thereby yielding

$$y(t) = b_0 x_1(t) + b_1 x_2(t) + \cdots + b_{n-1}x_n(t) + b_n x_n^{(1)}(t) \tag{3.57}$$

and substituting Eq. (3.56d) into Eq. (3.57), the output equation is expressed as follows:

$$y(t) = (b_0 - a_0 b_n)x_1(t) + (b_1 - a_1 b_n)x_2(t) + \cdots + (b_{n-1} - a_{n-1}b_n)x_n(t) + b_n u(t) \tag{3.58}$$

The matrix form of the state-equations is the following:

$$\begin{bmatrix} x_1^{(1)}(t) \\ x_2^{(1)}(t) \\ \vdots \\ x_{n-1}^{(1)}(t) \\ x_n^{(1)}(t) \end{bmatrix} = \begin{bmatrix} 0 & 1 & 0 & \cdots & 0 \\ 0 & 0 & 1 & \cdots & 0 \\ \vdots & \vdots & \vdots & \vdots & \vdots \\ 0 & 0 & 0 & \cdots & 1 \\ -a_0 & -a_1 & -a_2 & \cdots & -a_{n-1} \end{bmatrix} \begin{bmatrix} x_1(t) \\ x_2(t) \\ \vdots \\ x_{n-1}(t) \\ x_n(t) \end{bmatrix} + \begin{bmatrix} 0 \\ 0 \\ \vdots \\ 0 \\ 1 \end{bmatrix} u(t) \tag{3.59}$$

and the output equation in matrix form is as follows:

$$y(t) = [(b_0 - a_0 b_n) \ (b_1 - a_1 b_n) \dots (b_{n-1} - a_{n-1}b_n)] \begin{bmatrix} x_1(t) \\ x_2(t) \\ \vdots \\ x_n(t) \end{bmatrix} + b_n u(t) \tag{3.60}$$

Comparing Eqs. (3.59) and (3.60) with the matrix equations Eqs. (3.33) and (3.34), respectively, we can easily find the coefficient matrices, $\mathbf{A}$, $\mathbf{B}$, $\mathbf{C}$, $\mathbf{D}$, of the state-space representation. Note that the matrix $\mathbf{A}$ of Eq. (3.59) has a particular structure: all elements except the *last row* and the *superdiagonal* (i.e. the diagonal *above* the main diagonal) are *zeros*. The superdiagonal elements are all *ones*, while the last row consists of the coefficients with a negative sign, $-a_0, -a_1, \ldots, -a_{n-1}$. Taking the Laplace transform of Eq. (3.53), you can verify that the coefficients $a_0, a_1, \ldots, a_{n-1}$ are the *coefficients of the characteristic polynomial* of the system (i.e. the denominator polynomial of the transfer function, $Y(s)/U(s)$) given by $s^n + a_{n-1}s^{n-1} + \cdots + a_1 s + a_0$. The matrix $\mathbf{B}$ of Eq. (3.59) has all elements zeros, except the last row (which equals 1). Such a state-space representation has a name: *the controller companion form*. It is thus called because it has a special place in the design of controllers, which we will see in Chapter 5.

Another *companion form*, called the *observer companion form*, is obtained as follows for the system obeying Eq. (3.53). In Eq. (3.53) the terms involving derivatives of $y(t)$ and $u(t)$ of the same order are collected, and the equation is written as follows:

$$[y^{(n)}(t) - b_n u^{(n)}(t)] + [a_{n-1}y^{(n-1)}(t) - b_{n-1}u^{(n-1)}(t)]$$
$$+ \cdots + [a_1 y^{(1)}(t) - b_1 u^{(1)}(t)] + [a_0 y(t) - b_0 u(t)] = 0 \qquad (3.61)$$

On taking the Laplace transform of Eq. (3.61) subject to zero initial conditions, we get the following:

$$s^n[Y(s) - b_n U(s)] + s^{n-1}[a_{n-1}Y(s) - b_{n-1}U(s)]$$
$$+ \cdots + s[a_1 Y(s) - b_1 U(s)] + [a_0 Y(s) - b_0 U(s)] = 0 \qquad (3.62)$$

Dividing Eq. (3.61) by s^n leads to

$$Y(s) = b_n U(s) + [b_{n-1}U(s) - a_{n-1}Y(s)]/s + \cdots + [b_1 U(s) - a_1 Y(s)]/s^{n-1}$$
$$+ [b_0 U(s) - a_0 Y(s)]/s^n \qquad (3.63)$$

We can draw a schematic diagram for Eq. (3.63), using the fact that the multiplication factor $1/s$ in the Laplace domain represents an integration in time. Therefore, according to Eq. (3.63), $[b_{n-1}U(s) - a_{n-1}Y(s)]$ must pass through *one integrator* before contributing to the output, $Y(s)$. Similarly, $[b_1 U(s) - a_1 Y(s)]$ must pass through $(n-1)$ integrators, and $[b_0 U(s) - a_0 Y(s)]$ through n integrators in the schematic diagram. Figure 3.5 shows the schematic diagram of Eq. (3.63).

On comparing Figures 3.4 and 3.5, we see that both the figures have a series of n integrators, but the feedback paths from the output, $y(t)$, to the integrators are in *opposite* directions in the two figures. If we select the outputs of the integrators as state variables beginning from the *left* of Figure 3.5, we get the following state-equations:

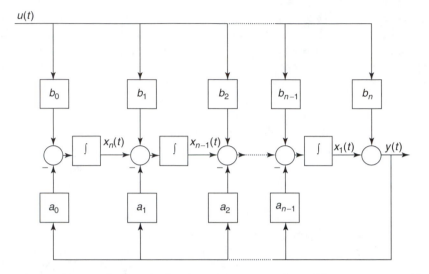

Figure 3.5   Schematic diagram for the observer companion form of the system in Example 3.6

$$
\begin{aligned}
x_1^{(1)}(t) &= -a_0 x_n(t) + (b_0 - a_0 b_n)u(t) \\
x_2^{(1)}(t) &= x_1(t) - a_1 x_n(t) + (b_1 - a_1 b_n)u(t) \\
&\;\;\vdots \\
x_{n-1}^{(1)}(t) &= x_{n-2}(t) - a_{n-2}x_n(t) + (b_{n-2} - a_{n-2}b_n)u(t) \\
x_n^{(1)}(t) &= -a_{n-1}x_n(t) + (b_{n-1} - a_{n-1}b_n)
\end{aligned}
\tag{3.64}
$$

and the output equation is

$$
y(t) = x_n(t) + b_n u(t)
\tag{3.65}
$$

Therefore, the state coefficient matrices, **A**, **B**, **C**, and **D**, of the observer companion form are written as follows:

$$
\mathbf{A} =
\begin{bmatrix}
0 & 0 & \cdots & 0 & -a_0 \\
1 & 0 & \cdots & 0 & -a_1 \\
0 & 1 & \cdots & 0 & -a_2 \\
\cdot & \cdot & \cdots & \cdot & \cdot \\
\cdot & \cdot & \cdots & \cdot & \cdot \\
\cdot & \cdot & \cdots & \cdot & \cdot \\
0 & 0 & \cdots & 1 & -a_{n-2} \\
0 & 0 & \cdots & 0 & -a_{n-1}
\end{bmatrix}
; \quad
\mathbf{B} =
\begin{bmatrix}
(b_0 - a_0 b_n) \\
(b_1 - a_1 b_n) \\
(b_2 - a_2 b_n) \\
\cdot \\
\cdot \\
\cdot \\
(b_{n-2} - a_{n-2}b_n) \\
(b_{n-1} - a_{n-1}b_n)
\end{bmatrix}
$$

$$
\mathbf{C} = [0 \;\; 0 \;\; \cdots \;\; 0 \;\; 1]; \quad \mathbf{D} = b_n
\tag{3.66}
$$

Note that the **A** matrix of the observer companion form is the transpose of the **A** matrix of the controller companion form. Also, the **B** matrix of the observer companion form is the transpose of the **C** matrix of the controller companion form, and vice versa. The **D** matrices of the both the companion forms are the same. Now we know why these state-space representations are called *companion forms*: they can be obtained from one another merely by taking the transpose of the coefficient matrices. The procedure used in this example for obtaining the companion forms of a single-input, single-output system can be extended to multi-variable systems.

Thus far, we have only considered examples having single inputs. Let us take up an example with multi-inputs.

Example 3.7

Consider the electrical network for an amplifier-motor shown in Figure 3.6. It is desired to change the angle, $\theta(t)$, and angular velocity, $\theta^{(1)}(t)$, of a load attached to the motor by changing the input voltage, $e(t)$, in the presence of a torque, $T_L(t)$, applied by the load on the motor. The governing differential equations for the amplifier-motor are the following:

$$Li^{(1)}(t) + (R + R_0)i(t) + a\theta^{(1)}(t) = K_A e(t) \qquad (3.67)$$

$$J\theta^{(2)}(t) + b\theta^{(1)}(t) - ai(t) = -T_L(t) \qquad (3.68)$$

where J, R, L, and b are the moment of inertia, resistance, self-inductance, and viscous damping-coefficient of the motor, respectively, and a is a machine constant. R_0 and K_A are the resistance and voltage amplification ratio of the amplifier.

Since the loading torque, $T_L(t)$, acts as a disturbance to the system, we can consider it as an additional input variable. The input vector is thus given by $\mathbf{u}(t) = [e(t); T_L(t)]^T$. The output vector is given by $\mathbf{y}(t) = [\theta(t); \theta^{(1)}(t)]^T$. We see from Eqs. (3.67) and (3.68) that the system is of third order. Hence, we need three state variables for the state-space representation of the system. Going by the desirable convention of choosing state variables to be physical quantities, let us select the state variables as $x_1(t) = \theta(t)$, $x_2(t) = \theta^{(1)}(t)$, and $x_3(t) = i(t)$. Then the state-equations can be written as follows:

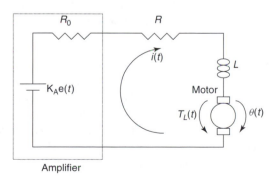

Figure 3.6 Amplifier-motor circuit of Example 3.7

$$x_1^{(1)}(t) = x_2(t) \tag{3.69}$$

$$x_2^{(1)}(t) = -(b/J)x_2(t) + (a/J)x_3(t) - T_\mathrm{L}(t)/J \tag{3.70}$$

$$x_3^{(1)}(t) = -(a/L)x_2(t) - [(R + R_0)/L]x_3(t) + (K_\mathrm{A}/L)e(t) \tag{3.71}$$

and the output equations are

$$\theta(t) = x_1(t) \tag{3.72}$$

$$\theta^{(1)(t)} = x_2(t) \tag{3.73}$$

In matrix form, the state-equation and output equations are written as Eqs. (3.33) and (3.34), respectively, with the state-vector, $\mathbf{x}(t) = [x_1(t); x_2(t); x_3(t)]^T$ and the following coefficient matrices:

$$\mathbf{A} = \begin{bmatrix} 0 & 1 & 0 \\ 0 & -b/J & a/J \\ 0 & -a/L & -(R + R_0)/L \end{bmatrix}; \quad \mathbf{B} = \begin{bmatrix} 0 & 0 \\ 0 & -1/J \\ K_\mathrm{A}/L & 0 \end{bmatrix};$$

$$\mathbf{C} = \begin{bmatrix} 1 & 0 & 0 \\ 0 & 1 & 0 \end{bmatrix}; \quad \mathbf{D} = \begin{bmatrix} 0 & 0 \\ 0 & 0 \end{bmatrix} \tag{3.74}$$

3.2 Linear Transformation of State-Space Representations

Since the state-space representation of a system is not unique, we can always find another state-space representation for the same system by the use of a *state transformation*. State transformation refers to the act of producing another state-space representation, starting from a given state-space representation. If a system is linear, the state-space representations also are linear, and the state transformation is a *linear transformation* in which the original state-vector is *pre-multiplied* by a *constant transformation matrix* yielding a new state-vector. Suppose $\mathbf{T}$ is such a *transformation matrix* for a linear system described by Eqs. (3.33) and (3.34). Let us find the new state-space representation in terms of $\mathbf{T}$ and the coefficient matrices, $\mathbf{A}$, $\mathbf{B}$, $\mathbf{C}$, $\mathbf{D}$. The transformed state-vector, $\mathbf{x}'(t)$, is expressed as follows:

$$\mathbf{x}'(t) = \mathbf{T}\mathbf{x}(t) \tag{3.75}$$

Equation (3.75) is called a *linear state-transformation* with transformation matrix, $\mathbf{T}$. Note that for a system of order n, $\mathbf{T}$ must be a square matrix of size $(n \times n)$, because order of the system remains unchanged in the transformation from $\mathbf{x}(t)$ to $\mathbf{x}'(t)$. Let us assume that it is possible to transform the new state-vector, $\mathbf{x}'(t)$, back to the original state-vector, $\mathbf{x}(t)$, with the use of the following *inverse transformation*:

$$\mathbf{x}(t) = \mathbf{T}^{-1}\mathbf{x}'(t) \tag{3.76}$$

Equation (3.76) requires that the inverse of the transformation matrix, $\mathbf{T}^{-1}$, should *exist* (in other words, $\mathbf{T}$ should be *nonsingular*). Equation (3.70) is obtained by *pre-multiplying* both sides of Eq. (3.75) by $\mathbf{T}^{-1}$, and noting that $\mathbf{T}\,\mathbf{T}^{-1} = \mathbf{I}$. To find the transformed state-equation, let us differentiate Eq. (3.76) with time and substitute the result, $\mathbf{x}^{(1)}(t) = \mathbf{T}^{-1}\mathbf{x}'^{(1)}(t)$, along with Eq. (3.76), into Eq. (3.33), thereby yielding

$$\mathbf{T}^{-1}\mathbf{x}'^{(1)}(t) = \mathbf{A}\mathbf{T}^{-1}\mathbf{x}'(t) + \mathbf{B}\mathbf{u}(t) \tag{3.77}$$

Pre-multiplying both sides of Eq. (3.77) by $\mathbf{T}$, we get the transformed state-equation as follows:

$$\mathbf{x}'^{(1)}(t) = \mathbf{T}\mathbf{A}\mathbf{T}^{-1}\mathbf{x}'(t) + \mathbf{T}\mathbf{B}\mathbf{u}(t) \tag{3.78}$$

We can write the transformed state-equation Eq. (3.78) in terms of the new coefficient matrices $\mathbf{A}'$, $\mathbf{B}'$, $\mathbf{C}'$, $\mathbf{D}'$, as follows:

$$\mathbf{x}'^{(1)}(t) = \mathbf{A}'\mathbf{x}'(t) + \mathbf{B}'\mathbf{u}(t) \tag{3.79}$$

where $\mathbf{A}' = \mathbf{T}\mathbf{A}\mathbf{T}^{-1}$, and $\mathbf{B}' = \mathbf{T}\mathbf{B}$. Similarly, substituting Eq. (3.76) into Eq. (3.34) yields the following transformed output equation:

$$\mathbf{y}(t) = \mathbf{C}'\mathbf{x}'(t) + \mathbf{D}'\mathbf{u}(t) \tag{3.80}$$

where $\mathbf{C}' = \mathbf{C}\mathbf{T}^{-1}$, and $\mathbf{D}' = \mathbf{D}$.

There are several reasons for transforming one state-space representation into another, such as the utility of a particular form of state-equations in control system design (the *controller* or *observer companion form*), the requirement of transforming the state variables into those that are *physically meaningful* in order to implement a control system, and sometimes, the need to *decouple* the state-equations so that they can be easily solved. We will come across such state-transformations in the following chapters.

Example 3.8

We had obtained two different state-space representations for the same electrical network in Examples 3.4 and 3.5. Let us find the state-transformation matrix, $\mathbf{T}$, which transforms the state-space representation given by Eqs. (3.42) and (3.43) to that given by Eqs. (3.51) and (3.52), respectively. In this case, the original state-vector is $\mathbf{x}(t) = [i_2(t); i_2^{(1)}(t) - R_3 e(t)/\{L(R_1 + R_3)\}]^T$, whereas the transformed state-vector is $\mathbf{x}'(t) = [LC\{e(t) - i_2(t)(R_1 R_3 + R_1 R_2 + R_2 R_3)/R_3 - Li_2^{(1)}(t)$ $(R_1 + R_3)/R_3\}; i_2(t)L(R_1 + R_3)/R_3]^T$. The state-transformation matrix, $\mathbf{T}$, is of size (2×2). From Eq. (3.69), it follows that

$$\begin{bmatrix} LC\{e(t) - i_2(t)(R_1 R_3 + R_1 R_2 + R_2 R_3)/R_3 - Li_2^{(1)}(t)(R_1 + R_3)/R_3\} \\ i_2(t)L(R_1 + R_3)/R_3 \end{bmatrix}$$

$$= \begin{bmatrix} T_{11} & T_{12} \\ T_{21} & T_{22} \end{bmatrix} \begin{bmatrix} i_2(t) \\ i_2^{(1)}(t) - R_3 e(t)/\{L(R_1 + R_3)\} \end{bmatrix} \tag{3.81}$$

where T_{11}, T_{12}, T_{21}, and T_{22} are the *unknown* elements of **T**. We can write the following *two* scalar equations out of the matrix equation, Eq. (3.81):

$$LCe(t) - LCi_2(t)(R_1R_3 + R_1R_2 + R_2R_3)/R_3 - L^2Ci_2^{(1)}(t)(R_1 + R_3)/R_3$$

$$= T_{11}i_2(t) + T_{12}[i_2^{(1)}(t) - R_3e(t)/\{L(R_1 + R_3)\}] \tag{3.82}$$

$$i_2(t)L(R_1 + R_3)/R_3 = T_{21}i_2(t) + T_{22}[i_2^{(1)}(t) - R_3e(t)/\{L(R_1 + R_3)\}] \tag{3.83}$$

Equating the coefficients of $i_2(t)$ on both sides of Eq. (3.82), we get

$$T_{11} = -LC(R_1R_3 + R_1R_2 + R_2R_3)/R_3 \tag{3.84}$$

Equating the coefficients of $e(t)$ on both sides of Eq. (3.82), we get

$$T_{12} = -L^2C(R_1 + R_3)/R_3 \tag{3.85}$$

Note that the same result as Eq. (3.85) is obtained if we equate the coefficients of $i_2^{(1)}(t)$ on both sides of Eq. (3.82). Similarly, equating the coefficients of corresponding variables on both sides of Eq. (3.83) we get

$$T_{21} = L(R_1 + R_3)/R_3 \tag{3.86}$$

and

$$T_{22} = 0 \tag{3.87}$$

Therefore, the required state-transformation matrix is

$$\mathbf{T} = \begin{bmatrix} -LC(R_1R_3 + R_1R_2 + R_2R_3)/R_3 & -L^2C(R_1 + R_3)/R_3 \\ L(R_1 + R_3)/R_3 & 0 \end{bmatrix} \tag{3.88}$$

With the transformation matrix of Eq. (3.88), you may verify that the state-space coefficient matrices of Example 3.5 are related to those of Example 3.4 according to Eqs. (3.79) and (3.80).

Example 3.9

For a linear, time-invariant state-space representation, the coefficient matrices are as follows:

$$\mathbf{A} = \begin{bmatrix} 1 & 2 \\ -3 & -1 \end{bmatrix}; \quad \mathbf{B} = \begin{bmatrix} 1 & 0 \\ 0 & 1 \end{bmatrix}; \quad \mathbf{C} = [1 \quad 2]; \quad \mathbf{D} = [0 \quad 0] \tag{3.89}$$

If the state-transformation matrix is the following:

$$\mathbf{T} = \begin{bmatrix} -1 & 1 \\ -1 & -1 \end{bmatrix} \tag{3.90}$$

let us find the transformed state-space representation. The first thing to do is to check whether **T** is singular. The determinant of **T**, $|\mathbf{T}| = 2$. Hence, **T** is nonsingular and its inverse can be calculated as follows (for the definitions of *determinant*, *inverse*, and other matrix operations see Appendix B):

$$\mathbf{T}^{-1} = \mathrm{adj}(\mathbf{T})/|\mathbf{T}| = (1/2)\begin{bmatrix} -1 & 1 \\ -1 & -1 \end{bmatrix}^T = (1/2)\begin{bmatrix} -1 & -1 \\ 1 & -1 \end{bmatrix} \tag{3.91}$$

Then the transformed state coefficient matrices, $\mathbf{A}'$, $\mathbf{B}'$, $\mathbf{C}'$, $\mathbf{D}'$, of Eqs. (3.79) and (3.80) are then calculated as follows:

$$\mathbf{A}' = \mathbf{TAT}^{-1} = (1/2)\begin{bmatrix} -1 & 1 \\ -1 & -1 \end{bmatrix}\begin{bmatrix} 1 & 2 \\ -3 & -1 \end{bmatrix}\begin{bmatrix} -1 & -1 \\ 1 & -1 \end{bmatrix} = \begin{bmatrix} 1/2 & 7/2 \\ -3/2 & -1/2 \end{bmatrix} \tag{3.92}$$

$$\mathbf{B}' = \mathbf{TB} = \begin{bmatrix} -1 & 1 \\ -1 & -1 \end{bmatrix}\begin{bmatrix} 1 & 0 \\ 0 & 1 \end{bmatrix} = \begin{bmatrix} -1 & 1 \\ -1 & -1 \end{bmatrix} \tag{3.93}$$

$$\mathbf{C}' = \mathbf{CT}^{-1} = (1/2)[1 \quad 2]\begin{bmatrix} -1 & -1 \\ 1 & -1 \end{bmatrix} = [1/2 \quad -3/2] \tag{3.94}$$

$$\mathbf{D}' = \mathbf{D} = [0 \quad 0] \tag{3.95}$$

It will be appreciated by anyone who has tried to invert, multiply, or find the determinant of matrices of size larger than (2×2) by hand, that doing so can be a tedious process. Such calculations are easily done using MATLAB, as the following example will illustrate.

Example 3.10

Consider the following state-space representation of the linearized longitudinal dynamics of an aircraft depicted in Figure 2.25:

$$\begin{bmatrix} v^{(1)}(t) \\ \alpha^{(1)}(t) \\ \theta^{(1)}(t) \\ q^{(1)}(t) \end{bmatrix} = \begin{bmatrix} -0.045 & 0.036 & -32 & -2 \\ -0.4 & -3 & -0.3 & 250 \\ 0 & 0 & 0 & 1 \\ 0.002 & -0.04 & 0.001 & -3.2 \end{bmatrix}\begin{bmatrix} v(t) \\ \alpha(t) \\ \theta(t) \\ q(t) \end{bmatrix}$$

$$+ \begin{bmatrix} 0 & 0.1 \\ -30 & 0 \\ 0 & 0 \\ -10 & 0 \end{bmatrix}\begin{bmatrix} \delta(t) \\ \mu(t) \end{bmatrix} \tag{3.96}$$

$$\begin{bmatrix} y_1(t) \\ y_2(t) \end{bmatrix} = \begin{bmatrix} 0 & 0 & 1 & 0 \\ 0 & 0 & 0 & 1 \end{bmatrix}\begin{bmatrix} v(t) \\ \alpha(t) \\ \theta(t) \\ q(t) \end{bmatrix} + \begin{bmatrix} 0 & 0 \\ 0 & 0 \end{bmatrix}\begin{bmatrix} \delta(t) \\ \mu(t) \end{bmatrix} \tag{3.97}$$

where the *elevator deflection*, $\delta(t)$ (Figure 2.25) and *throttle position*, $\mu(t)$ (not shown in Figure 2.25) are the two inputs, whereas the change in the *pitch angle*, $\theta(t)$, and the *pitch-rate*, $q(t) = \theta^{(1)}(t)$, are the two outputs. The state-vector selected to represent the dynamics in Eqs. (3.96) and (3.97) is $\mathbf{x}(t) = [v(t); \alpha(t); \theta(t); q(t)]^T$, where $v(t)$ represents a change in the *forward speed*, and $\alpha(t)$ is the change in the *angle of attack*. All the changes are measured from an initial equilibrium state of the aircraft given by $\mathbf{x}(0) = \mathbf{0}$. Let us transform the state-space representation using the following transformation matrix:

$$\mathbf{T} = \begin{bmatrix} 1 & -1 & 0 & 0 \\ 0 & 0 & -2 & 0 \\ -3.5 & 1 & 0 & -1 \\ 0 & 0 & 2.2 & 3 \end{bmatrix} \tag{3.98}$$

You may verify that $\mathbf{T}$ is nonsingular by finding its determinant by hand, or using the MATLAB function *det*. The state-transformation can be easily carried out using the intrinsic MATLAB functions as follows:

```
>>A=[-0.045 0.036 -32 -2; -0.4 -3 -0.3 250; 0 0 0 1; 0.002 -0.04 0.001
   -3.2]; <enter>

>>B=[0 0.1;-30 0;0 0;-10 0]; C=[0 0 1 0; 0 0 0 1]; D=zeros(2,2); <enter>

>>T=[1 -1 0 0; 0 0 -2 0; -3.5 1 0 -1; 0 0 2.2 3]; <enter>

>>Aprime=T*A*inv(T), Bprime=T*B, Cprime=C*inv(T) <enter>

Aprime =

      -4.3924      -77.0473       -1.3564      -84.4521
       0            -0.7333        0            -0.6667
       4.4182        40.0456       1.3322        87.1774
       0.1656        -2.6981       0.0456        -2.4515


Bprime =
      30.0000        0.1000
       0             0
     -20.0000       -0.3500
     -30.0000        0

Cprime =
       0           -0.5000       0           0
       0            0.3667       0           0.3333
```

and $\mathbf{D}'$ is, of course, just $\mathbf{D}$. The transformed state coefficient matrices can be obtained in one step by using the MATLAB Control System Toolbox (CST) command *ss2ss*. First, a state-space LTI object is created using the function *ss* as follows:

```
>> sys1=ss(A,B,C,D) <enter>

a =
                  x1            x2            x3         x4
           x1     -0.045        0.036         -32        -2
           x2     -0.4          -3            -0.3       250
```

```
        x3          0            0            0    1
        x4          0.002       -0.04         0.001 -3.2

b =
                    u1           u2
        x1          0            0.1
        x2         -30           0
        x3          0            0
        x4        -10            0

c =
                    x1     x2     x3     x4
        y1          0      0      1      0
        y2          0      0      0      1

d =
                    u1           u2
        y1          0            0
        y2          0            0
```

Continuous-time model.

Then, the function *ss2ss* is used to transform the LTI object, *sys1*, to another state-space representation, *sys2*:

```
>>sys2 = ss2ss (sys1,T) <enter>

a =
                    x1           x2           x3          x4
        x1         -4.3924      -77.047      -1.3564     -84.452
        x2          0           -0.73333      0          -0.66667
        x3          4.4182       40.046        1.3322     87.177
        x4          0.1656       -2.6981       0.0456     -2.4515

b =
                    u1           u2
        x1          30           0.1
        x2          0            0
        x3        -20           -0.35
        x4        -30            0

c =
                    x1           x2           x3          x4
        y1          0           -0.5          0           0
        y2          0            0.36667      0           0.33333

d =
                    u1     u2
        y1          0      0
        y2          0      0
```

Continuous-time model.

Since a system's characteristics do not change when we express the same system by different state-space representations, the linear state transformations are also called *similarity transformations*. Let us now see how we can obtain information about a system's characteristics – locations of poles, performance, stability, etc. – from its state-space representation.

3.3 System Characteristics from State-Space Representation

In Chapter 2, we defined the characteristics of a system by its characteristic equation, whose roots are the poles of the system. We also saw how the locations of the poles indicate a system's performance – such as natural frequency, damping factor, system type – as well as whether the system is stable. Let us see how a system's characteristic equation can be derived from its state-space representation.

The characteristic equation was defined in Chapter 2 to be the denominator polynomial of the system's transfer function (or transfer matrix) equated to zero. Hence, we should first obtain an expression for the transfer matrix in terms of the state-space coefficient matrices, $\mathbf{A}$, $\mathbf{B}$, $\mathbf{C}$, $\mathbf{D}$. Recall that the transfer matrix is obtained by taking the Laplace transform of the governing differential equations, for zero initial conditions. Taking the Laplace transform of both sides of the matrix state-equation, Eq. (3.33), assuming zero initial conditions (i.e. $\mathbf{x}(0) = \mathbf{0}$) yields the following result:

$$s\mathbf{X}(s) = \mathbf{A}\mathbf{X}(s) + \mathbf{B}\mathbf{U}(s) \tag{3.99}$$

where $\mathbf{X}(s) = \mathcal{L}[\mathbf{x}(t)]$, and $\mathbf{U}(s) = \mathcal{L}[\mathbf{u}(t)]$. Rearranging Eq. (3.99), we can write

$$(s\mathbf{I} - \mathbf{A})\mathbf{X}(s) = \mathbf{B}\mathbf{U}(s) \tag{3.100}$$

or

$$\mathbf{X}(s) = (s\mathbf{I} - \mathbf{A})^{-1}\mathbf{B}\mathbf{U}(s) \tag{3.101}$$

Similarly, taking the Laplace transform of the output equation, Eq. (3.34), with $\mathbf{Y}(s) = \mathcal{L}[\mathbf{y}(t)]$, yields

$$\mathbf{Y}(s) = \mathbf{C}\mathbf{X}(s) + \mathbf{D}\mathbf{U}(s) \tag{3.102}$$

Substituting Eq. (3.101) into Eq. (3.102) we get

$$\mathbf{Y}(s) = \mathbf{C}(s\mathbf{I} - \mathbf{A})^{-1}\mathbf{B}\mathbf{U}(s) + \mathbf{D}\mathbf{U}(s) = [\mathbf{C}(s\mathbf{I} - \mathbf{A})^{-1}\mathbf{B} + \mathbf{D}]\mathbf{U}(s) \tag{3.103}$$

From Eq. (3.103), it is clear that the transfer matrix, $\mathbf{G}(s)$, defined by $\mathbf{Y}(s) = \mathbf{G}(s)\mathbf{U}(s)$, is the following:

$$\mathbf{G}(s) = \mathbf{C}(s\mathbf{I} - \mathbf{A})^{-1}\mathbf{B} + \mathbf{D} \tag{3.104}$$

Equation (3.104) tells us that the transfer matrix is a sum of the *rational matrix* (i.e. a matrix whose elements are ratios of polynomials in s), $\mathbf{C}(s\mathbf{I} - \mathbf{A})^{-1}\mathbf{B}$, and the matrix **D**. Thus, **D** represents a *direct connection* between the input, $\mathbf{U}(s)$, and the output, $\mathbf{Y}(s)$, and is called the *direct transmission matrix*. Systems having $\mathbf{D} = \mathbf{0}$ are called *strictly proper*, because the numerator polynomials of the elements of $\mathbf{G}(s)$ are smaller in degree than the corresponding denominator polynomials (see the discussion following Eq. (2.1) for the definition of strictly proper single variable systems). In Example 2.28, we had obtained the characteristic equation of a multivariable system from the denominator polynomial of $\mathbf{G}(s)$. Hence, the characteristic polynomial of the system must be related to the denominator polynomial resulting from the matrix, $\mathbf{C}(s\mathbf{I} - \mathbf{A})^{-1}\mathbf{B}$.

Example 3.11

For a linear system described by the following state coefficient matrices, let us determine the transfer function and the characteristic equation:

$$\mathbf{A} = \begin{bmatrix} 1 & 2 \\ -2 & 1 \end{bmatrix}; \quad \mathbf{B} = \begin{bmatrix} 1 \\ 0 \end{bmatrix}; \quad \mathbf{C} = [1 \quad 1]; \quad \mathbf{D} = 0 \qquad (3.105)$$

The inverse $(s\mathbf{I} - \mathbf{A})^{-1}$ is calculated as follows:

$$(s\mathbf{I} - \mathbf{A})^{-1} = \mathrm{adj}(s\mathbf{I} - \mathbf{A})/|(s\mathbf{I} - \mathbf{A})| \qquad (3.106)$$

where the *determinant* $|(s\mathbf{I} - \mathbf{A})|$ is given by

$$|(s\mathbf{I} - \mathbf{A})| = \begin{bmatrix} (s-1) & -2 \\ 2 & (s-1) \end{bmatrix} = (s-1)^2 + 4 = s^2 - 2s + 5 \qquad (3.107)$$

and the *adjoint*, $\mathrm{adj}(s\mathbf{I} - \mathbf{A})$, is given by

$$\mathrm{adj}(s\mathbf{I} - \mathbf{A}) = \begin{bmatrix} (s-1) & -2 \\ 2 & (s-1) \end{bmatrix}^T = \begin{bmatrix} (s-1) & 2 \\ -2 & (s-1) \end{bmatrix} \qquad (3.108)$$

(See Appendix B for the definitions of the inverse, adjoint, and determinant.) Substituting Eqs. (3.107) and (3.108) into Eq. (3.106) we get

$$(s\mathbf{I} - \mathbf{A})^{-1} = [1/(s^2 - 2s + 5)]\begin{bmatrix} (s-1) & 2 \\ -2 & (s-1) \end{bmatrix} \qquad (3.109)$$

Then the transfer matrix is calculated as follows:

$$\mathbf{C}(s\mathbf{I} - \mathbf{A})^{-1} = [1/(s^2 - 2s + 5)][1 \quad 1]\begin{bmatrix} (s-1) & 2 \\ -2 & (s-1) \end{bmatrix}$$

$$= [1/(s^2 - 2s + 5)][(s-3) \quad (s+1)] \qquad (3.110)$$

$$\mathbf{C}(s\mathbf{I} - \mathbf{A})^{-1}\mathbf{B} = [1/(s^2 - 2s + 5)][(s-3); \quad (s+1)]\begin{bmatrix} 1 \\ 0 \end{bmatrix}$$

$$= (s-3)/(s^2 - 2s + 5) \qquad (3.111)$$

The conversion of a system's state-space representation into its transfer matrix is easily carried out with the MATLAB Control System Toolbox's (CST) LTI object function *tf* as follows:

```
>>sys = tf(sys) <enter>
```

Example 3.12

Let us convert the state-space representation of the aircraft longitudinal dynamics (Example 3.10) given by Eqs. (3.96) and (3.97) into the transfer matrix of the system, as follows:

```
>>sys1=tf(sys1) <enter>

Transfer function from input 1 to output...
            -10s^2-29.25s-1.442
#1: ---------------------------
    s^4+6.245s^3+19.9s^2+1.003s+0.7033


    -10s^3-29.25s^2-1.442s-5.293e-016
#2: ---------------------------
    s^4+6.245s^3+19.9s^2+1.003s+0.7033


Transfer function from input 2 to output...
            0.0002s+0.0022
#1: ---------------------------
    s^4+6.245s^3+19.9s^2+1.003s+0.7033


    0.0002s^2+0.0022s+3.671e-019
#2: ---------------------------
    s^4+6.245s^3+19.9s^2+1.003s+0.7033
```

Note that the single-input, single-output system of Example 3.11, the transfer function has a denominator polynomial $s^2 - 2s + 5$, which is also the *characteristic polynomial* of the system (see Chapter 2). The denominator polynomial is equal to $|(s\mathbf{I} - \mathbf{A})|$ (Eq. (3.107)). Thus, the poles of the transfer function are the roots of the characteristic equation, $|(s\mathbf{I} - \mathbf{A})| = 0$. This is also true for the multivariable system of Example 3.12, where all the elements of the transfer matrix have the same denominator polynomial. Using linear algebra, the characteristic equation of a general, linear time-invariant system is obtained from the following *eigenvalue problem* for the system:

$$\mathbf{A}\mathbf{v}_k = \lambda_k\mathbf{v}_k \qquad (3.112)$$

where λ_k is the kth *eigenvalue* of the matrix $\mathbf{A}$, and $\mathbf{v}_k$ is the *eigenvector* associated with the eigenvalue, λ_k (see Appendix B). Equation (3.112) can be written as follows:

$$(\lambda\mathbf{I} - \mathbf{A})\mathbf{v} = \mathbf{0} \qquad (3.113)$$

For the nontrivial solution of Eq. (3.113) (i.e. $\mathbf{v} \neq \mathbf{0}$), the following must be true:

$$|(\lambda \mathbf{I} - \mathbf{A})| = 0 \qquad (3.114)$$

Equation (3.114) is another way of writing the characteristic equation, whose the roots are the eigenvalues, λ. Hence, the *poles of the transfer matrix* are the *same as the eigenvalues of the matrix*, $\mathbf{A}$. Since $\mathbf{A}$ contains information about the characteristic equation of a system, it influences all the properties such as stability, performance and robustness of the system. For this reason, $\mathbf{A}$ is called the system's *state-dynamics matrix*.

Example 3.13

For the state-space representation of the electrical network derived in Examples 3.4 and 3.5, let us substitute the numerical values from Example 2.8 ($R_1 = R_3 = 10$ ohms, $R_2 = 25$ ohms, $L = 1$ henry, $C = 10^{-6}$ farad) and calculate the transfer functions and eigenvalues for both the state-space representations.

The state-space representation of Example 3.4 yields the following state coefficient matrices:

$$\mathbf{A} = \begin{bmatrix} 0 & 1 \\ -10^6 & -30 \end{bmatrix}; \quad \mathbf{B} = \begin{bmatrix} 0.5 \\ -15 \end{bmatrix}; \quad \mathbf{C} = [-0.5 \quad 0]; \quad \mathbf{D} = 0.05 \quad (3.115)$$

while the state-space representation of Example 3.5 has the following coefficient matrices:

$$\mathbf{A}' = \begin{bmatrix} 0 & 1 \\ -10^6 & -30 \end{bmatrix}; \quad \mathbf{B}' = \begin{bmatrix} 0 \\ 1 \end{bmatrix}; \quad \mathbf{C}' = [0 \quad -0.25]; \quad \mathbf{D}' = 0.05$$
$$(3.116)$$

Either using Eq. (3.104) by hand, or using the CST LTI object function, *tf*, we can calculate the respective transfer functions as follows:

```
>>A=[0 1;-1e6 -30]; B=[0.5; -15]; C=[-0.5 0]; D=0.05;sys1=ss(A,B,C,D);
  sys1=tf(sys1) <enter>


Transfer function:
0.05s^2+1.25s+5e004
-----------------
  s^2+30s+1e006


>> A=[0 1;-1e6 -30]; B=[0; 1]; C=[0 -0.25]; D=0.05;sys2=ss(A,B,C,D);
  sys2=tf(sys1) <enter>


Transfer function:
0.05s^2+1.25s+5e004
-----------------
  s^2+30s+1e006
```

Note that the two transfer functions are identical, as expected, because the two state-space representations are for the *same system*. The characteristic equation is obtained by equating the denominator polynomial to zero, i.e. $s^2 + 30s + 10^6 = 0$. Solving the characteristic equation, we get the poles of the system as follows:

```
>>roots([1 30 1e6]) <enter>

ans =
-1.5000e+001 +9.9989e+002i
-1.5000e+001 -9.9989e+002i
```

which agree with the result of Example 2.8. These poles should be the same as the eigenvalues of the matrix, $\mathbf{A}(= \mathbf{A}')$, obtained using the intrinsic MATLAB function *eig* as follows:

```
>>eig([0 1;-1e6 -30]) <enter>

ans =
-1.5000e+001 +9.9989e+002i
-1.5000e+001 -9.9989e+002i
```

Example 3.13 shows that the system's characteristics are unchanged by using different state-space representations.

Example 3.14

Consider the following two-input, two-output *turbo-generator* system [2]:

$$\mathbf{A} = \begin{bmatrix} -18.4456 & 4.2263 & -2.2830 & 0.2260 & 0.4220 & -0.0951 \\ -4.0977 & -6.0706 & 5.6825 & -0.6966 & -1.2246 & 0.2873 \\ 1.4449 & 1.4336 & -2.6477 & 0.6092 & 0.8979 & -0.2300 \\ -0.0093 & 0.2302 & -0.5002 & -0.1764 & -6.3152 & 0.1350 \\ -0.0464 & -0.3489 & 0.7238 & 6.3117 & -0.6886 & 0.3645 \\ -0.0602 & -0.2361 & 0.2300 & 0.0915 & -0.3214 & -0.2087 \end{bmatrix}$$

$$\mathbf{B} = \begin{bmatrix} -0.2748 & 3.1463 \\ -0.0501 & -9.3737 \\ -0.1550 & 7.4296 \\ 0.0716 & -4.9176 \\ -0.0814 & -10.2648 \\ 0.0244 & 13.7943 \end{bmatrix}$$

$$\mathbf{C} = \begin{bmatrix} 0.5971 & -0.7697 & 4.8850 & 4.8608 & -9.8177 & -8.8610 \\ 3.1013 & 9.3422 & -5.6000 & -0.7490 & 2.9974 & 10.5719 \end{bmatrix}$$

$$\mathbf{D} = \begin{bmatrix} 0 & 0 \\ 0 & 0 \end{bmatrix} \tag{3.117}$$

The eigenvalues and the associated natural frequencies and damping factors of the system are found by using the MATLAB command *damp(A)* as follows:

```
>>damp(A) <enter>

Eigenvalue                      Damping          Freq. (rad/sec)
-2.3455e-001                    1.0000e+000      2.3455e-001
-3.4925e-001+6.3444e+000i       5.4966e-002      6.3540e+000
-3.4925e-001-6.3444e+000i       5.4966e-002      6.3540e+000
-1.0444e+000                    1.0000e+000      1.0444e+000
-1.0387e+001                    1.0000e+000      1.0387e+001
-1.5873e+001                    1.0000e+000      1.5873e+001
```

Note that there are four real eigenvalues, and a pair of complex conjugate eigenvalues. All the eigenvalues (i.e. poles) have negative real parts, implying an *asymptotically stable system* from the stability criteria of Chapter 2. Also from Chapter 2, the damping factors associated with all real eigenvalues with negative real parts are 1.0, since such eigenvalues represent *exponentially decaying* responses. Only complex conjugate eigenvalues have damping factors less than 1.0. These eigenvalues represent an oscillatory response. If it were possible to *decouple* the state-equations by the use of a state transformation, such that the each of the transformed state-equations is in terms of *only one* state variable, then each eigenvalue would represent a particular *mode* in which the system can respond. Hence, there are *six modes* in this sixth order system, consisting of four real (or *first order*) modes, and a *second order* mode defined by a pair of complex conjugate eigenvalues. Note that the second order mode has a relatively small damping factor (0.055). The transfer matrix, $\mathbf{G}(s)$, defined by $\mathbf{Y}(s) = \mathbf{G}(s)\mathbf{U}(s)$, of this two-input, two-output system, is written as follows:

$$\mathbf{G}(s) = [\mathbf{G}_1(s) \ \mathbf{G}_2(s)] \tag{3.118}$$

where $\mathbf{Y}(s) = \mathbf{G}_1(s)U_1(s) + \mathbf{G}_2(s)U_2(s)$, with $U_1(s)$ and $U_2(s)$ being the two inputs. $\mathbf{G}_1(s)$ and $\mathbf{G}_2(s)$ are usually obtained using the CST LTI objects *ss* and *tf* as follows:

```
>>syst=ss(A,B,C,D); syst=tf(syst) <enter>

Transfer function from input 1 to output...
    0.04829s^5+1.876s^4+1.949s^3-1228s^2-5762s-2385
#1:---------------------------------------------------
    s^6+28.24s^5+258.3s^4+1468s^3+8214s^2+8801s+1631


    -0.4919s^5+9.483s^4-49.05s^3+551.6s^2-939.6s+907.6
#2:---------------------------------------------------
    s^6+28.24s^5+258.3s^4+1468s^3+8214s^2+8801s+1631


Transfer function from input 2 to output...
    0.02915s^5+1.289s^4-0.3041s^3-2.388e004s^2-8.29e005s
                        -9.544e005
```

```
#1:-------------------------------------------------------
     s^6+28.24s^5+258.3s^4+1468s^3+8214s^2+8801s+1631

     -0.6716s^5+804.5s^4+2.781e004s^3+8.085e004s^2+1.214e006s
                                                +1.082e006
#2:-------------------------------------------------------
     s^6+28.24s^5+258.3s^4+1468s^3+8214s^2+8801s+1631
```

Therefore,

$\mathbf{G}_1(s)$

$$= \begin{bmatrix} (0.04829s^5 + 1.876s^4 + 1.949s^3 - 1228s^2 - 5762s - 2385)/d(s) \\ (-0.4919s^5 + 9.483s^4 - 49.05s^3 + 551.6s^2 - 939.6s + 907.6)/d(s) \end{bmatrix}$$

$\mathbf{G}_2(s)$

$$= \begin{bmatrix} (0.02915s^5 + 1.289s^4 - 0.3041s^3 - 2.388 \times 10^4 s^2 - 8.29 \times 10^5 s - 9.544 \times 10^5)/d(s) \\ (-0.6716s^5 + 804.5s^4 + 27810s^3 + 8085s^2 + 1.214 \times 10^6 s + 1.082 \times 10^6)/d(s) \end{bmatrix}$$

$$(3.119)$$

where

$$d(s) = (s^6 + 28.24s^5 + 258.3s^4 + 1468s^3 + 8214s^2 + 8801s + 1631) \quad (3.120)$$

For brevity, the coefficients have been rounded off to four significant digits. If the MATLAB's *long* format is used to report the results, a greater accuracy is possible. Note that all the elements of the transfer matrix $\mathbf{G}(s)$ have a common denominator polynomial, whose roots are the poles (or eigenvalues) of the system. This is confirmed by using the coefficients of $d(s)$ rounded to five significant digits with the intrinsic MATLAB function *roots* as follows:

```
>>roots([1 28.238 258.31 1467.9 8214.5 8801.2 1630.6]) <enter>

ans =
-15.8746
-10.3862
-0.3491+6.3444i
-0.3491-6.3444i
-1.0444
-0.2346
```

3.4 Special State-Space Representations: The Canonical Forms

In Section 3.1 we saw how some special state-space representations can be obtained, such as the controller and observer companion forms. The companion forms are members of a special set of state-space representations, called *canonical forms*. In addition to the companion forms, another canonical form is the *Jordan canonical form*, which is derived from the partial fraction expansion of the system's transfer matrix as described below. In Section 3.3, we saw how the transfer matrix can be obtained from a state-space representation. Now we will address the inverse problem, namely deriving special state-space representations from the transfer matrix. For simplicity, consider a single-input, single-output system with the transfer function given by the following partial fraction expansion:

$$Y(s)/U(s) = k_0 + k_1/(s - p_1) + k_2/(s - p_2) + \cdots + k_i/(s - p_i) + k_{i+1}/(s - p_i)^2$$
$$+ \cdots + k_{i+m}/(s - p_i)^{m+1} + \cdots + k_n/(s - p_n) \tag{3.121}$$

where n is the order of the system, and all poles, except $s = p_i$, are simple poles. The pole, $s = p_i$, is of multiplicity $(m + 1)$, i.e. $(s - p_i)$ occurs as a power $(m + 1)$ in the transfer function. Let us select the state-variables of the system as follows:

$$X_1(s)/U(s) = 1/(s - p_1); \; X_2(s)/U(s) = 1/(s - p_2); \ldots;$$
$$X_i(s)/U(s) = 1/(s - p_i); \; X_{i+1}(s)/U(s) = 1/(s - p_i)^2; \ldots;$$
$$X_{i+m}(s)/U(s) = 1/(s - p_i)^{m+1}; \ldots; \; X_n(s)/U(s) = 1/(s - p_n) \tag{3.122}$$

Taking the inverse Laplace transform of Eq. (3.122), we get the following state-equations:

$$x_1^{(1)}(t) = p_1 x_1(t) + u(t)$$
$$x_2^{(1)}(t) = p_2 x_2(t) + u(t)$$
$$\vdots$$
$$x_i^{(1)}(t) = p_i x_i(t) + u(t)$$
$$x_{i+1}^{(1)}(t) = x_i(t) + p_i x_{i+1}(t)$$
$$\vdots$$
$$x_{i+m}^{(1)}(t) = x_{i+m-1}(t) + p_i x_{i+m}(t) \tag{3.123}$$
$$\vdots$$
$$x_n^{(1)}(t) = p_n x_n(t) + u(t)$$

and the output equation is given by

$$y(t) = k_1 x_1(t) + k_2 x_2(t) + \cdots + k_i x_i(t) + k_{i+1} x_{i+1}(t)$$
$$+ \cdots + k_{i+m} x_{i+m}(t) + \cdots + k_n x_n(t) + k_0 u(t) \tag{3.124}$$

(Note that in deriving the state-equations corresponding to the repeated pole, $s = p_i$, we have used the relationship $X_{i+1}(s) = X_i/(s - p_i)$, $X_{i+2}(s) = X_{i+1}/(s - p_i)$, and so on.) The state-space representation given by Eqs. (3.123) and (3.124) is called the *Jordan canonical form*. The state coefficient matrices of the Jordan canonical form are, thus, the following:

$$
\mathbf{A} =
\begin{bmatrix}
p_1 & 0 & \cdots & 0 & 0 & 0 & \cdots & 0 & 0 & \cdots & 0 \\
0 & p_2 & \cdots & 0 & 0 & 0 & \cdots & 0 & 0 & \cdots & 0 \\
0 & 0 & \cdots & 0 & 0 & 0 & \cdots & 0 & 0 & \cdots & 0 \\
 & & \cdot & & \cdot & & \cdot & & \cdot & & \cdot \\
0 & 0 & \cdots & p_i & 0 & 0 & \cdots & 0 & 0 & \cdots & 0 \\
0 & 0 & \cdots & 1 & p_i & 0 & \cdots & 0 & 0 & \cdots & 0 \\
0 & 0 & \cdots & 0 & 1 & p_i & \cdots & 0 & 0 & \cdots & 0 \\
 & & \cdot & & \cdot & & \cdot & & \cdot & & \cdot \\
0 & 0 & \cdots & 0 & 0 & 0 & \cdots & 1 & p_i & \cdots & 0 \\
 & & \cdot & & \cdot & & \cdot & & \cdot & & \cdot \\
0 & 0 & \cdots & 0 & 0 & 0 & \cdots & 0 & 0 & \cdots & p_n
\end{bmatrix}
$$

$$
\mathbf{B} = \begin{bmatrix} 1 & 1 & \cdots & 1 & 0 & \cdots & 0 & 1 & \cdots & 1 \end{bmatrix}^T
$$

$$
\mathbf{C} = \begin{bmatrix} k_1 & k_2 & \cdots & k_i & k_{i+1} & \cdots & k_{i+m} & \cdots & k_n \end{bmatrix}
$$

$$
\mathbf{D} = k_0 \tag{3.125}
$$

Note the particular structure of the **A** matrix in Eq. (3.125). The system's poles (i.e. the eigenvalues of **A**) occur on the main diagonal of **A**, with the repeated pole, p_i, occurring as many times as the multiplicity of the pole. A square block associated with the repeated pole, p_i, is marked by a dashed border, and is known as the *Jordan block* of pole, p_i. The diagonal *below* the main diagonal – called the *subdiagonal* – of this block has all elements equal to 1. All other elements of **A** are zeros. The matrix **B** also has a particular structure: all elements associated with the simple poles are *ones*, the first element of the Jordan block of repeated pole, p_i – shown in dashed border – is one, while the remaining elements of the Jordan block of p_i are zeros. The elements of matrix **C** are simply the residues corresponding to the poles in the partial fraction expansion (Eq. (1.121)), while the matrix **D** is equal to the direct term, k_0, in the partial fraction expansion. If a system has more than one repeated poles, then there is a Jordan block associated with each repeated pole of the same structure as in Eq. (3.125). If none of the poles are repeated, then **A** is a diagonal matrix. The Jordan canonical form can be also obtained similarly for multi-input, multi-output systems.

Example 3.15

Let us find the Jordan canonical form of the following system:

$$
Y(s)/U(s) = (s + 1)/[(s - 1)^2 (s - 3)] \tag{3.126}
$$

The partial fraction expansion of Eq. (3.126) is the following:

$$Y(s)/U(s) = 1/(s - 3) - 1/(s - 1) - 1/(s - 1)^2 \qquad (3.127)$$

Comparing Eq. (3.127) with Eq. (3.121) and using Eq. (3.125), we get the following Jordan canonical form:

$$\mathbf{A} = \begin{bmatrix} 3 & 0 & 0 \\ 0 & 1 & 0 \\ 0 & 1 & 1 \end{bmatrix}; \quad \mathbf{B} = \begin{bmatrix} 1 \\ 1 \\ 0 \end{bmatrix}$$

$$\mathbf{C} = \begin{bmatrix} 1 & -1 & -1 \end{bmatrix}; \quad \mathbf{D} = 0 \qquad (3.128)$$

The Jordan block associated with the repeated pole, $s = 1$, is shown in dashed borders.

While Jordan canonical form is an easy way of obtaining the state-space representation of a system, it has a major drawback: for the Jordan canonical form to be a *practical* representation, all the poles of the system (i.e. eigenvalues of $\mathbf{A}$) must be *real*. After all, the purpose of having state-space representations is to practically implement a control system, using electrical circuits or mechanical devices. A state-space representation with complex coefficient matrices cannot be implemented in a hardware (have you ever heard of a *complex electrical resistance*, or a spring with *complex stiffness*?). To make some sense out of Jordan canonical form for a system with complex poles, we can combine the partial fractions corresponding to each pair of complex conjugate poles, $p_{1,2} = \sigma \pm i\omega$, into a second order *real* sub-system as follows:

$$k_1/(s - p_1) + k_2/(s - p_2) = 2[\alpha s - (\alpha\sigma + \beta\omega)]/(s^2 - 2\sigma s + \sigma^2 + \omega^2) \qquad (3.129)$$

where $k_{1,2} = \alpha \pm i\beta$ are the residues corresponding to the poles, $p_{1,2}$. Remember that the residues corresponding to complex conjugate poles are also complex conjugates. Since the complex poles *always* occur as complex conjugates, their combination into real second order sub-systems using Eq. (3.129) will lead to a real state-space representation. From Eq. (3.129), it can be shown that the real Jordan block in $\mathbf{A}$ corresponding to a pair of complex conjugate poles, $p_{1,2} = \sigma \pm i\omega$, is a 2×2 block with real parts on the diagonal, and the imaginary parts off the diagonal as follows:

$$\text{Jordan block of } p_{1,2} = \sigma \pm i\omega \text{ in } \mathbf{A} = \begin{bmatrix} \sigma & \omega \\ -\omega & \sigma \end{bmatrix} \qquad (3.130)$$

The MATLAB Control System Toolbox function *canon* provides an easy derivation of the canonical forms, using the methodology presented above. The function is used as follows:

```
>>[csys,T] = canon(sys,'type') <enter>
```

where *sys* is an LTI object of the system (either transfer matrix, or state-space), *'type'* is either *'modal'* for Jordan canonical form, or *'companion'* for the observer companion form, *csys* is the returned canonical form, and **T** is the returned state-transformation matrix which transforms the state-space representation from *sys* to *csys* (**T** is meaningful only if *sys* is a state-space representation, and not the transfer matrix). The matrix **A** of the Jordan canonical form obtained using *canon* has each pair of complex eigenvalues in a real Jordan block given by Eq. (3.130).

Example 3.16

For the system of Example 3.14, let us obtain the canonical forms. For the Jordan canonical form, the MATLAB (CST) command *canon* is used as follows:

```
>> syst=ss(A, B, C, D); [jsyst,T] = canon(syst, 'modal') <enter>
a =

           x1          x2          x3          x4          x5
   x1    -15.873       0           0           0           0
   x2     0          -10.387       0           0           0
   x3     0           0          -0.34925     6.3444       0
   x4     0           0          -6.3444     -0.34925      0
   x5     0           0           0           0          -1.0444
   x6     0           0           0           0           0

           x6
   x1     0
   x2     0
   x3     0
   x4     0
   x5     0
   x6    -0.23455

b =

           u1          u2
   x1     0.50702     -20.055
   x2    -0.36131      30.035
   x3     0.092163     -5.577
   x4     0.13959      13.23
   x5    -0.17417       8.7113
   x6     0.021513     14.876

c =

           x1          x2          x3          x4          x5
   y1    0.86988      2.3105      2.7643      6.459       2.8803
   y2   -7.9857     -11.128      -0.19075    -0.78991     3.2141

           x6
   y1    -9.885
   y2    10.406

d =

           u1          u2
   y1     0           0
   y2     0           0

Continuous-time model.
```

```
T =
Columns 1 through 2
-1.67367664062877        0.82071823417532
0.93271689676861        -1.37011949477365
0.00703777816771         0.01235017857500
-0.01678546516033        0.05675504485937
0.01434336645020         0.32743224909474
0.00550594309107        -0.03421867428744


Columns 3 through 4
-0.64877657888967        0.02185375479665
1.30954638127904         0.00863637157274
0.06144994008648         1.42410332092227
-0.12409538851455        0.03844790932243
1.10820137234454         0.07909786976484
0.01382978986462        -0.04137902672467


Columns 5 through 6
0.15933965118530        -0.03863614709032
-0.32626410503493        0.08854860917358
-0.00622411675595        0.07244625298095
-1.41411419523105        0.02977656775124
-0.08538437917204        0.21852854369639
-0.01967206771160        1.01709518519537
```

Note the 2×2 real Jordan block in **A** corresponding to the complex eigenvalues $p_{3,4} = -0.349\,25 \pm 6.3444i$. Also, note that the corresponding terms in the matrices **B** and **C** are also real. The transformation matrix, **T**, has been reported in long format for greater accuracy in calculations. Next, we calculate the observer canonical form of the system as follows:

```
>>[csyst, T] = canon(syst, 'companion') <enter>

a =
        x1          x2          x3          x4          x5
   x1  9.6034e-015  -1.3017e-013  1.6822e-012  -4.8772e-011  1.5862e-009
   x2  1            5.6899e-014   1.6964e-012  -3.4476e-011  7.7762e-010
   x3  7.7716e-016  1             2.6068e-013  -4.6541e-012  1.0425e-010
   x4  4.5103e-017  2.4997e-015   1            -8.3311e-013  1.9014e-011
   x5  1.5179e-018  2.7864e-016   5.5303e-015  1             2.1529e-012
   x6  5.421e-020   1.0103e-017   1.9602e-016  -3.2543e-015  1

        x6
   x1  -1630.6
   x2  -8801.2
   x3  -8214.5
   x4  -1467.9
   x5  -258.31
   x6  -28.238

b =
        u1          u2
   x1  1           945.61
   x2  0           1128.8
   x3  0           201.9
```

```
   x4  0  36.481
   x5  0  4.0669
   x6  0  0.1451


c =

        x1            x2        x3       x4       x5
   y1  0.04829     0.51209  -24.985  -725.97  20044
   y2  -0.49194    23.374   -581.99  11670    -2.1041e+005


        x6
   y1  -3.488e+005
   y2  3.5944e+006

d =

        u1            u2
   y1  0             0
   y2  0             0


Continuous-time model.


T =

Columns 1 through 2
0.36705413565042   -1.48647885543276
0.48941400292950    0.46746500786628
0.06992034388180    0.10572273608335
0.01406952155565    0.02165321186650
0.00129963404010    0.00361303652580
0.00003944900770    0.00015766993850


Columns 3 through 4
3.02790751537344   -2.47039343978461
10.58180159205530  -2.47081762640110
1.83987823178029   -0.61525593744681
0.34386037047772   -0.10310276871196
0.04003837042044   -0.00922166413447
0.00145527317768   -0.00028916486064


Columns 5 through 6
-1.39030096046515   63.91103372168656
-2.11561239460537   73.88483216381728
-0.39195667361013   13.19012284392485
-0.09863522809937    2.36076797272696
-0.01208541334668    0.26314010184181
-0.00043627032173    0.00940569645931
```

In the **A** matrix of the computed observer companion form, all the elements except those in the last column and the subdiagonal are negligible, and can be assumed to be zeros. We can also derive the controller companion form merely by taking the transposes of **A**, **B**, and **C** of the observer form computed above. The controller companion form is thus denoted by the coefficient set $(\mathbf{A}^T, \mathbf{C}^T, \mathbf{B}^T, \mathbf{D})$.

The Jordan canonical form is useful for *decoupling* the state-equations of systems with distinct eigenvalues; such systems have a diagonal **A** matrix of the Jordan canonical form.

The companion forms are useful in designing control systems. However, a great disadvantage of the companion forms (both controller and observer) is that they are *ill-conditioned*, which means that the eigenvalues and eigenvectors of the matrix **A** are very sensitive to *perturbations* in the elements of **A**. This results in large inaccuracies in the computed eigenvalues (and eigenvectors), even if there is a small error in calculating **A**. Since a system's characteristics are governed by the eigenvalues and eigenvectors of **A**, an ill-conditioned **A** matrix is undesirable. The ill-conditioning of companion forms generally gets worse as the order of the system increases. Hence, we should normally avoid using the companion forms as state-space representations, especially for large order systems. MATLAB assigns a *condition number* to each square matrix. The condition number indicates how close a matrix is to being *singular* (i.e. determinant of the matrix being zero). A larger condition number means that the matrix is closer to being singular. With MATLAB we can assess the condition number of a square matrix, **A**, using the command *cond(A)*. If *cond(A)* is small, it indicates that **A** is *well-conditioned*. If *cond(A)* is very large, it implies an *ill-conditioned* **A**. Whenever we try to invert an ill-conditioned matrix, MATLAB issues a warning that the matrix is ill-conditioned and the results may be inaccurate.

Example 3.17

Let us compare the condition numbers for the Jordan canonical form and the observer companion form derived in Example 3.16. The condition number for the Jordan canonical form is calculated by first retrieving the state coefficient matrices from the LTI object *jsyst* using the Control System Toolbox (CST) function *ssdata*, and then applying *cond* to matrix **A** as follows:

```
>> [Aj,Bj,Cj,Dj]=ssdata(jsyst); cond(Aj) <enter>

ans =
        67.6741
```

while the condition number for the companion form is the following:

```
>> [Ac,Bc,Cc,Dc]=ssdata(csyst); cond(Ac) <enter>

ans =
        9.1881e+004
```

The condition number for the companion form is, thus, very large in comparison to that of the Jordan canonical form, confirming that the former is ill-conditioned. Why is a companion form ill-conditioned while the Jordan canonical form is not? The answer lies in all the diagonal elements, except the last, being zeros in the matrix **A** of the companion forms. In contrast, the Jordan form's matrix **A** has a *populated* diagonal, i.e. none of the diagonal elements are zeros.

For conversion of a transfer matrix into a state-space representation, you can use the MATLAB (CST) LTI object *ss*. However, the state-space conversion of a transfer matrix with *ss* results in the controller companion form, which we know to be ill-conditioned. Hence, we should avoid converting a transfer matrix to state-space representation using the command *ss*, unless we are dealing with a low order system.

3.5 Block Building in Linear, Time-Invariant State-Space

Control systems are generally interconnections of various sub-systems. If we have a state-space representation for each sub-system, we should know how to obtain the state-space representation of the entire system. Figure 2.55 shows three of the most common types of interconnections, namely the *series*, *parallel*, and *feedback* arrangement. Rather than using the transfer matrix description of Figure 2.55, we would like to depict the three common arrangements in state-space, as shown in Figure 3.7.

The series arrangement in Figure 3.7(a) is described by the following matrix equations:

$$\mathbf{x}_1^{(1)}(t) = \mathbf{A}_1\mathbf{x}_1(t) + \mathbf{B}_1\mathbf{u}(t) \tag{3.131}$$

$$\mathbf{y}_1(t) = \mathbf{C}_1\mathbf{x}_1(t) + \mathbf{D}_1\mathbf{u}(t) \tag{3.132}$$

$$\mathbf{x}_2^{(1)}(t) = \mathbf{A}_2\mathbf{x}_2(t) + \mathbf{B}_2\mathbf{y}_1(t) \tag{3.133}$$

$$\mathbf{y}(t) = \mathbf{C}_2\mathbf{x}_2(t) + \mathbf{D}_2\mathbf{y}_1(t) \tag{3.134}$$

where the state-space representation of the first sub-system is $(\mathbf{A}_1, \mathbf{B}_1, \mathbf{C}_1, \mathbf{D}_1)$, while that of the second subsystem is $(\mathbf{A}_2, \mathbf{B}_2, \mathbf{C}_2, \mathbf{D}_2)$. The input to the system, $\mathbf{u}(t)$, is also

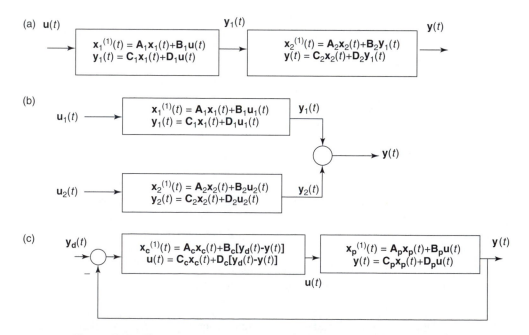

Figure 3.7   Three common arrangements of sub-systems models in state-space

the input to the first sub-system, while the system's output, $\mathbf{y}(t)$, is the output of the second sub-system. The output of the first sub-system, $\mathbf{y}_1(t)$, is the input to the second sub-system. Substitution of Eq. (3.132) int Eq. (3.133) yields:

$$\mathbf{x}_2^{(1)}(t) = \mathbf{A}_2\mathbf{x}_2(t) + \mathbf{B}_2\mathbf{C}_1\mathbf{x}_1(t) + \mathbf{B}_2\mathbf{D}_1\mathbf{u}(t) \tag{3.135}$$

and substituting Eq. (3.132) into Eq. (3.134), we get

$$\mathbf{y}(t) = \mathbf{C}_2\mathbf{x}_2(t) + \mathbf{D}_2\mathbf{C}_1\mathbf{x}_1(t) + \mathbf{D}_2\mathbf{D}_1\mathbf{u}(t) \tag{3.136}$$

If we define the state-vector of the system as $\mathbf{x}(t) = [\mathbf{x}_1^T(t); \mathbf{x}_2^T(t)]^T$, Eqs. (3.131) and (3.135) can be expressed as the following state-equation of the system:

$$\mathbf{x}^{(1)}(t) = \begin{bmatrix} \mathbf{A}_1 & \mathbf{0} \\ \mathbf{B}_2\mathbf{C}_1 & \mathbf{A}_2 \end{bmatrix} \mathbf{x}(t) + \begin{bmatrix} \mathbf{B}_1 \\ \mathbf{B}_2\mathbf{D}_1 \end{bmatrix} \mathbf{u}(t) \tag{3.137}$$

and the output equation is Eq. (3.136), re-written as follows:

$$\mathbf{y}(t) = [\mathbf{D}_2\mathbf{C}_1 \quad \mathbf{C}_2]\mathbf{x}(t) + \mathbf{D}_2\mathbf{D}_1\mathbf{u}(t) \tag{3.138}$$

The MATLAB (CST) command *series* allows you to connect two sub-systems in series using Eqs. (3.137) and (3.138) as follows:

```
>>sys = series(sys1,sys2) <enter>
```

The command *series* allows connecting the sub-systems when only some of the outputs of the first sub-system are going as inputs into the second sub-system (type *help series* ⟨*enter*⟩ for details; also see Example 2.28). Note that the sequence of the sub-systems is crucial. We will get an *entirely different* system by switching the sequence of the sub-systems in Figure 3.7(a), unless the two sub-systems are identical.

Deriving the state and output equations for the parallel connection of sub-systems in Figure 3.7(b) is left to you as an exercise. For connecting two parallel sub-systems, MATLAB (CST) has the command *parallel*, which is used in a manner similar to the command *series*.

The feedback control system arrangement of Figure 3.7(c) is more complicated than the series or parallel arrangements. Here, a controller with state-space representation $(\mathbf{A}_c, \mathbf{B}_c, \mathbf{C}_c, \mathbf{D}_c)$ is connected in series with the plant $(\mathbf{A}_p, \mathbf{B}_p, \mathbf{C}_p, \mathbf{D}_p)$ and the feedback loop from the plant output, $\mathbf{y}(t)$, to the summing junction is closed. The input to the closed-loop system is the desired output, $\mathbf{y}_d(t)$. The input to the controller is the error $[\mathbf{y}_d(t) - \mathbf{y}(t)]$, while its output is the input to the plant, $\mathbf{u}(t)$. The state and output equations of the plant and the controller are, thus, given by

$$\mathbf{x}_p^{(1)}(t) = \mathbf{A}_p\mathbf{x}_p(t) + \mathbf{B}_p\mathbf{u}(t) \tag{3.139}$$

$$\mathbf{y}(t) = \mathbf{C}_p\mathbf{x}_p(t) + \mathbf{D}_p\mathbf{u}(t) \tag{3.140}$$

$$\mathbf{x}_c^{(1)}(t) = \mathbf{A}_c\mathbf{x}_c(t) + \mathbf{B}_c[\mathbf{y}_d(t) - \mathbf{y}(t)] \tag{3.141}$$

$$\mathbf{u}(t) = \mathbf{C}_c\mathbf{x}_c(t) + \mathbf{D}_c[\mathbf{y}_d(t) - \mathbf{y}(t)] \tag{3.142}$$

Substituting Eq. (3.142) into Eqs. (3.139) and (3.140) yields the following:

$$\mathbf{x_p}^{(1)}(t) = \mathbf{A_p}\mathbf{x_p}(t) + \mathbf{B_p}\mathbf{C_c}\mathbf{x_c}(t) + \mathbf{B_p}\mathbf{D_c}[\mathbf{y_d}(t) - \mathbf{y}(t)] \quad (3.143)$$

$$\mathbf{y}(t) = \mathbf{C_p}\mathbf{x_p}(t) + \mathbf{D_p}\mathbf{C_c}\mathbf{x_c}(t) + \mathbf{D_p}\mathbf{D_c}[\mathbf{y_d}(t) - \mathbf{y}(t)] \quad (3.144)$$

Equation (3.144) can be expressed as

$$\mathbf{y}(t) = (\mathbf{I} + \mathbf{D_p}\mathbf{D_c})^{-1}[\mathbf{C_p}\mathbf{x_p}(t) + \mathbf{D_p}\mathbf{C_c}\mathbf{x_c}(t)] + (\mathbf{I} + \mathbf{D_p}\mathbf{D_c})^{-1}\mathbf{D_p}\mathbf{D_c}\mathbf{y_d}(t) \quad (3.145)$$

provided the square matrix $(\mathbf{I} + \mathbf{D_p}\mathbf{D_c})$ is non-singular. Substituting Eq. (3.145) into Eq. (3.143) yields the following state-equation of the closed-loop system:

$$\mathbf{x}^{(1)}(t) = \mathbf{A}\mathbf{x}(t) + \mathbf{B}\mathbf{y_d}(t) \quad (3.146)$$

and the output equation of the closed-loop system is Eq. (3.145) re-written as:

$$\mathbf{y}(t) = \mathbf{C}\mathbf{x}(t) + \mathbf{D}\mathbf{y_d}(t) \quad (3.147)$$

where

$$\mathbf{x}(t) = \begin{bmatrix} \mathbf{x_p}(t) \\ \mathbf{x_c}(t) \end{bmatrix};$$

$$\mathbf{A} = \begin{bmatrix} \mathbf{A_p} - \mathbf{B_p}\mathbf{D_c}(\mathbf{I} + \mathbf{D_p}\mathbf{D_c})^{-1}\mathbf{C_p} & \mathbf{B_p}\mathbf{C_c} - \mathbf{B_p}\mathbf{D_c}(\mathbf{I} + \mathbf{D_p}\mathbf{D_c})^{-1}\mathbf{D_p}\mathbf{C_c} \\ -\mathbf{B_c}(\mathbf{I} + \mathbf{D_p}\mathbf{D_c})^{-1}\mathbf{C_p} & \mathbf{A_c} - \mathbf{B_c}(\mathbf{I} + \mathbf{D_p}\mathbf{D_c})^{-1}\mathbf{D_p}\mathbf{C_c} \end{bmatrix}$$

$$\mathbf{C} = (\mathbf{I} + \mathbf{D_p}\mathbf{D_c})^{-1}\begin{bmatrix} \mathbf{C_p} & \mathbf{D_p}\mathbf{C_c} \end{bmatrix};$$

$$\mathbf{D} = (\mathbf{I} + \mathbf{D_p}\mathbf{D_c})^{-1}\mathbf{D_p}\mathbf{D_c}; \quad \mathbf{B} = \begin{bmatrix} \mathbf{B_p}\mathbf{D_c}(\mathbf{I} - \mathbf{D}) \\ \mathbf{B_c}(\mathbf{I} - \mathbf{D}) \end{bmatrix} \quad (3.148)$$

Using MATLAB (CST), the closed-loop system given by Eqs. (3.146)–(3.148) can be derived as follows:

```
>>sys0 = series(sysc,sysp) % series connection of LTI blocks sysc
  and sysp <enter>

>>sys1=ss(eye(size(sys0))) % state-space model (A=B=C=0, D=I) of
  the feedback block, sys1 <enter>

>>sysCL= feedback(sys0, sys1) % negative feedback from output to
  input of sys0 <enter>
```

where *sys0* is the state-space representation of the controller, *sysc*, in series with the plant, *sysp*, *sys1* is the state-space representation ($\mathbf{A} = \mathbf{B} = \mathbf{C} = 0, \mathbf{D} = \mathbf{I}$) of the feedback block in Figure 3.7(c), and *sysCL* is the state-space representation of the closed-loop system. Note that *sys0* is the *open-loop system* of Figure 3.7(c), i.e. the system when the feedback loop is absent.

Example 3.18

Let us derive the state-space representation of an interesting system, whose block-diagram is shown in Figure 3.8. The system represents a missile tracking

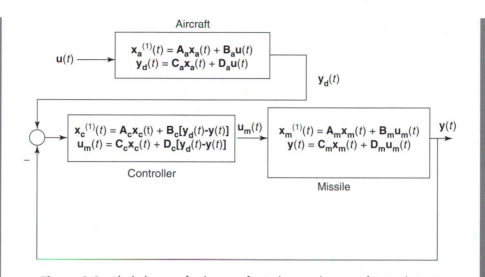

Figure 3.8 Block diagram for the aircraft-missile control system of Example 3.18

a maneuvering aircraft. The pilot of the aircraft provides an input vector, $\mathbf{u}(t)$, to the aircraft represented as $(\mathbf{A_a}, \mathbf{B_a}, \mathbf{C_a}, \mathbf{D_a})$. The input vector, $\mathbf{u}(t)$, consists of the aircraft pilot's deflection of the *rudder* and the *aileron*. The motion of the aircraft is described by the vector, $\mathbf{y_d}(t)$, which is the desired output of the missile, i.e. the missile's motion – described by the output vector, $\mathbf{y}(t)$ – should closely follow that of the aircraft. The output vector, $\mathbf{y}(t)$, consists of the missile's *linear* and *angular velocities* with respect to *three mutually perpendicular axes* attached to the missile's *center of gravity* – a total of six output variables. The state-space representation for the missile is $(\mathbf{A_m}, \mathbf{B_m}, \mathbf{C_m}, \mathbf{D_m})$. The missile is controlled by a feedback controller with the state-space representation $(\mathbf{A_c}, \mathbf{B_c}, \mathbf{C_c}, \mathbf{D_c})$ whose task is to ultimately make $\mathbf{y}(t) = \mathbf{y_d}(t)$, i.e. cause the missile to hit the maneuvering aircraft.

The matrices representing the aircraft, missile, and controller are as follows:

$$\mathbf{A_a} = \begin{bmatrix} -0.0100 & -0.1000 & 0 & 0 & 0 & 0 & 0 \\ 0 & -0.4158 & 1.0250 & 0 & 0 & 0 & 0 \\ 0 & 0.0500 & -0.8302 & 0 & 0 & 0 & 0 \\ 0 & 0 & 0 & -0.5600 & -1.0000 & 0.0800 & 0.0400 \\ 0 & 0 & 0 & 0.6000 & -0.1200 & -0.3000 & 0 \\ 0 & 0 & 0 & -3.0000 & 0.4000 & -0.4700 & 0 \\ 0 & 0 & 0 & 0 & 0.0800 & 1.0000 & 0 \end{bmatrix}$$

$$\mathbf{B_a} = \begin{bmatrix} 0 & 0 \\ 0 & 0 \\ 0 & 0 \\ 0.0730 & 0.0001 \\ -4.8000 & 1.2000 \\ 1.5000 & 10.0000 \\ 0 & 0 \end{bmatrix}$$

$$\mathbf{C_a} = \begin{bmatrix} 1 & 0 & 0 & 0 & 0 & 0 & 0 \\ 0 & 0 & 0 & 250 & 0 & 0 & 0 \\ 0 & 250 & 0 & 0 & 0 & 0 & 0 \\ 0 & 0 & 0 & 0 & 0 & 1 & 0 \\ 0 & 0 & 1 & 0 & 0 & 0 & 0 \\ 0 & 0 & 0 & 0 & 1 & 0 & 0 \end{bmatrix}; \quad \mathbf{D_a} = \begin{bmatrix} 0 & 0 \\ 0 & 0 \\ 0 & 0 \\ 0 & 0 \\ 0 & 0 \\ 0 & 0 \end{bmatrix}$$

$$\mathbf{A_m} = \begin{bmatrix} 0.4743 & 0 & 0.0073 & 0 & 0 & 0 \\ 0 & -0.4960 & 0 & 0 & 0 & 0 \\ -0.0368 & 0 & -0.4960 & 0 & 0 & 0 \\ 0 & -0.0015 & 0 & -0.0008 & 0 & 0.0002 \\ 0 & 0 & -0.2094 & 0 & -0.0005 & 0 \\ 0 & -0.2094 & 0 & 0 & 0 & -0.0005 \end{bmatrix}$$

$$\mathbf{B_m} = \begin{bmatrix} 0 & 0 & 0 \\ 191.1918 & 0 & 0 \\ 0 & 191.1918 & 0 \\ 0 & 0 & 1.0000 \\ 0 & 232.5772 & 0 \\ 232.5772 & 0 & 0 \end{bmatrix}; \quad \mathbf{C_m} = \begin{bmatrix} 1 & 0 & 0 & 0 & 0 & 0 \\ 0 & 1 & 0 & 0 & 0 & 0 \\ 0 & 0 & 1 & 0 & 0 & 0 \\ 0 & 0 & 0 & 1 & 0 & 0 \\ 0 & 0 & 0 & 0 & 1 & 0 \\ 0 & 0 & 0 & 0 & 0 & 1 \end{bmatrix}$$

$$\mathbf{D_m} = \begin{bmatrix} 0 & 0 & 0 \\ 0 & 0 & 0 \\ 0 & 0 & 0 \\ 0 & 0 & 0 \\ 0 & 0 & 0 \\ 0 & 0 & 0 \end{bmatrix}$$

$$\mathbf{A_c} = \begin{bmatrix} 0 & 0 & 0 & 1.0 & 0 & 0 \\ 0 & 0 & 0 & 0 & 1.0 & 0 \\ 0 & 0 & 0 & 0 & 0 & 1.0 \\ -1.0 & 0 & 0 & -0.3 & 0 & 0 \\ 0 & -1.0 & 0 & 0 & -0.3 & 0 \\ 0 & 0 & -1.0 & 0 & 0 & -0.3 \end{bmatrix}$$

$$\mathbf{B_c} = \begin{bmatrix} 0 & 0 & 0 & 0 & 0 & 0 \\ 0 & 0 & 0 & 0 & 0 & 0 \\ 0 & 0 & 0 & 0 & 0 & 0 \\ 0.0000 & 0.0000 & 0.0000 & 0.0000 & 0.0000 & 0.0001 \\ 0.6774 & 0.0000 & 0.0052 & 0.0000 & -0.0001 & 0.0000 \\ 0.0000 & 0.0000 & 0.0000 & 0.0000 & 0.0000 & 0.0000 \end{bmatrix}$$

$$\mathbf{C_c} = \begin{bmatrix} 1 & 0 & 0 & 0 & 0 & 0 \\ 0 & 1 & 0 & 0 & 0 & 0 \\ 0 & 0 & 1 & 0 & 0 & 0 \end{bmatrix}; \quad \mathbf{D_c} = \begin{bmatrix} 0 & 0 & 0 & 0 & 0 & 0 \\ 0 & 0 & 0 & 0 & 0 & 0 \\ 0 & 0 & 0 & 0 & 0 & 0 \end{bmatrix}$$

Note that the aircraft is a *seventh* order sub-system, while the missile and the controller are *sixth* order sub-systems. The state-space representation of the entire system is obtained as follows:

```
>>sysc=ss(Ac,Bc,Cc,Dc); sysm=ss(Am,Bm,Cm,Dm); sys0 = series(sysc,sysm);
  <enter>

>>sys1=ss(eye(size(sys0))); <enter>

>>sysCL=feedback(sys0,sys1); <enter>

>>sysa=ss(Aa,Ba,Ca,Da); syst=series(sysa, sysCL) <enter>
```

a =

	x1	x2	x3	x4	x5
x1	0.4743	0	0.0073258	0	0
x2	0	-0.49601	0	0	0
x3	-0.036786	0	-0.49601	0	0
x4	0	-0.0015497	0	-0.00082279	0
x5	0	0	-0.20939	0	-0.00048754
x6	0	-0.20939	0	-8.2279e-006	0
x7	0	0	0	0	0
x8	0	0	0	0	0
x9	0	0	0	0	0
x10	0	0	0	0	0
x11	-0.6774	0	-0.0052	0	0.0001
x12	0	0	0	0	0
x13	0	0	0	0	0
x14	0	0	0	0	0
x15	0	0	0	0	0
x16	0	0	0	0	0
x17	0	0	0	0	0
x18	0	0	0	0	0
x19	0	0	0	0	0

	x6	x7	x8	x9	x10
x1	0	0	0	0	0
x2	0	191.19	0	0	0
x3	0	0	191.19	0	0
x4	0.00017749	0	0	1	0
x5	0	0	232.58	0	0
x6	-0.00048754	232.58	0	0	0
x7	0	0	0	0	1
x8	0	0	0	0	0
x9	0	0	0	0	0
x10	-0.0001	-1	0	0	-0.3
x11	0	0	-1	0	0
x12	0	0	0	-1	0
x13	0	0	0	0	0
x14	0	0	0	0	0
x15	0	0	0	0	0
x16	0	0	0	0	0
x17	0	0	0	0	0
x18	0	0	0	0	0
x19	0	0	0	0	0

	x11	x12	x13	x14	x15
x1	0	0	0	0	0
x2	0	0	0	0	0
x3	0	0	0	0	0
x4	0	0	0	0	0
x5	0	0	0	0	0
x6	0	0	0	0	0
x7	0	0	0	0	0
x8	1	0	0	0	0
x9	0	1	0	0	0
x10	0	0	0	0	0
x11	-0.3	0	0.6774	1.3	-0.0001
x12	0	-0.3	0	0	0
x13	0	0	-0.01	-0.1	0
x14	0	0	0	-0.4158	1.025
x15	0	0	0	0.05	-0.8302
x16	0	0	0	0	0
x17	0	0	0	0	0
x18	0	0	0	0	0
x19	0	0	0	0	0

	x16	x17	x18	x19
x1	0	0	0	0
x2	0	0	0	0
x3	0	0	0	0
x4	0	0	0	0
x5	0	0	0	0
x6	0	0	0	0
x7	0	0	0	0
x8	0	0	0	0
x9	0	0	0	0
x10	0	0.0001	0	0
x11	0	0	0	0
x12	0	0	0	0
x13	0	0	0	0
x14	0	0	0	0
x15	0	0	0	0
x16	-0.56	-1	0.08	0.04
x17	0.6	-0.12	-0.3	0
x18	-3	0.4	-0.47	0
x19	0	0.08	1	0

b =

	u1	u2
x1	0	0
x2	0	0
x3	0	0
x4	0	0
x5	0	0
x6	0	0
x7	0	0
x8	0	0
x9	0	0
x10	0	0
x11	0	0

x12	0	0
x13	0	0
x14	0	0
x15	0	0
x16	0.073	0.0001
x17	-4.8	1.2
x18	1.5	10
x19	0	0

c =

	x1	x2	x3	x4	x5
y1	1	0	0	0	0
y2	0	1	0	0	0
y3	0	0	1	0	0
y4	0	0	0	1	0
y5	0	0	0	0	1
y6	0	0	0	0	0

	x6	x7	x8	x9	x10
y1	0	0	0	0	0
y2	0	0	0	0	0
y3	0	0	0	0	0
y4	0	0	0	0	0
y5	0	0	0	0	0
y6	1	0	0	0	0

	x11	x12	x13	x14	x15
y1	0	0	0	0	0
y2	0	0	0	0	0
y3	0	0	0	0	0
y4	0	0	0	0	0
y5	0	0	0	0	0
y6	0	0	0	0	0

	x16	x17	x18	x19
y1	0	0	0	0
y2	0	0	0	0
y3	0	0	0	0
y4	0	0	0	0
y5	0	0	0	0
y6	0	0	0	0

d =

	u1	u2
y1	0	0
y2	0	0
y3	0	0
y4	0	0
y5	0	0
y6	0	0

Continuous-time model.

The total system, *syst*, is of order 19, which is the sum of the individual orders of the sub-systems. If the entire system, *syst*, is asymptotically stable, the missile will ultimately hit the aircraft, irrespective of the pilot's inputs to the aircraft. To analyze whether the pilot can escape the missile by maneuvering the aircraft with the help of rudder and aileron inputs, let us find the eigenvalues of the entire system as follows:

```
>> [a,b,c,d]=ssdata(syst); damp(a) <enter>
```

Eigenvalue	Damping	Freq. (rad/s)
3.07e-001+1.02e+000i	-2.89e-001	1.06e+000
3.07e-001-1.02e+000i	-2.89e-001	1.06e+000
-7.23e-004	1.00e+000	7.23e-004
-8.23e-004	1.00e+000	8.23e-004
-1.00e-002	1.00e+000	1.00e-002
-1.55e-002	1.00e+000	1.55e-002
-1.59e-002	1.00e+000	1.59e-002
-2.74e-002+1.13e+000i	2.42e-002	1.13e+000
-2.74e-002-1.13e+000i	2.42e-002	1.13e+000
-1.39e-001+9.89e-001i	1.39e-001	9.99e-001
-1.39e-001-9.89e-001i	1.39e-001	9.99e-001
-1.50e-001+9.89e-001i	1.50e-001	1.00e+000
-1.50e-001-9.89e-001i	1.50e-001	1.00e+000
-3.16e-001	1.00e+000	3.16e-001
-3.82e-001	1.00e+000	3.82e-001
-5.04e-001	1.00e+000	5.04e-001
-5.38e-001	1.00e+000	5.38e-001
-9.30e-001	1.00e+000	9.30e-001
-1.09e+000	1.00e+000	1.09e+000

The complex conjugate eigenvalues $0.307 \pm 1.02i$ with a positive real part indicate that the system is unstable. Hence, it is possible for the pilot to ultimately escape the missile. The controller, *sysc*, must be re-designed to enable a hit by making the entire system asymptotically stable.

Exercises

3.1. Derive a state-space representation for each of the systems whose governing differential equations are the following, with outputs and inputs denoted by $y_i(t)$ and $u_i(t)$ (if $i > 1$), respectively:

(a) $17d^3y(t)/dt^3 + 10dy(t)/dt - 2y(t) = 2du(t)/dt + 5u(t)$.

(b) $d^2y_1(t)/dt^2 + 3dy_1(t)/dt - 6y_2(t) = -u_1(t)/7$; $-2d^2y_2(t)/dt^2 + 9y_2(t) - dy_1(t)/dt = 5du_1(t)/dt - u_2(t)$.

(c) $100d^4y(t)/dt^4 - 33d^3y(t)/dt^3 + 12d^2y(t)/dt^2 + 8y(t) = 27du_1(t)/dt - u_1(t) + 5u_2(t)$.

(d) $d^5y_1(t)/dt + 9y_1(t) - 7d^2y_2(t)/dt^2 + dy_2(t)/dt = 2d^3u(t)/dt^3 - 16d^2u(t)/dt^2 + 47du(t)/dt + 3u(t)$.

3.2. Derive a state-space representation for the systems whose transfer matrices are the following:

(a) $Y(s)/U(s) = (s^2 - 3s + 1)/(s^5 + 4s^3 + 3s^2 - s + 5)$.

(b) $\mathbf{Y}(s)/\mathbf{U}(s) = [(s + 1)/(s^2 + 2s + 3) \quad s/(s + 3) \quad 1/(s^3 + 5)]$.

(c) $\mathbf{Y}(s)/\mathbf{U}(s) = \begin{bmatrix} 1/(s + 2) & 0 \\ -1/(s^2 + 3s) & (s + 4)/(s + 7) \end{bmatrix}$

3.3. Derive a state-space representation for a satellite orbiting a planet (Example 2.3). Linearize the nonlinear state-space representation for small deviations from a circular orbit.

3.4. For a missile guided by *beam-rider* guidance law (Eq. (2.19)), derive a state-space representation considering the commanded missile acceleration, $a_{\mathrm{Mc}}(t)$, as the input, and the missile's angular position, $\theta_{\mathrm{M}}(t)$, as the output.

3.5. For the closed-loop *beam-rider* guidance of a missile shown in Figure 2.8, derive a state-space representation if the target's angular position, $\theta_{\mathrm{T}}(t)$, is the input, and the missile's angular position, $\theta_{\mathrm{M}}(t)$, is the output.

3.6. For a missile guided by the *command line-of-sight* guidance law (Eq. (2.20)), derive a state-space representation considering the commanded missile acceleration, $a_{\mathrm{Mc}}(t)$, as the input, and the missile's angular position, $\theta_{\mathrm{M}}(t)$, as the output.

3.7. For the closed-loop *command line-of-sight* guidance of a missile shown in Figure 2.9, derive a state-space representation if the target's angular position, $\theta_{\mathrm{T}}(t)$, is the input, and the missile's angular position, $\theta_{\mathrm{M}}(t)$, is the output. Can the state-space representation be linearized about an equilibrium point?

3.8. Derive a state-space representation for the longitudinal dynamics of an aircraft (Example 2.10) with elevator deflection, $\delta(t)$, as the input, and $[v(t) \quad \alpha(t) \quad \theta(t)]^T$ as the output vector. Convert the state-space representation into:

(a) the Jordan canonical form,

(b) the controller companion form,

(c) the observer companion form.

3.9. Derive a state-space representation for the compensated closed-loop chemical plant of Example 2.25, with the closed-loop transfer function given by Eq. (2.159). Convert the state-space representation into:

(a) the Jordan canonical form,

(b) the controller companion form,

(c) the observer canonical form.

3.10. For the closed-loop multivariable chemical process of Example 2.29, derive a state-space representation. Transform the state-space representation into:

    (a) the Jordan canonical form,

    (b) the controller companion form,

    (c) the observer canonical form.

3.11. For the aircraft longitudinal dynamics of Example 3.10 derive:

    (a) the Jordan canonical form,

    (b) the controller companion form,

    (c) the observer canonical form.

3.12. For the nonlinear electrical network of Exercise 2.3, derive a state-space representation with input as the voltages, $v_1(t)$ and $v_2(t)$, and the output as the current $i_2(t)$. Linearize the state-space representation about the equilibrium point falling in the *dead-zone*, $-a < v_1(t) < a$. Use $L = 1000$ henry, $R_1 = 100$ ohm, $R_2 = 200$ ohm, $C_1 = 2 \times 10^{-5}$ farad, and $C_2 = 3 \times 10^{-5}$ farad. Is the electrical network stable about the equilibrium point?

3.13. Repeat Exercise 2.29 using a state-space representation for each of the multivariable systems.

3.14. For the multivariable closed-loop system of Exercise 2.30, derive a state-space representation, and convert it into the Jordan canonical form.

References

1. Nise, N.S. *Control Systems Engineering*. Addison-Wesley, 1995.
2. Maciejowski, J.M. *Multivariable Feedback Design*. Addison-Wesley, 1989, pp. 406–407.

4

Solving the State-Equations

4.1 Solution of the Linear Time Invariant State Equations

We learnt in Chapter 3 how to represent the governing differential equation of a system by a set of first order differential equations, called state-equations, whose number is equal to the order of the system. Before we can begin designing a control system based on the state-space approach, we must be able to solve the state-equations. To see how the state-equations are solved, let us consider the following single first order differential equation:

$$x^{(1)}(t) = ax(t) + bu(t) \qquad (4.1)$$

where $x(t)$ is the state variable, $u(t)$ is the input, and a and b are the constant coefficients. Equation (4.1) represents a first order system. Let us try to solve this equation for $t > t_0$ with the *initial condition*, $x(t_0) = x_0$. (Note that since the differential equation, Eq. (4.1), is of first order we need only *one* initial condition to obtain its solution). The solution to Eq. (4.1) is obtained by multiplying both sides of the equation by $\exp\{-a(t - t_0)\}$ and re-arranging the resulting equation as follows:

$$\exp\{-a(t - t_0)\}x^{(1)}(t) - \exp\{-a(t - t_0)\}ax(t) = \exp\{-a(t - t_0)\}bu(t) \qquad (4.2)$$

We recognize the term on the left-hand side of Eq. (4.2) as $d/dt[\exp\{-a(t - t_0)\}x(t)]$. Therefore, Eq. (4.2) can be written as

$$d/dt[\exp\{-a(t - t_0)\}x(t)] = \exp\{-a(t - t_0)\}bu(t) \qquad (4.3)$$

Integrating both sides of Eq. (4.3) from t_0 to t, we get

$$\exp\{-a(t - t_0)\}x(t) - x(t_0) = \int_{t_0}^{t} \exp\{-a(\tau - t_0)\}bu(\tau)d\tau \qquad (4.4)$$

Applying the initial condition, $x(t_0) = x_0$, and multiplying both sides of Eq. (4.4) by $\exp\{a(t - t_0)\}$, we get the following expression for the state variable, $x(t)$:

$$x(t) = \exp\{a(t - t_0)\}x_0 + \int_{t_0}^{t} e^{a(t - \tau)}bu(\tau)d\tau; \quad (t \geq t_0) \qquad (4.5)$$

Note that Eq. (4.5) has two terms on the right-hand side. The first term, $\exp\{a(t - t_0)\}x_0$, depends upon the initial condition, x_0, and is called the *initial response* of the system. This will be the only term present in the response, $x(t)$, if the applied input, $u(t)$, is zero. The integral term on the right-hand side of Eq. (4.5) is independent of the initial condition, but depends upon the input. Note the similarity between this integral term and the *convolution integral* given by Eq. (2.120), which was derived as the response of a linear system to an arbitrary input by linearly superposing the individual impulse responses. The lower limit of the integral in Eq. (4.5) is t_0 (instead of $-\infty$ in Eq. (2.120)), because the input, $u(t)$, *starts acting* at time t_0 onwards, and is assumed to be zero at all times $t < t_0$. (Of course, one could have an *ever-present* input, which starts acting on the system at $t = -\infty$; in that case, $t_0 = -\infty$). If the coefficient, a, in Eq. (4.1) is *negative*, then the system given by Eq. (4.1) is *stable* (why?), and the response given by Eq. (4.5) will reach a *steady-state* in the limit $t \to \infty$. Since the initial response of the stable system decays to zero in the limit $t \to \infty$, the integral term is the only term remaining in the response of the system in the steady-state limit. Hence, the integral term in Eq. (4.5) is called the *steady-state response* of the system. All the system responses to singularity functions with zero initial condition, such as the *step response* and the *impulse response*, are obtained form the steady-state response. Comparing Eqs. (2.120) and (4.5), we can say that for this first order system the *impulse response*, $g(t - t_0)$, is given by

$$g(t - t_0) = \exp\{a(t - t_0)\}b \qquad (4.6)$$

You may verify Eq. (4.6) by deriving the impulse response of the first order system of Eq. (4.1) using the Laplace transform method of Chapter 2 for $u(t) = \delta(t - t_0)$ and $x(t_0) = 0$. The step response, $s(t)$, of the system can be obtained as the time integral of the impulse response (see Eqs. (2.104) and (2.105)), given by

$$s(t) = \int_{t_0}^{t} e^{a(t-\tau)}b d\tau = [\exp\{a(t - t_0)\} - 1]/a \qquad (4.7)$$

Note that Eq. (4.7) can also be obtained directly from Eq. (4.5) by putting $u(t) = u_s(t - t_0)$ and $x(t_0) = 0$.

To find the response of a general system of order n, we should have a solution for each of the n state-equations in a form similar to Eq. (4.5). However, since the state-equations are usually *coupled*, their solutions cannot be obtained *individually*, but *simultaneously* as a *vector solution*, $\mathbf{x}(t)$, to the following matrix state-equation:

$$\mathbf{x}^{(1)}(t) = \mathbf{A}\mathbf{x}(t) + \mathbf{B}u(t) \qquad (4.8)$$

Before considering the general matrix state-equation, Eq. (4.8), let us take the special case of a system having *distinct eigenvalues*. We know from Chapter 3 that for such systems, the state-equations can be *decoupled* through an appropriate state transformation. Solving decoupled state-equations is a simple task, consisting of individual application of Eq. (4.5) to each decoupled state-equation. This is illustrated in the following example.

Example 4.1

Consider a system with the following state-space coefficient matrices:

$$A = \begin{bmatrix} -3 & 0 \\ 0 & -2 \end{bmatrix}; \quad B = \begin{bmatrix} 1 \\ -1 \end{bmatrix} \tag{4.9}$$

Let us solve the state-equations for $t > 0$ with the following initial condition:

$$x(0) = \begin{bmatrix} 1 \\ 0 \end{bmatrix} \tag{4.10}$$

The individual scalar state-equations can be expressed from Eq. (4.8) as follows:

$$x_1^{(1)}(t) = -3x_1(t) + u(t) \tag{4.11}$$

$$x_2^{(1)}(t) = -2x_2(t) - u(t) \tag{4.12}$$

where $x_1(t)$ and $x_2(t)$ are the state variables, and $u(t)$ is the input defined for $t \geq 0$. Since both Eqs. (4.11) and (4.12) are decoupled, they are solved independently of one another, and their solutions are given by Eq. (4.5) as follows:

$$x_1(t) = e^{-3t} + \int_0^t e^{-3(t-\tau)}u(\tau)d\tau; \quad (t \geq 0) \tag{4.13}$$

$$x_2(t) = -\int_0^t e^{-2(t-\tau)}u(\tau)d\tau; \quad (t \geq 0) \tag{4.14}$$

Example 4.1 illustrates the ease with which the decoupled state-equations are solved. However, only systems with distinct eigenvalues can be decoupled. For systems having repeated eigenvalues, we must be able to solve the coupled state-equations given by Eq. (4.8).

To solve the general state-equations, Eq. (4.8), let us first consider the case when the input vector, $u(t)$, is *always zero*. Then Eq. (4.8) becomes a *homogeneous* matrix state-equation given by

$$x^{(1)}(t) = Ax(t) \tag{4.15}$$

We are seeking the vector solution, $x(t)$, to Eq. (4.15) subject to the initial condition, $x(t_0) = x_0$. The solution to the *scalar counterpart* of Eq. (4.15) (i.e. $x^{(1)}(t) = ax(t)$) is just the initial response given by $x(t) = \exp\{a(t - t_0)\}x_0$, which we obtain from Eq. (4.5) by setting $u(t) = 0$. Taking a hint from the scalar solution, let us write the vector solution to Eq. (4.15) as

$$x(t) = \exp\{A(t - t_0)\}x(t_0) \tag{4.16}$$

In Eq. (4.16) we have introduced a strange beast, $\exp\{\mathbf{A}(t - t_0)\}$, which we will call *the matrix exponential of* $\mathbf{A}(t - t_0)$. This beast is somewhat like the Loch Ness monster, whose existence has been conjectured, but not proven. Hence, it is a figment of our imagination. Everybody has seen and used the scalar exponential, $\exp\{a(t - t_0)\}$, but talking about a *matrix* raised to the power of a *scalar*, e, appears to be stretching our credibility beyond its limits! Anyhow, since Eq. (4.16) tells us that the matrix exponential can help us in solving the general state-equations, let us see how this animal can be defined.

We know that the Taylor series expansion of the scalar exponential, $\exp\{a(t - t_0)\}$, is given by

$$\exp\{a(t - t_0)\} = 1 + a(t - t_0) + a^2(t - t_0)^2/2!$$
$$+ a^3(t - t_0)^3/3! + \cdots + a^k(t - t_0)^k/k! + \cdots \qquad (4.17)$$

Since the matrix exponential behaves exactly like the scalar exponential in expressing the solution to a first order differential equation, we conjecture that it must also have the same expression for its Taylor series as Eq. (4.17) with the scalar, a, replaced by the matrix, **A**. Therefore, we *define* the matrix exponential, $\exp\{\mathbf{A}(t - t_0)\}$, as a matrix that has the following Taylor series expansion:

$$\exp\{\mathbf{A}(t - t_0)\} = \mathbf{I} + \mathbf{A}(t - t_0) + \mathbf{A}^2(t - t_0)^2/2!$$
$$+ \mathbf{A}^3(t - t_0)^3/3! + \cdots + \mathbf{A}^k(t - t_0)^k/k! + \cdots \qquad (4.18)$$

Equation (4.18) tells us that the matrix exponential is of the same size as the matrix **A**. Our definition of $\exp\{\mathbf{A}(t - t_0)\}$ must satisfy the homogeneous matrix state-equation, Eq. (4.15), whose solution is given by Eq. (4.16). To see whether it does so, let us differentiate Eq. (4.16) with time, t, to yield

$$\mathbf{x}^{(1)}(t) = d/dt[\exp\{\mathbf{A}(t - t_0)\}\mathbf{x}(t_0)] = d/dt[\exp\{\mathbf{A}(t - t_0)\}]\mathbf{x}(t_0) \qquad (4.19)$$

The term $d/dt[\exp\{\mathbf{A}(t - t_0)\}]$ is obtained by differentiating Eq. (4.17) with respect to time, t, as follows:

$d/dt[\exp\{\mathbf{A}(t - t_0)\}]$

$$= \mathbf{A} + \mathbf{A}^2(t - t_0) + \mathbf{A}^3(t - t_0)^2/2! + \mathbf{A}^4(t - t_0)^3/3! + \cdots + \mathbf{A}^{k+1}(t - t_0)^k/k! + \cdots$$
$$= \mathbf{A}[\mathbf{I} + \mathbf{A}(t - t_0) + \mathbf{A}^2(t - t_0)^2/2! + \mathbf{A}^3(t - t_0)^3/3! + \cdots + \mathbf{A}^k(t - t_0)^k/k! + \cdots]$$
$$= \mathbf{A}\exp\{\mathbf{A}(t - t_0)\} \qquad (4.20)$$

(Note that the right-hand side of Eq. (4.20) can also be expressed as $[\mathbf{I} + \mathbf{A}(t - t_0) + \mathbf{A}^2(t - t_0)^2/2! + \mathbf{A}^3(t - t_0)^3/3! + \cdots + \mathbf{A}^k(t - t_0)^k/k! + \cdots]\mathbf{A} = \exp\{\mathbf{A}(t - t_0)\}\mathbf{A}$, which implies that $\mathbf{A}\exp\{\mathbf{A}(t - t_0)\} = \exp\{\mathbf{A}(t - t_0)\}\mathbf{A}$.) Substituting Eq. (4.20) into Eq. (4.19), and using Eq. (4.16), we get

$$\mathbf{x}^{(1)}(t) = \mathbf{A}\exp\{\mathbf{A}(t - t_0)\}\mathbf{x}_0 = \mathbf{A}\mathbf{x}(t) \qquad (4.21)$$

which is the same as Eq. (4.15). Hence, our definition of the matrix exponential by Eq. (4.18) does satisfy Eq. (4.15). In Eq. (4.20) we saw that the matrix exponential, $\exp\{\mathbf{A}(t - t_0)\}$, *commutes* with the matrix, $\mathbf{A}$, i.e. $\mathbf{A}\exp\{\mathbf{A}(t - t_0)\} = \exp\{\mathbf{A}(t - t_0)\}\mathbf{A}$. This is a special property of $\exp\{\mathbf{A}(t - t_0)\}$, because only rarely do two matrices commute with one another (see Appendix B). Looking at Eq. (4.16), we see that the matrix exponential, $\exp\{\mathbf{A}(t - t_0)\}$, performs a *linear transformation* on the initial state-vector, $\mathbf{x}(t_0)$, to give the state-vector at time t, $\mathbf{x}(t)$. Hence, $\exp\{\mathbf{A}(t - t_0)\}$, is also known as the *state-transition matrix*, as it *transitions* the system given by the homogeneous state-equation, Eq. (4.15), from the state, $\mathbf{x}(t_0)$, at time, t_0, to the state $\mathbf{x}(t)$, at time, t. Thus, using the state-transition matrix we can find the state at any time, t, if we know the state at any previous time, $t_0 < t$. Table 4.1 shows some important properties of the state-transition matrix, which you can easily verify from the definition of $\exp\{\mathbf{A}(t - t_0)\}$, Eq. (4.18).

Now that we know how to solve for the initial response (i.e. response when $\mathbf{u}(t) = \mathbf{0}$) of the system given by Eq. (4.8), let us try to obtain the general solution, $\mathbf{x}(t)$, when the input vector, $\mathbf{u}(t)$, is non-zero for $t \geq t_0$. Again, we will use the steps similar to those for the scalar state-equation, i.e. Eqs. (4.1)–(4.5). However, since now we are dealing with matrix equation, we have to be careful with the sequence of matrix multiplications. Pre-multiplying Eq. (4.8) by $\exp\{-\mathbf{A}(t - t_0)\}$, we get

$$\exp\{-\mathbf{A}(t - t_0)\}\mathbf{x}^{(1)}(t) = \exp\{-\mathbf{A}(t - t_0)\}\mathbf{A}\mathbf{x}(t) + \exp\{-\mathbf{A}(t - t_0)\}\mathbf{B}\mathbf{u}(t) \quad (4.22)$$

Bringing the terms involving $\mathbf{x}(t)$ to the left-hand side, we can write

$$\exp\{-\mathbf{A}(t - t_0)\}[\mathbf{x}^{(1)}(t) - \mathbf{A}\mathbf{x}(t)] = \exp\{-\mathbf{A}(t - t_0)\}\mathbf{B}\mathbf{u}(t) \quad (4.23)$$

From Table 4.1 we note that $d/dt[\exp\{-\mathbf{A}(t - t_0)\}] = -\exp\{-\mathbf{A}(t - t_0)\}\mathbf{A}$. Therefore, the left-hand side of Eq. (4.23) can be expressed as follows:

$$\exp\{-\mathbf{A}(t - t_0)\}\mathbf{x}^{(1)}(t) - \exp\{-\mathbf{A}(t - t_0)\}\mathbf{A}\mathbf{x}(t)$$
$$= \exp\{-\mathbf{A}(t - t_0)\}\mathbf{x}^{(1)}(t) + d/dt[\exp\{-\mathbf{A}(t - t_0)\}]\mathbf{x}(t)$$
$$= d/dt[\exp\{-\mathbf{A}(t - t_0)\}\mathbf{x}(t)] \quad (4.24)$$

Hence, Eq. (4.23) can be written as

$$d/dt[\exp\{-\mathbf{A}(t - t_0)\}\mathbf{x}(t)] = \exp\{-\mathbf{A}(t - t_0)\}\mathbf{B}\mathbf{u}(t) \quad (4.25)$$

Table 4.1 Some important properties of the state-transition matrix

S. No.	Property	Expression
1	Stationarity	$\exp\{\mathbf{A}(t_0 - t_0)\} = \mathbf{I}$
2	Commutation with $\mathbf{A}$	$\mathbf{A}\exp\{\mathbf{A}(t - t_0)\} = \exp\{\mathbf{A}(t - t_0)\}\mathbf{A}$
3	Differentiation with time, t	$d/dt[\exp\{\mathbf{A}(t - t_0)\}] = \exp\{\mathbf{A}(t - t_0)\}\mathbf{A}$
4	Inverse	$[\exp\{\mathbf{A}(t - t_0)\}]^{-1} = \exp\{\mathbf{A}(t_0 - t)\}$
5	Time-marching	$\exp\{\mathbf{A}(t - t_1)\}\exp\{\mathbf{A}(t_1 - t_0)\} = \exp\{\mathbf{A}(t - t_0)\}$

Integrating Eq. (4.25) with respect to time, from t_0 to t, we get

$$\exp\{-\mathbf{A}(t - t_0)\}\mathbf{x}(t) - \mathbf{x}(t_0) = \int_{t_0}^{t} \exp\{-\mathbf{A}(\tau - t_0)\}\mathbf{B}\mathbf{u}(\tau)d\tau \qquad (4.26)$$

Pre-multiplying both sides of Eq. (4.26) by $\exp\{\mathbf{A}(t - t_0)\}$, and noting from Table 4.1 that $\exp\{-\mathbf{A}(t - t_0)\} = [\exp\{\mathbf{A}(t - t_0)\}]^{-1}$, we can write the solution state-vector as follows:

$$\mathbf{x}(t) = \exp\{\mathbf{A}(t - t_0)\}\mathbf{x}(t_0) + \int_{t_0}^{t} \exp\{\mathbf{A}(t - \tau)\}\mathbf{B}\mathbf{u}(\tau)d\tau; \quad (t \geq t_0) \qquad (4.27)$$

Note that the matrix equation, Eq. (4.27), is of the same form as the scalar equation, Eq. (4.5). Using Eq. (4.27), we can calculate the solution to the general matrix state-equation, Eq. (4.8), for $t \geq t_0$. However, we do not yet know how to calculate the state-transition matrix, $\exp\{\mathbf{A}(t - t_0)\}$.

4.2 Calculation of the State-Transition Matrix

If we can calculate the state-transition matrix, $\exp\{\mathbf{A}(t - t_0)\}$, when the state-dynamics matrix, $\mathbf{A}$, and the times, t_0 and $t \geq t_0$, are specified, our task of solving the linear state-equations will simply consist of plugging $\exp\{\mathbf{A}(t - t_0)\}$ into Eq. (4.27) and getting the solution $\mathbf{x}(t)$, provided we know the initial state-vector, $\mathbf{x}(t_0)$, and the input vector, $\mathbf{u}(t)$, for $t \geq t_0$. As stated at the beginning of Section 4.1, the easiest way to solve a matrix state-equation is by decoupling the individual scalar state-equations, which is possible only if the system has distinct eigenvalues. First, let us calculate the state-transition matrix for such a system.

For a linear system of order n, having n *distinct eigenvalues*, $\lambda_1, \lambda_2, \ldots, \lambda_n$, the *eigenvalue problem* (see Chapter 3) is written as follows:

$$\mathbf{A}\mathbf{v}_k = \lambda_k\mathbf{v}_k; \quad (k = 1, 2, \ldots, n) \qquad (4.28)$$

We know from Chapter 3 that such a system can be decoupled (or *diagonalized*) by using the following state-transformation:

$$\mathbf{x}'(t) = \mathbf{T}\mathbf{x}(t); \quad \mathbf{T} = [\mathbf{v}_1; \mathbf{v}_2; \ldots; \mathbf{v}_n]^{-1} \qquad (4.29)$$

and the state-dynamics matrix then becomes diagonalized as follows:

$$\mathbf{A}' = \mathbf{T}\mathbf{A}\mathbf{T}^{-1} = \begin{bmatrix} \lambda_1 & 0 & 0 & \ldots & 0 \\ 0 & \lambda_2 & 0 & \ldots & 0 \\ \cdot & \cdot & \cdot & \ldots & \cdot \\ 0 & 0 & 0 & \ldots & \lambda_n \end{bmatrix} \qquad (4.30)$$

You can easily show from the definition of the state-transition matrix, Eq. (4.18), that the state-transition matrix for the decoupled system is given by

$$\exp\{\mathbf{A}(t - t_0)\} = \begin{bmatrix} \exp\{\lambda_1(t - t_0)\} & 0 & 0 & \cdots & 0 \\ 0 & \exp\{\lambda_2(t - t_0)\} & 0 & \cdots & 0 \\ \cdot & \cdot & \cdot & \cdots & \cdot \\ 0 & 0 & 0 & \cdots & \exp\{\lambda_n(t - t_0)\} \end{bmatrix}$$

$$(4.31)$$

Equation (4.31) shows that the state-transition matrix for a decoupled system is a diagonal matrix. In general, the state-transition matrix for any transformed system, $\mathbf{x}'(t) = \mathbf{Tx}(t)$, can be expressed as

$$\exp\{\mathbf{A}(t - t_0)\}$$

$$= \mathbf{I} + \mathbf{A}'(t - t_0) + (\mathbf{A}')^2(t - t_0)^2/2! + (\mathbf{A}')^3(t - t_0)^3/3! + \cdots + (\mathbf{A}')^k(t - t_0)^k/k! + \cdots$$

$$= \mathbf{TT}^{-1} + (\mathbf{TAT}^{-1})(t - t_0) + (\mathbf{TAT}^{-1})^2(t - t_0)^2/2!$$

$$+ (\mathbf{TAT}^{-1})^3(t - t_0)^3/3! + \cdots + (\mathbf{TAT}^{-1})^k(t - t_0)^k/k! + \cdots$$

$$= \mathbf{T}[\mathbf{I} + \mathbf{A}(t - t_0) + \mathbf{A}^2(t - t_0)^2/2! + \mathbf{A}^3(t - t_0)^3/3! + \cdots + \mathbf{A}^k(t - t_0)^k/k! + \cdots]\mathbf{T}^{-1}$$

$$= \mathbf{T}\exp\{\mathbf{A}(t - t_0)\}\mathbf{T}^{-1} \qquad (4.32)$$

Example 4.2

Let us calculate the state-transition matrix of the following system, and then solve for the state-vector if the initial condition is $\mathbf{x}(0) = [1; \quad 0]^T$ and the applied input is $u(t) = 0$:

$$\mathbf{A} = \begin{bmatrix} -1 & 2 \\ -1 & -3 \end{bmatrix}; \quad \mathbf{B} = \begin{bmatrix} 0 \\ -1 \end{bmatrix} \qquad (4.33)$$

The eigenvalues of the system are obtained by solving the following characteristic equation:

$$|\lambda\mathbf{I} - \mathbf{A}| = \begin{vmatrix} (\lambda + 1) & -2 \\ 1 & (\lambda + 3) \end{vmatrix} = (\lambda + 1)(\lambda + 3) + 2 = \lambda^2 + 4\lambda + 5 = 0$$

$$(4.34)$$

which gives the following eigenvalues:

$$\lambda_{1,2} = -2 \pm i \qquad (4.35)$$

Note that the negative real parts of both the eigenvalues indicate an asymptotically stable system. Since the eigenvalues are distinct, the system can be decoupled using the state-transformation given by Eq. (4.29). The eigenvectors, $\mathbf{v}_1 = [v_{11}; v_{21}]^T$ and $\mathbf{v}_2 = [v_{12}; v_{22}]^T$ are calculated from Eq. (4.28). The equation $\mathbf{Av}_1 = \lambda_1\mathbf{v}_1$ yields the following scalar equations:

$$\lambda_1 v_{11} = -v_{11} + 2v_{21} \qquad (4.36a)$$

$$\lambda_1 v_{21} = -v_{11} - 3v_{21} \qquad (4.36b)$$

Note that Eqs. (4.36a) and (4.36b) are *linearly dependent*, i.e. we cannot get the *two unknowns*, v_{11}, and v_{21}, by solving these two equations. You may verify this fact by trying to solve for v_{11} and v_{21}. (This behavior of the eigenvector equations is true for a general system of order n; only $(n - 1)$ equations relating the eigenvector elements are linearly independent). The best we can do is *arbitrarily specify* one of the two unknowns, and use *either* Eq. (4.36a) *or* Eq. (4.36b) – since both give us the same relationship between v_{11} and v_{21} – to get the remaining unknown. Let us arbitrarily choose $v_{11} = 1$. Then either Eq. (4.36a) or Eq. (4.36b) gives us $v_{21} = (1 + \lambda_1)/2 = (-1 + i)/2$. Hence, the first eigenvector is $\mathbf{v}_1 = [\, 1; \quad (-1 + i)/2\,]^T$. Similarly, the second eigenvector is obtained by 'solving' $\mathbf{A}\mathbf{v}_2 = \lambda_2\mathbf{v}_2$, yielding the second eigenvector as $\mathbf{v}_2 = [\, 1; \quad (-1 - i)/2\,]^T$. Plugging the two eigenvectors in Eq. (4.29), we get the state-transformation matrix, $\mathbf{T}$, as

$$\mathbf{T} = \begin{bmatrix} v_{11} & v_{12} \\ v_{21} & v_{22} \end{bmatrix}^{-1} = \begin{bmatrix} 1 & 1 \\ (-1 + i)/2 & (-1 - i)/2 \end{bmatrix}^{-1} = \begin{bmatrix} (1 - i)/2 & -i \\ (1 + i)/2 & i \end{bmatrix}$$

(4.37)

Then the diagonalized state-dynamics matrix, $\mathbf{A}'$, is given by

$$\mathbf{A}' = \mathbf{T}\mathbf{A}\mathbf{T}^{-1} = \begin{bmatrix} \lambda_1 & 0 \\ 0 & \lambda_2 \end{bmatrix} = \begin{bmatrix} (-2 + i) & 0 \\ 0 & (-2 - i) \end{bmatrix}$$

(4.38)

and the state-transition matrix for the transformed system is

$$e^{\mathbf{A}'t} = \begin{bmatrix} \exp(\lambda_1 t) & 0 \\ 0 & \exp(\lambda_2 t) \end{bmatrix} = \begin{bmatrix} e^{(-2+i)t} & 0 \\ 0 & e^{(-2-i)t} \end{bmatrix}$$

(4.39)

Note that $t_0 = 0$ in this example. Then from Eq. (4.32) the state-transition matrix for the original system is given by

$$e^{\mathbf{A}t} = \mathbf{T}^{-1} e^{\mathbf{A}'t} \mathbf{T}$$

$$= \begin{bmatrix} 1 & 1 \\ (-1 + i)/2 & (-1 - i)/2 \end{bmatrix} \begin{bmatrix} e^{(-2+i)t} & 0 \\ 0 & e^{(-2-i)t} \end{bmatrix} \begin{bmatrix} (1 - i)/2 & -i \\ (1 + i)/2 & i \end{bmatrix}$$

$$= \begin{bmatrix} [(1 - i)e^{(-2+i)t} + (1 + i)e^{(-2-i)t}]/2 & i(e^{(-2-i)t} - e^{(-2+i)t}) \\ i(e^{(-2+i)t} - e^{(-2-i)t})/2 & [(1 + i)e^{(-2+i)t} + (1 - i)e^{(-2-i)t}]/2 \end{bmatrix}$$

(4.40)

Those with a taste for complex algebra may further simplify Eq. (4.40) by using the identity $e^{a+ib} = e^a[\cos(b) + i\sin(b)]$, where a and b are real numbers. The resulting expression for $e^{\mathbf{A}t}$ is as follows:

$$e^{\mathbf{A}t} = \begin{bmatrix} e^{-2t}[\cos(t) + \sin(t)] & 2e^{-2t}\sin(t) \\ -e^{-2t}\sin(t) & e^{-2t}[\cos(t) - \sin(t)] \end{bmatrix}; \quad (t \geq 0) \quad (4.41)$$

The solution, $\mathbf{x}(t)$, is then given by Eq. (4.27) with $\mathbf{u}(t) = 0$ as follows:

$$\mathbf{x}(t) = [x_1(t); x_2(t)]^T = e^{\mathbf{A}t}\mathbf{x}(0) = \begin{bmatrix} e^{-2t}[\cos(t) + \sin(t)] \\ -e^{-2t}\sin(t) \end{bmatrix}; \quad (t \geq 0) \quad (4.42)$$

The state variables, $x_1(t)$ and $x_2(t)$, given by Eq. (4.42) are plotted in Figure 4.1. Note that both the state variables shown in Figure 4.1 decay to zero in about 3 s, thereby confirming that the system is asymptotically stable.

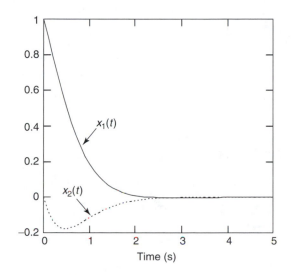

Figure 4.1   The calculated state variables, $x_1(t)$ and $x_2(t)$, for Example 4.2

The method presented in Example 4.2 for calculating the state-transition matrix is restricted to those systems which have distinct eigenvalues. The intrinsic MATLAB function *expm3* lets you use the diagonalization method for the calculation of the matrix exponential for systems with distinct eigenvalues as follows:

```
>>eP = expm3(P) <enter>
```

where **P** is a square matrix of which the matrix exponential, **eP**, is to be calculated. Alternatively, you can use the intrinsic MATLAB function *eig* as follows to calculate the eigenvector matrix, **V**, and the diagonalized matrix, **D**, with eigenvalues of **P** as its diagonal elements:

```
>>[V,D] = eig(P) <enter>
```

Then use the MATLAB function *exp* as follows to calculate the matrix exponential of **D**:

```
>>eD = exp(D) - rot90(eye(size(D))) <enter>
```

The MATLAB function *exp(D)* calculates a matrix whose elements are exponentials of the corresponding elements of the matrix **D**. The matrix exponential of **D** is obtained by subtracting the *off-diagonal* elements of *exp(D)* (which are all ones) from *exp(D)*; this is done by forming a matrix whose diagonal elements are *zeros* and whose off-diagonal elements are all *ones* – an *identity matrix* of same size as **D**, *eye(size(D)), rotated by 90* *degrees* using the command *rot90(eye(size(D)))*. Finally, the matrix exponential of **P** can be obtained as follows, using $\mathbf{eP = VeDV^{-1}}$:

```
>>eP = V*eD*inv(V) <enter>
```

Example 4.3

Using MATLAB, let us calculate the state-transition matrix for the system in Example 4.2 for $t_0 = 0$ and $t = 2$ s. First, let us use the command *expm3* as follows:

```
>>A=[-1 2; -1 -3]; eAt = expm3(A*2) <enter>

eAt =
9.0324e-003    3.3309e-002
-1.6654e-002   -2.4276e-002
```

Now let us use the alternative approach with the command *eig* as follows:

```
>>[V, D] = eig(A*2) <enter>

V =

8.1650e-001                    8.1650e-001
-4.0825e-001 +4.0825e-001i   -4.0825e-001 -4.0825e-001i

D =

-4.0000e+000+ 2.0000e+000i        0

     0          -4.0000e+000- 2.0000e+000i
```

Then the state-transition matrix of the diagonalized system is calculated as follows:

```
>>eD = exp(D) - rot90(eye(size(D))) <enter>

eD =
-7.6220e-003 +1.6654e-002i    0
     0                        -7.6220e-003 -1.6654e-002i
```

Finally, using the inverse state-transformation from the diagonalized system to the original system, we get the state-transition matrix, $e^{\mathbf{A}t}$, as follows:

```
>>eAt = V*eD*inv(V) <enter>

eAt =

9.0324e-003    3.3309e-002    -2.2646e-018i
-1.6654e-002   -2.4276e-002   +1.6505e-018i
```

which is the same result as that obtained using *expm3* (ignoring the negligible imaginary parts). You may verify the accuracy of the computed value of $e^{\mathbf{A}t}$ by comparing it with the exact result obtained in Eq. (4.41) for $t = 2$ s.

For systems with repeated eigenvalues, a general method of calculating the state-transition matrix is the Laplace transform method, in which the Laplace transform is taken of the homogeneous state-equation, Eq. (4.15) subject to the initial condition, $\mathbf{x}(0) = \mathbf{x}_0$ as follows:

$$s\mathbf{X}(s) - \mathbf{x}(0) = \mathbf{A}\mathbf{X}(s) \tag{4.43}$$

where $\mathbf{X}(s) = \mathcal{L}[\mathbf{x}(t)]$. Collecting the terms involving $\mathbf{X}(s)$ to the left-hand side of Eq. (4.43), we get

$$(s\mathbf{I} - \mathbf{A})\mathbf{X}(s) = \mathbf{x}(0) \tag{4.44}$$

$$\mathbf{X}(s) = (s\mathbf{I} - \mathbf{A})^{-1}\mathbf{x}(0) = (s\mathbf{I} - \mathbf{A})^{-1}\mathbf{x}_0 \tag{4.45}$$

Taking the inverse Laplace transform of Eq. (4.45), we get the state-vector, $\mathbf{x}(t)$, as

$$\mathbf{x}(t) = \mathcal{L}^{-1}[\mathbf{X}(s)] = \mathcal{L}^{-1}[(s\mathbf{I} - \mathbf{A})^{-1}\mathbf{x}_0] = \mathcal{L}^{-1}[(s\mathbf{I} - \mathbf{A})^{-1}]\mathbf{x}_0 \tag{4.46}$$

Comparing Eq. (4.46) with Eq. (4.16), we obtain the following expression for the state-transition matrix:

$$e^{\mathbf{A}t} = \mathcal{L}^{-1}[(s\mathbf{I} - \mathbf{A})^{-1}] \tag{4.47}$$

Thus, Eq. (4.47) gives us a general method for calculating the state-transition matrix for $t > 0$. The matrix $(s\mathbf{I} - \mathbf{A})^{-1}$ is called the *resolvent* because it helps us in solving the state-equation by calculating $e^{\mathbf{A}t}$. If the initial condition is specified at $t = t_0$, we would be interested in the state-transition matrix, $\exp\{\mathbf{A}(t - t_0)\}$, for $t > t_0$, which is obtained from Eq. (4.47) merely by substituting t by $(t - t_0)$.

Example 4.4

Consider a system with the following state-dynamics matrix:

$$\mathbf{A} = \begin{bmatrix} -2 & 1 & 5 \\ 0 & 0 & -3 \\ 0 & 0 & 0 \end{bmatrix} \tag{4.48}$$

Let us calculate the state-transition matrix and the initial response, if the initial condition is $\mathbf{x}(0) = [\,0; \quad 0; \quad 1\,]^T$. The eigenvalues of the system are calculated as follows:

$$|\lambda\mathbf{I} - \mathbf{A}| = \begin{vmatrix} (\lambda + 2) & -1 & -5 \\ 0 & \lambda & 3 \\ 0 & 0 & \lambda \end{vmatrix} = \lambda^2(\lambda + 2) = 0 \tag{4.49}$$

From Eq. (4.49) it follows that the eigenvalues of the system are $\lambda_1 = \lambda_2 = 0$, and $\lambda_3 = -2$. Since the first two eigenvalues are repeated, the system cannot be decoupled, and the approach of Example 4.2 for calculating the state-transition matrix is inapplicable. Let us apply the Laplace transform approach given by Eq. (4.47). First, the *resolvent* $(s\mathbf{I} - \mathbf{A})^{-1}$ is calculated as follows:

$$(s\mathbf{I} - \mathbf{A})^{-1} = \text{adj}((s\mathbf{I} - \mathbf{A})/|s\mathbf{I} - \mathbf{A}| = 1/[s^2(s + 2)]$$

$$\times \begin{bmatrix} s^2 & 0 & 0 \\ s & s(s+2) & 0 \\ (5s+3) & -3(s+2) & s(s+2) \end{bmatrix}^T$$

$$= \begin{bmatrix} 1/(s+2) & 1/[s(s+2)] & (5s+3)/[s^2(s+2)] \\ 0 & 1/s & -3/s^2 \\ 0 & 0 & 1/s \end{bmatrix} \quad (4.50)$$

Taking the inverse Laplace transform of Eq. (4.50) with the help of partial fraction expansions for the elements of $(s\mathbf{I} - \mathbf{A})^{-1}$ and using Table 2.1, we get the state-transition matrix as follows:

$$e^{\mathbf{A}t} = \begin{bmatrix} e^{-2t} & (1-e^{-2t})/2 & 7(1-e^{-2t})/4 + 3t/2 \\ 0 & 1 & -3t \\ 0 & 0 & 1 \end{bmatrix}; \quad (t > 0) \quad (4.51)$$

Note that the inverse Laplace transform of $1/s$ is $u_s(t)$ from Table 2.1. However, since we are interested in finding the state-transition matrix and the response only for $t > 0$ (because the response at $t = 0$ is known from the initial condition, $\mathbf{x}(0)$) we can write $\mathcal{L}^{-1}(1/s) = 1$ for $t > 0$, which has been used in Eq. (4.51). The initial response is then calculated as follows:

$$\mathbf{x}(t) = \begin{bmatrix} x_1(t) \\ x_2(t) \\ x_3(t) \end{bmatrix} = e^{\mathbf{A}t}\mathbf{x}(0) = \begin{bmatrix} 7(1-e^{-2t})/4 + 3t/2 \\ -3t \\ 1 \end{bmatrix}; \quad (t > 0) \quad (4.52)$$

Note that the term $3t/2$ makes $x_1(t)$ keep on increasing with time, $t > 0$. Similarly, $x_2(t)$ keeps on increasing with time. This confirms that the system is unstable. A plot of $x_1(t)$, $x_2(t)$, and $x_3(t)$ is shown in Figure 4.2.

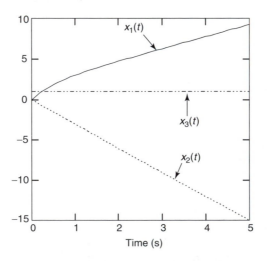

Figure 4.2 The calculated state variables, $x_1(t)$, $x_2(t)$, and $x_3(t)$, for Example 4.4

4.3 Understanding the Stability Criteria through the State-Transition Matrix

In Section 2.6 we listed the criteria by which we can judge whether a system is stable. The state-transition matrix allows us to understand the stability criteria. We saw in Example 4.2 that the elements of e^{At} are *linear combinations* of $\exp(\lambda_k t)$, where λ_k for $k = 1, 2, \ldots, n$ are the *distinct* eigenvalues of the system (Eq. (4.40)). Such elements can be expressed as $\exp(a_k t)$ multiplied by oscillatory terms, $\sin(b_k t)$ and $\cos(b_k t)$, where a_k and b_k are real and imaginary parts of the kth eigenvalue, $\lambda_k = a_k + ib_k$ (Eq. (4.41)). If the real parts, a_k, of all the eigenvalues are *negative* – as in Example 4.2 – the initial responses of *all* the state-variables will decay to zero as time, t, becomes large due to the presence of $\exp(a_k t)$ as the factor in all the elements of e^{At}. Hence, a system with *all* eigenvalues having negative real parts is *asymptotically stable*. This is the first stability criterion. By the same token, if any eigenvalue λ_k, has a *positive* real part, a_k, then the corresponding factor $\exp(a_k t)$ will diverge to infinity as time, t, becomes large, signifying an *unstable* system. This is the second stability criterion.

In Example 4.4, we saw that if a zero eigenvalue is repeated twice, it leads to the presence of terms such as ct, where c is a constant, in the elements of e^{At} (Eq. (4.51)). More generally, if an eigenvalue, λ_k, which is repeated *twice* has *zero real part* (i.e. $\lambda_k = ib_k$), then e^{At} will have terms such as $t \sin(b_k t)$ and $t \cos(b_k t)$ – and their combinations – in its elements. If an eigenvalue with zero real part is repeated *thrice*, then e^{At} will have combinations of $t^2 \sin(b_k t)$, $t^2 \cos(b_k t)$, $t \sin(b_k t)$, and $t \cos(b_k t)$ in its elements. Similarly, for eigenvalues with zero real parts repeated larger number of times, there will be *higher powers* of t present as *coefficients* of the oscillatory terms in the elements of e^{At}. Hence, if any eigenvalue, λ_k, having zero real part is *repeated two or more times*, the presence of powers of t as coefficients of the oscillatory terms, $\sin(b_k t)$ and $\cos(b_k t)$, causes elements of e^{At} to blow-up as time, t, increases, thereby indicating an *unstable* system. This is the third stability criterion.

Note that individual initial responses to a *specific initial condition* may not be sufficient to tell us whether a system is stable. This is seen in the following example.

Example 4.5

Reconsider the system of Example 4.4 with the initial condition, $\mathbf{x}(0) = [\, 1; \quad 0; \quad 0\,]^T$. Substituting the initial condition into Eq. (4.16), we get the following initial response for the system:

$$\mathbf{x}(t) = \begin{bmatrix} x_1(t) \\ x_2(t) \\ x_3(t) \end{bmatrix} = e^{At} \; \mathbf{x}(0) = \begin{bmatrix} e^{-2t} \\ 0 \\ 0 \end{bmatrix}; \quad (t > 0) \qquad (4.53)$$

Equation (4.53) indicates that $\mathbf{x}(t) \to 0$ as $t \to \infty$. Thus, a system we *know* to be *unstable* from Example 4.4 (and from the third stability criterion), has an initial response decaying asymptotically to zero when the initial condition is $\mathbf{x}(0) = [\, 1; \quad 0; \quad 0\,]^T$, which is the characteristic of an asymptotically stable system.

Example 4.5 illustrates that we can be fooled into believing that a system is stable if we look at its initial response to only *some specific* initial conditions. A true mirror of the system's stability is its state-transition matrix, which reflects the three stability criteria. If *any element* of the state-transition matrix grows to infinity as time becomes large, the system is *unstable*. Then it is possible to find at least *one* initial condition that leads to an unbounded initial response. The state-transition matrix contains information about how a system will respond to an *arbitrary* initial condition. Hence, the stability of a system is deduced from *all possible* initial conditions (i.e. the state-transition matrix), rather than from only some *specific* ones.

4.4 Numerical Solution of Linear Time-Invariant State-Equations

In the previous sections, we saw two methods for calculating the state-transition matrix, which is required for the solution of the linear state-equations. The diagonalization method works only if the system eigenvalues are distinct. Calculating the state-transition matrix by the inverse Laplace transform method of Eq. (4.47) is a tedious process, taking into account the matrix inversion, partial fraction expansion, and inverse Laplace transformation of each element of the resolvent, as Example 4.4 illustrates. While the partial fraction expansions can be carried out using the intrinsic MATLAB function *residue*, the other steps must be performed by hand. Clearly, the utility of Eq. (4.47) is limited to small order systems. Even for the systems which allow easy calculation of the state-transition matrix, the calculation of the steady-state response requires time integration of the input terms (Eq. (4.27)), which is no mean task if the inputs are arbitrary functions of time.

The definition of the matrix exponential, $\exp\{\mathbf{A}(t - t_0)\}$, by the Taylor series expansion of Eq. (4.18) gives us another way of calculating the state-transition matrix. However, since Eq. (4.18) requires evaluation of an *infinite* series, the *exact* calculation of $\exp\{\mathbf{A}(t - t_0)\}$ is impossible by this approach. Instead, we use an *approximation* to $\exp\{\mathbf{A}(t - t_0)\}$ in which only a *finite* number of terms are retained in the series on the right-hand side of Eq. (4.18):

$$\exp\{\mathbf{A}(t - t_0)\} \approx \mathbf{I} + \mathbf{A}(t - t_0) + \mathbf{A}^2(t - t_0)^2/2!$$
$$+ \mathbf{A}^3(t - t_0)^3/3! + \cdots + \mathbf{A}^N(t - t_0)^N/N! \tag{4.54}$$

Note that the approximation given by Eq. (4.54) consists of powers of $\mathbf{A}(t - t_0)$ up to N. In Eq. (4.54), we have neglected the following infinite series, called the *remainder series*, $\mathbf{R}_N$, which is also the *error* in our approximation of $\exp\{\mathbf{A}(t - t_0)\}$:

$$\mathbf{R}_N = \mathbf{A}^{N+1}(t - t_0)^{N+1}/(N + 1)! + \mathbf{A}^{N+2}(t - t_0)^{N+2}/(N + 2)!$$
$$+ \mathbf{A}^{N+3}(t - t_0)^{N+3}/(N + 3)! + \cdots = \sum_{k=N+1}^{\infty} \mathbf{A}^k(t - t_0)^k/k! \tag{4.55}$$

Clearly, the accuracy of the approximation in Eq. (4.54) depends upon how large is the error, $\mathbf{R}_N$, given by Eq. (4.55). Since $\mathbf{R}_N$ is a matrix, when we ask how large is the error, we mean how large is each element of $\mathbf{R}_N$. The magnitude of the matrix, $\mathbf{R}_N$, is a matrix consisting of *magnitudes* of the elements of $\mathbf{R}_N$. However, it is quite useful to assign a scalar quantity, called the *norm*, to measure the magnitude of a matrix. There are several ways in which the norm of a matrix can be defined, such as the sum of the magnitudes of all the elements, or the square-root of the sum of the squares of all the elements. Let us assign such a scalar norm to measure the magnitude of the error matrix, $\mathbf{R}_N$, and denote it by the symbol $\|\mathbf{R}_N\|$, which is written as follows:

$$\|\mathbf{R}_N\| = \left\| \sum_{k=N+1}^{\infty} \mathbf{A}^k (t-t_0)^k / k! \right\| \leq \sum_{k=N+1}^{\infty} \|\mathbf{A}^k\| (t-t_0)^k / k \tag{4.56}$$

The inequality on the right-hand side of Eq. (4.56) is due to the well known *triangle inequality*, which implies that if a and b are real numbers, then $|a+b| \leq |a| + |b|$. Now, for our approximation of Eq. (4.54) to be accurate, the first thing we require is that the magnitude of error, $\|\mathbf{R}_N\|$, be a *finite quantity*. Secondly, the error magnitude should be *small*. The first requirement is met by noting that the Taylor series of Eq. (4.18) is *convergent*, i.e. the successive terms of the series become smaller and smaller. The inifinite series on the extreme right-hand side of Eq. (4.56) – which is a part of the Taylor series – is also finite. Hence, irrespective of the value of N, $\|\mathbf{R}_N\|$ is always finite. From Eq. (4.56), we see that the approximation error can be made small in two ways: (a) by *increasing* N, and (b) by *decreasing* $(t-t_0)$. The implementation of Eq. (4.54) in a computer program can be done using an *algorithm* which selects the highest power, N, based on the desired accuracy, i.e. the error given by Eq. (4.56). MATLAB uses a similar algorithm in its function named *expm2* which computes the matrix exponential using the finite series approximation. Other algorithms based on the finite series approximation to the matrix exponential are given in Golub and van Loan [1] and Moler and van Loan [2]. The accuracy of the algorithms varies according to their implementation. The MATLAB functions *expm* and *expm1* use two different algorithms for the computation of the matrix exponential based on Laplace transform of the finite-series of Eq. (4.54) – which results in each element of the matrix exponential being approximated by a rational polynomial in s, called the *Padé approximation*. Compared to *expm2* – which directly implements the finite Taylor series approximation – *expm* and *expm1* are more accurate.

There is a limit to which the number of terms in the approximation can be increased. Therefore, for a given N, the accuracy of approximation in Eq. (4.54) can be increased by making $(t-t_0)$ small. How small is small enough? Obviously, the answer depends upon the system's dynamics matrix, $\mathbf{A}$, as well as on N. If $(t-t_0)$ is chosen to be small, how will we evaluate the state-transition matrix for large time? For this purpose, we will use the *time-marching* approach defined by the following property of the state-transition matrix (Table 4.1):

$$\exp\{\mathbf{A}(t-t_0)\} = \exp\{\mathbf{A}(t-t_1)\} \exp\{\mathbf{A}(t_1-t_0)\} \tag{4.57}$$

where $t_0 < t_1 < t$. The *time-marching* approach for the computation of the state-transition matrix consists of evaluating $e^{\mathbf{A}\Delta t}$ as follows using Eq. (4.54):

$$e^{\mathbf{A}\Delta t} \approx \mathbf{I} + \mathbf{A}\Delta t + \mathbf{A}^2 (\Delta t)^2 / 2! + \mathbf{A}^3 (\Delta t)^3 / 3! + \cdots + \mathbf{A}^N (\Delta t)^N / N! \tag{4.58}$$

where Δt is a *small time-step*, and then marching ahead in time – like an army marches on with fixed footsteps – using Eq. (4.57) with $t = t_0 + n\Delta t$ and $t_1 = t_0 + (n-1)\Delta t$ as follows:

$$\exp\{\mathbf{A}(t_0 + n\Delta t)\} = e^{\mathbf{A}\Delta t} \exp\{\mathbf{A}[t_0 + (n-1)\Delta t]\} \qquad (4.59)$$

Equation (4.59) allows successive evaluation of $\exp\{\mathbf{A}(t_0 + n\Delta t)\}$ for $n = 1, 2, 3, \ldots$ until the final time, t, is reached. The time-marching approach given by Eqs. (4.58) and (4.59) can be easily programmed on a digital computer. However, instead of finding the state-transition matrix at time, $t > t_0$, we are more interested in obtaining the solution of the state-equations, Eq. (4.8), $\mathbf{x}(t)$, when initial condition is specified at time t_0. To do so, let us apply the time-marching approach to Eq. (4.27) by substituting $t = t_0 + n\Delta t$, and writing the solution after n time-steps as follows:

$$\mathbf{x}(t_0 + n\Delta t) = e^{\mathbf{A}n\Delta t}\mathbf{x}(t_0) + \int_{t_0}^{t_0+n\Delta t} \exp\{\mathbf{A}(t_0 + n\Delta t - \tau)\}\mathbf{B}\mathbf{u}(\tau)d\tau; \quad (n = 1, 2, 3, \ldots)$$

$$(4.60)$$

For the first time step, i.e. $n = 1$, Eq. (4.60) is written as follows:

$$\mathbf{x}(t_0 + \Delta t) = e^{\mathbf{A}\Delta t}\mathbf{x}(t_0) + \int_{t_0}^{t_0+\Delta t} \exp\{\mathbf{A}(t_0 + \Delta t - \tau)\}\mathbf{B}\mathbf{u}(\tau)d\tau \qquad (4.61)$$

The integral term in Eq. (4.61) can be expressed as

$$\int_{t_0}^{t_0+\Delta t} \exp\{\mathbf{A}(t_0 + \Delta t - \tau)\}\mathbf{B}\mathbf{u}(\tau)d\tau = e^{\mathbf{A}\Delta t} \int_{0}^{\Delta t} e^{-\mathbf{A}T}\mathbf{B}\mathbf{u}(t_0 + T)dT \qquad (4.62)$$

where $T = \tau - t_0$. Since the time step, Δt, is small, we can *assume* that the integrand vector $e^{-\mathbf{A}T}\mathbf{B}\mathbf{u}(t_0 + T)$ is *essentially constant* in the interval $0 < T < \Delta t$, and is equal to $e^{-\mathbf{A}\Delta t}\mathbf{B}\mathbf{u}(t_0 + \Delta t)$. Thus, we can *approximate* the integral term in Eq. (4.62) as follows:

$$e^{\mathbf{A}\Delta t} \int_{0}^{\Delta t} e^{-\mathbf{A}T}\mathbf{B}\mathbf{u}(t_0 + T)dT \approx e^{\mathbf{A}\Delta t}e^{-\mathbf{A}\Delta t}\mathbf{B}\mathbf{u}(t_0 + \Delta t)\Delta t = \mathbf{B}\mathbf{u}(t_0 + \Delta t)\Delta t = \mathbf{B}\mathbf{u}(t_0)\Delta t$$

$$(4.63)$$

Note that in Eq. (4.63), we have used $\mathbf{u}(t_0 + \Delta t) = \mathbf{u}(t_0)$, because the input vector is assumed to be *constant* in the interval $t_0 < t < t_0 + \Delta t$. Substituting Eq. (4.63) into Eq. (4.61), the approximate solution after the first time step is written as

$$\mathbf{x}(t_0 + \Delta t) \approx e^{\mathbf{A}\Delta t}\mathbf{x}(t_0) + \mathbf{B}\mathbf{u}(t_0)\Delta t \qquad (4.64)$$

For the next time step, i.e. $n = 2$ and $t = t_0 + 2\Delta t$, we can use the solution after the first time step, $\mathbf{x}(t_0 + \Delta t)$, which is already known from Eq. (4.64), as the initial condition and, assuming that the input vector is constant in the interval $t_0 + \Delta t < t < t_0 + 2\Delta t$, the solution can be written as follows:

$$\mathbf{x}(t_0 + 2\Delta t) \approx e^{\mathbf{A}\Delta t}\mathbf{x}(t_0 + \Delta t) + \mathbf{B}\mathbf{u}(t_0 + \Delta t)\Delta t \qquad (4.65)$$

The process of time-marching, i.e. using the solution after the previous time step as the initial condition for calculating the solution after the next time step, is continued and the solution after n time steps can be approximated as follows:

$$\mathbf{x}(t_0 + n\Delta t) \approx e^{\mathbf{A}\Delta t}\mathbf{x}(t_0 + (n-1)\Delta t) + \mathbf{B}\mathbf{u}(t_0 + (n-1)\Delta t)\Delta t; \quad (n = 1, 2, 3, \ldots)$$

(4.66)

A special case of the system response is to the *unit impulse inputs*, i.e. $\mathbf{u}(t) = \delta(t - t_0)[1; \quad 1; \quad \ldots; \quad 1]^T$. In such a case, the integral in Eq. (4.62) is exactly evaluated as follows, using the *sampling property* of the unit impulse function, $\delta(t - t_0)$, given by Eq. (2.24):

$$\int_{t_0}^{t_0+\Delta t} \exp\{\mathbf{A}(t_0 + \Delta t - \tau)\}\mathbf{B}\mathbf{u}(\tau)d\tau = e^{\mathbf{A}\Delta t}\mathbf{B}$$

(4.67)

which results in the following solution:

$$\mathbf{x}(t_0 + n\Delta t) = e^{\mathbf{A}\Delta t}[\mathbf{x}(t_0 + (n-1)\Delta t) + \mathbf{B}\Delta t]; \quad (n = 1, 2, 3, \ldots)$$

(4.68)

Note that the solution given by Eq. (4.68) is an exact result, and is valid only if all the inputs are unit impulse functions applied at time $t = t_0$.

By the time-marching method of solving the state-equations we have essentially converted the *continuous-time* system, given by Eq. (4.8), to a *discrete-time* (or *digital*) system given by Eq. (4.66) (or, in the special case of unit impulse inputs, by Eq. (4.67)). The difference between the two is enormous, as we will see in Chapter 8. While in a continuous-time system the time is smoothly changing and can assume *any real value*, in a digital system the time can *only be an integral multiple of the time step*, Δt (i.e. Δt multiplied by an integer). The continuous-time system is clearly the limiting case of the digital system in the limit $\Delta t \to 0$. Hence, the accuracy of approximating a continuous-time system by a digital system is crucially dependent on the size of the time step, Δt; the accuracy improves as Δt becomes smaller. The *state-equation* of a linear, time-invariant, digital system with $t_0 = 0$ can be written as

$$\mathbf{x}(n\Delta t) = \mathbf{A}_d\mathbf{x}((n-1)\Delta t) + \mathbf{B}_d\mathbf{u}((n-1)\Delta t); \quad (n = 1, 2, 3, \ldots)$$

(4.69)

where $\mathbf{A}_d$ and $\mathbf{B}_d$ are the digital state coefficient matrices. Comparing Eqs. (4.66) and (4.69) we find that the solution of a continuous-time state-equation is approximated by the solution of a digital state-equation with $\mathbf{A}_d = e^{\mathbf{A}\Delta t}$ and $\mathbf{B}_d = \mathbf{B}\Delta t$ when the initial condition is specified at time $t = 0$. The digital solution, $\mathbf{x}(n\Delta t)$, is simply obtained from Eq. (4.69) using the time-marching method starting from the initial condition, $\mathbf{x}(0)$, and assuming that the input vector is constant during each time step. The digital solution of Eq. (4.69) is easily implemented on a *digital computer*, which itself works with a *non-zero time step* and input signals that are specified over each time step (called *digital signals*).

The assumption of a constant input vector during each time step, used in Eq. (4.63), results in a *staircase* like approximation of $\mathbf{u}(t)$ (Figure 4.3), and is called a *zero-order hold*, i.e. a zero-order linear interpolation of $\mathbf{u}(t)$ during each time step, $(n-1)\Delta t < t < n\Delta t$. The zero-order hold approximates $\mathbf{u}(t)$ by a step function in each time step. It is a good approximation even with a large time step, Δt, if $\mathbf{u}(t)$ itself is a *step* like function in continuous-time, such as a *square wave*. However, if $\mathbf{u}(t)$ is a smooth function in continuous-time, then it is more accurate to use a higher order interpolation to approximate $\mathbf{u}(t)$ in each time step, rather than using the zero-order hold. One such approximation is

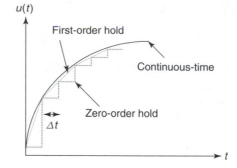

Figure 4.3 The zero-order and first-order hold digital approximations of a continuous-time input, $u(t)$

the *first-order hold* which approximates $\mathbf{u}(t)$ as a *ramp function* (i.e. a first-order linear interpolation) in each time step $(n-1)\Delta t < t < n\Delta t$ (Figure 4.3).

The conversion of a continuous-time system to the corresponding digital approximation using the zero-order hold for the input vector (Eq. (4.66)) is performed by the MATLAB Control System Toolbox (CST) function *c2d*, which calculates $e^{A\Delta t}$ using the intrinsic MATLAB function *expm*. The command *c2d* is employed as follows:

```
>>sysd = c2d(sysc,Ts,'method') <enter>
```

where *sysc* is the continuous-time state-space LTI object, Ts is the specified time step, (Δt), and *sysd* is the resulting digital state-space approximation of Eq. (4.69). The *'method'* allows a user to select among zero-order hold (*'zoh'*), first-order hold (*'foh'*), or higher-order interpolations for the input vector, called *Tustin* (or *bilinear*) approximation (*'tustin'*), and Tustin interpolation with *frequency prewarping* (*'prewarp'*). The Tustin approximation involves a *trapezoidal* approximation for $\mathbf{u}^{(1)}(t)$ in each time step (we will discuss the *Tustin* approximation a little more in Chapter 8). Tustin interpolation with frequency prewarping (*'prewarp'*) is a more accurate interpolation than plain *'tustin'*. An alternative to *c2d* is the CST function *c2dm*, which lets the user work directly with the state coefficient matrices rather than the LTI objects of the continuous time and digital systems as follows:

```
>>[Ad,Bd,Cd,Dd] = c2dm(A,B,C,D,Ts,'method') <enter>
```

where **A, B, C, D** are the continuous-time state-space coefficient matrices, **Ad, Bd, Cd, Dd** are the returned *digital* state-space coefficient matrices, and Ts and *'method'* are the same as in *c2d*. For more information on these MATLAB (CST) commands, you may refer to the *Users' Guide* for MATLAB *Control System Toolbox* [3].

How large should be the step size, Δt, selected in obtaining the digital approximation given by Eq. (4.69)? This question is best answered by considering how fast the system is likely to respond to a given initial condition, or to an applied input. Obviously, a fast changing response will not be captured very accurately by using a large Δt. Since the state-transition matrix, $e^{A\Delta t}$, has elements which are combinations of $\exp(\lambda_k \Delta t)$, where λ_k, $k = 1, 2$, etc., are the eigenvalues of **A**, it stand to reason that the time step,

Δt, should be small enough to accurately evaluate the *fastest changing element* of $e^{A\Delta t}$, which is represented by the eigenvalue, λ_k, corresponding to the largest *natural frequency*. Recall from Chapter 2 that the natural frequency is associated with the *imaginary part*, b, of the eigenvalue, $\lambda_k = a + bi$, which leads to *oscillatory* terms such as $\sin(b\Delta t)$ and $\cos(b\Delta t)$ in the elements of $e^{A\Delta t}$ (Example 4.2). Hence, we should select Δt such that $\Delta t < 1/|b|_{max}$ where $|b|_{max}$ denotes the largest imaginary part magnitude of all the eigenvalues of A. To be on the safe-side of accuracy, it is advisable to make the time step smaller than the *ten times* the reciprocal of the largest imaginary part magnitude, i.e. $\Delta t < 0.1/|b|_{max}$. If all the eigenvalues of a system are real, then the oscillatory terms are absent in the state-transition matrix, and one can choose the time step to be smaller than the reciprocal of the largest *real part* magnitude of all the eigenvalues of the system, i.e. $\Delta t < 1/|a|_{max}$.

Once a digital approximation, Eq. (4.69), to the linear, time-invariant, continuous-time system is available, the MATLAB (CST) command *ltitr* can be used to solve for $x(n\Delta t)$ using time-marching with $n = 1, 2, 3, \ldots$, given the initial condition, $x(0)$, and the input vector, $u(t)$, at the time points, $t = (n-1)\Delta t$, $n = 1, 2, 3, \ldots$ as follows:

```
>>x = ltitr(Ad,Bd,u,x0) <enter>
```

where **Ad, Bd**, are the digital state-space coefficient matrices, **x0** is the initial condition vector, **u** is a matrix having as many columns as there are inputs, and the ith row of **u** corresponds to the ith time point. **x** is the returned matrix with as many columns as there

Table 4.2 Listing of the M-file *march.m*

---

march.m

```
function [y,X] = march(A,B,C,D,X0,t,u,method)
% Time-marching solution of linear, time-invariant
% state-space equations using the digital approximation.
% A= state dynamics matrix; B= state input coefficient matrix;
% C= state output coefficient matrix;
% D= direct transmission matrix;
% X0= initial state vector; t= time vector.
% u=matrix with the ith input stored in the ith column, and jth row
% corresponding to the jth time point.
% y= returned output matrix with ith output stored in the ith column,
% and jth row corresponding to the jth time point.
% X= returned state matrix with ith state variable stored in the ith
% column, and jth row corresponding to the jth time point.
% method= method of digital interpolation for the inputs(see 'c2dm')
% copyright(c)2000 by Ashish Tewari
n=size(t,2);
dt=t(2)-t(1);
% digital approximation of the continuous-time system:-
[ad,bd,cd,dd]=c2dm(A,B,C,D,dt,method);
% solution of the digital state-equation by time-marching:-
X=ltitr(ad,bd,u,X0);
% calculation of the outputs:-
y=X*C'+u*D';
```

are state variables, and with the *same number* of rows as **u**, with the ith row corresponding to the ith time-point (the first row of **x** consists of the elements of **x0**).

The entire solution procedure for the state-space equation using the digital approximation of Eq. (4.66) can be programmed in a new M-file named *march*, which is tabulated in Table 4.2. This M-file can be executed as follows:

```
>>[y,x] = march(A,B,C,D,x0,t,u,'method') <enter>
```

where **A, B, C, D, x0, u, x**, and '*method*' are the same as those explained previously in the usage of the MATLAB (CST) command *c2dm*, while **t** is the time vector containing the equally spaced time-points at which the input, **u**, is specified, and **y** is the returned matrix containing the outputs of the system in its columns, with each row of **y** corresponding to a different time-point.

Example 4.6

Using the time-marching approach, let us calculate the response of the system given in Example 4.1 when $u(t) = u_s(t)$. Since the largest eigenvalue is -3, we can select the time step to be $\Delta t < 1/3$, or $\Delta t < 0.333$. Selecting $\Delta t = 0.1$, we can generate the time vector regularly spaced from $t = 0$ to $t = 2$ s, and specify the unit step input, u, as follows:

```
>>t=0:0.1:2; u=ones(size(t,2),1); <enter>
```

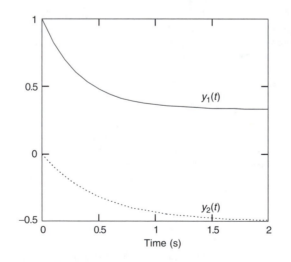

Figure 4.4 The calculated outputs, $y_1(t)$ and $y_2(t)$, for Example 4.6

Then, the M-file *march.m* is executed with zero-order hold, after specifying the state coefficient matrices and the initial condition vector (it is assumed that the outputs are the state variables themselves, i.e. $\mathbf{C} = \mathbf{I}$, $\mathbf{D} = \mathbf{0}$) as follows:

```
>>A= [-3 0; 0 -2]; B = [1; -1]; C = eye(2); D = zeros(2,1); X0 = [1; 0]; <enter>

>>[y,X] = march(A,B,C,D,X0,t,u,'zoh'); <enter>
```

The outputs, which are also the two state variables, $x_1(t)$ and $x_2(t)$, are plotted against the time vector, **t**, in Figure 4.4 as follows:

```
>>plot(t,y) <enter>
```

You may verify that the result plotted in Figure 4.4 is almost indistinguishable from the analytical result obtained in Eqs. (4.13) and (4.14), which for a unit step input yield the following:

$$x_1(t) = (2e^{-3t} + 1)/3; \quad x_2(t) = (e^{-2t} - 1)/2 \qquad (4.70)$$

Example 4.7

In Examples 2.10 and 3.10, we saw how the linearized longitudinal motion of an aircraft can be represented by appropriate transfer functions and state-space representations. These examples had involved the assumption that the structure of the aircraft is *rigid*, i.e. the aircraft does not get *deformed* by the *air-loads* acting on it. However, such an assumption is invalid, because most aircraft have rather *flexible* structures. Deformations of a flexible aircraft under changing air-loads caused by the aircraft's motion, result in a complex dynamics, called *aeroelasticity*. Usually, the *short-period mode* has a frequency closer to that of the elastic motion, while the *phugoid mode* has little aeroelastic effect. The longitudinal motion of a *flexible bomber* aircraft is modeled as a second order short-period mode, a second-order *fuselage bending mode*, and two first-order *control-surface actuators*. The sixth order system is described by the following linear, time-invariant, state-space representation:

$$\mathbf{A} = \begin{bmatrix} 0.4158 & 1.025 & -0.00267 & -0.0001106 & -0.08021 & 0 \\ -5.5 & -0.8302 & -0.06549 & -0.0039 & -5.115 & 0.809 \\ 0 & 0 & 0 & 1.0 & 0 & 0 \\ -1040 & -78.35 & -34.83 & -0.6214 & -865.6 & -631 \\ 0 & 0 & 0 & 0 & -75 & 0 \\ 0 & 0 & 0 & 0 & 0 & -100 \end{bmatrix}$$

$$\mathbf{B} = \begin{bmatrix} 0 & 0 \\ 0 & 0 \\ 0 & 0 \\ 0 & 0 \\ 75 & 0 \\ 0 & 100 \end{bmatrix}$$

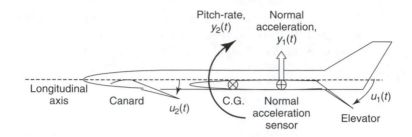

Pitch-rate, $y_2(t)$ Normal acceleration, $y_1(t)$

Longitudinal axis   Canard $u_2(t)$ C.G.   Normal acceleration sensor $u_1(t)$ Elevator

Figure 4.5 Inputs and outputs for the longitudinal dynamics of a flexible bomber aircraft

$$\mathbf{C} = \begin{bmatrix} -1491 & -146.43 & -40.2 & -0.9412 & -1285 & -564.66 \\ 0 & 1.0 & 0 & 0 & 0 & 0 \end{bmatrix}$$

$$\mathbf{D} = \begin{bmatrix} 0 & 0 \\ 0 & 0 \end{bmatrix} \tag{4.71}$$

The inputs are the *desired elevator deflection* (rad.), $u_1(t)$, and the *desired canard deflection* (rad.), $u_2(t)$, while the outputs are the sensor location's *normal acceleration* (m/s^2), $y_1(t)$, and the *pitch-rate* (rad./s), $y_2(t)$. See Figure 4.5 for a description of the inputs and outputs.

Let us calculate the response of the system if the initial condition and the input vector are the following:

$$\mathbf{x}(0) = [\, 0.1; \quad 0; \quad 0; \quad 0; \quad 0; \quad 0\,]^T; \quad \mathbf{u}(t) = \begin{bmatrix} -0.1\sin(10t) \\ \sin(12t) \end{bmatrix} \tag{4.72}$$

First, let us select a proper time step for solving the state-equations. The system's eigenvalues are calculated using the command *damp* as follows:

```
>>damp(A) <enter>
```

```
Eigenvalue                 Damping        Freq. (rad/sec)
-4.2501e-001+1.8748e+000i  2.2109e-001    1.9224e+000
-4.2501e-001-1.8748e+000i  2.2109e-001    1.9224e+000
-5.0869e-001+6.0289e+000i  8.4077e-002    6.0503e+000
-5.0869e-001-6.0289e+000i  8.4077e-002    6.0503e+000
-7.5000e+001               1.0000e+000    7.5000e+001
-1.0000e+002               1.0000e+000    1.0000e+002
```

The largest imaginary part magnitude of the eigenvalues is 6.03, while the largest real part magnitude is 100. Therefore, from our earlier discussion, the time step should be selected such that $\Delta t < 0.1/6$ s and $\Delta t < 1/100$ s. Clearly, selecting the smaller of the two numbers, i.e. $\Delta t < 1/100$ s, will satisfy both the inequalities.

Hence, we select $\Delta t = 1/150$ s $= 6.6667e - 003$ s. The time vector, **t**, the input matrix, **u**, and initial condition vector, **x0**, are then specified as follows:

```
>>t = 0:6.6667e-3:5; u = [-0.05*sin(10*t); 0.05*sin(12*t)]';
  X0 = [0.1 zeros(1,5)]'; <enter>
```

Then the M-file *march.m* is used with a first-order hold for greater accuracy (since the inputs are smoothly varying) as follows:

```
>>[yf,Xf] = march(A,B,C,D,X0,t,u,'foh'); <enter>
```

To see how much is the difference in the computed outputs if a less accurate zero-order hold is used, we re-compute the solution using *march.m* with '*zoh*' as the '*method*':

```
>>[yz,Xz] = march(A,B,C,D,X0,t,u,'zoh'); <enter>
```

The computed outputs, $y_1(t)$, and $y_2(t)$, are plotted in Figures 4.6 and 4.7, respectively, for the zero-order and first-order holds. It is observed in Figure 4.6 that the output $y_1(t)$ calculated using the first-order hold has slightly lower peaks when compared to that calculated using the zero-order hold. Figure 4.7 shows virtually no difference between the values of $y_2(t)$ calculated by zero-order and first-order holds.

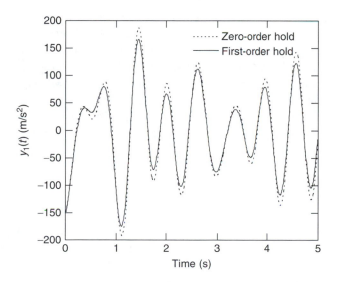

Figure 4.6 The normal acceleration output, $y_1(t)$, for the flexible bomber aircraft of Example 4.7

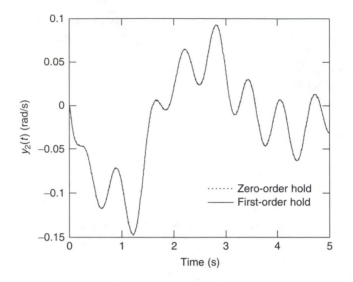

Figure 4.7 The pitch-rate output, $y_2(t)$, of the flexible bomber aircraft of Example 4.7

The MATLAB (CST) function *lsim* is an alternative to *march* for solving the state-equations by digital approximation, and is used as follows:

```
>>[y,t,x] = lsim(sys,u,t,x0,'method'); <enter>
```

where *sys* is an LTI object of the system, while the arguments u, t, 'method' (either '*zoh*', or '*foh*') and the returned output matrix, y, and state solution matrix, x, are defined in the same manner as in the M-file *march*. The user *need not* specify which interpolation method between '*zoh*' and '*foh*' has to be used in *lsim*. If a '*method*' is not specified, the function *lsim* checks the shape of the input, and applies a zero-order hold to the portions which have step-like changes, and the first-order hold to the portions which are smooth functions of time. In this regard, *lsim* is more efficient than *march*, since it optimizes the interpolation of the input, $\mathbf{u}(t)$, portion by portion, instead of applying a user specified interpolation to the entire input done by *march*. However, *lsim* can be confused if there are rapid changes between smooth and step-like portions of the input. Hence, there is a need for selecting a small time step for rapidly changing inputs in *lsim*.

Example 4.8

For the system in Example 4.7, compare the solutions obtained using the MATLAB (CST) command *lsim* and the M-file *march* when the initial condition is $\mathbf{x0} = [\,0.1; \quad 0; \quad 0; \quad 0; \quad 0; \quad 0\,]^T$ when the elevator input, $u_1(t)$ is a rectangular pulse applied at $t = 0$ with amplitude 0.05 rad. and duration 0.05 s, while the canard input, $u_2(t)$, is a sawtooth pulse applied at $t = 0$ with amplitude 0.05 rad. and duration 0.05 s. The rectangular pulse and the sawtooth pulse are defined in Examples 2.5 and 2.7, respectively.

Mathematically, we can express the input vector as follows, using Eqs. (2.32) and (2.33):

$$\mathbf{u}(t) = \begin{bmatrix} 0.05[u_s(t+0.05) - u_s(t)] \\ r(t) - r(t-0.05) - 0.05u_s(t-0.05) \end{bmatrix} \tag{4.73}$$

where $u_s(t)$ and $r(t)$ are the unit step and unit ramp functions, respectively. Using MATLAB, the inputs are generated as follows:

```
>>dt=6.6667e-3; t=0:dt:5; u1=0.05*(t<0.05+dt); i=find(t<0.05+dt);
  u2=u1; u2(i)=t(i); u = [u1' u2']; <enter>
```

Assuming that **A, B, C, D, x0** are already available in the MATLAB workspace from Example 4.7, we can calculate the response of the system using *march* as follows:

```
>>[Y1,X1] = march(A,B,C,D,X0,t,u,'zoh'); <enter>
```

where **Y1** is the returned output matrix for zero-order hold. For comparison, the solution is also obtained using *lsim* and the output is stored in matrix **Y2** as follows:

```
>>sys=ss(A,B,C,D); [Y2,t,X2] = lsim(sys,u,t,X0); <enter>
```

The computed outputs, $y_1(t)$ and $y_2(t)$, by the two different methods are compared in Figures 4.8 and 4.9, respectively. Note that the responses calculated using *march* with zero-order hold and *lsim* are indistinguishable.

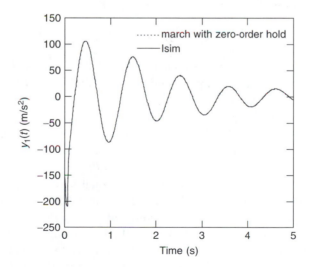

Figure 4.8 Normal acceleration output for the flexible bomber aircraft in Example 4.8 with rectangular pulse elevator input and sawtooth pulse canard input

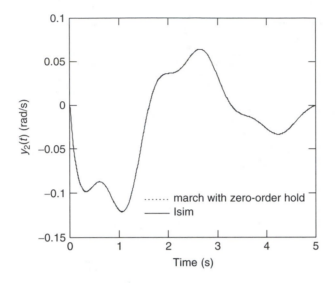

Figure 4.9 Pitch-rate output for the flexible bomber aircraft in Example 4.8 with rectangular pulse elevator input and sawtooth pulse canard input

When the inputs are impulse functions, the M-files *march* and *lsim* cannot be used directly, because it is impossible to describe the impulse function (which, by definition, goes to infinity in almost zero time) by an input vector. Instead, the digital approximation is obtained using an equation similar to Eq. (4.68), which gives the solution if all the inputs are unit impulse functions applied at $t = t_0$. For a more general case, i.e. when the input vector is given by $\mathbf{u}(t) = \delta(t - t_0)[\, c_1; \quad c_2; \quad \ldots; \quad c_m\,]^T$, where $c_1, c_2, \ldots, c_m$ are constants, we can write the solution to the state-equation as follows:

$$\mathbf{x}(t_0 + n\Delta t) = e^{\mathbf{A}\Delta t}[\mathbf{x}(t_0 + (n-1)\Delta t) + \mathbf{Bc}\Delta t]; \quad (n = 1, 2, 3, \ldots) \tag{4.74}$$

where $\mathbf{c} = [\, c_1; \quad c_2; \quad \ldots; \quad c_m\,]^T$. Comparing Eqs. (4.68) and (4.74), we find that the digital approximation for impulse inputs is given by $\mathbf{A}_d = e^{\mathbf{A}\Delta t}$ and $\mathbf{B}_d = e^{\mathbf{A}\Delta t}\mathbf{B}\Delta t$, if and only if the input vector is given by $\mathbf{u}(t) = u_s(t)\mathbf{c}$. For calculating the response to general impulse inputs of this type applied at $t = 0$, we can write a MATLAB M-file in a manner similar to *march.m*. Such an M-file, called *impgen.m* is given in Table 4.3. MATLAB (CST) does have a standard M-file for calculating the impulse response called *impulse.m*, which, however, is limited to the special case when *all* of the inputs are simultaneous *unit* impulses, i.e. all elements of the vector $\mathbf{c}$ are equal to 1. Clearly, *impgen* is more versatile than *impulse*. (MATLAB (CST) also has a dedicated function for calculating the step response, called *step*, which also considers *all inputs to be simultaneous unit step functions*.) The advantage of using the MATLAB command *impulse* (and *step*) lies in quickly checking a new control design, without having to generate the time vector, because the time vector is automatically generated. Also, for systems that are *not* strictly proper (i.e. $\mathbf{D} \neq \mathbf{0}$) the CST function *impulse* disregards the impulse in the response at

Table 4.3 Listing of the M-file *impgen.m*

---

impgen.m

```
function [y,X] = impgen(A,B,C,D,X0,t,c)
% Time-marching solution of linear, time-invariant
% state-space equations using the digital approximation when the
% inputs are impulse functions scaled by constants. The scaling
% constants for the impulse inputs are contained in vector 'c'.
% A= state dynamics matrix; B= state input coefficient matrix;
% C= state output coefficient matrix;
% D= direct transmission matrix;
% X0= initial state vector; t= time vector.
% y= returned output matrix with ith output stored in the ith
% column, and jth row corresponding to the jth time point.
% X= returned state matrix with ith state variable stored in the
% ith column, and jth row corresponding to the jth time point.
% copyright(c)2000 by Ashish Tewari
n=size(t,2);
m=size(c,2);
dt=t(2)-t(1);
% digital approximation of the continuous-time system:-
[ad,bd,cd,dd]=c2dm(A,B,C,D,dt,'zoh');
Bd=ad*bd;
u=ones(n,1)*c';
% time-marching solution of the digital state equation:-
X=ltitr(ad,Bd,u,X0);
% calculation of the outputs:-
y=X*C'+u*D';
```

$t = 0$ (see Eq. (2.115)). For details on the usage of these specialized CST functions, use the MATLAB *help* command.

Example 4.9

For the flexible bomber aircraft of Example 4.7, let us determine the response if the initial condition is $\mathbf{x0} = [\,0.1; \quad 0; \quad 0; \quad 0; \quad 0; \quad 0\,]^T$ and the input vector is given by:

$$\mathbf{u}(t) = \begin{bmatrix} 0.1\delta(t) \\ -0.1\delta(t) \end{bmatrix} \tag{4.75}$$

First, the time vector, **t**, and the coefficient vector, **c**, are specified as follows:

```
>>dt=6.6667e-3; t=0:dt:10; c=[0.1; -0.1]; <enter>
```

Then, *impgen* is invoked as follows, assuming **A**, **B**, **C**, **D**, **x0** have been already computed and stored in the MATLAB workspace:

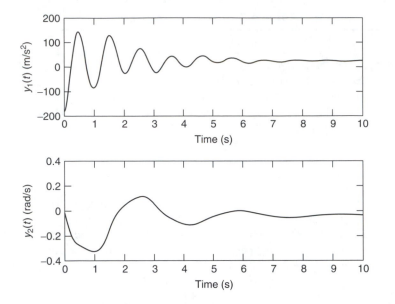

Figure 4.10   Response of the flexible bomber aircraft of Example 4.9 to elevator and canard impulse inputs of same magnitude but opposite signs

```
>>[y,X]=impgen(A,B,C,D,X0,t,c); <enter>
```

The computed outputs, $y_1(t)$ and $y_2(t)$, which are the first and second columns of the returned matrix, y, respectively, are plotted in Figure 4.10. Note that we could not get this response from the CST function *impulse*, which is confined to the case of $\mathbf{c} = [\, 1; \quad 1\,]^T$.

4.5 Numerical Solution of Linear Time-Varying State-Equations

There is no general analytical procedure of solving the state-equations for linear *time-varying* systems (i.e. when the state coefficient matrices, $\mathbf{A},\mathbf{B},\mathbf{C},\mathbf{D}$, are functions of time). Thus, numerical solution procedures using digital approximation methods, similar to those of the previous section, are required for solving the state-equations of such systems. For special time-varying systems in which the state-dynamics matrix, $\mathbf{A}$, is a constant and the matrices $\mathbf{B}$, $\mathbf{C}$, and $\mathbf{D}$ are functions of time, the analytical solution given by Eq. (4.27) is valid with the following modification:

$$\mathbf{x}(t) = \exp\{\mathbf{A}(t - t_0)\}\mathbf{x}(t_0) + \int_{t_0}^{t} e^{A(t-\tau)}\mathbf{B}(\tau)\mathbf{u}(\tau)d\tau; \quad (t \geq t_0) \qquad (4.76)$$

where $\mathbf{x}(t_0)$ is the initial condition and $\mathbf{B}(\tau)$ indicates the controls coefficient matrix evaluated at time $t = \tau$. For linear, time-varying systems the output equation is given by:

$$\mathbf{y}(t) = \mathbf{C}(t)\mathbf{x}(t) + \mathbf{D}(t)\mathbf{u}(t) \qquad (4.77)$$

For systems with a constant $\mathbf{A}$ matrix, the output can be calculated by Eq. (4.77), in which $\mathbf{x}(t)$ is calculated using Eq. (4.76). A digital approximation of Eq. (4.76) is given by

$$\mathbf{x}(n\Delta t) = \mathbf{A}_d\mathbf{x}((n-1)\Delta t) + \mathbf{B}_d((n-1)\Delta t)\mathbf{u}((n-1)\Delta t); \; (n = 1, 2, 3, \ldots) \quad (4.78)$$

where $\mathbf{A}_d = e^{A\Delta t}$ and $\mathbf{B}_d((n-1)\Delta t) = \mathbf{B}((n-1)\Delta t)\Delta t$. The solution given by Eq. (4.78) can be implemented in a computer program, such as *march.m*. However, since time-varying systems with a constant $\mathbf{A}$ matrix are rarely encountered, such a program will be rarely used. Instead, we need a general solution procedure which is applicable to a general time-varying system given by:

$$\mathbf{x}^{(1)}(t) = \mathbf{A}(t)\mathbf{x}(t) + \mathbf{B}(t)\mathbf{u}(t) \quad (4.79)$$

A practical procedure for solving Eq. (4.79) is the application of the time-marching approach of Section 4.5, assuming that $\mathbf{A}(t)$ is constant *during each time step*, Δt, but

Table 4.4 Listing of the M-file *vmarch.m*

---

vmarch.m

```
function [y,X] = vmarch(tvfun,X0,t,u,method)
% Time-marching solution of linear, time-varying
% state-space equations using the matrix exponential.
% X0= initial state vector; t= time vector;
% u=matrix with the ith input stored in the ith column, and jth row

% corresponding to the jth time point
% y= returned matrix with the ith output stored in the ith column,
  and jth row
% corresponding to the jth time point
% X= returned matrix with the ith state variable stored in the ith column,
% and jth row corresponding to the jth time point
% method= method of digital interpolation for the inputs (see 'c2dm')
% copyright(c)2000 by Ashish Tewari
n=size(t,2);
dt=t(2)-t(1);
% initial condition:-
X(1,:)=X0';
% function evaluation of time varying state coefficient matrices
% using the M-file 'tvfun.m' for initial time t=t(1):-
[A,B,C,D]=feval(tvfun,t(1));
% outputs for t=t(1):-
y(1,:)=X(1,:)*C'+u(1,:)*D';
% beginning of the time-loop:-
for i=1:n-1
% function evaluation of time varying state coefficient matrices
% using the M-file 'tvfun.m' for t=t(i):-
[A,B,C,D]=feval(tvfun,t(i));
% digital approximation of the continuous-time system:-
[ad,bd,cd,dd]=c2dm(A,B,C,D,dt,method);
% solution of the digital state and output equations:-
X(i+1,:)=X(i,:)*ad'+u(i,:)*bd';
y(i+1,:)=X(i+1,:)*C'+u(i,:)*D';
end
```

varies as we proceed from *one time step to the next*. In essence, this procedure applies a *zero-order hold* to $\mathbf{A}(t)$. Thus, the approximate digital solution for a general time-varying system can be written as follows:

$$\mathbf{x}(n\Delta t) = \mathbf{A}_d((n-1)\Delta t)\mathbf{x}((n-1)\Delta t) + \mathbf{B}_d((n-1)\Delta t)\mathbf{u}((n-1)\Delta t); \quad (n = 1, 2, 3, \ldots)$$
$$(4.80)$$

where $\mathbf{A}_d = \exp\{\mathbf{A}((n-1)\Delta t)\Delta t\}$ and $\mathbf{B}_d((n-1)\Delta t) = \mathbf{B}((n-1)\Delta t)\Delta t$. A MATLAB M-file can be written based on Eqs. (4.80) and (4.77) for computing the time marching output of a general linear, time-varying system. Such an M-file, called *vmarch.m* is given in Table 4.4.

Example 4.10

Consider the following linear, time-varying system:

$$\mathbf{A}(t) = \begin{bmatrix} -0.1\sin(t) & 0 \\ 0 & -0.7\cos(t) \end{bmatrix}; \quad \mathbf{B}(t) = \begin{bmatrix} 0.2\sin(t) \\ 0 \end{bmatrix} \qquad (4.81)$$

$$\mathbf{C}(t) = [1 \quad -1]; \quad \mathbf{D}(t) = -0.05\cos(t) \qquad (4.82)$$

Let us calculate the output, $\mathbf{y}(t)$, when the initial condition is $\mathbf{x}(0) = [0; \quad -1]^T$ and the input is a unit setp function, $\mathbf{u}(t) = u_s(t)$. The time vector, initial condition, and input are specified by the following MATLAB command:

```
>>t = 0:0.1:10; X0 = [0 -1]'; u = ones(size(t,2), 1); <enter>
```

The time-varying coefficient matrices are calculated by the M-file *timv.m* tabulated in Table 4.5. Then, the solution to the time-varying state-equations and the output are obtained by calling *vmarch.m* as follows:

```
>>[y,X] = vmarch('timv',X0,t,u,'zoh'); <enter>
```

The resulting output, $y(t)$, is plotted in Figure 4.11.

Table 4.5   Listing of the M-file *timv.m*

---

timv.m

```
function [A,B,C,D]=timv(t);
% Linear time-varying state coefficient matrices for Example 4.10.
A=[-0.1*sin(t) 0;0 -0.7*cos(t)];
B=[0.2*sin(t);0];
C=[1 -1];
D=-0.05*cos(t);
```

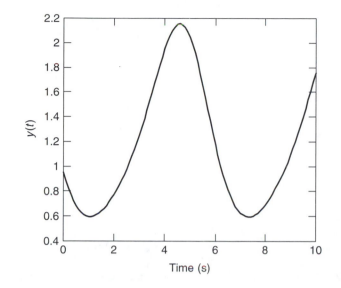

Figure 4.11   Output of the linear, time-varying system of Example 4.10

Example 4.11

Let us obtain the response of an interesting time-varying system. Consider a *surface-to-air missile* propelled by a rocket engine. Once launched, the rocket consumes the fuel rapidly, thereby changing the *mass, center-of-gravity*, and *moments of inertia* of the missile. Also, as the fuel is burned, the missile accelerates and changes altitude. The aerodynamic forces and moments acting on the missile are functions of the flight velocity and the altitude; therefore, the aerodynamic properties of the missile also keep changing with time. The motion of the missile is described by the following state variables: velocities, U, V, and W, along *three mutually perpendicular axes*, x, y, z, respectively, passing through the missile's *center of gravity*, and the rotation rates, p, q, and r, about x, y, and z, respectively. The state-vector is thus $\mathbf{x}(t) = [U(t); V(t); W(t); p(t); q(t); r(t)]^T$. All the state-variables are assumed to be the outputs of the system. A diagram of the missile showing the state-variables and the inputs is given in Figure 4.12. One input to the missile is the *rolling-moment*, ΔL, about the longitudinal axis of the missile caused by the deflection of an aerodynamic *control surface*. In addition, the *thrust* from the rocket engine can be *vectored* (i.e. deflected) by small angles, α and β, in the *longitudinal* (X, Z) plane and *lateral* (Y, Z) plane, respectively, of the missile (Figure 4.12). These thrust deflection angles constitute two additional inputs to the missile. The input vector is $\mathbf{u}(t) = [\alpha; \beta; \Delta L]^T$. The thrust of the missile is *assumed constant* until the rocket motor *burns-out* at $t = 20$ s, after which time the thrust is zero. The time-varying state-coefficient matrices representing the

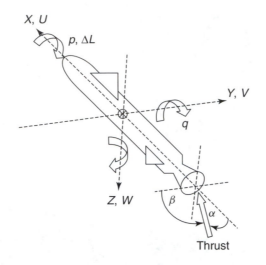

Figure 4.12 State-variables and inputs for a surface-to-air missile equipped with thrust vectoring and aerodynamic roll control surfaces

linearized motion of the missile are calculated by the M-file *misstimv.m* tabulated in Table 4.6.

Let us calculate the response of the missile if the initial condition is zero, and the inputs are given by

$$\mathbf{u}(t) = \begin{bmatrix} 0.01\sin(t) \\ 0 \\ 0.01\cos(t) \end{bmatrix} \tag{4.83}$$

The time vector, initial condition, and input are specified by the following MATLAB command:

```
>>t = 0:0.2:30; X0 = zeros(6,1); u = [0.01*sin(0.1*t)' 0.01*cos(0.1*t)'
  zeros(size(t,2),1)]; <enter>
```

The output and state solution are then calculated by the following call to *vmarch.m*:

```
>>[y,X] = vmarch('misstimv',X0,t,u,'zoh'); <enter>
```

The calculated outputs, $U(t)$, $V(t)$, $W(t)$, and $p(t)$, are plotted in Figure 4.13, while the outputs $q(t)$ and $r(t)$ are shown in Figure 4.14. Note the discontinuity in the slopes of all the responses at $t = 20$ s, which is the rocket *burn-out* time.

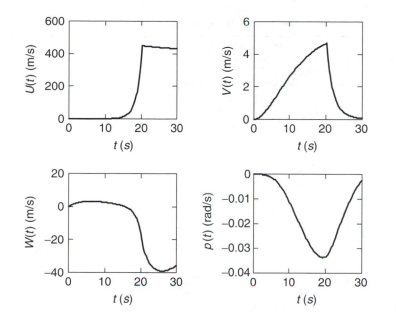

Figure 4.13 The outputs $U(t)$, $V(t)$, $W(t)$, and $p(t)$, for the missile of Example 4.11

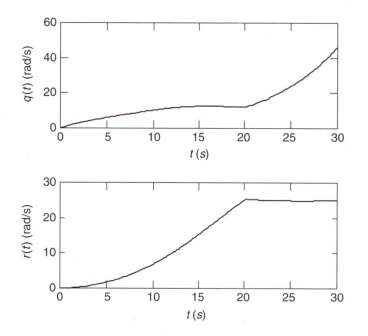

Figure 4.14 The outputs $q(t)$ and $r(t)$ for the missile of Example 4.11

Table 4.6 Listing of the M-file *misstimv.m*

misstimv.m

```
function [A,B,C,D]=misstimv(t);
% Linear, time-varying state coefficient matrices for a missile
% x = [u v w p q r]';
% A= state dynamics matrix; B= state input coefficient matrix
% C= state output coefficient matrix; D= direct transmission
matrix
% t= time after launch
% copyright(c)2000 by Ashish Tewari
% Thrust as a function of time:-
if t<=20
    Th=1.2717e+005;
else
    Th=0;
end
% inertia properties as functions of time:-
m=983.6716-29.4208*t-0.5812*t^2+0.0338*t^3;
iy=1000*(2.5475-0.0942*t+0.0022*t^2);ix=iy/10;
xcg=3.6356-0.0217*t-0.0008*t^2;
% aerodynamic and propulsive properties as functions of time:-
Zw=-(246.44+21.9038*t-1.5996*t*t+0.0244*t^3);
Mw=-(872.95-52.7448*t-0.0006*t^2+0.0368*t^3);
Mq=(-3-1.39*t+0.08*t^2)/10;
Lr=0.0134+0.0029*t-0.0001*t^2;
Lp=-0.0672-0.0143*t+0.0006*t^2;
Lv=-0.1159-0.0317*t+0.0015*t^2;
Zu=-9.5383-2.592*t+0.1209*t^2-0.0011*t^3;
Xw=1.9067+0.5186*t-0.0242*t^2+0.0002*t^3;
Md=1e5*(1.3425-2.3946*t+0.1278*t^2-0.0017*t^3);
Zd=1e4*(-2.0143-3.6649*t+0.1854*t^2-0.0023*t^3);
Yv=Zw;Nv=Mw;Nr=Mq;Np=Lp/10;Xu=Th/400+Zu/10;ci=(iy-ix)/iy;
% the state coefficient matrices:-
A=[Xu/m 0 Xw/m 0 0 0;0 Yv/m 0 0 0 0;Zu/m 0 Zw/m 0 0 0;
  0 Lv/ix 0 Lp/ix 0 Lr/ix;0 0 Mw/iy 0 Mq/iy 0;
  0 Nv/iy 0 Np/iy 0 Nr/iy];
B=[0 0 0;Th/m 0 0;0 Th/m 0;0 0 1;0 Th*xcg/iy 0;Th*xcg/iy 0 0];
C=eye(6);D=zeros(6,3);
```

4.6 Numerical Solution of Nonlinear State-Equations

The nonlinear state-equations are the most difficult to solve. As for time-varying systems, there is no analytical solution for a set of general nonlinear state-equations. There are several numerical schemes available for solving a set of nonlinear, first-order differential equations using the digital approximation to the continuous-time differential equations. Since the state-equations of a nonlinear system are also a set of nonlinear, first-order differential equations, we can use such numerical schemes to solve the state-equations of nonlinear systems. Due to the nonlinear nature of the differential equations, the solution procedure often is more complicated than merely marching forward in time, as we did for linear systems in the previous two sections. Instead, an *iterative* solution procedure may be required at each time step, which means that we assume a starting solution, and then go back and keep on changing the assumed solution until the solution *converges* (i.e. stops changing appreciably with the steps of the iteration).

A general set of nonlinear state-equations can be expressed as follows:

$$\mathbf{x}^{(1)}(t) = f(\mathbf{x}(t), \mathbf{u}(t), t) \tag{4.84}$$

where $\mathbf{x}(t)$ is the state-vector, $\mathbf{u}(t)$ is the input vector, and $f(\mathbf{x}(t), \mathbf{u}(t), t)$ denotes a nonlinear function involving the state variables, the inputs, and time, t. The solution, $\mathbf{x}(t)$, of Eq. (4.84) with the initial condition, $\mathbf{x}(t_0) = \mathbf{x}_0$ may not always exist. The existence of solution of nonlinear differential equations requires that the nonlinear function, $f(\mathbf{x}(t), \mathbf{u}(t), t)$, should be defined and *continuous* for all *finite* times, $t \geq t_0$. Also, it is required that $f(\mathbf{x}(t), \mathbf{u}(t), t)$ must satisfy the following condition, known as the *Lipschitz condition*:

$$|f(\mathbf{x}(t), \mathbf{u}(t), t) - f(\mathbf{x}^*(t), \mathbf{u}(t), t)| \leq K|\mathbf{x}(t) - \mathbf{x}^*(t)| \tag{4.85}$$

where $\mathbf{x}^*(t)$ is a vector different from $\mathbf{x}(t)$, K is a constant, and $|\mathbf{V}|$ denotes a vector consisting of the absolute value of each element of the vector $\mathbf{V}$. For greater details on the existence of solution of ordinary nonlinear differential equations see a textbook on ordinary differential equations, such as Henrici [4]. In this book, we will assume that we are dealing with nonlinear system which have a solution to their state-equations. Owing to the nonlinear nature of the differential equations, the numerical procedure cannot be a one-shot (i.e. an open-loop) process, such as that used for linear differential equations. Instead, an *iterative* solution procedure is required for nonlinear systems, which consists of repeatedly evaluating the solution in a *loop* (such as the feedback loop) at each time step, until the solution meets certain desirable conditions. Hence, a nonlinear solution procedure itself is a *closed-loop system*.

The digital solution procedures for Eq. (4.84) can be divided into *single-step methods*, *multi-step methods*, and *hybrid methods*. The single-step methods obtain the approximate solution vector by using the state-vector and input vector *only at the previous time step*. The time marching solution methods of the previous two sections are single-step methods for linear state-equations. For nonlinear systems, examples of single-step methods are *Runge–Kutta* and *Adams* methods. The multi-step methods use information from *more than one* previous time steps to obtain the approximate solution at a given time step. The *predictor-corrector* methods are examples of multi-step methods. The *hybrid methods* are those that either do not fall into the categories of single- and multi-step methods, or those that use information from previous time steps as well as future (extrapolated) time steps. The *Euler method* falls in the hybrid category. For more information on the numerical solution methods for nonlinear differential equations see Ralston and Rabinowitz [5].

While choosing which method to use for solving a nonlinear set of differential equations one should consider *numerical accuracy* (how large is the digital approximation error), *efficiency* (how fast is the algorithm when implemented on a computer), *numerical stability* (whether the algorithm *converges* to a solution), and *starting problem* (how the algorithm can be started). While the multi-step and hybrid methods offer a greater efficiency for a comparable accuracy than the single-step methods, they are usually very difficult to start, and special attention must be paid in changing the time steps to avoid stability problems. Such issues make multi-step or hybrid methods more complicated than the single-step methods. Complexity of an algorithm often results in a reduced efficiency when implemented in a computer program. A single-step method which is simple to implement, and which provides good accuracy in a wide variety of problems is the *Runge–Kutta* method. The Runge–Kutta method uses the following digital approximation to Eq. (4.84):

$$\mathbf{x}(t_n) - \mathbf{x}(t_{n-1}) = \sum_{i=1}^{p} w_i \mathbf{k}_i \tag{4.86}$$

where t_n and t_{n-1} are the nth and $(n-1)$th time steps, respectively, w_i are constants, and

$$\mathbf{k}_i = \Delta t_n f\left(\mathbf{x}(t_{n-1}) + \sum_{j=1}^{i=1} \beta_{ij} \mathbf{k}_j, \mathbf{u}(t_{n-1}), t_{n-1} + \alpha_i \Delta t_n\right) \tag{4.87}$$

where $\Delta t_n = t_n - t_{n-1}$, α_i and β_{ij} are constants, with $\alpha_1 = 0$. The time step size, Δt_n, can be variable. The constants α_i and β_{ij} are evaluated by equating the right-hand side of Eq. (4.86), with the following Taylor series expansion:

$$\mathbf{x}(t_n) - \mathbf{x}(t_{n-1}) = \sum_{k=1}^{\infty} \Delta t_n \mathbf{x}^{(k)}(t_{n-1})/k! \tag{4.88}$$

However, since we cannot numerically evaluate an infinite series, the right-hand side of Eq. (4.88) is approximated by a finite series of m terms as follows:

$$\mathbf{x}(t_n) - \mathbf{x}(t_{n-1}) \approx \sum_{k=1}^{m} \Delta t_n \mathbf{x}^{(k)}(t_{n-1})/k! \tag{4.89}$$

The approximation given by Eq. (4.89) leads to a Runge–Kutta method of *order m*. The higher the number of terms in the series of Eq. (4.89), the greater will be the accuracy of the approximation. Comparing Eqs. (4.86) and (4.89), it can be shown that the *largest number* of terms that can be retained in the series of Eq. (4.89) is $m = p$. Usually, when $m = 4$, the resulting *fourth order Runge–Kutta method* is accurate enough for most practical purposes. It can be shown [5] that substituting Eq. (4.89) into Eq. (4.86), and making use of the exact differential equation, Eq. (4.84), results in the following relationships for the parameters of the fourth order Runge-Kutta method:

$$\alpha_i = \sum_{j=1}^{i-1} \beta_{ij}; \quad (i = 2, 3, 4) \tag{4.90}$$

$$\sum_{i=1}^{4} w_i = 1; \quad \sum_{i=1}^{4} w_i \alpha_i = 1/2; \quad \sum_{i=1}^{4} w_i \alpha_i^2 = 1/3; \quad \sum_{i=1}^{4} w_i \alpha_i^3 = 1/4;$$

$$w_3 \alpha_2 \beta_{32} + w_4(\alpha_2 \beta_{42} + \alpha_3 \beta_{43}) = 1/6; \quad w_3 \alpha_2^2 \beta_{32} + w_4(\alpha_2^2 \beta_{42} + \alpha_3^2 \beta_{43}) = 1/12;$$

$$w_3 \alpha_2 \alpha_3 \beta_{32} + w_4 \alpha_4(\alpha_2 \beta_{42} + \alpha_3 \beta_{43}) = 1/8; \quad w_4 \alpha_2 \beta_{32} \beta_{43} = 1/24 \tag{4.91}$$

Equations (4.90) and (4.91) represent 11 equations and 13 unknowns. Hence, we can obtain the solution of any 11 unknowns in terms of the remaining two unknowns, which we choose to be α_2 and α_3. The Runge–Kutta parameters are thus the following:

$$w_1 = 1/2 + [1 - 2(\alpha_2 + \alpha_3)]/(12\alpha_2\alpha_3); \quad w_2 = (2\alpha_3 - 1)/[12\alpha_2(\alpha_3 - \alpha_2)(1 - \alpha_2)]$$

$$w_3 = (1 - 2\alpha_2)/[12\alpha_3(\alpha_3 - \alpha_2)(1 - \alpha_3)]; \quad w_4 = 1/2 + [2(\alpha_2 + \alpha_3) - 3]/[12(1 - \alpha_2)(1 - \alpha_3)]$$

$$\beta_{32} = \beta_{23} = \alpha_3(\alpha_3 - \alpha_2)/[2\alpha_2(1 - \alpha_2)]; \quad \alpha_4 = 1 \tag{4.92}$$

$$\beta_{42} = \beta_{24} = (1 - \alpha_2)[\alpha_2 + \alpha_3 - 1 - (2\alpha_3 - 1)^2]/\{2\alpha_2(\alpha_3 - \alpha_2)[6\alpha_2\alpha_3 - 4(\alpha_2 + \alpha_3) + 3]\}$$

$$\beta_{43} = \beta_{34} = (1 - 2\alpha_2)(1 - \alpha_2)(1 - \alpha_3)/\{\alpha_3(\alpha_3 - \alpha_2)[6\alpha_2\alpha_3 - 4(\alpha_2 + \alpha_3) + 3]\}$$

Obviously, we should take care to avoid selecting those values for the two parameters, α_2 and α_3, which lead to the denominators of the expressions in Eq. (4.92) becoming zero. A popular choice of these two parameters is $\alpha_2 = \alpha_3 = 1/2$. However, the choice which *minimizes* the *approximation error* in the fourth order Runge-Kutta method is $\alpha_2 = 0.4$, and $\alpha_3 = 7/8 - (3/16)\sqrt{5}$.

The Runge–Kutta algorithm consists of marching in time using Eq. (4.86) with a variable time step size, Δt_n. The *truncation-error* (i.e. error of approximating Eq. (4.88) by Eq. (4.89)) is estimated using the *matrix norm* (such as the one defined in Section 4.5) after each time step. If the error is acceptable, then the solution is updated; if not, the time step size is reduced, and the error re-calculated until the error becomes acceptable. This process is repeated for the next time step, using the solution from the previous step, and so on until the final time is reached. Using MATLAB, a Runge–Kutta algorithm can be easily programmed. Fortunately, MATLAB comes with intrinsic nonlinear functions *ode23* and *ode45*, which are based on third and fifth order Runge–Kutta algorithms, respectively. Other MATLAB functions for solving nonlinear equations are *ode113*, *ode15s*, *ode23s*, *ode23t*, and *ode23tb*. The function *ode113* uses a variable order integration method for nonlinear equations. The functions with names ending with the letters *s*, *t*, or *tb* are specially suited for solving *stiff equations*. Stiff equations [5] are a set of first-order nonlinear equations with a large difference in their *time scales* (e.g. solution to each equation may have a significantly different time for reaching a *steady state*). The normal solution procedure that takes into account only the shortest time scale of stiff equations may either fail to converge, or may require very large number of time steps to arrive at a steady state. Hence, stiff equations require special solution procedures [5]. We will consider the more common variety of nonlinear equations (i.e. non-stiff equations) that can be solved using *ode23*, *ode113*, and *ode45*. These functions are used as follows to obtain a solution to a set of nonlinear state-equations:

```
>>[t,X] = ode23(@fun,tspan,X0,options); <enter>
```

where *@fun* denotes a user supplied M-file, *fun.m*, in which the time derivative of the state-vector, $\mathbf{x}^{(1)}(t)$, is evaluated using Eq. (4.84), $tspan = [ti\ t1\ t2\ t3\ \ldots tf]$ is a row vector containing the initial time, ti, at which the initial condition vector, $\mathbf{x0}$, is specified, any intermediate times, $t1$, $t2$, $t3$, $\ldots$, at which the solution is desired (optional), and the final time, tf, and $\mathbf{t}$ is a vector containing the time points at which the returned solution,

x, is obtained. The returned matrix **x** contains as many rows as there are the time points, and each column of **x** corresponds to a state variable. The first executable statement of the M-file *fun.m* should be the following:

```
>>function xdot = (t,x)
```

and the remaining statements of *fun.m* should evaluate the derivative of the state-vector, **xdot**, based on the state-vector, **x**, and time, t. Note that the input vector, $\mathbf{u}(t)$, is internal to the M-file *fun.m*, i.e. it is not used directly in *ode23*, *ode113* or *ode45*, but only indirectly through **xdot**. The fourth input argument, *options*, can be used to specify *relative error* and *absolute error tolerances*, *Reltol* (a scalar) and *Abstol* (a vector of the same size as **x**), respectively, for convergence through the function *odeset*. This ensures that the error in the ith component of the solution vector, x_i, does not exceed the greater number between $Reltol|x_i|$ and $Abstol(i)$. If options are not specified, then the default values of relative tolerance of 0.001 and absolute tolerance of 10^{-6} are used. For more information on the *ode* functions use the MATLAB's help command.

Example 4.12

Consider a double-pendulum (Figure 1.5). A choice of the state variables for this fourth order system is $x_1(t) = \theta_1(t)$, $x_2(t) = \theta_2(t)$, $x_3(t) = \theta_1^{(1)}(t)$; $x_4(t) = \theta_2^{(1)}(t)$, which results in the following state-equations:

$$x_1^{(1)}(t) = x_3(t)$$

$$x_2^{(1)}(t) = x_4(t)$$

$$x_3^{(1)}(t) = [m_2 L_1 x_3^2(t) \sin(x_2(t) - x_1(t)) \cos(x_2(t) - x_1(t))$$

$$+ m_2 L_2 x_4^2(t) \sin(x_2(t) - x_1(t)) + m_2 g \sin(x_2(t)) \cos(x_2(t) - x_1(t))$$

$$- (m_1 + m_2) g \sin(x_1(t))]/[L_1(m_1 + m_2) - m_2 L_1 \cos^2(x_2(t) - x_1(t))]$$

$$x_4^{(1)}(t) = -[g \sin(x_2(t)) + L_1 x_3^2(t) \sin(x_2(t) - x_1(t)) + L_1 x_3^{(1)}(t) \cos(x_2(t)$$

$$- x_1(t))]/L_2 + u(t)/(m_2 L_2^2) \tag{4.93}$$

where $u(t)$ is the input torque applied on the mass, m_2. Note that the last state-equation has a term involving $x_3^{(1)}(t)$ on the right-hand side. This has been done for the sake of brevity (you can substitute $x_3^{(1)}(t)$ from the previous state-equation into the last state-equation to obtain the state-equations in explicit form). It is desired to obtain the solution of Eq. (4.93) for the initial condition $\mathbf{x}(0) = [0.7$ rad.; 1.4 rad.; 0 rad./s; 0 rad./s$]^T$ and input, $u(t) = 0.01 \sin(5t) N$-m. The function M-file for evaluating the time derivative of the state-vector, $\mathbf{x}^{(1)}(t)$, is called *doub.m* and is tabulated in Table 4.7. Note that the input, $u(t)$, must be specified within the function file *doub.m*, while the initial condition is specified in the call to the Runge–Kutta solver (either *ode23* or *ode45*).

Table 4.7 Listing of the M-file *doub.m*

doub.m

```
function xp=doub(t,x)
% Nonlinear state-equations for a double-pendulum, excited
% by input, u(t), torque acting on mass m2.
% x=[theta1 theta2 thetadot1 thetadot2]
% xp is the time derivative of the state-vector, x.
% copyright(c)2000 by Ashish Tewari
m1=1;m2=2;l1=1;l2=2;g=9.8;
u=0.01*sin(5*t);
xp(1,1)=x(3);
xp(2,1)=x(4);
x21=x(2)-x(1);
xp(3,1)=(m2*l1*x(3)*x(3)*sin(x21)*cos(x21)+m2*l2*x(4)*x(4)*sin(x21)...
  + m2*g*sin(x(2))*cos(x21)-(m1+m2)*g*sin(x(1)))/((m1+m2)*l1-...
  m2*l1*cos(x21)*cos(x21));
xp(4,1)=-(g*sin(x(2))+l1*x(3)*x(3)*sin(x21)+l1*xp(3)*cos(x21))/l2
+u/(m2*l2*l2);
```

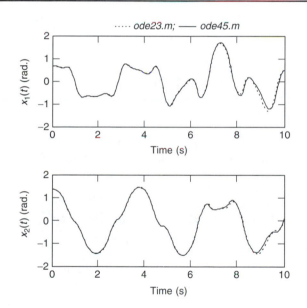

Figure 4.15 The calculated state variables, $x_1(t) = \theta_1(t)$ and $x_2(t) = \theta_2(t)$ for the double-pendulum, Example 4.12

Let us compare the solutions obtained using *ode23* and *ode45*, as follows:

```
>>[t1,x1]= ode23(@doub, [0 10], [0.7 1.4 0 0]'); <enter>

>>[t2,x2]= ode45(@doub, [0 10], [0.7 1.4 0 0]'); <enter>

>>subplot(211), plot(t1,x1(:,1),t2,x2(:,1)), hold on, subplot(212),
  plot(t1,x1(:,2),t2,x2(:,2)) <enter>
```

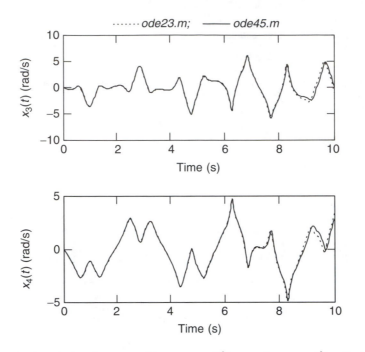

Figure 4.16 The calculated state variables, $x_3(t) = \theta_1^{(1)}(t)$, and $x_4(t) = \theta_2^{(1)}(t)$ for the double-pendulum, Example 4.12

The resulting plots of $x_1(t)$ and $x_2(t)$ are shown in Figure 4.15. Figure 4.16 plotting $x_3(t)$ and $x_4(t)$ is obtained by the following commands:

```
>>newfig <enter>

>>subplot(211), plot(t1,x1(:,3),t2,x2(:,3)), hold on, subplot(212),
    plot(t1,x1(:,4),t2,x2(:,4)) <enter>
```

Note that in Figures 4.15 and 4.16, a very small difference is observed between the state variables calculated by *ode23.m* and those calculated by *ode45.m*. This difference is seen to increase with time, indicating a larger truncation error for the third order Runge–Kutta method of *ode23.m* when compared to the fifth order Runge–Kutta method of *ode45*. Since the truncation error is added up after each time step, there is an error accumulation as time increases. The double-pendulum falls into a special category of nonlinear systems, called *chaotic systems*, which were discussed in Section 1.3. Figure 1.6 compared the state variable, $x_2(t)$, calculated for two very slightly different initial conditions and a zero input, and was generated using *ode45*. Figure 1.6 showed a large difference in the response, $x_2(t)$, when the initial conditions differed by a very small amount, which is the hallmark of a chaotic system.

Example 4.13

Let us consider another interesting nonlinear system, called the *wing-rock phenomenon*. *Wing-rock* is a special kind of *rolling* and *yawing* motion observed in modern fighter type aircraft when operating at a large angle of attack (defined as the angle made by the *longitudinal axis* of the aircraft and the *direction of flight*). The nonlinear state-equations modeling the wing-rock dynamics of a fighter aircraft are the following [6]:

$$x_1^{(1)}(t) = x_2(t)$$

$$x_2^{(1)}(t) = -\omega^2 x_1(t) + \mu_1 x_2(t) + \mu_2 x_1^2(t)x_2(t) + b_1 x_2^3(t) + b_2 x_1(t)x_2^2(t)$$

$$+ L_\delta x_3(t) + L_\beta x_4(t) - L_r x_5(t)$$

$$x_3^{(1)}(t) = -kx_3(t) + ku(t)$$

$$x_4^{(1)}(t) = x_5(t)$$

$$x_5^{(1)}(t) = -N_p x_2(t) - N_\beta x_4(t) - N_r x_5(t) \tag{4.94}$$

where the state variables are, $x_1(t)$: *bank angle* (rad.), $x_2(t)$: *roll-rate* (rad./s), $x_3(t)$: *aileron deflection angle* (rad.), $x_4(t)$: *sideslip angle* (rad.), and $x_5(t)$: *sideslip-rate* (rad./s). The input, $u(t)$, is the *desired aileron deflection* (rad.). The constants ω, μ_1, μ_2, b_1, b_2, L_δ, L_β, L_r, N_p, N_β, and N_r depend on the *inertial* and *aerodynamic* properties of the aircraft, while the constant k is the aileron-actuator's *time-constant* (the aileron is an aerodynamic *control surface* which is deployed using a first order actuator). Let us obtain the solution to Eq. (4.94) when the initial condition is $\mathbf{x}(0) = [1.0 \text{ rad.}; 0.5 \text{ rad./s}; 0 \text{ rad.}; 0 \text{ rad.}; 0 \text{ rad./s}]^T$ and the input is zero. The time derivative of state-vector, $\mathbf{x}^{(1)}(t)$, is evaluated using the M-file called *wrock.m*, which is tabulated in Table 4.8.

Using *ode45*, the initial response (i.e. response when $u(t) = 0$) is obtained as follows:

```
>>[t,x]= ode45(@wrock, [0 700], [0.2 0 0 0 0]'); <enter>
```

The plot of the bank angle, $x_1(t)$, from $t = 0$ s to $t = 700$ s is shown in Figure 4.17. Note that instead of decaying to zero in the limit $t \to \infty$, the initial response keeps on oscillating with a constant time period, and an amplitude which becomes constant in the limit $t \to \infty$. Such a motion is called a *limit cycle motion*. Note that while the system is not unstable (i.e. the response *does not* tend to infinity in the limit $t \to \infty$), a limit cycle response is undesirable from weapons aiming and delivery considerations, and also because it may lead to structural fatigue in the aircraft (or other mechanical systems) thereby causing the wings to come-off. Figure 4.18 shows a plot of $x_2(t)$ against $x_1(t)$. Such a plot in which the time derivative of a variable $(x_2(t) = x_1^{(1)}(t)$ is plotted against the variable itself $(x_1(t))$ is called a *phase-plane plot*. Figure 4.18 shows that the limit cycle motion corresponds to a limiting outer boundary in the phase-plane plot, indicating that the amplitude

Table 4.8 Listing of the M-file *wrock.m*

wrock.m

```
function xdot=wrock(t,x)
% nonlinear state-equations for the wing-rock problem;
% including first order aileron actuator
% xdot is the time derivative of the state-vector, x.
% copyright(c)2000 by Ashish Tewari
a=[-0.05686 0.03254 0.07334 -0.3597 1.4681];
% pure-rolling mode natural-frequency squared:-
w=-0.354*a(1);
% aileron-actuator time-constant:-
k=1/0.0495;
% linear aerodynamic coefficients:-
lbet=-0.02822;lr=0.1517;np=-0.0629;nbet=1.3214;nr=-0.2491;ldelt=1;
% nonlinear inertial and aerodynamic coefficients:-
u(1)=0.354*a(2)-0.001;
u(2)=0.354*a(4);
b(1)=0.354*a(3);
b(2)=0.354*a(5);
% desired aileron deflection as the input, 'f':-
f=0;
% the nonlinear state-equations:-
xdot(1,1)=x(2);
xdot(2,1)=-w*x(1)+u(1)*x(2)+b(1)*x(2)^3+u(2)*x(2)*x(1)^2
        +b(2)*x(1)*x(2)^2...+ldelt*x(3)+lbet*x(4)-lr*x(5);
xdot(3,1)=-k*x(3)+k*f;
xdot(4,1)=x(5);
xdot(5,1)=-nbet*x(4)+nr*x(5)-np*x(2);
```

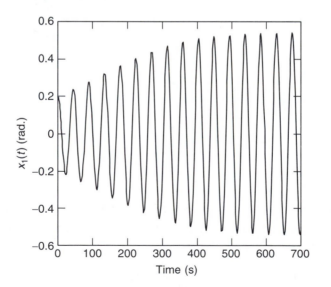

Figure 4.17 Initial response of bank angle, $x_1(t)$, for the wing-rock problem of Example 4.13 showing a *limit cycle motion* (i.e. constant amplitude oscillation in the limit $t \to \infty$)

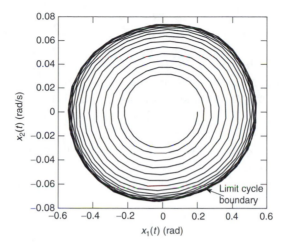

Figure 4.18   Phase-plane plot of $x_2(t)$ vs. $x_1(t)$ for the wing-rock problem of Example 4.13 showing the limit cycle boundary

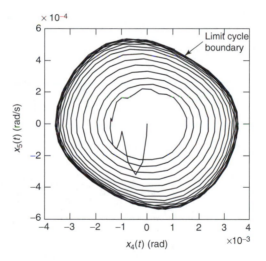

Figure 4.19   Phase-plane plot of $x_5(t)$ vs. $x_1(t)$ for the wing-rock problem of Example 4.13 showing the limit cycle boundary

of the motion has become constant. Similarly, Figure 4.19 shows the phase-plane plot of $x_5(t)$ against $x_4(t)$, displaying the limit cycle boundary.

4.7 Simulating Control System Response with SIMULINK

SIMULINK is very handy in quickly obtaining solutions to linear or nonlinear state-equations resulting from control systems with many sub-systems, using the ordinary differential equation solvers of MATLAB. SIMULINK allows the representation of each system by a set of linear and nonlinear *blocks*, and lets inputs and outputs to be modeled as special blocks called *sources* and *sinks*, respectively. A multivariable system can be represented either using the individual transfer functions between scalar inputs and outputs, or more conveniently by state-space model blocks. Copying the blocks from the SIMULINK *block library* makes simulating a control system very easy. Refer to Appendix B, or SIMULINK *User's Guide* [7] for more information on SIMULINK. Since SIMULINK works seamlessly with MATLAB and the Control System Toolbox (CST), you can draw upon the functions libraries of MATLAB and CST (and any other toolboxes that you may happen to have on your computer). The default simulation of SIMULINK uses the MATLAB function *ode45*, which is a fifth order, variable time-step Runge–Kutta solver (see Section 4.6). This allows a simulation of both linear and nonlinear state-equations.

Example 4.14

Let us simulate the step response of the flexible bomber aircraft (Example 4.7) using SIMULINK. The SIMULINK block diagram and the resulting simulation are shown in Figure 4.20. Note the *state-space* block modeling the aircraft dynamics, which requires that the state-coefficient matrices **A**, **B**, **C**, **D** of the aircraft (Eq. (4.71)) be available in the MATLAB work-space. The source *step input* block applies simultaneous unit step inputs, $u_1(t)$ and $u_2(t)$, to the aircraft, and the resulting output

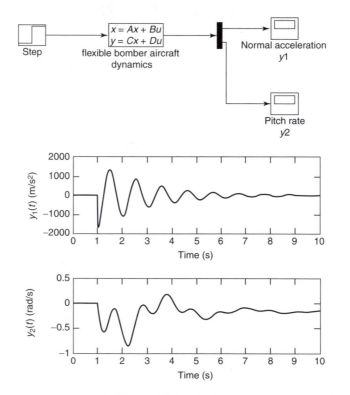

Figure 4.20 Simulation of the flexible bomber's step response using SIMULINK

vector is split into its two scalar elements, $y_1(t)$ (normal acceleration), and $y_2(t)$ (pitch-rate), by a *demux* block. The two simulated outputs are displayed individually on the scope blocks, as shown in Figure 4.20. The simulation is run with default parameters (variable time-step *ode45* solver with relative error tolerance of 0.001 and absolute tolerance of 10^{-6} per element).

A useful feature of SIMULINK is the availability of many nonlinear blocks, such as *dead-zone, backlash, saturation, rate-limiter, switch, relay, coulomb* and *viscous friction,* etc., which are commonly encountered in modeling many practical control systems. The following example illustrates the use of SIMULINK for a nonlinear control system.

Example 4.15

Let us simulate a nonlinear control system for controlling the roll dynamics of a fighter aircraft (Example 2.24). The *aileron input*, $\delta(t)$, to the aircraft is modeled

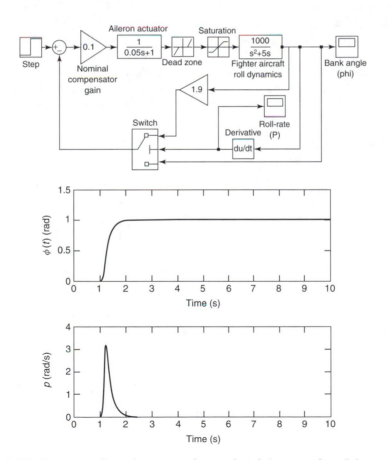

Figure 4.21 Simulation of a nonlinear control system for a fighter aircraft's roll dynamics using SIMULINK

by a *first-order actuator* with a *time-constant* of 0.05 s, and a *dead-zone* in the interval $-0.05 < t < 0.05$. The actuator is saturated at $\delta(t) = \pm 0.3$ rad. A nonlinear feedback control system is devised using a switch which allows a feedback gain of 1.9 when the roll-rate, $p(t) = \phi^{(1)}(t)$, is greater than or equal to 1 rad/s, and a unity feedback otherwise. A nominal cascade compensator gain of 0.1 is used to achieve desirable response (which is different from that designed in Example 2.24 due to the additional actuator dynamics). Figure 4.21 shows the SIMULINK block diagram and the resulting step response of the closed-loop system. Note that a zero overshoot, a zero steady-state error, zero oscillation, and a settling time of less than two seconds have been achieved, despite the nonlinear control system dynamics. A maximum roll-rate of about 3 rad/s is observed. This performance is representative of modern fighter aircraft, where the ability to achieve a large steady bank angle quickly and without any overshoots or oscillations is an important measure of the aircraft's *dog-fighting* maneuverability.

SIMULINK has many advanced features for simulating a complex control system, such as the creation of new sub-system blocks and *masking blocks* through M-files, C programs, or SIMULINK block diagrams, for effortless integration in your system's model. This allows an extension of the SIMULINK graphical functions to suit your own needs of analysis and design. The SIMULINK *demos* and the User's Guide for SIMULINK [7] are very helpful in explaining the advanced usage and extension of SIMULINK block library.

Exercises

4.1. A homogeneous linear, time-invariant system is described by the following state-dynamics matrix:

$$\mathbf{A} = \begin{bmatrix} -2 & -3 & 0 \\ 0 & 0 & 1 \\ 0 & -2 & -4 \end{bmatrix} \tag{4.95}$$

(a) Find the state-transition matrix of the system.

(b) Find the eigenvalues of the system. Is the system stable?

(c) Calculate the response, $\mathbf{x}(t)$, of the system to the initial condition, $\mathbf{x}(0) = [1; -0.1; 0.5]^T$.

4.2. For a linear, time-invariant system described by the following state-space representation:

$$\mathbf{A} = \begin{bmatrix} -1.5 & 0.2 \\ 0.13 & 0 \end{bmatrix}; \quad \mathbf{B} = \begin{bmatrix} 1 & 0 \\ 1 & -1 \end{bmatrix} \tag{4.96}$$

calculate the response, $\mathbf{x}(t)$, if the initial condition is $\mathbf{x}(0) = [10; 2]^T$, and the input vector is $\mathbf{u}(t) = [t; 1]^T$.

4.3. The lateral dynamics of an aircraft are described by the following state-equation:

$$\begin{bmatrix} p^{(1)}(t) \\ r^{(1)}(t) \\ \beta^{(1)}(t) \\ \phi^{(1)}(t) \end{bmatrix} = \begin{bmatrix} -15 & 0 & -15 & 0 \\ 0 & -0.8 & 10 & 0 \\ 0 & -1 & -0.8 & 0 \\ 1 & 0 & 0 & 0 \end{bmatrix} \begin{bmatrix} p(t) \\ r(t) \\ \beta(t) \\ \phi(t) \end{bmatrix} + \begin{bmatrix} 25 & 3 \\ 0 & -3.5 \\ 0 & 0 \\ 0 & 0 \end{bmatrix} \begin{bmatrix} \delta_A(t) \\ \delta_R(t) \end{bmatrix} \quad (4.97)$$

(a) Determine the step response, $\mathbf{x}(t) = [p(t); r(t); \beta(t); \phi(t)]^T$, to the *aileron* input, $\delta_A(t) = u_s(t)$.

(b) Determine the step response, $\mathbf{x}(t) = [p(t); r(t); \beta(t); \phi(t)]^T$, to the *rudder* input, $\delta_R(t) = u_s(t)$.

(c) Determine the response, $\mathbf{x}(t) = [p(t); r(t); \beta(t); \phi(t)]^T$, if the initial condition is $p(0) = 0.1$ rad/s,
$r(0) = \beta(0) = \phi(0) = 0$, and the two inputs are zeros, $\delta_A(t) = \delta_R(t) = 0$.

(d) Determine the response, $\mathbf{y}(t) = [p(t); \phi(t)]^T$, if the initial condition is zero, $\mathbf{x}(0) = \mathbf{0}$, and the input vector is $\mathbf{u}(t) = [\delta_A(t); \delta_R(t)]^T = [0.1\delta(t); -0.1\delta(t)]^T$.

4.4. For the compensated closed-loop chemical plant of Example 2.25 (for which a state-space representation was obtained in Exercise 3.9), determine the output, $y(t)$, and the plant input, $u(t)$, if the desired output, $y_d(t)$, is a *unit impulse function*, and the initial condition is zero.

4.5. For the multivariable closed-loop system of Exercise 2.30 (for which a state-space representation was obtained in Exercise 3.13) with controller parameter $K = 1$, determine the output, $\mathbf{y}(t)$, and the plant input, $\mathbf{u}(t)$, if the desired output is $\mathbf{y}_d(t) = [u_s(t); -u_s(t)]^T$, and the initial condition is zero.

4.6 A linear time-varying system has the following state-space representation:

$$\mathbf{A}(t) = \begin{bmatrix} -5/13 & 1 \\ 1 & -5/13 \end{bmatrix}; \mathbf{B}(t) = \begin{bmatrix} e^{t/2} \\ e^{-t} \end{bmatrix} \quad (4.98)$$

Calculate the response, $\mathbf{x}(t)$, if the input is a unit step function and the initial condition is $\mathbf{x}(0) = [-1; 1]^T$.

4.7. For the missile of Example 4.11, compute:

(a) the step response to $\alpha(t) = u_s(t)$.

(b) the step response to $\Delta L(t) = u_s(t)$.

(c) the initial response, $\mathbf{x}(t)$, to the initial condition, $\mathbf{x}(0) = [100; -10; 100; 0; 0; 0]^T$ and zero input, $\mathbf{u}(t) = \mathbf{0}$.

4.8. Solve the *van der Pol* equation (Eq. (2.204)) of Exercise 2.2, if $a = 5$ and $b = 3$, and the initial condition is given by $x^{(1)}(0) = -0.1$, and $x(0) = 0.5$.

4.9. Solve for the motion of the double-pendulum (Example 4.12) if the initial condition is zero, $\mathbf{x}(0) = \mathbf{0}$, and the input torque is a unit step function, $u(t) = u_s(t)$.

4.10. For the wing-rock dynamics of Example 4.13, solve for the *bank angle*, $x_1(t)$, and *roll-rate*, $x_2(t)$, if the initial condition is zero, $\mathbf{x}(0) = \mathbf{0}$, and the *desired aileron input* is a unit step function, $u(t) = u_s(t)$.

References

1. Golub, G.H. and van Loan, C.F. *Matrix Computation*. Johns Hopkins University Press, 1983.
2. Moler, C.B. and van Loan, C.F. Nineteen dubious ways to compute the exponential of a matrix. *SIAM Review*, **20**, 801–836, 1979.
3. *Control System Toolbox-5.0 for Use with MATLAB®-User's Guide*, The Math Works Inc., Natick, MA, USA, 2000.
4. Henrici P., *Discrete Variable Methods in Ordinary Differential Equations*. John Wiley & Sons, New York, 1962.
5. Ralston, A. and Rabinowitz, P. *A First Course in Numerical Analysis*. McGraw-Hill, New York, 1988.
6. Tewari, A. Nonlinear optimal control of wing rock including yawing motion. Paper No. AIAA-2000-4251, *Proceedings of AIAA Guidance, Navigation, and Controls Conference*, Denver, CO, August 14–17, 2000.
7. *SIMULINK-4.0 User's Guide*. The Math Works Inc., Natick, MA, USA, 2000.

5

Control System Design in State-Space

5.1 Design: Classical vs. Modern

A fashion designer tailors the apparel to meet the tastes of fashionable people, keeping in mind the desired fitting, season and the occasion for which the clothes are to be worn. Similarly, a control system engineer *designs* a control system to meet the desired objectives, keeping in mind issues such as where and how the control system is to be implemented. We need a control system because we do not like the way a plant behaves, and by *designing* a control system we try to *modify* the behavior of the plant to suit our needs. *Design* refers to the process of changing a control system's *parameters* to meet the specified stability, performance, and robustness objectives. The design *parameters* can be the unknown constants in a *controller's* transfer function, or its state-space representation. In Chapter 1 we compared open- and closed-loop control systems, and saw how a closed-loop control system has a better chance of achieving the desired performance. In Chapter 2 we saw how the classical transfer function approach can be used to design a closed-loop control system, i.e. the use of graphical methods such as Bode, Nyquist, and root-locus plots. Generally, the classical design consists of varying the controller transfer function until a desired closed-loop performance is achieved. The classical indicators of the closed-loop performance are the closed-loop frequency response, or the locations of the closed-loop poles. For a large order system, by varying a limited number of constants in the controller transfer function, we can vary in a pre-specified manner the locations of only a few of the closed-loop poles, *but not all of them*. This is a major limitation of the classical design approach. The following example illustrates some of the limitations of the classical design method.

> *Example 5.1*
>
> Let us try to design a closed-loop control system for the following plant transfer function in order to achieve a zero steady-state error when the desired output is the unit step function, $u_s(t)$:
>
> $$G(s) = (s + 1)/[(s - 1)(s + 2)(s + 3)] \qquad (5.1)$$

The single-input, single-output plant, $G(s)$, has poles located at $s = 1$, $s = -2$, and $s = -3$. Clearly, the plant is unstable due to a pole, $s = 1$, in the right-half s-plane. Also, the plant is of type 0. For achieving a zero steady-state error, we need to do *two things*: make the closed-loop system stable, and make type of the closed-loop system at least unity. Selecting a closed-loop arrangement of Figure 2.32, both of these requirements are apparently met by the following choice of the controller transfer function, $H(s)$:

$$H(s) = K(s - 1)/s \qquad (5.2)$$

Such a controller would apparently cancel the plant's unstable pole at $s = 1$ by a zero at the same location in the closed-loop transfer function, and make the system of type 1 by having a pole at $s = 0$ in the open-loop transfer function. The open-loop transfer function, $G(s)H(s)$, is then the following:

$$G(s)H(s) = K(s + 1)/[s(s + 2)(s + 3)] \qquad (5.3)$$

and the closed-loop transfer function is given by

$$Y(s)/Y_{\mathrm{d}}(s) = G(s)H(s)/[1 + G(s)H(s)]$$
$$= K(s + 1)/[s(s + 2)(s + 3) + K(s + 1)] \qquad (5.4)$$

From Eq. (5.4), it is apparent that the closed-loop system can be made stable by selecting those value of the design parameter, K, such that all the closed-loop poles lie in the left-half s-plane. The root-locus of the closed-loop system is plotted in Figure 5.1 as K

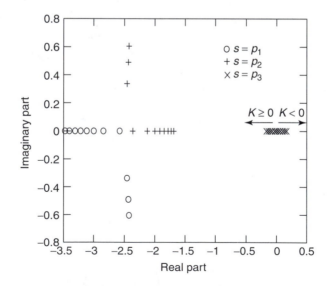

Figure 5.1   Apparent loci of the closed-loop poles of Example 5.1 as the classical design parameter, K, is varied from -1 to 1

is varied from -1 to 1. Apparently, from Figure 5.1, the closed-loop system is stable for $K \geq 0$ and unstable for $K < 0$. In Figure 5.1, it appears that the pole $s = p_3$ determines the stability of the closed-loop system, since it is the pole which crosses into the right-half s-plane for $K < 0$. This pole is called the *dominant pole* of the system, because being closest to the imaginary axis it *dominates* the system's response (recall from Chapter 4 that the smaller the real part magnitude of a pole, the longer its contribution to the system's response persists). By choosing an appropriate value of K, we can place only the dominant pole, $s = p_3$, at a desired location in the left-half s-plane. The locations of the other two poles, $s = p_1$ and $s = p_2$, would then be governed by such a choice of K. In other words, by choosing the sole parameter K, the locations of all the three poles cannot be chosen *independently of each other*. Since all the poles contribute to the closed-loop performance, the classical design approach may fail to achieve the desired performance objectives when only a few poles are being directly affected in the design process.

Furthermore, the chosen design approach of Example 5.1 is misleading, because it *fails to even stabilize* the closed-loop system! Note that the closed-loop transfer function given by Eq. (5.4) is of *third order*, whereas we expect that it should be of *fourth order*, because the closed-loop system is obtained by combining a third order plant with a first order controller. This discrepancy in the closed-loop transfer function's order has happened due to our attempt to *cancel* a pole with a zero at the same location. Such an attempt is, however, doomed to fail as shown by a state-space analysis of the closed-loop system.

Example 5.2

Let us find a state-space representation of the closed-loop system designed using the classical approach in Example 5.1. Since the closed-loop system is of the configuration shown in Figure 3.7(c), we can readily obtain its state-space representation using the methods of Chapter 3. The Jordan canonical form of the plant, $G(s)$, is given by the following state coefficient matrices:

$$\mathbf{A_p} = \begin{bmatrix} 1 & 0 & 0 \\ 0 & -2 & 0 \\ 0 & 0 & -3 \end{bmatrix}; \quad \mathbf{B_p} = \begin{bmatrix} 0 \\ 1/3 \\ -1/2 \end{bmatrix}$$

$$\mathbf{C_p} = [1 \quad 1 \quad 1]; \quad \mathbf{D_p} = 0 \tag{5.5}$$

A state-space representation of the controller, $H(s)$, is the following:

$$\mathbf{A_c} = 0; \quad \mathbf{B_c} = K; \quad \mathbf{C_c} = -1; \quad \mathbf{D_c} = K \tag{5.6}$$

Therefore, on substituting Eqs. (5.5) and (5.6) into Eqs. (3.146)–(3.148), we get the following state-space representation of the closed-loop system:

$$\mathbf{x}^{(1)}(t) = \mathbf{A}\mathbf{x}(t) + \mathbf{B}\mathbf{y_d}(t) \tag{5.7}$$

$$\mathbf{y}(t) = \mathbf{C}\mathbf{x}(t) + \mathbf{D}\mathbf{y_d}(t) \tag{5.8}$$

where

$$\mathbf{A} = \begin{bmatrix} 1 & 0 & 0 & 0 \\ -K/3 & (-2-K/3) & -K/3 & -1/3 \\ K/2 & K/2 & (-3+K/2) & 1/2 \\ -K & -K & -K & 0 \end{bmatrix}; \quad \mathbf{B} = \begin{bmatrix} 0 \\ K/3 \\ -K/2 \\ K \end{bmatrix}$$

$$\mathbf{C} = [1 \quad 1 \quad 1 \quad 0]; \quad \mathbf{D} = 0 \tag{5.9}$$

The closed-loop system is of fourth order, as expected. The closed-loop poles are the eigenvalues of $\mathbf{A}$, i.e. the solutions of the following characteristic equation:

$$|\lambda\mathbf{I} - \mathbf{A}| = \begin{vmatrix} (\lambda - 1) & 0 & 0 & 0 \\ K/3 & (\lambda + 2 + K/3) & K/3 & 1/3 \\ -K/2 & -K/2 & (\lambda + 3 - K/2) & -1/2 \\ K & K & K & \lambda \end{vmatrix} = 0$$

$$\tag{5.10}$$

It is evident from Eq. (5.10) that, irrespective of the value of K, one of the eigenvalues of $\mathbf{A}$ is $\lambda = 1$, which corresponds to a closed-loop pole at $s = 1$. Hence, irrespective of the design parameter, K, we have an *unstable* closed-loop system, which means that the chosen design approach of cancelling an unstable pole with a zero *does not work*. More importantly, even though we have an *unconditionally* unstable closed-loop system, the closed-loop transfer function given by Eq. (5.4) *fools us into believing* that we can stabilize the closed-loop system by selecting an *appropriate value* for K. Such a system which remains unstable irrespective of the values of the control design parameters is called an *unstabilizable system*. The classical design approach of Example 5.1 gave us an unstabilizable closed-loop system, and we didn't even know it! Stabilizability of a system is a consequence of an important property known as *controllability*, which we will consider next. (Although we considered a closed-loop system in Example 5.2, the properties *controllability* and *stabilizability* are more appropriately defined for a plant.)

5.2 Controllability

When as children we sat in the back seat of a car, our collective effort to move the car by pushing on the front seat always ended in failure. This was because the input we provided to the car in this manner, *no matter how large*, did not affect the overall motion of the car. There was something known as the *third law of Newton*, which physically prevented us from achieving our goal. Hence, for us the car was *uncontrollable when we were sitting in the car*. The same car could be moved, however, by stepping out and giving a hefty push to it from the outside; then it became a *controllable* system for our purposes. *Controllability* can be defined as the property of a system when it is possible to take the system from *any initial state*, $\mathbf{x}(t_0)$, to *any final state*, $\mathbf{x}(t_f)$, in a *finite* time, $(t_f - t_0)$, by means of the input vector, $\mathbf{u}(t)$, $t_0 \leq t \leq t_f$. It is important to stress the words *any* and *finite*, because it may be possible to move an uncontrollable system from *some initial states* to *some final states*, or take an *infinite* amount of time in moving the uncontrollable system, using the input vector, $\mathbf{u}$ (t). Controllability of a system can be easily determined if we can *decouple* the state-equations of a system. Each decoupled scalar state-equation corresponds to a *sub-system*. If any of the decoupled state-equations of the system is *unaffected* by the input vector, then it is not possible to change the corresponding state variable using the input, and hence, the sub-system is *uncontrollable*. If any sub-system is uncontrollable, i.e. if any of the state variables is unaffected by the input vector, then it follows that the entire system is uncontrollable.

Example 5.3

Re-consider the closed-loop system of Example 5.2. The state-equations of the closed-loop system (Eqs. (5.7)–(5.9)) can be expressed in scalar form as follows:

$$x_1^{(1)}(t) = x_1(t) \tag{5.11a}$$

$$x_2^{(1)}(t) = -Kx_1(t)/3 - (2 + K/3)x_2(t) - Kx_3(t)/3$$
$$- x_4(t)/3 + Ky_d(t)/3 \tag{5.11b}$$

$$x_3^{(1)}(t) = Kx_1(t)/2 + Kx_2(t)/2 + (-3 + K/2)x_3(t)$$
$$+ x_4(t)/2 - Ky_d(t)/2 \tag{5.11c}$$

$$x_4^{(1)}(t) = -Kx_1(t) - Kx_2(t) - Kx_3(t) + Ky_d(t) \tag{5.11d}$$

On examining Eq. (5.11a), we find that the equation is decoupled from the other state-equations, and does not contain the input to the closed-loop system, $y_d(t)$. Hence, the state variable, $x_1(t)$, is entirely unaffected by the input, $y_d(t)$, which implies that the system is *uncontrollable*. Since the uncontrollable sub-system described by Eq. (5.11a) is also unstable (it corresponds to the eigenvalue $\lambda = 1$), there is no way we can stabilize the closed-loop system by changing the controller design parameter, K. Hence, the system is *unstabilizable*. In fact, the plant of this system given by the state-space representation of Eq. (5.5) is itself unstabilizable, because of the zero in the matrix $\mathbf{B_p}$ corresponding to the sub-system having eigen-value $\lambda = 1$. The unstabilizable plant leads to an unstabilizable closed-loop system.

Example 5.3 shows how a decoupled state-equation indicating an *uncontrollable* and *unstable* sub-system implies an *unstabilizable* system.

Example 5.4

Let us analyze the controllability of the following system:

$$\mathbf{A} = \begin{bmatrix} 0 & 0 & 1 & 0 \\ 0 & 0 & 0 & 1 \\ 0 & 0 & 0 & 0 \\ 0 & 0 & 0 & 0 \end{bmatrix}; \quad \mathbf{B} = \begin{bmatrix} 0 \\ 0 \\ -1 \\ 1 \end{bmatrix} \tag{5.12}$$

The system is *unstable*, with four zero eigenvalues. Since the state-equations of the system are coupled, we cannot directly deduce controllability. However, some of the state-equations can be decoupled by transforming the state-equations using the transformation $\mathbf{z}(t) = \mathbf{Tx}(t)$, where

$$\mathbf{T} = \begin{bmatrix} 1 & 1 & 0 & 0 \\ 1 & -1 & 0 & 0 \\ 0 & 0 & 1 & 1 \\ 0 & 0 & 1 & -1 \end{bmatrix} \tag{5.13}$$

The transformed state-equations can be written in the following scalar form:

$$z_1^{(1)}(t) = x_3'(t) \tag{5.14a}$$

$$z_2^{(1)}(t) = x_4'(t) \tag{5.14b}$$

$$z_3^{(1)}(t) = 0 \tag{5.14c}$$

$$z_4^{(1)}(t) = -2u(t) \tag{5.14d}$$

Note that the state-equation, Eq. (5.14c) denotes an uncontrollable sub-system in which the state variable, $z_3(t)$, is unaffected by the input, $u(t)$. Hence, the system is uncontrollable. However, since the only uncontrollable sub-system denoted by Eq. (5.14c) is *stable* (its eigenvalue is, $\lambda = 0$), we can safely *ignore* this sub-system and *stabilize* the remaining sub-systems denoted by Eqs. (5.14a),(5.14b), and (5.14d), using a feedback controller that modifies the control input, $u(t)$. An uncontrollable system all of whose uncontrollable sub-systems are stable is thus said to be *stabilizable*. The process of stabilizing a stabilizable system consists of ignoring all uncontrollable but stable sub-systems, and designing a controller based on the remaining (controllable) sub-systems. Such a control system will be successful, because each ignored sub-system will be stable.

In the previous two examples, we could determine controllability, only because certain state-equations were decoupled from the other state-equations. Since decoupling state-equations is a cumbersome process, and may not be always possible, we need another criterion for testing whether a system is controllable. The following *algebraic controllability test theorem* provides an easy way to check for controllability.

Theorem
A linear, time-invariant system described by the matrix state-equation, $\mathbf{x}^{(1)}(t) = \mathbf{Ax}(t) + \mathbf{Bu}(t)$ *is controllable* if and only if the *controllability test matrix*

$$\mathbf{P} = [\mathbf{B}; \quad \mathbf{AB}; \quad \mathbf{A}^2\mathbf{B}; \quad \mathbf{A}^3\mathbf{B}; \quad \ldots; \quad \mathbf{A}^{n-1}\mathbf{B}]$$

is of rank n, *the order of the system.*

(The *rank* of a matrix, $\mathbf{P}$, is defined as the dimension of the *largest non-zero* determinant formed out of the matrix, $\mathbf{P}$ (see Appendix B). If $\mathbf{P}$ is a square matrix, the largest determinant formed out of $\mathbf{P}$ is $|\mathbf{P}|$. If $\mathbf{P}$ is not a square matrix, the largest determinant formed out of $\mathbf{P}$ is either the determinant formed by taking all the rows and equal number of columns, or all the columns and equal number of rows of $\mathbf{P}$. See Appendix B for an illustration of the rank of a matrix. Note that for a system of order n with r inputs, the size of the controllability test matrix, $\mathbf{P}$, is $(n \times nr)$. The largest non-zero determinant of $\mathbf{P}$ can be of dimension n. Hence, the rank of $\mathbf{P}$ can be *either less than or equal* to n.)

A rigourous proof of the algebraic controllability test theorem can be found in Friedland [2]. An analogous form of algebraic controllability test theorem can be obtained for linear, time-varying systems [2]. Alternatively, we can form a *time-varying* controllability test matrix as

$$\mathbf{P}(t) = [\mathbf{B}(t); \quad \mathbf{A}(t)\mathbf{B}(t); \quad \mathbf{A}^2(t)\mathbf{B}(t); \quad \mathbf{A}^3(t)\mathbf{B}(t); \quad \ldots; \quad \mathbf{A}^{n-1}(t)\mathbf{B}(t)] \quad (5.15)$$

and check the rank of $\mathbf{P}(t)$ for all times, $t \geq t_0$, for a linear, time-varying system. If at any instant, t, the rank of $\mathbf{P}(t)$ is less than n, the system is uncontrollable. However, we must use the time-varying controllability test matrix of Eq. (5.15) with great *caution*, when the state-coefficient matrices are rapidly changing with time, because the test can be practically applied at *discrete time step* – rather than at all possible times (see Chapter 4) – and there may be some time intervals (smaller than the time steps) in which the system may be uncontrollable.

Example 5.5

Using the controllability test theorem, let us find whether the following system is controllable:

$$\mathbf{A} = \begin{bmatrix} -2 & 1 \\ -1 & -3 \end{bmatrix}; \quad \mathbf{B} = \begin{bmatrix} 1 \\ 0 \end{bmatrix} \quad (5.16)$$

The controllability test matrix is the following:

$$\mathbf{P} = [\mathbf{B}; \quad \mathbf{AB}] = \begin{bmatrix} 1 & -2 \\ 0 & -1 \end{bmatrix} \quad (5.17)$$

The largest determinant of $\mathbf{P}$ is $|\mathbf{P}| = -1 \neq 0$, Hence the rank of $\mathbf{P}$ is equal to 2, the order of the system. Thus, by the controllability test theorem, the system is *controllable*.

Applying the algebraic controllability test involves finding the rank of $\mathbf{P}$, and checking whether it is equal to n. This involves forming all possible determinants of dimension n out of the matrix $\mathbf{P}$, by removing some of the columns (if $m > 1$), and checking whether all of those determinants are non-zero. By any account, such a process is cumbersome if performed by hand. However, MATLAB provides us the command *rank(P)* for finding the rank of a matrix, $\mathbf{P}$. Moreover, MATLAB's Control System Toolbox (CST) lets you directly form the controllability test matrix, $\mathbf{P}$, using the command *ctrb* as follows:

```
>>P = ctrb(A, B) <enter>
```

```
or

>>P = ctrb(sys) <enter>
```

where **A** and **B** are the state coefficient matrices of the system whose LTI object is *sys*.

Example 5.6

Let us verify the uncontrollability of the system given in Example 5.4, using the controllability test. The controllability test matrix is constructed as follows:

```
>>A=[0 0 1 0; zeros(1,3)1; zeros(2,4)]; B=[0 0 -1 1]'; P=ctrb(A,B)
  <enter>

P =

    0  -1  0  0
    0   1  0  0
   -1   0  0  0
    1   0  0  0
```

Then the rank of **P** is found using the MATLAB command *rank:*

```
   >>rank(P) <enter>

   ans =

    2
```

Since the rank of **P** is *less than* 4, the order of the system, it follows from the controllability test theorem that the system is *uncontrollable*.

What are the causes of uncontrollability? As our childhood attempt of pushing a car while sitting inside it indicates, whenever we choose an input vector that does not affect *all* the state variables *physically*, we will have an uncontrollable system. An attempt to cancel a pole of the plant by a zero of the controller may also lead to an uncontrollable closed-loop system *even though the plant itself may be controllable*. Whenever you see a system in which pole-zero cancellations have occurred, the chances are high that such a system is uncontrollable.

Example 5.7

Let us analyze the controllability of the closed-loop system of configuration shown in Figure 2.32, in which the controller, $H(s)$, and plant, $G(s)$, are as follows:

$$H(s) = K(s-2)/(s+1); \quad G(s) = 3/(s-2) \tag{5.18}$$

The closed-loop transfer function in which a pole-zero cancellation has occurred at $s = 2$ is the following:

$$Y(s)/Y_d(s) = G(s)H(s)/[1 + G(s)H(s)] = 3K/(s + 3K + 1) \qquad (5.19)$$

The Jordan canonical form of the plant is the following:

$$\mathbf{A_p} = 2; \quad \mathbf{B_p} = 3; \quad \mathbf{C_p} = 1; \quad \mathbf{D_p} = 0 \qquad (5.20)$$

Note that the plant is controllable (the controllability test matrix for the plant is just $\mathbf{P} = \mathbf{B_p}$, which is of rank 1). The Jordan canonical form of the controller is the following:

$$\mathbf{A_c} = -1; \quad \mathbf{B_c} = K; \quad \mathbf{C_c} = -3; \quad \mathbf{D_c} = K \qquad (5.21)$$

The closed-loop state-space representation is obtained using Eqs. (3.146)–(3.148) as the following:

$$\mathbf{A} = \begin{bmatrix} (2 - 3K) & -9 \\ -K & -1 \end{bmatrix}; \quad \mathbf{B} = \begin{bmatrix} 3K \\ K \end{bmatrix}$$

$$\mathbf{C} = [1 \quad 0]; \quad \mathbf{D} = 0 \qquad (5.22)$$

The controllability test matrix for the closed-loop system is the following:

$$\mathbf{P} = [\mathbf{B} \quad \mathbf{AB}] = \begin{bmatrix} 3K & -(9K^2 + 3K) \\ K & -(3K^2 + K) \end{bmatrix} \qquad (5.23)$$

To see whether $\mathbf{P}$ is of rank 2 (i.e. whether $\mathbf{P}$ is non-singular) let us find its determinant as follows:

$$|\mathbf{P}| = \begin{vmatrix} 3K & -(9K^2 + 3K) \\ K & -(3K^2 + K) \end{vmatrix} = -9K^3 - 3K^2 + 9K^3 + 3K^2 = 0 \qquad (5.24)$$

Since $|\mathbf{P}| = 0$, $\mathbf{P}$ is singular, its rank is less than 2. Therefore, the closed-loop system is uncontrollable no matter what value of the controller design parameter, K, is chosen. Hence, a *controllable* plant has led to an *uncontrollable* closed-loop system in which a pole-zero cancellation has occurred.

Other causes of uncontrollability could be *mathematical*, such as using *superfluous* state variables (i.e. more state variables than the order of the system) when modeling a system; the superfluous state variables will be definitely unaffected by the inputs to the system, causing the state-space representation to be uncontrollable, even though the system may be physically controllable. A rare cause of uncontrollability is *too much symmetry* in the system's mathematical model. Electrical networks containing *perfectly balanced bridges* are examples of systems with too much symmetry. However, perfect symmetry almost never exists in the real world, or in its digital computer model.

Now that we know how to determine the controllability of a system, we can avoid the pitfalls of Examples 5.1 and 5.7, and are ready to design a control system using state-space methods.

5.3 Pole-Placement Design Using Full-State Feedback

In Section 5.1 we found that it may be required to change a plant's characteristics by using a closed-loop control system, in which a controller is designed to *place* the *closed-loop poles* at desired locations. Such a design technique is called the *pole-placement* approach. We also discussed in Section 5.1 that the classical design approach using a controller transfer function with a few design parameters is insufficient to place all the closed-loop poles at desired locations. The state-space approach using *full-state feedback* provides sufficient number of controller design parameters to move all the closed-loop poles independently of each other. *Full-state feedback* refers to a controller which generates the input vector, $\mathbf{u}(t)$, according to a *control-law* such as the following:

$$\mathbf{u}(t) = \mathbf{K}[\mathbf{x_d}(t) - \mathbf{x}(t)] - \mathbf{K_d}\mathbf{x_d}(t) - \mathbf{K_n}\mathbf{x_n}(t) \qquad (5.25)$$

where $\mathbf{x}(t)$ is the state-vector of the plant, $\mathbf{x_d}(t)$ is the *desired* state-vector, $\mathbf{x_n}(t)$ is the *noise* state-vector and $\mathbf{K}$, $\mathbf{K_d}$ and $\mathbf{K_n}$ are the *controller gain matrices*. The desired state-vector, $\mathbf{x_d}(t)$, and the noise state-vector, $\mathbf{x_n}(t)$, are generated by external processes, and act as inputs to the control system. The task of the controller is to achieve the desired state-vector in the steady state, while counteracting the affect of the noise. The input vector, $\mathbf{u}(t)$, generated by Eq. (5.25) is applied to the plant described by the following state and output equations:

$$\mathbf{x}^{(1)}(t) = \mathbf{A}\mathbf{x}(t) + \mathbf{B}\mathbf{u}(t) + \mathbf{F}\mathbf{x_n}(t) \qquad (5.26)$$

$$\mathbf{y}(t) = \mathbf{C}\mathbf{x}(t) + \mathbf{D}\mathbf{u}(t) + \mathbf{E}\mathbf{x_n}(t) \qquad (5.27)$$

where $\mathbf{F}$ and $\mathbf{E}$ are the noise coefficient matrices in the state and output equations, respectively. Designing a control system using full-state feedback requires that the plant described by Eq. (5.26) must be *controllable*, otherwise the control input generated using Eq. (5.25) will not affect all the state variables of the plant. Furthermore, Eq. (5.25) requires that the all the state variables of the system must be *measurable*, and capable of being fed back to the controller. The controller thus consists of physical *sensors*, which measure the state variables, and electrical or mechanical devices, called *actuators*, which provide inputs to the plant based on the desired outputs and the *control-law* of Eq. (5.25). Modern controllers invariably use digital electronic circuits to implement the control-law in a hardware. The controller gain matrices, $\mathbf{K}$, $\mathbf{K_d}$, and $\mathbf{K_n}$ are the *design parameters* of the control system described by Eqs. (5.25)–(5.27). Note that the order of the full-state feedback closed-loop system is the *same* as that of the plant. A schematic diagram of the general control system with full-state feedback is shown in Figure 5.2.

Let us first consider control systems having $\mathbf{x_d}(t) = \mathbf{0}$. A control system in which the desired state-vector is zero is called a *regulator*. Furthermore, for simplicity let us assume that all the measurements are *perfect*, and that there is *no error* committed in *modeling*

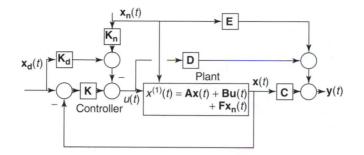

Figure 5.2 Schematic diagram of a general full-state feedback control system with desired state, $\mathbf{x_d}(t)$, and noise, $\mathbf{x_n}(t)$

the plant by Eqs. (5.26) and (5.27). These two assumptions imply that all undesirable inputs to the system in the form of noise, are absent, i.e. $\mathbf{x}_n(t) = \mathbf{0}$. Consequently, the control-law of Eq. (5.25) reduces to

$$\mathbf{u}(t) = -\mathbf{K}\mathbf{x}(t) \tag{5.28}$$

and the schematic diagram of a noiseless regulator is shown in Figure 5.3.

On substituting Eq. (5.28) into Eqs. (5.26) and (5.27), we get the closed-loop state and output equations of the regulator as follows:

$$\mathbf{x}^{(1)}(t) = (\mathbf{A} - \mathbf{BK})\mathbf{x}(t) \tag{5.29}$$

$$\mathbf{y}(t) = (\mathbf{C} - \mathbf{DK})\mathbf{x}(t) \tag{5.30}$$

Equations. (5.29) and (5.30) indicate that the regulator is a *homogeneous* system, described by the closed-loop state coefficient matrices $\mathbf{A}_{CL} = \mathbf{A} - \mathbf{BK}$, $\mathbf{B}_{CL} = \mathbf{0}$, $\mathbf{C}_{CL} = \mathbf{C} - \mathbf{DK}$, and $\mathbf{D}_{CL} = \mathbf{0}$. The closed-loop poles are the *eigenvalues* of $\mathbf{A}_{CL}$. Hence, by selecting the controller gain matrix, $\mathbf{K}$, we can place the closed-loop poles at desired locations. For a plant of order n with r inputs, the size of $\mathbf{K}$ is $(r \times n)$. Thus, we have a total of $r \cdot n$ scalar design parameters in our hand. For multi-input systems (i.e. $r > 1$), the number of design parameters are, therefore, *more than sufficient* for selecting the locations of n poles.

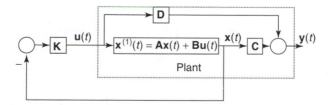

Figure 5.3 Schematic diagram of a full-state feedback regulator (i.e. control system with a zero desired state-vector) without any noise

Example 5.8

Let us design a full-state feedback regulator for the following plant such that the closed-loop poles are $s = -0.5 \pm i$:

$$\mathbf{A} = \begin{bmatrix} 1 & 0 \\ 0 & -2 \end{bmatrix}; \quad \mathbf{B} = \begin{bmatrix} 1 \\ -1 \end{bmatrix} \tag{5.31}$$

The plant, having poles at $s = 1$ and $s = -2$, is unstable. Also, the plant is controllable, because its decoupled state-space representation in Eq. (5.31) has no elements of $\mathbf{B}$ equal to zero. Hence, we can place closed-loop poles *at will* using the following full-state feedback gain matrix:

$$\mathbf{K} = [K_1; \quad K_2] \tag{5.32}$$

The closed-loop state-dynamics matrix, $\mathbf{A_{CL}} = \mathbf{A} - \mathbf{BK}$, is the following:

$$\mathbf{A_{CL}} = \mathbf{A} - \mathbf{BK} = \begin{bmatrix} (1 - K_1) & -K_2 \\ K_1 & (-2 + K_2) \end{bmatrix} \tag{5.33}$$

The closed-loop poles are the eigenvalues of $\mathbf{A_{CL}}$, which are calculated as follows:

$$|\lambda \mathbf{I} - \mathbf{A_{CL}}| = \begin{vmatrix} (\lambda - 1 + K_1) & K_2 \\ -K_1 & (\lambda + 2 - K_2) \end{vmatrix}$$

$$= (\lambda - 1 + K_1)(\lambda + 2 - K_2) + K_1 K_2 = 0 \tag{5.34}$$

The roots of the characteristic equation (Eq. (5.34)) are the closed-loop eigenvalues given by

$$\lambda_{1,2} = -0.5(K_1 - K_2 + 1) \pm 0.5(K_1^2 + K_2^2 - 2K_1 K_2 - 6K_1 - 2K_2 + 9)^{1/2}$$

$$= -0.5 \pm i \tag{5.35}$$

Solving Eq. (5.35) for the unknown parameters, K_1 and K_2, we get

$$K_1 = K_2 = 13/12 \tag{5.36}$$

Thus, the full-state feedback regulator gain matrix which moves the poles from $s = 1$, $s = -2$ to $s = -0.5 \pm i$ is $\mathbf{K} = [13/12; \quad 13/12]$.

5.3.1 Pole-placement regulator design for single-input plants

Example 5.8 shows that even for a single-input, second order plant, the calculation for the required regulator gain matrix, $\mathbf{K}$, by hand is rather involved, and is likely to get out of hand as the order of the plant increases beyond three. Luckily, if the plant is in the *controller companion form*, then such a calculation is greatly simplified for single-input plants. Consider a single-input plant of order n whose controller companion form is the following (see Chapter 3):

$$\mathbf{A} = \begin{bmatrix} -a_{n-1} & -a_{n-2} & -a_{n-3} & \cdots & -a_1 & -a_0 \\ 1 & 0 & 0 & \cdots & 0 & 0 \\ 0 & 1 & 0 & \cdots & 0 & 0 \\ 0 & 0 & 1 & \cdots & 0 & 0 \\ \cdot & \cdot & \cdot & \cdots & \cdot & \cdot \\ \cdot & \cdot & \cdot & \cdots & \cdot & \cdot \\ 0 & 0 & 0 & \cdots & 1 & 0 \\ 0 & 0 & 0 & \cdots & 0 & 1 \end{bmatrix} ; \quad \mathbf{B} = \begin{bmatrix} 1 \\ 0 \\ 0 \\ 0 \\ \cdot \\ \cdot \\ 0 \\ 0 \end{bmatrix} \qquad (5.37)$$

where $a_0, \ldots, a_{n-1}$ are the coefficients of the plant's characteristic polynomial $|s\mathbf{I} - \mathbf{A}| = s^n + a_{n-1}s^{n-1} + \ldots + a_1 s + a_0$. The full-state feedback regulator gain matrix is a row vector of n unknown parameters given by

$$\mathbf{K} = [K_1; \quad K_2; \quad \ldots; \quad K_n] \qquad (5.38)$$

It is desired to place the closed-loop poles such that the closed-loop characteristic polynomial is the following:

$$|s\mathbf{I} - \mathbf{A_{CL}}| = |s\mathbf{I} - \mathbf{A} + \mathbf{BK}| = s^n + \alpha_{n-1}s^{n-1} + \alpha_{n-2}s^{n-2} \ldots + \alpha_1 s + \alpha_0 \qquad (5.39)$$

where the closed-loop state dynamics matrix, $\mathbf{A_{CL}} = \mathbf{A} - \mathbf{BK}$, is the following:

$$\mathbf{A_{CL}} = \begin{bmatrix} (-a_{n-1} - K_1) & (-a_{n-2} - K_2) & (-a_{n-3} - K_3) & \cdots & (-a_1 - K_{n-1}) & (-a_0 - K_n) \\ 1 & 0 & 0 & \cdots & 0 & 0 \\ 0 & 1 & 0 & \cdots & 0 & 0 \\ 0 & 0 & 1 & \cdots & 0 & 0 \\ \cdot & \cdot & \cdot & \cdots & \cdot & \cdot \\ 0 & 0 & 0 & \cdots & 1 & 0 \\ 0 & 0 & 0 & \cdots & 0 & 1 \end{bmatrix} \qquad (5.40)$$

It is interesting to note that the closed-loop system is also in the controller companion form! Hence, from Eq. (5.40), the coefficients of the closed-loop characteristic polynomial must be the following:

$$\alpha_{n-1} = a_{n-1} + K_1; \quad \alpha_{n-2} = a_{n-2} + K_2; \quad \ldots; \quad \alpha_1 = a_1 + K_{n-1}; \quad \alpha_0 = a_0 + K_n \qquad (5.41)$$

or, the unknown regulator parameters are calculated simply as follows:

$$K_1 = \alpha_{n-1} - a_{n-1}; \quad K_2 = \alpha_{n-2} - a_{n-2}; \quad \ldots; \quad K_{n-1} = \alpha_1 - a_1; \quad K_n = \alpha_0 - a_0 \qquad (5.42)$$

In vector form, Eq. (5.42) can be expressed as

$$\mathbf{K} = \boldsymbol{\alpha} - \mathbf{a} \qquad (5.43)$$

where $\boldsymbol{\alpha} = [\alpha_{n-1}; \alpha_{n-2}; \ldots; \alpha_1; \alpha_0]$ and $\mathbf{a} = [a_{n-1}; a_{n-2}; \ldots; a_1; a_0]$. If the state-space representation of the plant is *not* in the controller companion form, a state-transformation can be used to transform the plant to the controller companion form as follows:

$$\mathbf{x}'(t) = \mathbf{T}\mathbf{x}(t); \mathbf{A}' = \mathbf{TAT}^{-1}; \mathbf{B}' = \mathbf{TB} \tag{5.44}$$

where $\mathbf{x}'(t)$ is the state-vector of the plant in the controller companion form, $\mathbf{x}(t)$ is the original state-vector, and $\mathbf{T}$ is the state-transformation matrix. The single-input regulator's control-law (Eq. (5.28)) can thus be expressed as follows:

$$u(t) = -\mathbf{K}\mathbf{x}(t) = -\mathbf{KT}^{-1}\mathbf{x}'(t) \tag{5.45}$$

Since $\mathbf{KT}^{-1}$ is the regulator gain matrix when the plant is in the controller companion form, it must be given by Eq. (5.43) as follows:

$$\mathbf{KT}^{-1} = \boldsymbol{\alpha} - \mathbf{a} \tag{5.46}$$

or

$$\mathbf{K} = (\boldsymbol{\alpha} - \mathbf{a})\mathbf{T} \tag{5.47}$$

Let us derive the state-transformation matrix, $\mathbf{T}$, which transforms a plant to its controller companion form. The controllability test matrix of the plant in its *original* state-space representation is given by

$$\mathbf{P} = [\mathbf{B}; \quad \mathbf{AB}; \quad \mathbf{A}^2\mathbf{B}; \quad \ldots; \quad \mathbf{A}^{n-1}\mathbf{B}] \tag{5.48}$$

Substitution of inverse transformation, $\mathbf{B} = \mathbf{T}^{-1}\mathbf{B}'$, and $\mathbf{A} = \mathbf{T}^{-1}\mathbf{A}'\mathbf{T}$ into Eq. (5.48) yields

$$\mathbf{P} = [\mathbf{T}^{-1}\mathbf{B}'; \quad (\mathbf{T}^{-1}\mathbf{A}'\mathbf{T})\mathbf{T}^{-1}\mathbf{B}'; \quad (\mathbf{T}^{-1}\mathbf{A}'\mathbf{T})^2\mathbf{T}^{-1}\mathbf{B}'; \quad \ldots; \quad (\mathbf{T}^{-1}\mathbf{A}'\mathbf{T})^{n-1}\mathbf{T}^{-1}\mathbf{B}']$$

$$= \mathbf{T}^{-1}[\mathbf{B}'; \quad \mathbf{A}'\mathbf{B}'; \quad (\mathbf{A}')^2\mathbf{B}'; \quad \ldots; \quad (\mathbf{A}')^{n-1}\mathbf{B}'] = \mathbf{T}^{-1}\mathbf{P}' \tag{5.49}$$

where $\mathbf{P}'$ is the controllability test matrix of the plant in *controller companion form*. Premultiplying both sides of Eq. (5.49) with $\mathbf{T}$, and then post-multiplying both sides of the resulting equation with $\mathbf{P}^{-1}$ we get the following expression for $\mathbf{T}$:

$$\mathbf{T} = \mathbf{P}'\mathbf{P}^{-1} \tag{5.50}$$

You can easily show that $\mathbf{P}'$ is the following *upper triangular* matrix (thus called because all the elements *below* its main diagonal are zeros):

$$\mathbf{P}' = \begin{bmatrix} 1 & -a_{n-1} & -a_{n-2} & \ldots & -a_2 & -a_1 \\ 0 & 1 & -a_{n-1} & \ldots & -a_3 & -a_2 \\ 0 & 0 & 1 & \ldots & -a_4 & -a_3 \\ \cdot & \cdot & \cdot & \ldots & \cdot & \cdot \\ \cdot & \cdot & \cdot & \ldots & \cdot & \cdot \\ 0 & 0 & 0 & \ldots & 1 & -a_{n-1} \\ 0 & 0 & 0 & \ldots & 0 & 1 \end{bmatrix} \tag{5.51}$$

Also note from Eq. (5.51) that the determinant of $\mathbf{P}'$ is unity, and that $(\mathbf{P}')^{-1}$ is obtained merely by replacing all the elements above the main diagonal of $\mathbf{P}'$ by their *negatives*. Substituting Eq. (5.50) into Eq. (5.47), the regulator gain matrix is thus given by

$$\mathbf{K} = (\boldsymbol{\alpha} - \mathbf{a})\mathbf{P}'\mathbf{P}^{-1} \tag{5.52}$$

Equation (5.52) is called the *Ackermann's pole-placement formula*. For a single-input plant considered here, both $\mathbf{P}$ and $\mathbf{P}'$ are square matrices of size $(n \times n)$. Note that if the plant is uncontrollable, $\mathbf{P}$ is singular, thus $\mathbf{T} = \mathbf{P}'\mathbf{P}^{-1}$ *does not exist*. This confirms our earlier requirement that for pole-placement, a plant must be controllable.

Example 5.9

Let us design a full-state feedback regulator for an inverted pendulum on a moving cart (Figure 2.59). A linear state-space representation of the plant is given by Eqs. (3.31) and (3.32), of which the state coefficient matrices are the following:

$$\mathbf{A} = \begin{bmatrix} 0 & 0 & 1 & 0 \\ 0 & 0 & 0 & 1 \\ (M+m)g/(ML) & 0 & 0 & 0 \\ -mg/M & 0 & 0 & 0 \end{bmatrix}; \quad \mathbf{B} = \begin{bmatrix} 0 \\ 0 \\ -1/(ML) \\ 1/M \end{bmatrix}$$

$$\mathbf{C} = \begin{bmatrix} 1 & 0 & 0 & 0 \\ 0 & 1 & 0 & 0 \end{bmatrix}; \quad \mathbf{D} = \begin{bmatrix} 0 \\ 0 \end{bmatrix} \tag{5.53}$$

The single-input, $u(t)$, is a force applied horizontally to the cart, and the two outputs are the angular position of the pendulum, $\theta(t)$, and the horizontal position of the cart, $x(t)$. The state-vector of this fourth order plant is $\mathbf{x}(t) = [\theta(t); x(t); \theta^{(1)}(t); x^{(1)}(t)]^T$. Let us assume the numerical values of the plant's parameters as follows: $M = 1$ kg, $m = 0.1$ kg, $L = 1$ m, and $g = 9.8$ m/s^2. Then the matrices $\mathbf{A}$ and $\mathbf{B}$ are the following:

$$\mathbf{A} = \begin{bmatrix} 0 & 0 & 1 & 0 \\ 0 & 0 & 0 & 1 \\ 10.78 & 0 & 0 & 0 \\ -0.98 & 0 & 0 & 0 \end{bmatrix}; \quad \mathbf{B} = \begin{bmatrix} 0 \\ 0 \\ -1 \\ 1 \end{bmatrix} \tag{5.54}$$

Let us first determine whether the plant is controllable. This is done by finding the controllability test matrix, $\mathbf{P}$, using the MATLAB (CST) command *ctrb* as follows:

```
>>P = ctrb(A,B)) <enter>

P =

0          -1.0000      0          -10.7800
0           1.0000      0            0.9800
-1.0000      0        -10.7800      0
1.0000      0          0.9800      0
```

The determinant of the controllability test matrix is then computed as follows:

```
>>det(P) <enter>

ans =
-96.0400
```

Since $|\mathbf{P}| \neq 0$, it implies that the plant is controllable. However, the *magnitude* of $|\mathbf{P}|$ depends upon the *scaling* of matrix $\mathbf{P}$, and is *not* a good indicator of how *far away* $\mathbf{P}$ is from being singular, and thus how *strongly* the plant is controllable. A better way of detecting the measure of controllability is the condition number, obtained using the MATLAB function *cond* as follows:

```
>>cond(p) <enter>

ans =
 12.0773
```

Since condition number of $\mathbf{P}$ is *small* in magnitude, the plant is *strongly* controllable. Thus, our pole-placement results are expected to be accurate. (Had the condition number of $\mathbf{P}$ been *large* in magnitude, it would have indicated a *weakly* controllable plant, and the inversion of $\mathbf{P}$ to get the feedback gain matrix would have been inaccurate.) The poles of the plant are calculated by finding the eigenvalues of the matrix $\mathbf{A}$ using the MATLAB command *damp* as follows:

```
>>damp(A) <enter>
```

Eigenvalue	Damping	Freq. (rad/sec)
3.2833	-1.0000	3.2833
0	-1.0000	0
0	-1.0000	0
-3.2833	1.0000	3.2833

The plant is *unstable* due to a pole with positive real-part (and also due to a pair of poles at $s = 0$). Controlling this unstable plant is like balancing a vertical stick on your palm. The task of the regulator is to stabilize the plant. Let us make the closed-loop system stable, by selecting the closed-loop poles as $s = -1 \pm i$, and $s = -5 \pm 5i$. The coefficients of the plant's characteristic polynomial can be calculated using the MATLAB command *poly* as follows:

```
>>a=poly(A) <enter>

a =

      1.0000   0.0000   -10.7800   0   0
```

which implies that the characteristic polynomial of the plant is $s^4 - 10.78s^2 = 0$. Hence, the polynomial coefficient vector, $\mathbf{a}$, is the following:

$$\mathbf{a} = [0; \quad -10.78; \quad 0; \quad 0] \tag{5.55}$$

The characteristic polynomial of the closed-loop system can also be calculated using the command *poly* as follows:

```
>>v = [-1+j; -1-j; -5+5*j; -5-5*j]; alpha = poly(v) <enter>

alpha =

  1 12 72 120 100
```

which implies that the closed-loop characteristic polynomial is $\alpha^4 + 12\alpha^3 + 72\alpha^2 + 120\alpha + 100$, and the vector $\boldsymbol{\alpha}$ is thus the following:

$$\boldsymbol{\alpha} = [12; \quad 72; \quad 120; \quad 100] \tag{5.56}$$

Note that the MATLAB function *poly* can be used to compute the characteristic polynomial either *directly* from a square matrix, or from the *roots* of the characteristic polynomial (i.e. the eigenvalues of a square matrix). It now remains to find the upper triangular matrix, $\mathbf{P}'$, by either Eq. (5.49) or Eq. (5.51). Since a controller companion form is generally ill-conditioned (see Chapter 3), we would like to avoid using Eq. (5.49) which involves higher powers of the ill-conditioned matrix, $\mathbf{A}'$. From Eq. (5.51), we get

$$\mathbf{P}' = \begin{bmatrix} 1 & 0 & 10.78 & 0 \\ 0 & 1 & 0 & 10.78 \\ 0 & 0 & 1 & 0 \\ 0 & 0 & 0 & 1 \end{bmatrix} \tag{5.57}$$

Finally, the regulator gain matrix is obtained through Eq. (5.52) as follows:

```
>>Pdash=[1 0 10.78 0; 0 1 0 10.78; 0 0 1 0; 0 0 0 1]; a=[0 -10.78 0 0];

            alpha=[12 72 120 100]; K = (alpha-a)*Pdash*inv(P) <enter>

K =

-92.9841   -10.2041   -24.2449   -12.2449
```

The regulator gain matrix is thus the following:

$$\mathbf{K} = [-92.9841; \quad -10.2041; \quad -24.2449; \quad -12.2449] \tag{5.58}$$

Let us confirm that the eigenvalues of the closed-loop state-dynamics matrix, $\mathbf{A}_{CL} = \mathbf{A} - \mathbf{BK}$, are indeed what we set out to achieve as follows:

```
>>ACL = A-B*K <enter>

ACL =
    0            0          1.0000       0
    0            0          0            1.0000
  -82.2041    -10.2041    -24.2449    -12.2449
   92.0041     10.2041     24.2449     12.2449
```

The closed-loop poles are then evaluated by the command *eig* as follows:

```
>>eig(ACL) <enter>

ans =
 -5.0000+5.0000i
 -5.0000-5.0000i
 -1.0000+1.0000i
 -1.0000-1.0000i
```

Hence, the desired locations of the closed-loop poles have been obtained.

The computational steps of Example 5.9 are programmed in the MATLAB (CST) function called *acker* for computing the regulator gain matrix for single-input plants using the Ackermann's formula (Eq. (5.52). The command *acker* is used as follows:

```
>>K = acker(A,B,V) <enter>
```

where A, B are the state coefficient matrices of the plant, V is a vector containing the desired closed-loop pole locations, and K is the returned regulator gain matrix. Since Ackermann's formula is based on transforming the plant into the controller companion form, which becomes ill-conditioned for large order plants, the computed regulator gain matrix may be inaccurate when n is greater than, say, 10. The command *acker* produces a warning, if the computed closed-loop poles are *more than* 10% *off* from their desired locations. A similar MATLAB (CST) function called *place* is also available for computing the pole-placement regulator gain for single-input plants. The function *place* also provides an output *ndigits*, which indicates the number of significant digits to which the closed-loop poles have been placed. The design of Example 5.9 is simply carried out by using the command *place* as follows:

```
>>V = [-1+j; -1-j; -5+5*j; -5-5*j]; K = place(A,B,V) <enter>

place: ndigits= 17

K =

-92.9841   -10.2041   -24.2449   -12.2449
```

The result is identical to that obtained in Example 5.9; $ndigits = 17$ indicates that the locations of the closed-loop poles match the desired values up to 17 significant digits.

The locations of closed-loop poles determine the performance of the regulator, such as the settling time, maximum overshoot, etc. (see Chapter 2 for performance parameters) when the system is disturbed by a non-zero initial condition. A design is usually specified in terms of such performance parameters, rather than the locations of the closed-loop poles themselves. It is the task of the designer to ensure that the desired performance is achieved by selecting an appropriate set of closed-loop poles. This is illustrated in the following example.

Example 5.10

For the inverted-pendulum on a moving cart of Example 5.9, let us design a regulator which achieves a 5% maximum overshoot and a settling time less than 1 second for both the outputs, when the cart is initially displaced by 0.01 m. The state coefficient matrices, **A**, **B**, **C**, and **D**, of the plant are given in Eq. (5.53). The initial condition vector has the perturbation to the cart displacement, $x(t)$, as the only non-zero element; thus, $\mathbf{x}(0) = [0; 0.01; 0; 0]^T$. Let us begin by testing whether the regulator designed in Example 5.9 meets the performance specifications. This is done by using the MATLAB (CST) function *initial* to find the initial response as follows:

```
>>t = 0:0.1:10; sysCL=ss(A-B*K, zeros(4,1),C,D); [y,t,X] = initial
  (sysCL,[0 0.01 0 0]',t); <enter>
```

where **y**, **X**, and **t** denote the returned output, state, and time vectors and *sysCL* is the state-space LTI model of the closed-loop system. The resulting outputs $\mathbf{y}(t) = [\theta(t); x(t)]^T$ are plotted in Figure 5.4.

In Figure 5.4, both the responses are seen to have acceptably small maximum overshoots, but settling-times in excess of 5 s, which is unacceptable. In order to *speed-up* the closed-loop response, let us move all the poles *deeper* inside the left-half plane by decreasing their real parts such that the new desired closed-loop poles are $s = -7.5 \pm 7.5i$, and $s = -10 \pm 10i$. Then, the new regulator gain matrix, the closed-loop dynamics matrix, and the initial response are obtained as follows:

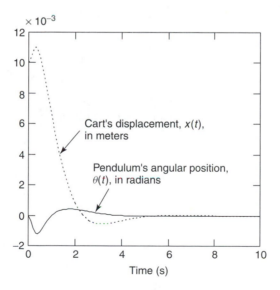

Figure 5.4 Closed-loop initial response of the regulated inverted pendulum on a moving cart to perturbation on cart displacement for the regulator gain matrix, **K** = [−92.9841; −10.2041; −24.2449; −12.2449]

```
>>V=[-7.5+7.5*j  -7.5-7.5*j  -10+10*j  -10-10*j]'; K = place(A,B,V) <enter>

place: ndigits= 19

K =

-2.9192e+003   -2.2959e+003   -5.7071e+002   -5.3571e+002
>>t = 0:0.01:2; sysCL=ss(A-B*K, zeros(4,1),C,D); [y,t,X] = initial(sysCL,
   [0 0.01 0 0]',t); <enter>
```

The resulting outputs are plotted in Figure 5.5, which indicates a maximum overshoot of the steady-state values less than 4%, and a settling time of less than 1 s
for both the responses.

How did we know that the new pole locations will meet our performance requirements?
We didn't. We tried for several pole configurations, until we hit upon the one that met
our requirements. This is the design approach in a nutshell. On comparing Figures 5.4
and 5.5, we find that by moving the closed-loop poles further inside the left-half plane,
we *speeded-up* the initial response *at the cost* of increased maximum overshoot. The
settling time and maximum overshoot are, thus, *conflicting requirements*. To decrease
one, we have to accept an increase in the other. Such a compromise, called a *trade-
off*, is a hallmark of control system design. Furthermore, there is another cost associated
with moving the poles deeper inside the left-half plane – that of the control input. Note
that the new regulator gain elements are *several times larger* than those calculated in
Example 5.9, which implies that the regulator must now apply an input which is much

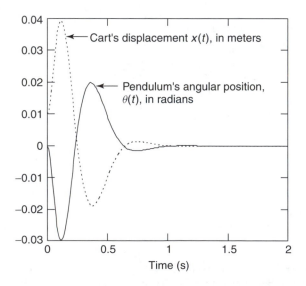

Figure 5.5 Closed-loop initial response of the regulated inverted pendulum on a moving cart to
perturbation on cart displacement for the regulator gain matrix, **K** = [−2919.2; −2295.9; −570.71;
−535.71]

larger in magnitude than that in Example 5.9. The input, $\mathbf{u}(t) = -\mathbf{K}\mathbf{x}(t)$, can be calculated from the previously calculated matrices, $\mathbf{K}$ and $\mathbf{x}$, as follows:

```
>>u = -K*X'; <enter>
```

The control inputs for the two values of the regulator gain matrix are compared in Figure 5.6. The control input, $u(t)$, which is a force applied to the cart, is seen to be *more than 200 times* in magnitude for the design of Example 5.10 than that of Example 5.9. The *actuator*, which applies the input force to the cart, must be physically able to generate this force for the design to be successful. The cost of controlling a plant is a function of the largest control input magnitude expected in *actual operating conditions*. For example, if the largest *expected* initial disturbance in cart displacement were 0.1 m instead of 0.01 m, a *ten times larger* control input would be required than that in Figure 5.6. The larger the control input magnitude, the bigger would be the energy spent by the actuator in generating the control input, and the higher would be the cost of control. It is possible to *minimize* the control effort required in controlling a plant by imposing conditions – other than pole-placement – on the regulator gain matrix, which we will see in Chapter 6. However, a rough method of ensuring that the performance requirements are met with the minimum control effort is to ensure that all the closed-loop poles are about the *same distance* from the imaginary axis in the left-half plane. The poles in the left-half plane that are *farthest* away from the imaginary axis dictate the control input magnitude, while the *speed* of response (i.e. the settling time of the transients) is governed by the poles with the *smallest* real parts, called the *dominant poles*. If some closed-loop poles are close to, and some are very far from the imaginary axis, it implies that *too much control energy* is being spent for a given settling time, and thus the design is *inefficient*. The most efficient closed-loop configuration thus appears to be the one where all the poles are placed in the

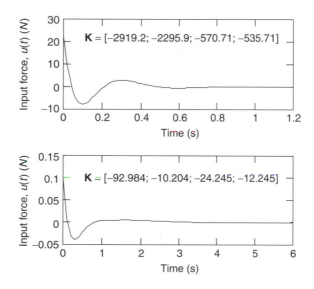

Figure 5.6 Control inputs of the regulated inverted pendulum on a moving cart for two designs of the full-state feedback regulator

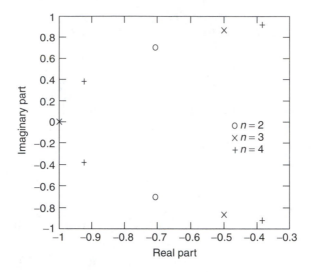

Figure 5.7 Butterworth pattern of poles in the left-half plane for $n = 2$, 3, and 4

left half plane, roughly the same distance from the imaginary axis. To increase the speed of the closed-loop response, one has to just increase this distance. One commonly used closed-loop pole configuration is the *Butterworth pattern*, in which the poles are placed on a circle of radius R *centered at the origin*, and are obtained from the solution of the following equation:

$$(s/R)^{2n} = (-1)^{n+1} \tag{5.59}$$

where n is the number of poles in the left-half plane (usually, we want all the closed-loop poles in the left-half plane; then n is the order of the system). For $n = 1$, the pole in the left-half plane satisfying Eq. (5.59) is $s = -R$. For $n = 2$, the poles in the left-half plane satisfying Eq. (5.59) are the solutions of $(s/R)^2 + (s/R)\sqrt{2} + 1 = 0$. The poles satisfying Eq. (5.59) in the left-half plane for $n = 3$ are the solutions of $(s/R)^3 + 2(s/R)^2 + 2(s/R) + 1 = 0$. For a given n, we can calculate the poles satisfying Eq. (5.59) by using the MATLAB function *roots*, and discard the poles having positive real parts. The Butterworth pattern for $n = 2$, 3, and 4 is shown in Figure 5.7. Note, however, that as n *increases*, the real part of the two Butterworth poles closest to the imaginary axis *decreases*. Thus for large n, it may be required to move these two poles further inside the left-half plane, in order to meet a given speed of response.

Example 5.11

Let us compare the closed-loop initial response and the input for the inverted pendulum on a moving cart with those obtained in Example 5.10 when the closed-loop poles are in a Butterworth pattern. For $n = 4$, the poles satisfying Eq. (5.59) in the left-half plane are calculated as follows:

```
>>z = roots([1 0 0 0 0 0 0 0 1]) <enter>

z =

  -0.9239+0.3827i
  -0.9239-0.3827i
  -0.3827+0.9239i
  -0.3827-0.9239i
   0.3827+0.9239i
   0.3827-0.9239i
   0.9239+0.3827i
   0.9239-0.3827i
```

The first four elements of z are the required poles in the left-half plane, i.e. $s/R = -0.9239 \pm 0.3827i$ and $s/R = -0.3827 \pm 0.9239i$. For obtaining a maximum over-shoot less than 5% and settling-time less than 1 s for the initial response (the design requirements of Example 5.10), let us choose $R = 15$. Then the closed-loop characteristic polynomial are obtained as follows:

```
>>i = find(real(z) < 0); p = poly(15*z(i)) <enter>

p =
 Columns 1 through 3
 1.0000e+000       3.9197e+001-3.5527e-015i       7.6820e+002-5.6843e-014i

 Columns 4 through 5
 8.8193e+003-3.1832e-012i       5.0625e+004-2.1654e-011i
```

Neglecting the small imaginary parts of **p**, the closed-loop characteristic polynomial is $s^4 + 39.197s^3 + 768.2s^2 + 8819.3s + 50625$, with the vector α given by

```
>>alpha=real(p(2:5)) <enter>

alpha =
  3.9197e+001   7.6820e+002   8.8193e+003   5.0625e+004
```

$$\alpha = [39.197; \quad 768.2; \quad 8819.3; \quad 50\,625] \qquad (5.60)$$

Then using the values of **a**, **P**, and **P**$'$ calculated in Example 5.9, the regulator gain matrix is calculated by Eq. (5.52) as follows:

```
>>K = (alpha-a)*Pdash*inv(P) <enter>

K =
 -5.9448e+003   -5.1658e+003   -9.3913e+002   -8.9993e+002
```

and the closed-loop state-dynamics matrix is obtained as

```
>>ACL=A-B*K <enter>

ACL =

0              0              1.0000e+000    0
0              0              0              1.0000e+000
-5.9340e+003   -5.1658e+003   -9.3913e+002   -8.9993e+002
5.9438e+003    5.1658e+003    9.3913e+002    8.9993e+002
```

The closed-loop eigenvalues are calculated as follows:

```
>>eig(ACL) <enter>

ans =
-5.7403e+000+1.3858e+001i
-5.7403e+000-1.3858e+001i
-1.3858e+001+5.7403e+000i
-1.3858e+001-5.7403e+000i
```

which are the required closed-loop Butterworth poles for $R = 15$. The initial response of the closed-loop system is calculated as follows, and is plotted in Figure 5.8:

```
>>t = 0:1.0753e-2:1.2; sysCL=ss(ACL,zeros(4,1),C,D); [y,t,X]=initial
  (sysCL,[0 0.01 0 0]',t); <enter>
```

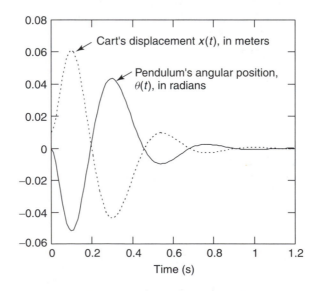

Figure 5.8 Initial response of the regulated inverted pendulum on a moving cart, for the closed-loop poles in a Butterworth pattern of radius, $R = 15$

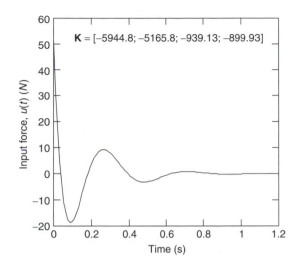

Figure 5.9 Control input for the regulated inverted pendulum on a moving cart, for closed-loop poles in a Butterworth pattern of radius, $R = 15$

Note from Figure 5.8 that the maximum overshoot for cart displacement is about 6% for both the outputs, and the settling time is greater than 1 s. The design is thus unacceptable. The *slow* closed-loop response is caused by the pair of *dominant poles* with real part -5.7403. If we try to increase the real part magnitude of the dominant poles by increasing R, we will have to pay for the increased speed of response in terms of *increased* input magnitude, because the poles furthest from the imaginary axis ($s/R = -0.9239 \pm 0.3827i$) will move still further away. The control input, $u(t)$, is calculated and plotted in Figure 5.9 as follows:

```
>>u = -K*X'; plot(t,u) <enter>
```

Figure 5.9 shows that the control input magnitude is much larger than that of the design in Example 5.10. The present pole configuration is unacceptable, because it does not meet the design specifications, and requires a large control effort. To reduce the control effort, we will try a Butterworth pattern with $R = 8.5$. To increase the speed of the response, we will move the *dominant poles* further inside the left-half plane than dictated by the Butterworth pattern, such that *all* the closed-loop poles have the *same* real parts. The selected closed-loop pole configuration is $s = -7.853 \pm 3.2528i$, and $s = -7.853 \pm 7.853i$. The regulator gain matrix which achieves this pole placement is obtained using MATLAB as follows:

```
>>format long e <enter>

>>v=[-7.853-3.2528i -7.853+3.2528i -7.853-7.853i -7.853+7.853i]';K=place
  (A,B,v) <enter>

place: ndigits= 18
```

```
K =
Columns 1 through 3
-1.362364050360232e+003  -9.093160795202226e+002  -3.448741667548096e+002

Column 4
-3.134621667548089e+002
```

Note that we have printed out **K** using the *long format*, because we will need this matrix later. A short format would have introduced unacceptable truncation errors. The closed-loop initial response is calculated and plotted in Figure 5.10 as follows:

```
>>sysCL=ss(A-B*K,zeros(4,1),C,D); [y,t,X] = initial(sysCL,
  [0 0.01 0 0]', t); <enter>
```

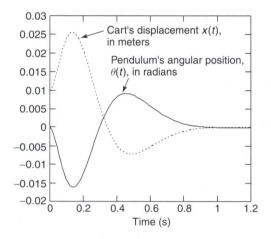

Figure 5.10 Initial response of the regulated inverted pendulum on a moving cart for the design of Example 5.11 with the closed-loop poles at $s = -7.853 \pm 3.2528i$, and $s = -7.853 \pm 7.853i$

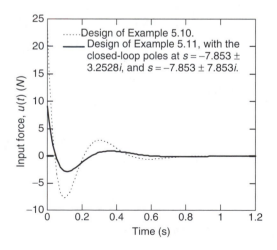

Figure 5.11 Comparison of the control input for the design of Example 5.10 with that of Example 5.11 with closed-loop poles at $s = -7.853 \pm 3.2528i$, and $s = -7.853 \pm 7.853i$

Figure 5.10 shows that the closed-loop response has a maximum overshoot of about 2.5% and a settling time of 1 s, which is a *better performance* than the design of Example 5.10. The control input of the present design is compared with that of Example 5.10 in Figure 5.11, which shows that the former is less than half of the latter. Hence, the present design results in a *better performance*, while requiring a *much smaller control effort*, when compared to Example 5.10.

5.3.2 Pole-placement regulator design for multi-input plants

For a plant having more than one input, the full-state feedback regulator gain matrix of Eq. (5.28) has $(r \times n)$ elements, where n is the order of the plant and r is the number of inputs. Since the number of poles that need to be placed is n, we have *more design parameters* than the number of poles. This over-abundance of design parameters allows us to specify *additional design conditions*, apart from the location of n poles. What can be these additional conditions? The answer depends upon the nature of the plant. For example, it is possible that a particular state variable is *not necessary* for generating the control input vector by Eq. (5.28); hence, the *column* corresponding to that state variable in **K** can be chosen as zero, and the pole-placement may yet be possible. Other conditions on **K** could be due to *physical relationships* between the inputs and the state variables; certain input variables could be *more closely related* to *some* state variables, requiring that the elements of **K** corresponding to the *other* state variables should be zeros. Since the structure of the regulator gain matrix for multi-input systems is *system specific*, we cannot derive a *general* expression for the regulator gain matrix, such as Eq. (5.52) for the single-input case. The following example illustrates the multi-input design process.

Example 5.12

Let us design a full-state feedback regulator for the following plant:

$$\mathbf{A} = \begin{bmatrix} 0 & 0 & 0 \\ 0 & 0.01 & 0 \\ 0 & 0 & -0.1 \end{bmatrix}; \quad \mathbf{B} = \begin{bmatrix} 1 & 0 \\ 0 & -1 \\ 0 & -2 \end{bmatrix}$$

$$\mathbf{C} = \begin{bmatrix} 1 & 0 \\ 0 & 0 \end{bmatrix}; \quad \mathbf{D} = \begin{bmatrix} 0 & 0 \\ 0 & 0 \end{bmatrix} \qquad (5.61)$$

The plant is unstable due to a pole at $s = 0.01$. The rank of the controllability test matrix of the plant is obtained as follows:

```
>>rank(ctrb(A, B)) <enter>

ans =

    3
```

Hence, the plant is controllable, and the closed-loop poles can be placed at will. The general regulator gain matrix is as follows:

$$\mathbf{K} = \begin{bmatrix} K_1 & K_2 & K_3 \\ K_4 & K_5 & K_6 \end{bmatrix} \quad (5.62)$$

and the closed-loop state dynamics matrix is the following:

$$\mathbf{A_{CL}} = \mathbf{A} - \mathbf{BK} = \begin{bmatrix} -K_1 & -K_2 & -K_3 \\ K_4 & (0.01 + K_5) & K_6 \\ 2K_4 & 2K_5 & (-0.1 + 2K_6) \end{bmatrix} \quad (5.63)$$

which results in the following closed-loop characteristic equation:

$$|s\mathbf{I} - \mathbf{A_{CL}}| = \begin{vmatrix} (s + K_1) & K_2 & K_3 \\ -K_4 & (s - 0.01 - K_5) & -K_6 \\ -2K_4 & -2K_5 & (s + 0.1 - 2K_6) \end{vmatrix} = 0 \quad (5.64)$$

or

$$(s + K_1)[(s - 0.01 - K_5)(s + 0.1 - 2K_6) - 2K_5K_6] + K_4[K_2(s + 0.1 - 2K_6)$$
$$+ 2K_3K_5] + 2K_4[K_2K_6 + K_3(s - 0.01 - K_5)] = 0 \quad (5.65)$$

or

$$s^3 + (0.09 - K_5 - 2K_6 + K_1)s^2 + (K_2K_4 + 2K_3K_4 + 0.09K_1 - 2K_1K_6$$
$$- K_1K_5 - 0.001 + 0.02K_6 - 0.1K_5)s + 0.1K_2K_4 + 0.02K_1K_6$$
$$- 0.001K_1 - 0.1K_1K_5 - 0.02K_3K_4 = 0 \quad (5.66)$$

Let us choose the closed-loop poles as $s = -1$, and $s = -0.045 \pm 0.5i$. Then the closed-loop characteristic equation must be $(s + 1)(s + 0.045 - 0.5i)(s + 0.045 + 0.5i) = s^3 + 1.09s^2 + 0.342s + 0.252 = 0$, and comparing with Eq. (5.66), it follows that

$$K_1 - K_5 - 2K_6 = 1;$$
$$K_2K_4 + 2K_3K_4 + 0.09K_1 - 2K_1K_6 - K_1K_5 + 0.02K_6 - 0.1K_5 = 0.343$$
$$0.1K_2K_4 + 0.02K_1K_6 - 0.001K_1 - 0.1K_1K_5 - 0.02K_3K_4 = 0.252 \quad (5.67)$$

which is a set of nonlinear algebraic equations to be solved for the regulator design parameters – apparently a hopeless task by hand. However, MATLAB (CST) again comes to our rescue by providing the function *place*, which allows placing the poles of multi-input plants. The function *place* employs an *eigenstructure assignment* algorithm [3], which specifies *additional conditions* to be satisfied by the regulator gain elements, provided the *multiplicity* of each pole to be placed *does not exceed* the number of inputs, and all complex closed-loop poles must appear in conjugate pairs. For the present example, the regulator gain matrix is determined using *place* as follows:

```
>>A=[0 0 0;0 0.01 0;0 0 -0.1];B=[1 0;0 -1;0 -2]; p=[-1 -0.045-0.5i
  -0.045+0.5I]; K=place(A,B,p) <enter>

place: ndigits= 16

K =

 0.9232  0.1570   -0.3052
 0.1780 -2.4595  1.1914

>>eig(A-B*K) <enter>

ans =

 -1.0000
 -0.0450+0.5000i
 -0.0450-0.5000i
```

You may verify that the computed values of the gain matrix satisfies Eq. (5.67). The *optimal control* methods of Chapter 6 offer an alternative design approach for regulators based on multi-input plants.

5.3.3 Pole-placement regulator design for plants with noise

In the previous two sections, we had ignored the presence of disturbances, or *noise*, in a plant when designing full-state feedback regulators. Designs that ignore noise in a plant are likely to fail when implemented in actual conditions where noise exists. Noise can be divided into two categories: *measurement noise*, or the noise caused by imperfections in the sensors that measure the output variables; and the *process noise*, or the noise which arises due to ignored dynamics when modeling a plant. Since neither the sensors nor a plant's mathematical model can be perfect, we should always expect some noise in a plant. The state-equation of a plant with noise vector, $\mathbf{x}_n(t)$, is the following:

$$\mathbf{x}^{(1)}(t) = \mathbf{A}\mathbf{x}(t) + \mathbf{B}\mathbf{u}(t) + \mathbf{F}\mathbf{x}_n(t) \tag{5.68}$$

where $\mathbf{F}$ is the noise coefficient matrix. To place the closed-loop poles at desired locations while counteracting the effect of the noise, a full-state feedback regulator is to be designed based on the following control-law:

$$\mathbf{u}(t) = -\mathbf{K}\mathbf{x}(t) - \mathbf{K}_n\mathbf{x}_n(t) \tag{5.69}$$

Substituting Eq. (5.69) into Eq. (5.68) yields the following state-equation of the closed-loop system:

$$\mathbf{x}^{(1)}(t) = (\mathbf{A} - \mathbf{B}\mathbf{K})\mathbf{x}(t) + (\mathbf{F} - \mathbf{B}\mathbf{K}_n)\mathbf{x}_n(t) \tag{5.70}$$

Note that Eq. (5.70) implies that the noise vector, $\mathbf{x}_n(t)$, acts as an *input vector* for the closed-loop system, whose state-dynamics matrix is $\mathbf{A}_{CL} = (\mathbf{A} - \mathbf{B}\mathbf{K})$. A schematic diagram of the full-state feedback regulator with noise is shown in Figure 5.12.

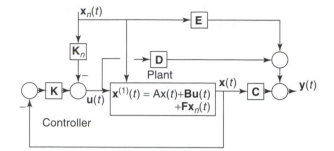

Figure 5.12 Schematic diagram of a full-state feedback regulator with noise, $\mathbf{x}_n(t)$

The regulator feedback gain matrix, $\mathbf{K}$, is selected, as before, to place the closed-loop poles (eigenvalues of $\mathbf{A_{CL}}$) at desired locations. While we may not know the exact process by which the noise, $\mathbf{x}_n(t)$, is generated (because it is usually a *stochastic process*, as discussed in Chapter 1), we can develop an *approximation* of how the noise affects the plant by deriving the noise coefficient matrix, $\mathbf{F}$, from experimental observations. Once $\mathbf{F}$ is known reasonably, the regulator noise gain matrix, $\mathbf{K}_n$, can be selected such that the effect of the noise vector, $\mathbf{x}_n(t)$, on the closed-loop system is minimized. It would, of course, be ideal if we can make $(\mathbf{F} - \mathbf{B}\mathbf{K}_n) = \mathbf{0}$, in which case there would be absolutely no influence of the noise on the closed-loop system. However, it may not be always possible to select the (rq) unknown elements of $\mathbf{K}_n$ to satisfy the (nq) scalar equations constituting $(\mathbf{F} - \mathbf{B}\mathbf{K}_n) = \mathbf{0}$, where n is the order of the plant, r is the number of inputs, and q is the number of noise variables in the noise vector, $\mathbf{x}_n(t)$. When $r < n$ (as it is usually the case), the number of unknowns in $(\mathbf{F} - \mathbf{B}\mathbf{K}_n) = \mathbf{0}$ is *less than* the number of scalar equations, and hence all the equations cannot be satisfied. If $r = n$, and the matrix $\mathbf{B}$ is non-singular, then we can uniquely determine the regulator noise gain matrix by $\mathbf{K}_n = -\mathbf{B}^{-1}\mathbf{F}$. In the rare event of $r > n$, the number of unknowns *exceed* the number of equations, and all the equations, $(\mathbf{F} - \mathbf{B}\mathbf{K}_n) = \mathbf{0}$, can be satisfied by appropriately selecting the unknowns, though not uniquely.

Example 5.13

Consider a fighter aircraft whose state-space description given by Eqs. (5.26) and (5.27) has the following coefficient matrices:

$$\mathbf{A} = \begin{bmatrix} -1.7 & 50 & 260 \\ 0.22 & -1.4 & -32 \\ 0 & 0 & -12 \end{bmatrix}; \quad \mathbf{B} = \begin{bmatrix} -272 \\ 0 \\ 14 \end{bmatrix}; \quad \mathbf{F} = \begin{bmatrix} 0.02 & 0.1 \\ -0.0035 & 0.004 \\ 0 & 0 \end{bmatrix}$$

$$\mathbf{C} = \mathbf{I}; \quad \mathbf{D} = \mathbf{0}; \quad \mathbf{E} = \mathbf{0} \tag{5.71}$$

The state variables of the aircraft model are normal acceleration in m/s^2, $x_1(t)$, pitch-rate in rad/s, $x_2(t)$, and elevator deflection in rad, $x_3(t)$, while the input, $u(t)$, is the desired elevator deflection in rad. (For a graphical description of the system's variables, see Figure 4.5.) The poles of the plant are calculated as follows:

```
>>A=[-1.7 50 260; 0.22 -1.4 -32; 0 0 -12]; damp(A) <enter>
Eigenvalue    Damping         Freq. (rad/sec)
   1.7700      -1.0000         1.7700
  -4.8700       1.0000         4.8700
 -12.0000       1.0000 12.0000
```

The plant is unstable due to a pole at $s = 1.77$. To stabilize the closed-loop system, it is desired to place the closed-loop poles at $s = -1 \pm i$ and $s = -1$. The following controllability test reveals a controllable plant, implying that pole-placement is possible:

```
>>B=[-272 0 14]'; rank(ctrb(A,B)) <enter>

ans =

     3
```

The regulator feedback gain matrix is thus obtained as follows:

```
>>v = [-1-i -1+i -1]; K = place(A,B,v) <enter>

place: ndigits= 19

K =

0.0006   -0.0244   -0.8519
```

and the closed-loop state dynamics matrix is the following:

```
>>ACL=A-B*K <enter>
ACL =

-1.5267       43.3608        28.2818
 0.2200       -1.4000       -32.0000
-0.0089        0.3417        -0.0733
```

To determine the remaining regulator matrix, $\mathbf{K}_n = [K_{n1} \quad K_{n2}]$, let us look at the matrix $(\mathbf{F} - \mathbf{BK}_n)$:

$$\mathbf{F} - \mathbf{BK}_n = \begin{bmatrix} 0.02 + 272K_{n1} & 0.1 + 272K_{n2} \\ -0.0035 & 0.004 \\ -14K_{n1} & -14K_{n2} \end{bmatrix} \tag{5.72}$$

Equation (5.72) tells us that it is *impossible* to make all the elements of $(\mathbf{F} - \mathbf{BK}_n)$ zeros, by selecting the two unknown design parameters, K_{n1} and K_{n2}. The next best thing to $(\mathbf{F} - \mathbf{BK}_n) = 0$ is making the *largest elements* of $(\mathbf{F} - \mathbf{BK}_n)$ zeros, and living with the other non-zero elements. This is done by selecting $K_{n1} = -0.02/272$ and $K_{n2} = -0.1/272$ which yields the following $(\mathbf{F} - \mathbf{BK}_n)$:

$$\mathbf{F} - \mathbf{BK}_n = \begin{bmatrix} 0 & 0 \\ -0.0035 & 0.004 \\ 0.00103 & 0.00515 \end{bmatrix} \tag{5.73}$$

With $(\mathbf{F} - \mathbf{BK}_n)$ given by Eq. (5.73), we are always going to have some effect of noise on the closed-loop system, which hopefully, will be small. The most satisfying thing about Eq. (5.73) is that the closed-loop system given by Eq. (5.70) is *uncontrollable with noise as the input* (you can verify this fact by checking the rank of *ctrb* $(\mathbf{A}_{\mathrm{CL}}, (\mathbf{F} - \mathbf{BK}_n))$). This means that the noise is not going to affect all the state variables of the closed-loop system. Let us see by what extent the noise affects our closed-loop design by calculating the system's response with a noise vector, $\mathbf{x}_n(t) = [1 \times 10^{-5}; -2 \times 10^{-6}]^T \sin(100t)$, which acts as an input to the closed-loop system given by Eq. (5.70), with zero initial conditions. Such a noise model is too simple; actual noise is *non-deterministic* (or *stochastic*), and consists of a combination of several frequencies, rather than only one frequency (100 rad/s) as assumed here. The closed-loop response to noise is calculated by using the MATLAB (CST) command *lsim* as follows:

```
>>t=0:0.01:5; xn=[1e-5 -2e-6]'*sin(100*t); Bn=[0 0;-3.5e-3 0.004;1.03e-3 5.15e-3];
  <enter>
```

```
>>sysCL=ss(ACL,Bn,eye(3),zeros(3,2)); [y,t,X]=lsim(sysCL,xn',t'); plot(t,X) <enter>
```

The resulting closed-loop state variables, $x_1(t)$, $x_2(t)$, and $x_3(t)$, are plotted in Figure 5.13, which shows oscillations with very small amplitudes. Since the amplitudes are very small, the effect of the noise on the closed-loop system can be said to be negligible. Let us see what may happen if we make the closed-loop system *excessively* stable. If the closed-loop poles are placed at $s = -100$, $s = -100 \pm 100i$, the resulting closed-loop response to the noise is shown in Figure 5.14. Note that the closed-loop response has increased by about 300 times in

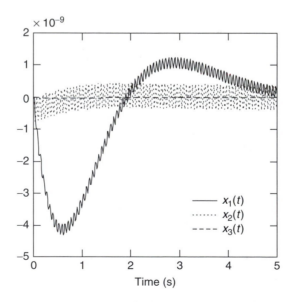

Figure 5.13 Closed-loop response of the regulated fighter aircraft to noise, when the closed-loop poles are $s = -1, s = -1 \pm i$

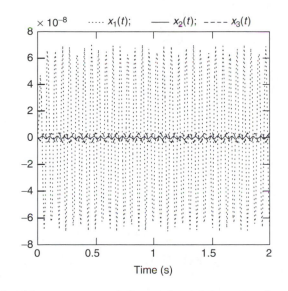

Figure 5.14 Closed-loop response of the regulated fighter aircraft to noise, when the closed-loop poles are $s = -100$, $s = -100 \pm 100i$

magnitude, compared with that of Figure 5.13. Therefore, moving the poles too far into the left-half plane has the effect of increasing the response of the system due to noise, which is undesirable. This kind of amplified noise effect is due to the resulting *high gain feedback*. High gain feedback is to be avoided in the frequency range of expected noise. This issue is appropriately dealt with by *filters* and *compensators* (Chapter 7).

The conflicting requirements of increasing the speed of response, and decreasing the effect of noise are met by a pole configuration that is neither too deep inside the left-half plane, nor too close to the imaginary axis. The *optimum* pole locations are obtained by trial and error, if we follow the pole-placement approach. However, the *optimal control* methods of Chapters 6 and 7 provide a more effective procedure of meeting both speed and noise attenuation requirements than the pole-placement approach.

5.3.4 Pole-placement design of tracking systems

Now we are in a position to extend the pole-placement design to *tracking systems*, which are systems in which the desired state-vector, $\mathbf{x}_d(t)$, is *non-zero*. Schematic diagram of a tracking system with noise was shown in Figure 5.2, with the plant described by Eqs. (5.26) and (5.27), and the control-law given by Eq. (5.25). The objective of the tracking system is to make the error, $\mathbf{e}(t) = (\mathbf{x}_d(t) - \mathbf{x}(t))$, zero in the steady-state, while counteracting the effect of the noise, $\mathbf{x}_n(t)$. If the process by which the desired state-vector is generated is *linear* and *time-invariant*, it can be represented by the following state-equation:

$$\mathbf{x}_d^{(1)}(t) = \mathbf{A}_d\mathbf{x}_d(t) \tag{5.74}$$

Note that Eq. (5.74) represents a homogeneous system, because the desired state vector is unaffected by the input vector, $\mathbf{u}(t)$. Subtracting Eq. (5.26) from Eq. (5.74), we can write the following plant state-equation in terms of the error:

$$\mathbf{x}_d^{(1)}(t) - \mathbf{x}^{(1)}(t) = \mathbf{A}_d\mathbf{x}_d(t) - \mathbf{A}\mathbf{x}(t) - \mathbf{B}\mathbf{u}(t) - \mathbf{F}\mathbf{x}_n(t) \tag{5.75}$$

or

$$\mathbf{e}^{(1)}(t) = \mathbf{A}\mathbf{e}(t) + (\mathbf{A}_d - \mathbf{A})\mathbf{x}_d(t) - \mathbf{B}\mathbf{u}(t) - \mathbf{F}\mathbf{x}_n(t) \tag{5.76}$$

and the control-law (Eq. (5.25)) can be re-written as follows:

$$\mathbf{u}(t) = \mathbf{K}\mathbf{e}(t) - \mathbf{K}_d\mathbf{x}_d(t) - \mathbf{K}_n\mathbf{x}_n(t) \tag{5.77}$$

Referring to Figure 5.2, we see that while $\mathbf{K}$ is a *feedback* gain matrix (because it multiplies the error signal which is generated by the fed back state-vector), $\mathbf{K}_d$ and $\mathbf{K}_n$ are *feedforward* gain matrices, which multiply the desired state-vector and the noise vector, respectively, and hence *feed* these two vectors *forward* into the control system. Substituting Eq. (5.77) into Eq. (5.76) yields the following state-equation for the tracking system:

$$\mathbf{e}^{(1)}(t) = (\mathbf{A} - \mathbf{B}\mathbf{K})\mathbf{e}(t) + (\mathbf{A}_d - \mathbf{A} + \mathbf{B}\mathbf{K}_d)\mathbf{x}_d(t) + (\mathbf{B}\mathbf{K}_n - \mathbf{F})\mathbf{x}_n(t) \tag{5.78}$$

The design procedure for the tracking system consists of determining the full-state feedback gain matrix, $\mathbf{K}$, such that the poles of the closed-loop system (i.e. eigenvalues of $\mathbf{A}_{CL} = \mathbf{A} - \mathbf{B}\mathbf{K}$) are placed at desired locations, and choose the gain matrices, $\mathbf{K}_d$ and $\mathbf{K}_n$, such that the error, $\mathbf{e}(t)$, is either *reduced to zero*, or *made as small as possible* in the *steady-state*, in the presence of the noise, $\mathbf{x}_n(t)$. Of course, the closed-loop system described by Eq. (5.78) must be *asymptotically stable*, i.e. all the closed-loop poles must be in the left-half plane, otherwise the error will not reach a steady-state *even in the absence of noise*. Furthermore, as seen in Example 5.13, there may not be enough design parameters (i.e. elements in $\mathbf{K}_d$ and $\mathbf{K}_n$) to make the error zero in the steady-state, in the presence of noise. If all the closed-loop poles are placed in the left-half plane, the tracking system is asymptotically stable, and the steady-state condition for the error is reached (i.e. the error becomes constant in the limit $t \to \infty$). Then the steady state condition is described by $\mathbf{e}^{(1)}(t) = \mathbf{0}$, and Eq. (5.78) becomes the following in the steady state:

$$\mathbf{0} = (\mathbf{A} - \mathbf{B}\mathbf{K})\mathbf{e}_{ss} + (\mathbf{A}_d - \mathbf{A} + \mathbf{B}\mathbf{K}_d)\mathbf{x}_{dss} + (\mathbf{B}\mathbf{K}_n - \mathbf{F})\mathbf{x}_{nss} \tag{5.79}$$

where $\mathbf{e}(t) \to \mathbf{e}_{ss}$ (the steady state error vector), $\mathbf{x}_d(t) \to \mathbf{x}_{dss}$, and $\mathbf{x}_n(t) \to \mathbf{x}_{nss}$ as $t \to \infty$. From Eq. (5.79), we can write the steady state error vector as follows:

$$\mathbf{e}_{ss} = (\mathbf{A} - \mathbf{B}\mathbf{K})^{-1}[(\mathbf{A} - \mathbf{B}\mathbf{K}_d - \mathbf{A}_d)\mathbf{x}_{dss} + (\mathbf{F} - \mathbf{B}\mathbf{K}_n)\mathbf{x}_{nss}] \tag{5.80}$$

Note that the closed-loop state-dynamics matrix, $\mathbf{A}_{CL} = \mathbf{A} - \mathbf{BK}$, is non-singular, because all its eigenvalues are in the left-half plane. Hence, $(\mathbf{A} - \mathbf{BK})^{-1}$ exists. For $\mathbf{e}_{ss}$ to be zero, irrespective of the values of $\mathbf{x}_{dss}$ and $\mathbf{x}_{nss}$, we should have $(\mathbf{A} - \mathbf{BK}_d - \mathbf{A}_d) = \mathbf{0}$ and $(\mathbf{F} - \mathbf{BK}_n) = \mathbf{0}$, by selecting the appropriate gain matrices, $\mathbf{K}_d$ and $\mathbf{K}_n$. However, as seen in Example 5.13, this is seldom possible, owing to the number of inputs to the plant, r, being *usually smaller* than the order of the plant, n. Hence, as in Example 5.13, the best one can usually do is to make *some elements* of $\mathbf{e}_{ss}$ zeros, and living with the other non-zero elements, provided they are small. *In the rare case of the plant having as many inputs as the plant's order*, i.e. $n = r$, we can uniquely determine $\mathbf{K}_d$ and $\mathbf{K}_n$ as follows, to make $\mathbf{e}_{ss} = \mathbf{0}$:

$$\mathbf{K}_d = \mathbf{B}^{-1}(\mathbf{A} - \mathbf{A}_d); \quad \mathbf{K}_n = \mathbf{B}^{-1}\mathbf{F} \qquad (5.81)$$

Example 5.14

For the fighter aircraft of Example 5.13, let us design a controller which makes the aircraft track a target, whose state-dynamics matrix, $\mathbf{A}_d$, is the following:

$$\mathbf{A}_d = \begin{bmatrix} -2.1 & 35 & 150 \\ 0.1 & -1.1 & -21 \\ 0 & 0 & -8 \end{bmatrix} \qquad (5.82)$$

The eigenvalues of $\mathbf{A}_d$ determine the poles of the target, which indicate how rapidly the desired state-vector, $\mathbf{x}_d(t)$, is changing, and are calculated as follows:

```
>>Ad = [-10.1 35 150; 0.1 -1.1 -21; 0 0 -8]; damp(Ad) <enter>

Eigenvalue     Damping      Freq. (rad/sec)
-0.7266        1.0000       0.7266
-8.0000        1.0000       8.0000
-10.4734       1.0000       10.4734
```

The target dynamics is asymptotically stable, with the pole closest to the imaginary axis being, $s = -0.7266$. This pole determines the settling time (or the speed) of the target's response. To track the target successfully, the closed-loop tracking system must be *fast enough*, i.e. the poles closest to the imaginary axis must have sufficiently small real parts, i.e. smaller than -0.7266. However, if the closed-loop dynamics is made *too fast* by increasing the negative real part magnitudes of the poles, there will be an *increased effect* of the noise on the system, as seen in Example 5.13. Also, recall that for an *efficient* design (i.e. smaller control effort), all the closed-loop poles must be about the same distance from the imaginary axis. Let us choose a closed-loop pole configuration as $s = -1$, $s = -1 \pm i$. The feedback gain matrix for this pole configuration was determined in Example 5.13 to be the following:

$$\mathbf{K} = [0.0006; \quad -0.0244; \quad -0.8519] \qquad (5.83)$$

with the closed-loop state-dynamics matrix given by

$$\mathbf{A}_{CL} = \mathbf{A} - \mathbf{BK} = \begin{bmatrix} -1.5267 & 43.3608 & 28.2818 \\ 0.2200 & -1.4000 & -32.0000 \\ -0.0089 & 0.3417 & -0.0733 \end{bmatrix}$$

(5.84)

The noise gain matrix, $\mathbf{K}_n$, was determined in Example 5.13 by making the largest elements of $(\mathbf{F} - \mathbf{BK}_n)$ vanish, to be the following:

$$\mathbf{K}_n = [-0.02/272; \quad -0.1/272]$$

(5.85)

It remains to find the feedforward gain matrix, $\mathbf{K}_d = [K_{d1}; \quad K_{d2}; \quad K_{d3}]$, by considering the steady state error, $\mathbf{e}_{ss}$, given by Eq. (5.81). Note from Eq. (5.80) that, since the target is asymptotically stable, it follows that $\mathbf{x}_{dss} = \mathbf{0}$, hence $\mathbf{K}_d$ will not affect the *steady state error*. However, the *transient error*, $\mathbf{e}(t)$, can be reduced by considering elements of the following matrix:

$$\mathbf{A} - \mathbf{A_d} - \mathbf{BK}_d = \begin{bmatrix} (8.4 + 272K_{d1}) & (15 + 272K_{d2}) & (110 + 272K_{d3}) \\ 0.12 & -0.3 & -11 \\ -14K_{d1} & -14K_{d2} & -4 - 14K_{d3} \end{bmatrix}$$

(5.86)

Since by changing $\mathbf{K}_d$ we can only affect the first and the third rows of $(\mathbf{A} - \mathbf{A_d} - \mathbf{BK}_d)$, let us select $\mathbf{K}_d$ such that the largest elements of $(\mathbf{A} - \mathbf{A_d} - \mathbf{BK}_d)$, which are in the first row, are minimized. By selecting $K_{d1} = -8.4/272$, $K_{d2} = -15/272$, and $K_{d3} = -110/272$, we can make the elements in the first row of $(\mathbf{A} - \mathbf{A_d} - \mathbf{BK}_d)$ zeros, and the resulting matrix is the following:

$$\mathbf{A} - \mathbf{A_d} - \mathbf{BK}_d = \begin{bmatrix} 0 & 0 & 0 \\ 0.12 & -0.3 & -11 \\ 0.432 & 0.772 & 1.704 \end{bmatrix}$$

(5.87)

and the required feedforward gain matrix is given by

$$\mathbf{K}_d = [-8.4/272; \quad -15/272; \quad -110/272]$$

(5.88)

The closed-loop error response to target initial condition, $\mathbf{x}_d(0) = [3; 0; 0]^T$, and noise given by $\mathbf{x}_n(t) = [1 \times 10^{-5}; \quad -2 \times 10^{-6}]^T \sin(100t)$, can be obtained by solving Eq. (5.78) with $\mathbf{x}_d(t)$ and $\mathbf{x}_n(t)$ as the known inputs. The noise vector, $\mathbf{x}_n(t)$, and the matrix $(\mathbf{BK}_n - \mathbf{F})$, are calculated for time upto 10 s as follows:

```
>>t = 0:0.01:10; Xn = [1e-5 -2e-6]'*sin(100*t); Bn = -[0 0; -3.5e-3
   0.004;1.03e-3 5.15e-3]; <enter>
```

The desired state-vector, $\mathbf{x}_d(t)$, is obtained by solving Eq. (5.74) using the MATLAB (CST) command *initial* as follows:

```
>>sysd=ss(Ad,zeros(3,1),eye(3),zeros(3,1)); [yd,t,Xd,] = initial(sysd,
   [3 0 0]',t); <enter>
```

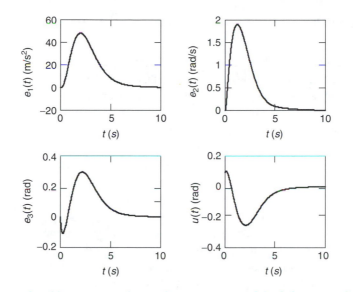

Figure 5.15 Closed-loop error and control input response of the fighter aircraft tracking a target with initial condition $\mathbf{X}_d(0) = [3; 0; 0]^T$

The closed-loop error dynamics given by Eq. (5.78) can be written as follows:

$$\mathbf{e}^{(1)}(t) = \mathbf{A}_{CL}\mathbf{e}(t) + \mathbf{B}_{CL}\mathbf{f}(t) \tag{5.89}$$

where $\mathbf{A}_{CL} = \mathbf{A} - \mathbf{BK}$, $\mathbf{B}_{CL} = [\,(\mathbf{A}_d - \mathbf{A} + \mathbf{BK}_d);\quad (\mathbf{BK}_n - \mathbf{F})\,]$, and the input vector, $\mathbf{f}(t) = [\,\mathbf{x}_d(t)^T;\quad \mathbf{x}_n(t)^T\,]^T$, which are calculated as follows:

```
>>ACL = A-B*K; BCL = [Ad-A+B*Kd Bn]; f = [Xd Xn']; <enter>
```

Finally, using the MATLAB command *lsim*, the closed-loop error response, $\mathbf{e}(t)$, is calculated as follows:

```
>>sysCL=ss(ACL,BCL,eye(3),zeros(3,5)); e = lsim(sysCL,f,t'); <enter>
```

The error, $\mathbf{e}(t) = [\,e_1(t);\quad e_2(t);\quad e_3(t)\,]^T$, and control input, $u(t) = \mathbf{K}\mathbf{e}(t) - \mathbf{K}_d\mathbf{x}_d(t) - \mathbf{K}_n\mathbf{x}_n(t)$, are plotted in Figure 5.15. Note that all the error transients decay to zero in about 10 s, with a negligible influence of the noise. The settling time of error could be made smaller than 10 s, but with a larger control effort and increased vulnerability to noise.

The controller design with gain matrices given by Eqs. (5.83), (5.85), and (5.88) is the best we can do with pole-placement, because there are not enough design parameters (controller gain elements) to make the steady state error identically zero. Clearly, this is a major drawback of the pole-placement method. A better design approach with full-state feedback is the optimal control method, which will be discussed in Chapters 6 and 7.

5.4 Observers, Observability, and Compensators

When we designed control systems using full-state feedback in the previous section, it was assumed that we can measure and feedback all the state variables of the plant using sensors. However, it is rarely possible to measure all the state variables. Some state variables are not even physical quantities. Even in such cases where all the state variables are physical quantities, accurate sensors may not be available, or may be too expensive to construct for measuring all the state variables. Also, some state variable measurements can be so noisy that a control system based on such measurements would be unsuccessful. Hence, it is invariably required to *estimate* rather than measure the state-vector of a system. How can one estimate the state-vector, if it cannot be measured? The answer lies in *observing* the output of the system for a known input and for a finite time interval, and then reconstructing the state-vector from the record of the output. The mathematical model of the process by which a state-vector is estimated from the measured output and the known input is called an *observer* (or *state estimator*). An observer is an essential part of modern control systems. When an observer estimates the *entire* state-vector, it is called a *full-order observer*. However, the state variables that can be measured need not be estimated, and can be directly deduced from the output. An observer which estimates only the unmeasurable state variables is called the *reduced-order observer*. A reduced-order observer results in a smaller order control system, when compared to the full-order observer. However, when the measured state variables are noisy, it is preferable to use a full-order observer to reduce the effect of noise on the control system. A controller which generates the control input to the plant based on the estimated state-vector is called a *compensator*. We will consider the design of observers and compensators below.

Before we can design an observer for a plant, the plant must be *observable*. *Observability* is an important property of a system, and can be defined as the property that makes it possible to determine *any initial state*, $\mathbf{x}(t_0)$, of an *unforced* system (i.e. when the input vector, $\mathbf{u}(t)$, is *zero*) by using a *finite record* of the output, $\mathbf{y}(t)$. The term *finite record* implies that the output is recorded for only a *finite time interval* beginning at $t = t_0$. In other words, observability is a property which enables us to determine what the system was doing at some time, t_0, after measuring its output for a finite time interval beginning at that time. The term *any initial state* is significant in the definition of observability; it may be possible to determine *some initial states* by recording the output, and the system may yet be *unobservable*. Clearly, observability requires that *all* the state variables must contribute to the output of the system, otherwise we cannot reconstruct *all possible* combinations of state variables (i.e. *any* initial state-vector) by measuring the output. The relationship between observability and the output is thus the *dual* of that between controllability and the input. For a system to be controllable, all the state variables must *be affected* by the input; for a system to be observable, all the state variables must *affect* the output. If there are some state variables which *do not* contribute to the output, then the system is *unobservable*. One way of determining observability is by looking at the decoupled state-equations, and the corresponding output equation of a system.

Example 5.15

Consider a system with the following scalar state-equations:

$$x_1^{(1)}(t) = 2x_1(t) + 3u(t)$$

$$x_2^{(1)}(t) = -x_2(t) \tag{5.90}$$

$$x_3^{(1)}(t) = 5x_3(t) - u(t)$$

The scalar output equations of the system are the following:

$$y_1(t) = x_1(t)$$

$$y_2(t) = 2x_2(t) + x_1(t) + u(t) \tag{5.91}$$

Equation (5.90) implies that the state variable, $x_3(t)$, is decoupled from the other two state variables, $x_1(t)$ and $x_2(t)$. Also, $x_3(t)$ does not affect either of the two output variables, $y_1(t)$ and $y_2(t)$. Since the state variable $x_3(t)$, does not contribute to the output vector, $\mathbf{y}(t) = [\, y_1(t); \quad y_2(t)\,]^T$, either *directly* or *indirectly* through $x_1(t)$ and $x_2(t)$, it follows that the system is *unobservable*.

As it is not always possible to decouple the state-equations, we need another way of testing for observability. Similar to the algebraic controllability test theorem, there is an *algebraic observability test theorem* for linear, time-invariant systems stated as follows.

Theorem
The unforced system, $\mathbf{x}^{(1)}(t) = \mathbf{A}\mathbf{x}(t)$, $\mathbf{y}(t) = \mathbf{C}\mathbf{x}(t)$, is observable if and only if the rank of the observability test matrix, $\mathbf{N} = [\,\mathbf{C}^T; \quad \mathbf{A}^T\mathbf{C}^T; \quad (\mathbf{A}^T)^2\mathbf{C}^T; \quad \dots; \quad (\mathbf{A}^T)^{n-1}\mathbf{C}^T\,]$, is equal to n, the order of the system.

The proof of this theorem, given in Friedland [2], follows from the definition of observability, and recalling from Chapter 4 that the output of an unforced (homogeneous) linear, time-invariant system is given by $\mathbf{y}(t) = \mathbf{C}\exp\{\mathbf{A}(t - t_0)\}\mathbf{x}(t_0)$, where $\mathbf{x}(t_0)$ is the initial state-vector.

Example 5.16

Let us apply the observability test theorem to the system of Example 5.15. The state coefficient matrices, $\mathbf{A}$ and $\mathbf{C}$, are the following:

$$\mathbf{A} = \begin{bmatrix} 1 & 0 & 0 \\ 0 & -1 & 0 \\ 0 & 0 & 5 \end{bmatrix}; \quad \mathbf{C} = \begin{bmatrix} 1 & 0 & 0 \\ 2 & 1 & 0 \end{bmatrix} \tag{5.92}$$

The observability test matrix, $\mathbf{N}$, is constructed as follows:

$$\mathbf{A}^T = \begin{bmatrix} 1 & 0 & 0 \\ 0 & -1 & 0 \\ 0 & 0 & 5 \end{bmatrix}; \quad \mathbf{C}^T = \begin{bmatrix} 1 & 2 \\ 0 & 1 \\ 0 & 0 \end{bmatrix}; \quad \mathbf{A}^T\mathbf{C}^T = \begin{bmatrix} 1 & 2 \\ 0 & -1 \\ 0 & 0 \end{bmatrix}$$

$$(\mathbf{A}^{\mathrm{T}})^2 \mathbf{C}^{\mathrm{T}} = \begin{bmatrix} 1 & 2 \\ 0 & 1 \\ 0 & 0 \end{bmatrix} \tag{5.93}$$

or

$$\mathbf{N} = \begin{bmatrix} 1 & 2 & 1 & 2 & 1 & 2 \\ 0 & 1 & 0 & -1 & 0 & 1 \\ 0 & 0 & 0 & 0 & 0 & 0 \end{bmatrix} \tag{5.94}$$

The entire third row of $\mathbf{N}$ consists of zeros; hence it is impossible to form a (3×3) sized, non-zero determinant out of the rows and columns of $\mathbf{N}$. Thus rank $(\mathbf{N}) < 3$ for this third order system, therefore the system is *unobservable*.

Rather than forming the observability test matrix, $\mathbf{N}$, by hand as in Example 5.16, which could be a tedious process for large order systems, we can use the MATLAB (CST) command *ctrb*, noting that $\mathbf{N}$ is the controllability test matrix in which $\mathbf{A}$ is replaced by $\mathbf{A}^{\mathrm{T}}$ and $\mathbf{B}$ is replaced by $\mathbf{C}^{\mathrm{T}}$. Thus, the command

```
>>N = ctrb(A',C') <enter>
```

will give us the observability test matrix.

The reasons for unobservability of a system are pretty much the same as those for uncontrollability, namely the use of superfluous state variables in state-space model, pole-zero cancellation in the system's transfer matrix, too much symmetry, and physical unobservability (i.e. selection of an output vector which is physically unaffected by one or more state variables). If the sub-systems which cause unobservability are *stable*, we can safely ignore those state variables that do not contribute to the output, and design an observer based on the remaining state variables (which would constitute an observable *sub-system*). Thus a stable, unobservable system is said to be *detectable*. If an unobservable sub-system is *unstable*, then the entire system is said to be *undetectable*, because an observer cannot be designed by ignoring the unobservable (and unstable) sub-system. In Example 5.15, the unobservable sub-system corresponding to the decoupled state variable, $x_3(t)$, is unstable (it has a pole at $s = 5$). Hence, the system of Example 5.15 is *undetectable*.

5.4.1 Pole-placement design of full-order observers and compensators

A full-order observer estimates the entire state-vector of a plant, based on the measured output and a known input. If the plant for which the observer is required is linear, the observer's dynamics would also be described by linear state-equations. Consider a noise-free, linear, time-invariant plant described by the following state and output equations:

$$\mathbf{x}^{(1)}(t) = \mathbf{A}\mathbf{x}(t) + \mathbf{B}\mathbf{u}(t) \tag{5.95}$$

$$\mathbf{y}(t) = \mathbf{C}\mathbf{x}(t) + \mathbf{D}\mathbf{u}(t) \tag{5.96}$$

The linear, time-invariant state-equation which describes the dynamics of a full-order observer can be expressed as follows:

$$\mathbf{x}_o^{(1)}(t) = \mathbf{A}_o\mathbf{x}_o(t) + \mathbf{B}_o\mathbf{u}(t) + \mathbf{L}\mathbf{y}(t) \tag{5.97}$$

where $\mathbf{x}_o(t)$ is the *estimated state-vector*, $\mathbf{u}(t)$ is the input vector, $\mathbf{y}(t)$ is the output vector, $\mathbf{A}_o$, $\mathbf{B}_o$ are the state-dynamics and control coefficient matrices of the observer, and $\mathbf{L}$ is the *observer gain matrix*. The matrices $\mathbf{A}_o$, $\mathbf{B}_o$, and $\mathbf{L}$ must be selected in a design process such that the *estimation error*, $\mathbf{e}_o(t) = \mathbf{x}(t) - \mathbf{x}_o(t)$, is brought to zero in the steady state. On subtracting Eq. (5.97) from Eq. (5.95), we get the following *error dynamics* state-equation:

$$\mathbf{e}_o^{(1)}(t) = \mathbf{A}_o\mathbf{e}_o(t) + (\mathbf{A} - \mathbf{A}_o)\mathbf{x}(t) + (\mathbf{B} - \mathbf{B}_o)\mathbf{u}(t) - \mathbf{L}\mathbf{y}(t) \tag{5.98}$$

Substitution of Eq. (5.96) into Eq. (5.98) yields

$$\mathbf{e}_o^{(1)}(t) = \mathbf{A}_o\mathbf{e}_o(t) + (\mathbf{A} - \mathbf{A}_o)\mathbf{x}(t) + (\mathbf{B} - \mathbf{B}_o)\mathbf{u}(t) - \mathbf{L}[\mathbf{C}\mathbf{x}(t) + \mathbf{D}\mathbf{u}(t)] \tag{5.99}$$

or

$$\mathbf{e}_o^{(1)}(t) = \mathbf{A}_o\mathbf{e}_o(t) + (\mathbf{A} - \mathbf{A}_o - \mathbf{L}\mathbf{C})\mathbf{x}(t) + (\mathbf{B} - \mathbf{B}_o - \mathbf{L}\mathbf{D})\mathbf{u}(t) \tag{5.100}$$

From Eq. (5.100), it is clear that estimation error, $\mathbf{e}_o(t)$, will go to zero in the steady state irrespective of $\mathbf{x}(t)$ and $\mathbf{u}(t)$, if all the *eigenvalues* of $\mathbf{A}_o$ are in the *left-half plane*, and the coefficient matrices of $\mathbf{x}(t)$ and $\mathbf{u}(t)$ are *zeros*, i.e. $(\mathbf{A} - \mathbf{A}_o - \mathbf{L}\mathbf{C}) = \mathbf{0}$, $(\mathbf{B} - \mathbf{B}_o - \mathbf{L}\mathbf{D}) = \mathbf{0}$. The latter requirement leads to the following expressions for $\mathbf{A}_o$ and $\mathbf{B}_o$:

$$\mathbf{A}_o = \mathbf{A} - \mathbf{L}\mathbf{C}; \quad \mathbf{B}_o = \mathbf{B} - \mathbf{L}\mathbf{D} \tag{5.101}$$

The error dynamics state-equation is thus the following:

$$\mathbf{e}_o^{(1)}(t) = (\mathbf{A} - \mathbf{L}\mathbf{C})\mathbf{e}_o(t) \tag{5.102}$$

The observer gain matrix, $\mathbf{L}$, must be selected to place all the eigenvalues of $\mathbf{A}_o$ (which are also the poles of the observer) at desired locations in the left-half plane, which implies that the estimation error dynamics given by Eq. (5.102) is *asymptotically stable* (i.e. $\mathbf{e}_o(t) \to \mathbf{0}$ as $t \to \infty$). On substituting Eq. (5.101) into Eq. (5.97), we can write the full-order observer's state-equation as follows:

$$\mathbf{x}_o^{(1)}(t) = (\mathbf{A} - \mathbf{L}\mathbf{C})\mathbf{x}_o(t) + (\mathbf{B} - \mathbf{L}\mathbf{D})\mathbf{u}(t) + \mathbf{L}\mathbf{y}(t) = \mathbf{A}\mathbf{x}_o(t) + \mathbf{B}\mathbf{u}(t)$$
$$+ \mathbf{L}[\mathbf{y}(t) - \mathbf{C}\mathbf{x}_o(t) - \mathbf{D}\mathbf{u}(t)] \tag{5.103}$$

Note that Eq. (5.103) approaches Eq. (5.95) in the steady state if $\mathbf{x}_o(t) \to \mathbf{x}(t)$ as $t \to \infty$. Hence, the observer *mirrors* the plant dynamics if the error dynamics is asymptotically stable. The term $[\mathbf{y}(t) - \mathbf{C}\mathbf{x}_o(t) - \mathbf{D}\mathbf{u}(t)]$ in Eq. (5.103) is called the *residual*, and can be expressed as follows:

$$[\mathbf{y}(t) - \mathbf{C}\mathbf{x}_o(t) - \mathbf{D}\mathbf{u}(t)] = \mathbf{C}\mathbf{x}(t) - \mathbf{C}\mathbf{x}_o(t) = \mathbf{C}\mathbf{e}_o(t) \tag{5.104}$$

From Eq. (5.104), it is clear that the residual is also forced to zero in the steady-state if the error dynamics is asymptotically stable.

CONTROL SYSTEM DESIGN IN STATE-SPACE

The observer design process merely consists of selecting $\mathbf{L}$ by pole-placement of the observer. For single-output plants, the pole-placement of the observer is carried out in a manner similar to the pole-placement of *regulators* for *single-input* plants (see Section 5.3.1). For a plant with the characteristic polynomial written as $|s\mathbf{I} - \mathbf{A}| = s^n + a_{n-1}s^{n-1} + \cdots + a_1s + a_0$, it can be shown by steps similar to Section 5.3.1 that the observer gain matrix, $\mathbf{L}$, which places the observer's poles such that the observer's characteristic polynomial is $|s\mathbf{I} - \mathbf{A_o}| = s^n + \beta_{n-1}s^{n-1} + \cdots + \beta_1s + \beta_0$ is given by

$$\mathbf{L} = [(\boldsymbol{\beta} - \mathbf{a})\mathbf{N}'\mathbf{N}^{-1}]^T \tag{5.105}$$

where $\boldsymbol{\beta} = [\beta_{n-1}; \quad \beta_{n-2}; \quad \ldots; \quad \beta_1; \quad \beta_0]$, $\mathbf{a} = [a_{n-1}; \quad a_{n-2}; \quad \ldots; \quad a_1; \quad a_0]$, $\mathbf{N}$ is the *observability test matrix* of the plant described by Eqs. (5.95) and (5.96), and $\mathbf{N}'$ is the observability test matrix of the plant when it is in the *observer companion form*. Since for single-input, single-output systems, the observer companion form can be obtained from the controller companion form merely by substituting $\mathbf{A}$ by $\mathbf{A}^T$, $\mathbf{B}$ by $\mathbf{C}^T$, and $\mathbf{C}$ by $\mathbf{B}^T$ (see Chapter 3), you can easily show that $\mathbf{N}' = \mathbf{P}'$, where $\mathbf{P}'$ is the *controllability test matrix* of the plant when it is in the *controller companion form*. Thus, we can write

$$\mathbf{L} = [(\boldsymbol{\beta} - \mathbf{a})\mathbf{P}'\mathbf{N}^{-1}]^T \tag{5.106}$$

Recall that $\mathbf{P}'$ is an upper triangular matrix, given by Eq. (5.51).

Example 5.17

Let us try to design a full-order observer for the inverted pendulum on a moving cart (Example 5.9). A state-space representation of the plant is given by Eq. (5.53), with the numerical values of $\mathbf{A}$ and $\mathbf{B}$ given by Eq. (5.54). For this single-input, two-output plant, let us try to design an observer using *only one of the outputs*. If we select the single output to be $y(t) = \theta(t)$, the angular position of the inverted pendulum, the matrices $\mathbf{C}$ and $\mathbf{D}$ are the following:

$$\mathbf{C} = [1; \quad 0; \quad 0; \quad 0]; \quad \mathbf{D} = 0 \tag{5.107}$$

The first thing to do is to check whether the plant is *observable* with this choice of the output. We do so by the following MATLAB command:

```
>>N = (ctrb(A',C'); rank(N) <enter>

ans =
       2
```

Since the rank of the observability test matrix, $\mathbf{N}$, is 2, i.e. *less than 4*, the order of the plant, the plant is *unobservable* with the angular position of the pendulum as the only output. Hence, we cannot design an observer using $y(t) = \theta(t)$. If we choose $y(t) = x(t)$, the cart's displacement, then the output coefficient matrices are as follows:

$$\mathbf{C} = [0; \quad 1; \quad 0; \quad 0]; \quad \mathbf{D} = 0 \tag{5.108}$$

On forming the observability test matrix, **N**, with this choice of output, and checking its rank we get

```
>>N = (ctrb(A',C')); rank(N) <enter>

ans =
      4
```

Since now rank (**N**) = 4, the order of the plant, the plant is observable with $y(t) = x(t)$, and an observer can be designed based on this choice of the output. Let us place the observer poles at $s = -10 \pm 10i$, and $s = -20 \pm 20i$. Then the observer's characteristic polynomial coefficients vector, β, is calculated as follows:

```
>>v = [-10-10i -10+10i -20-20i -20+20i]'; p = poly(v); beta = p(2:5)
   <enter>

beta =
      60    1800    24000    160000
```

The plant's characteristic polynomial coefficient vector, **a**, is calculated as follows:

```
>>p = poly(A); a = p(2:5) <enter>

a =
              0 -10.7800 0 0
```

and the matrix **P'** is evaluated using Eq. (5.51) as follows:

```
>>Pdash = [1 -a(1:3); 0 1 -a(1:2); 0 0 1 -a(1); 0 0 0 1] <enter>

Pdash =
          1.0000    0         10.7800    0
          0         1.000     0          10.7800
          0         0         1.0000     0
          0         0         0          1.0000
```

Finally, the observer gain matrix, **L**, is calculated using Eq. (5.106) as follows:

```
>>format long e; L = ((beta-a)*Pdash*inv(N))' <enter>

L =
  -2.514979591836735e+004
   6.000000000000000e+001
  -1.831838861224490e+005
   1.810780000000000e+003
```

Note that we have printed out **L** in the *long format*, since we need to store it for later calculations. Let us check whether the observer poles have been placed at desired locations, by calculating the eigenvalues of $\mathbf{A}_o = (\mathbf{A} - \mathbf{LC})$ as follows:

```
>>Ao = A-L*C; eig(Ao) <enter>

ans =
 -20.0000+20.0000i
 -20.0000-20.0000i
 -10.0000+10.0000i
 -10.0000-10.0000i
```

Hence, observer pole-placement has been accurately achieved.

Example 5.17 illustrates the ease by which single-output observers can be designed. However, it is impossible to design single-output observers for those plants which are *unobservable* with *any single* output. When *multi-output* observers are required, generally there are more design parameters (i.e. elements in the observer gain matrix, **L**) than the observer poles, hence all of these parameters cannot be determined by pole-placement alone. As in the design of regulators for multi-input plants (Section 5.3.2), additional conditions are required to be satisfied by multi-output observers, apart from pole-placement, to determine the observer gain matrix. These additional conditions are hard to come by, and thus pole-placement is not a good method of designing multi-output observers. A better design procedure in such cases is the *Kalman filter* approach of Chapter 7.

MATLAB's Control System Toolbox (CST) provides the command *estim* for constructing a state-space model, *syso*, of the observer with the observer gain matrix, **L**, and a state-space model, *sysp*, of the plant, with state coefficient matrices **A, B, C, D**, as follows:

```
>>sysp=ss[A,B,C,D]; sysp = estim(syso,L) <enter>
```

The input to the observer thus formed is the plant's output, $\mathbf{y}(t)$, while output vector of the observer is $[\{\mathbf{Cx_o}(t)\}^T; \quad \mathbf{x_o}(t)^T]^T$, where $\mathbf{x_o}(t)$ is the estimated state-vector.

Observers (also known as *estimators*) by themselves are very useful in estimating the plant dynamics from a limited number of outputs, and are employed in *parameter estimation, fault detection*, and other similar applications. The utility of an observer in a control system lies in feeding the estimated state-vector to a controller for generating input signals for the plant. The controllers which generate input signals for the plant based on the estimated state-vector (rather than the actual, fed back state-vector) are called *compensators*. However, design of compensators involves a dilemma. The estimated state-vector is obtained from an observer, which treats the plant's input vector as a *known quantity*, while the compensator is *yet to* generate the input vector based on the estimated state-vector. It is like the classic chicken and egg problem, since we do not know which came first: the control input on which the estimated state-vector is based, or the estimated state-vector on which the input is based! A practical way of breaking this vicious circle is the *separation principle*, which states that if we design an observer (assuming known input vector), and a compensator (assuming known estimated state-vector) *separately*, and then combine the two, we will end up with a control system *that works*. The separation principle thus allows us to design the observer and the controller *independently of each other*. The resulting control system can be a *regulator* or a *tracking system*, depending on the desired state-vector being *zero* or *non-zero*, respectively.

Let us consider a tracking system (i.e. a control system with a non-zero desired state-vector) based on a noise-free plant described by Eqs. (5.95) and (5.96), for which a full-order observer, given by Eq. (5.103) has been designed. Then a compensator can be designed to generate the input vector for the plant according to the following control-law:

$$\mathbf{u}(t) = \mathbf{K}[\mathbf{x}_d(t) - \mathbf{x}_o(t)] - \mathbf{K}_d\mathbf{x}_d(t) \qquad (5.109)$$

where $\mathbf{x}_o(t)$ is the *estimated state-vector*, $\mathbf{x}_d(t)$ is the *desired state-vector*, $\mathbf{K}$ is the *feedback gain matrix*, and $\mathbf{K}_d$ is the *feedforward gain matrix*. On substituting Eq. (5.109) into Eq. (5.103), the observer state-equation becomes

$$\mathbf{x}_o^{(1)}(t) = (\mathbf{A} - \mathbf{LC} - \mathbf{BK} + \mathbf{LDK})\mathbf{x}_o(t) + (\mathbf{B} - \mathbf{LD})(\mathbf{K} - \mathbf{K}_d)\mathbf{x}_d(t) + \mathbf{Ly}(t) \quad (5.110)$$

On substituting the output equation, Eq. (5.96), into Eq. (5.110), and again substituting Eq. (5.109), we get the following state-equation for the compensator:

$$\mathbf{x}_o^{(1)}(t) = (\mathbf{A} - \mathbf{LC} - \mathbf{BK})\mathbf{x}_o(t) + \mathbf{B}(\mathbf{K} - \mathbf{K}_d)\mathbf{x}_d(t) + \mathbf{LCx}(t) \qquad (5.111)$$

The plant's state-equation, Eq. (5.95), when the input is given by Eq. (5.109), becomes the following:

$$\mathbf{x}^{(1)}(t) = \mathbf{Ax}(t) - \mathbf{BKx}_o(t) + \mathbf{B}(\mathbf{K} - \mathbf{K}_d)\mathbf{x}_d(t) \qquad (5.112)$$

Equations. (5.111) and (5.112) are the state-equations of the closed-loop system, and can be expressed as follows:

$$\begin{bmatrix} \mathbf{x}^{(1)}(t) \\ \mathbf{x}_o^{(1)}(t) \end{bmatrix} = \begin{bmatrix} \mathbf{A} & -\mathbf{BK} \\ \mathbf{LC} & (\mathbf{A} - \mathbf{LC} - \mathbf{BK}) \end{bmatrix} \begin{bmatrix} \mathbf{x}(t) \\ \mathbf{x}_o(t) \end{bmatrix} + \begin{bmatrix} \mathbf{B}(\mathbf{K} - \mathbf{K}_d) \\ \mathbf{B}(\mathbf{K} - \mathbf{K}_d) \end{bmatrix} \mathbf{x}_d(t) \quad (5.113)$$

The closed-loop tracking system is thus of order $2n$, where n is the order of the plant. The input to the closed-loop system is the desired state-vector, $\mathbf{x}_d(t)$. A schematic diagram of the tracking system is shown in Figure 5.16. Note that this control system is essentially

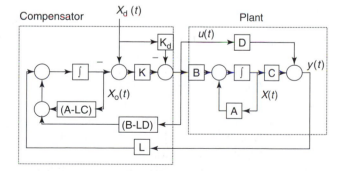

Figure 5.16 Closed-loop tracking system with a full-order compensator

based on the feedback of the output vector, $\mathbf{y}(t)$, to the compensator, which generates the input vector, $\mathbf{u}(t)$, for the plant.

To obtain the state-equation for the estimation error, $\mathbf{e}_o(t) = \mathbf{x}(t) - \mathbf{x}_o(t)$, let us write Eq. (5.112) as follows:

$$\mathbf{x}^{(1)}(t) = (\mathbf{A} - \mathbf{BK})\mathbf{x}(t) + \mathbf{BKe}_o(t) + \mathbf{B}(\mathbf{K} - \mathbf{K}_d)\mathbf{x}_d(t) \qquad (5.114)$$

On subtracting Eq. (5.111) from Eq. (5.114) we get

$$\mathbf{e}_o^{(1)}(t) = (\mathbf{A} - \mathbf{LC})\mathbf{e}_o(t) \qquad (5.115)$$

which is the same as Eq. (5.102). The state-equation for the tracking error, $\mathbf{e}(t) = \mathbf{x}_d(t) - \mathbf{x}(t)$, is obtained by subtracting Eq. (5.114) from Eq. (5.74), which results in

$$\mathbf{e}^{(1)}(t) = (\mathbf{A} - \mathbf{BK})\mathbf{e}(t) + (\mathbf{A}_d - \mathbf{A} + \mathbf{BK}_d)\mathbf{x}_d(t) - \mathbf{BKe}_o(t) \qquad (5.116)$$

The tracking system's error dynamics is thus represented by Eqs. (5.115) and (5.116), which can be expressed together as follows:

$$\begin{bmatrix} \mathbf{e}^{(1)}(t) \\ \mathbf{e}_o^{(1)}(t) \end{bmatrix} = \begin{bmatrix} (\mathbf{A} - \mathbf{BK}) & -\mathbf{BK} \\ \mathbf{0} & (\mathbf{A} - \mathbf{LC}) \end{bmatrix} \begin{bmatrix} \mathbf{e}(t) \\ \mathbf{e}_o(t) \end{bmatrix} + \begin{bmatrix} (\mathbf{A}_d - \mathbf{A} + \mathbf{BK}_d) \\ \mathbf{0} \end{bmatrix} \mathbf{x}_d(t) \quad (5.117)$$

Note that Eq. (5.117) represents the closed-loop tracking system in a *decoupled* state-space form. The closed-loop poles must be the eigenvalues of the following closed-loop state-dynamics matrix, $\mathbf{A}_{CL}$:

$$\mathbf{A}_{CL} = \begin{bmatrix} (\mathbf{A} - \mathbf{BK}) & \mathbf{0} \\ \mathbf{0} & (\mathbf{A} - \mathbf{LC}) \end{bmatrix} \qquad (5.118)$$

Equation (5.117) implies that the closed-loop poles are the eigenvalues of $\mathbf{A}_{CL}$, i.e. the roots of the characteristic equation $|s\mathbf{I} - \mathbf{A}_{CL}| = 0$, which can be written as $|[s\mathbf{I} - (\mathbf{A} - \mathbf{BK})][s\mathbf{I} - (\mathbf{A} - \mathbf{LC})]| = 0$, resulting in $|s\mathbf{I} - (\mathbf{A} - \mathbf{BK})| = 0$ and $|s\mathbf{I} - (\mathbf{A} - \mathbf{LC})| = 0$. Hence, the *closed-loop poles* are the *eigenvalues of* $(\mathbf{A} - \mathbf{BK})$ and *eigenvalues of* $(\mathbf{A} - \mathbf{LC})$, which are also the *poles of* the full-state feedback *regulator* and the *observer*, respectively. Note from Eq. (5.117) that for the estimation error, $\mathbf{e}_o(t)$, to go to zero in the steady state, all the eigenvalues of $(\mathbf{A} - \mathbf{LC})$ must be in the left-half plane. Also, for the tracking error, $\mathbf{e}(t)$, to go to zero in the steady state, irrespective of the desired state-vector, $\mathbf{x}_d(t)$, all the eigenvalues of $(\mathbf{A} - \mathbf{BK})$ must be in the left-half plane, and the coefficient matrix multiplying $\mathbf{x}_d(t)$ must be zero, $(\mathbf{A}_d - \mathbf{A} + \mathbf{BK}_d) = \mathbf{0}$. Recall from Section 5.3 that $(\mathbf{A} - \mathbf{BK})$ is the state-dynamics matrix of the *full-state feedback regulator*, and from Eq. (5.103) that $(\mathbf{A} - \mathbf{LC})$ is the state-dynamics matrix of the *full-order observer*. Hence, the compensator design process consists of *separately* deriving the feedback gain matrices $\mathbf{L}$ and $\mathbf{K}$, by pole-placement of the observer and the full-state feedback regulator, respectively, and selecting $\mathbf{K}_d$ to satisfy $(\mathbf{A}_d - \mathbf{A} + \mathbf{BK}_d) = \mathbf{0}$. Usually, it is impossible to satisfy $(\mathbf{A}_d - \mathbf{A} + \mathbf{BK}_d) = \mathbf{0}$ by selecting the feedforward gain matrix, $\mathbf{K}_d$. Alternatively, it may be possible to satisfy $(\mathbf{A}_d - \mathbf{A} + \mathbf{BK}_d)\mathbf{x}_d(t) = \mathbf{0}$ when some elements of $\mathbf{x}_d(t)$ are zeros. Hence, the steady state tracking error can generally be reduced to zero only for some values of the desired state-vector. In the above steps, we have assumed that the desired state-vector, $\mathbf{x}_d(t)$, is available for measurement. In many cases, it is possible to measure only a desired output, $\mathbf{y}_d(t) = \mathbf{C}_d\mathbf{x}_d(t)$, rather than $\mathbf{x}_d(t)$ itself. In such cases, an observer can be designed to estimate $\mathbf{x}_d(t)$ based on the measurement of the desired output. It is left to you as an exercise to derive the state-equations for the compensator when $\mathbf{x}_d(t)$ is not measurable.

Example 5.18

Let us design a compensator for the inverted pendulum on a moving cart (Example 5.9), when it is desired to move the cart by 1 m, while not letting the pendulum fall. Such a tracking system is representative of a *robot*, which is bringing to you an inverted champagne bottle precariously balanced on a finger! The plant is clearly unstable (as seen in Example 5.9). The task of the compensator is to stabilize the inverted pendulum, while moving the cart by the desired displacement. The desired state-vector is thus a constant, consisting of the desired angular position of the inverted pendulum, $\theta_d(t) = 0$, desired cart displacement, $x_d(t) = 1$ m, desired angular velocity of the pendulum, $\theta_d^{(1)}(t) = 0$, and desired cart velocity, $x_d^{(1)}(t) = 0$. Hence, $\mathbf{x}_d(t) = [0; 1; 0; 0]^T$. Since $\mathbf{x}_d(t)$ is constant, it implies that $\mathbf{x}_d^{(1)}(t) = \mathbf{0}$, and from Eq. (5.74), $\mathbf{A}_d = \mathbf{0}$. By the separation principle, we can design a tracking system *assuming full-state feedback*, and then combine it with a full-order observer, which estimates the plant's state-vector. A full-state feedback regulator has already been designed for this plant in Example 5.11, which places the eigenvalues of the regulator state-dynamics matrix, $(\mathbf{A} - \mathbf{BK})$, at $s = -7.853 \pm 3.2528i$, and $s = -7.853 \pm 7.853i$ using the following feedback gain matrix:

$\mathbf{K}$

$$= [\,-1362.364050360232; \quad -909.3160795202226; \quad -344.8741667548096; \quad -313.4621667548089\,]$$
$$(5.119)$$

We have also designed a full-order observer for this plant using the cart displacement, $x(t)$, as the output in Example 5.17. The observer poles, i.e. the eigenvalues of $(\mathbf{A} - \mathbf{LC})$, were selected to be at $s = -10 \pm 10i$, and $s = -20 \pm 20i$, and the observer gain matrix which achieved this observer pole configuration was obtained to be the following:

$$\mathbf{L} = [\,-25\,149.79591836735; \quad 60.0; \quad -183\,183.8861224490; \quad 1810.780\,]^T$$
$$(5.120)$$

The separation principle allows us to combine the separately designed observer and regulator into a compensator. However, it remains for us to determine the feedforward gain matrix, $\mathbf{K}_d$. The design requirement of zero tracking error in the steady state is satisfied if $(\mathbf{A}_d - \mathbf{A} + \mathbf{BK}_d)\mathbf{x}_d(t) = \mathbf{0}$ in Eq. (5.117). The elements of

$\mathbf{K}_d = [\, K_{d1}; \quad K_{d2}; \quad K_{d3}; \quad K_{d4} \,]$ are thus determined as follows:

$$(\mathbf{A}_d - \mathbf{A} + \mathbf{BK}_d)\mathbf{x}_d(t) = \begin{bmatrix} 0 \\ 0 \\ K_{d2} \\ K_{d2} \end{bmatrix} = \begin{bmatrix} 0 \\ 0 \\ 0 \\ 0 \end{bmatrix} \tag{5.121}$$

Equation (5.121) is exactly satisfied by selecting $K_{d2} = 0$. What about the other elements of $\mathbf{K}_d$? There are no conditions placed on the other elements of $\mathbf{K}_d$, and thus we can *arbitrarily* take them to be zeros. Therefore, by choosing $\mathbf{K}_d = \mathbf{0}$, we are able to meet the zero tracking error requirement in the steady state. On substituting the designed values of the gain matrices, $\mathbf{K}$, $\mathbf{L}$, and $\mathbf{K}_d$ into Eq. (5.113), we can get the closed-loop state-equations for the tracking system in terms of the plant's state-vector, $\mathbf{x}(t)$, and the estimated state-vector, $\mathbf{x}_o(t)$, and then solve them to get the closed-loop response. This is done using MATLAB as follows:

```
>>K=[-1362.364050360232 -909.3160795202226 -344.8741667548096 -313.46216675
    48089]; <enter>

>>L=[-25149.79591836735 60.0 -183183.8861224490 1810.780]'; Kd=zeros(1,4);
    <enter>

>>ACL = [A -B*K; L*C (A-L*C-B*K)]; BCL = [B*(K-Kd); B*(K-Kd)];<enter>
```

Let us confirm that the eigenvalues of $\mathbf{A}_{CL}$ are the poles of the regulator designed in Example 5.11 and the observer designed in Example 5.17 as follows:

```
>>eig(ACL) <enter>

ans =
-20.0000+20.0000i
-20.0000-20.0000i
-10.0000+10.0000i
-10.0000-10.0000i
-7.8530+7.8530i
-7.8530-7.8530i
-7.8530+3.2528i
-7.8530-3.2528i
```

which indeed they are. Finally, the closed-loop response to the desired state-vector is calculated as follows:

```
>>t = 0:1.0753e-2:1.2; n=size(t,2); for i=1:n; Xd(i,:) = [0 1 0 0]; end
    <enter>

>>sysCL=ss(ACL, BCL,[C zeros(1,4)],zeros(1,4)); [y,t,X] = lsim(sysCL,Xd,t');
    <enter>
```

The closed-loop cart's displacement, $x(t)$, and pendulum's angular position, $\theta(t)$, are plotted in Figure 5.17, as follows:

```
>>plot(t,X(:,1:2)) <enter>
```

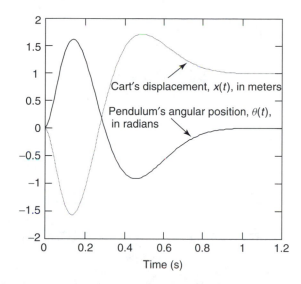

Figure 5.17 Response of the compensator based tracking system for inverted- pendulum on a moving cart, with desired angular position, $\theta_d(t) = 0$, and desired cart's displacement, $x_d(t) = 1$ m, when the regulator poles are $s = -7.853 \pm 3.2528i$, and $s = -7.853 \pm 7.853i$

The closed-loop transient response for $x(t)$ and $\theta(t)$ is seen in Figure 5.17 to settle to their respective desired values in about 1 s, with maximum overshoots of 1.65 m and 1.57 rad., respectively. However, an overshoot of 1.57 rad. corresponds to 90°, which implies that the pendulum *has been allowed to fall* and then brought back up to the inverted position, $\theta(t) = 0°$. If the inverted pendulum represents a drink being brought to you by a robot (approximated by the moving cart), clearly this compensator design would be unacceptable, and it will be necessary to reduce the maximum overshoot to an angle less than 90° by suitably modifying the closed-loop poles. Recall from Example 3.3 that the linearized state-space model of the system given by Eq. (5.53) is *invalid* when the pendulum sways by a large angle, $\theta(t)$, and the results plotted in Figure 5.17 are thus *inaccurate*. Hence, the regulator design that was adequate for stabilizing the plant in the presence of a *small* initial disturbance in cart displacement, is *unsatisfactory* for moving the cart by a *large* displacement. Note that the location of the regulator poles, i.e. the eigenvalues of $(\mathbf{A} - \mathbf{BK})$, governs the closed-loop response of the plant's state-vector, $\mathbf{x}(t)$. By moving the regulator poles closer to the imaginary axis, it would be possible to reduce the maximum overshoot *at the cost of increased settling time*. Let us select the new regulator poles as $s = -0.7853 \pm 3.25328i$ and $s = -0.7853 \pm 0.7853i$. The new feedback gain matrix, $\mathbf{K}$, is calculated as follows:

```
>>v=[-0.7853+3.25328i -0.7853-3.25328i -0.7853+0.7853i -0.7853-0.7853i]';
  K=place(A,B,v)
place: ndigits= 16

K =
 -27.0904    -1.4097    -5.1339    -1.9927
```

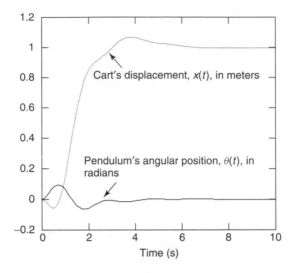

Figure 5.18 Response of the compensator based tracking system for inverted- pendulum on a moving cart, with desired angular position, $\theta_d(t) = 0$, and desired cart's displacement, $x_d(t) = 1$ m, when regulator poles are $s = -0.7853 \pm 3.25328i$ and $s = -0.7853 \pm 0.7853i$

and the new closed-loop response is plotted in Figure 5.18, which shows that the maximum overshoots have been reduced to less than 1.1 m and 0.1 rad. (5.7°) for $x(t)$ and $\theta(t)$, respectively, but the settling time is increased to about 7 s. Since the pendulum now sways by small angles, the linearized model of Eq. (5.53) is valid, and the compensator design is acceptable. However, the robot now takes 7 seconds in bringing your drink placed 1 m away! You may further refine the design by experimenting with the regulator pole locations.

Let us see how well the compensator estimates the state-vector by looking at the estimation error vector, $\mathbf{e}_o(t) = \mathbf{x}(t) - \mathbf{x}_o(t)$. The elements of the estimation error vector, $e_{o1}(t) = \theta_d(t) - \theta(t)$, $e_{o2}(t) = x_d(t) - x(t)$, $e_{o3}(t) = \theta_d^{(1)}(t) - \theta^{(1)}(t)$, and $e_{o4}(t) = x_d^{(1)}(t) - x^{(1)}(t)$ are plotted in Figure 5.19 as follows:

```
>>plot(t,X(:,1)-X(:,5),t,X(:,2)-X(:,6),t,X(:,3)-X(:,7),t,X(:,4)-X(:,8))
 <enter>
```

Figure 5.19 shows that the largest estimation error magnitude is about 1.5×10^{-9} rad/s for estimating the pendulum's angular velocity, $\theta^{(1)}(t)$, and about 5×10^{-10} rad. for estimating the pendulum's angular position, $\theta(t)$. Since the observer is based on the measurement of the cart's displacement, $x(t)$, the estimation error magnitudes of $x(t)$ and $x^{(1)}(t)$ are seen to be negligible in comparison with those of $\theta(t)$ and $\theta^{(1)}(t)$. All the estimation errors decay to zero in about 7 s, which is the same time as the settling time of the closed-loop response for the state-vector, $\mathbf{x}(t)$. The observer poles are therefore at acceptable locations. Note that we can move the observer poles as much inside the left-half plane as we want, because there is no control input cost associated with the observer. However, if the measurements of the output are noisy, there will be an increased influence

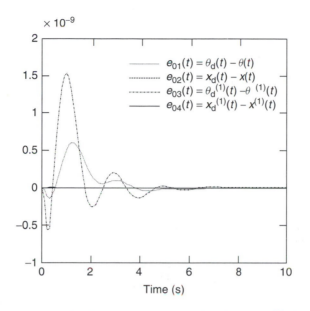

Figure 5.19 Estimation errors for the compensator based tracking system for inverted- pendulum on a moving cart, with desired angular position, $\theta_d(t) = 0$, and desired cart's displacement, $x_d(t) = 1$ m, when regulator poles are $s = -0.7853 \pm 3.25328i$ and $s = -0.7853 \pm 0.7853i$

of noise on the closed-loop system if the observer poles are too far inside the left-half plane.

5.4.2 Pole-placement design of reduced-order observers and compensators

When some of the state variables of a plant can be measured, it is unnecessary to estimate those state variables. Hence, a *reduced-order observer* can be designed which estimates only those state variables that cannot be measured. Suppose the state-vector of a plant, $\mathbf{x}(t)$, can be partitioned into a vector containing measured state variables, $\mathbf{x}_1(t)$, and unmeasurable state variables, $\mathbf{x}_2(t)$, i.e. $\mathbf{x}(t) = [\mathbf{x}_1(t)^T; \quad \mathbf{x}_2(t)^T]^T$. The measured output vector, $\mathbf{y}(t)$, may either be equal to the vector, $\mathbf{x}_1(t)$ – implying that all the state variables constituting $\mathbf{x}_1(t)$ can be *directly* measured – or it may be equal to a *linear combination* of the state variables constituting $\mathbf{x}_1(t)$. Hence, the output equation can be generally expressed as

$$\mathbf{y}(t) = \mathbf{C}\mathbf{x}_1(t) \tag{5.122}$$

where $\mathbf{C}$ is a constant, *square* matrix, indicating that there are as many outputs as the number of elements in $\mathbf{x}_1(t)$. When $\mathbf{x}_1(t)$ can be directly measured, $\mathbf{C} = \mathbf{I}$. The plant's state-equation (Eq. (5.95)) can be expressed in terms of the partitioned state-vector,

$\mathbf{x}(t) = [\mathbf{x}_1(t)^T; \quad \mathbf{x}_2(t)^T]^T$, as follows:

$$\mathbf{x}_1^{(1)}(t) = \mathbf{A}_{11}\mathbf{x}_1(t) + \mathbf{A}_{12}\mathbf{x}_2(t) + \mathbf{B}_1\mathbf{u}(t) \tag{5.123}$$

$$\mathbf{x}_2^{(1)}(t) = \mathbf{A}_{12}\mathbf{x}_1(t) + \mathbf{A}_{22}\mathbf{x}_2(t) + \mathbf{B}_2\mathbf{u}(t) \tag{5.124}$$

where

$$\mathbf{A} = \begin{bmatrix} \mathbf{A}_{11} & \mathbf{A}_{12} \\ \mathbf{A}_{21} & \mathbf{A}_{22} \end{bmatrix}; \quad \mathbf{B} = \begin{bmatrix} \mathbf{B}_1 \\ \mathbf{B}_2 \end{bmatrix} \tag{5.125}$$

Let the order of the plant be n, and the number of measured state variables (i.e. the dimension of $\mathbf{x}_1(t)$) be k. Then a reduced-order observer is required to estimate the vector $\mathbf{x}_2(t)$, which is of dimension $(n - k)$. Hence, the estimated state-vector is simply given by

$$\mathbf{x}_o(t) = \begin{bmatrix} \mathbf{x}_1(t) \\ \mathbf{x}_{o2}(t) \end{bmatrix} = \begin{bmatrix} \mathbf{C}^{-1}\mathbf{y}(t) \\ \mathbf{x}_{o2}(t) \end{bmatrix} \tag{5.126}$$

where $\mathbf{x}_{o2}(t)$ is the *estimation* of the vector $\mathbf{x}_2(t)$. Note that Eq. (5.126) requires that $\mathbf{C}$ should be a *non-singular* matrix, which implies that the plant should be *observable* with the output given by Eq. (5.122). If the plant is *unobservable* with the output given by Eq. (5.122), $\mathbf{C}$ would be *singular*, and a reduced-order observer *cannot* be designed.

The observer state-equation should be such that the *estimation error*, $\mathbf{e}_{o2}(t) = \mathbf{x}_2(t) - \mathbf{x}_{o2}(t)$, is always brought to zero in the steady state. A possible observer state-equation would appear to be the extension of the full-order observer state-equation (Eq. (5.103)) for the reduced-order observer, written as follows:

$$\mathbf{x}_o^{(1)}(t) = \mathbf{A}\mathbf{x}_o(t) + \mathbf{B}\mathbf{u}(t) + \mathbf{L}[\mathbf{y}(t) - \mathbf{C}\mathbf{x}_o(t)] \tag{5.127}$$

where the observer gain matrix, $\mathbf{L}$, would determine the estimation error dynamics. On substituting Eq. (5.126) into Eq. (5.127), and subtracting the resulting state-equation for $\mathbf{x}_{o2}(t)$ from Eq. (5.124), we can write the estimation error state-equation as follows:

$$\mathbf{e}_{o2}^{(1)}(t) = \mathbf{A}_{22}\mathbf{e}_{o2}(t) \tag{5.128}$$

However, Eq. (5.128) indicates that the estimation error is *unaffected* by the observer gain matrix, $\mathbf{L}$, and solely depends upon the plant's sub-matrix, $\mathbf{A}_{22}$. If $\mathbf{A}_{22}$ turns out to be a matrix having eigenvalues with *positive* real parts, we will be stuck with an estimation error that goes to *infinity* in the steady state! Clearly, the observer state-equation given by Eq. (5.127) is *unacceptable*. Let us try the following reduced-order observer dynamics:

$$\mathbf{x}_{o2}(t) = \mathbf{L}\mathbf{y}(t) + \mathbf{z}(t) \tag{5.129}$$

where $\mathbf{z}(t)$ is the solution of the following state-equation:

$$\mathbf{z}^{(1)}(t) = \mathbf{Fz}(t) + \mathbf{Hu}(t) + \mathbf{Gy}(t) \qquad (5.130)$$

Note that the reduced-order observer gain matrix, $\mathbf{L}$, defined by Eq. (5.129), is of size $[(n-k) \times k]$, whereas the full-order observer gain matrix would be of size $(n \times k)$. On differentiating Eq. (5.129) with respect to time, subtracting the result from Eq. (5.124), and substituting $\mathbf{z}(t) = \mathbf{x_{o2}}(t) - \mathbf{Ly}(t) = \mathbf{x_2}(t) - \mathbf{e_{o2}}(t) - \mathbf{LCx_1}(t)$, the state-equation for estimation error is written as follows:

$$\mathbf{e_{o2}^{(1)}}(t) = \mathbf{Fe_{o2}}(t) + (\mathbf{A_{21}} - \mathbf{LCA_{11}} + \mathbf{FLC})\mathbf{x_1}(t) + (\mathbf{A_{22}} - \mathbf{LCA_{12}} - \mathbf{F})\mathbf{x_2}(t)$$
$$+ (\mathbf{B_2} - \mathbf{LCB_1} - \mathbf{H})\mathbf{u}(t) \qquad (5.131)$$

Equation (5.131) implies that for the estimation error, $\mathbf{e_{o2}}(t)$, to go to zero in the steady state, irrespective of $\mathbf{x_1}(t)$, $\mathbf{x_2}(t)$, and $\mathbf{u}(t)$, the coefficient matrices multiplying $\mathbf{x_1}(t)$, $\mathbf{x_2}(t)$, and $\mathbf{u}(t)$ *must vanish*, and $\mathbf{F}$ must have *all* eigenvalues in the left-half plane. Therefore, it follows that

$$\mathbf{F} = \mathbf{A_{22}} - \mathbf{LCA_{12}}; \quad \mathbf{H} = \mathbf{B_2} - \mathbf{LCB_1}; \quad \mathbf{G} = \mathbf{FL} + (\mathbf{A_{21}} - \mathbf{LCA_{11}})\mathbf{C^{-1}} \quad (5.132)$$

The reduced-order observer design consists of selecting the observer gain matrix, $\mathbf{L}$, such that all the eigenvalues of $\mathbf{F}$ are in the left-half plane.

Example 5.19

Let us design a reduced-order observer for the inverted pendulum on a moving cart (Example 5.9), based on the measurement of the cart displacement, $x(t)$. The first step is to partition the state-vector into measurable and unmeasurable parts, i.e. $\mathbf{x}(t) = [\mathbf{x_1}(t)^T; \; \mathbf{x_2}(t)^T]^T$, where $\mathbf{x_1}(t) = x(t)$, and $\mathbf{x_2}(t) = [\theta(t); \; \theta^{(1)}(t); \; x^{(1)}(t)]^T$. However, in Example 5.9, the state-vector was expressed as $[\theta(t); \; x(t); \; \theta^{(1)}(t); \; x^{(1)}(t)]^T$. We must therefore rearrange the state coefficient matrices (Eq. (5.54)) such that the state-vector is $\mathbf{x}(t) = [x(t); \; \theta(t); \; \theta^{(1)}(t); \; x^{(1)}(t)]^T$ and partition them as follows:

$$\mathbf{A} = \begin{bmatrix} 0 & 0 & 0 & 1 \\ 0 & 0 & 1 & 0 \\ 0 & 10.78 & 0 & 0 \\ 0 & -0.98 & 0 & 0 \end{bmatrix}; \quad \mathbf{B} = \begin{bmatrix} 0 \\ 0 \\ -1 \\ 1 \end{bmatrix} \qquad (5.133)$$

From Eq. (5.133) it is clear that

$$\mathbf{A_{11}} = 0; \quad \mathbf{A_{12}} = [0 \;\; 0 \;\; 1]; \quad \mathbf{B_1} = 0$$

$$\mathbf{A_{21}} = \begin{bmatrix} 0 \\ 0 \\ 0 \end{bmatrix}; \quad \mathbf{A_{22}} = \begin{bmatrix} 0 & 1 & 0 \\ 10.78 & 0 & 0 \\ -0.98 & 0 & 0 \end{bmatrix}; \quad \mathbf{B_2} = \begin{bmatrix} 0 \\ -1 \\ 1 \end{bmatrix} \qquad (5.134)$$

Since the measured output is $\mathbf{x_1}(t) = x(t)$, the output equation is $\mathbf{y}(t) = \mathbf{C}\mathbf{x_1}(t)$, where $\mathbf{C} = 1$. We have to select an observer gain-matrix, $\mathbf{L}$, such that the eigenvalues of $\mathbf{F} = (\mathbf{A_{22}} - \mathbf{LCA_{12}})$ are in the left-half plane. Let us select the observer poles, i.e. the eigenvalues of $\mathbf{F}$, to be $s = -20$, $s = -20 \pm 20i$. Then $\mathbf{L}$ is calculated by pole-placement as follows:

```
>>A12 = [0 0 1]; A22 = [0 1 0; 10.78 0 0; -0.98 0 0]; C = 1; <enter>

>>v = [-20 -20+20i -20-20i]'; L = (place(A22',A12'*C',v))' <enter>

L =
-1.6437e+003
-1.6987e+004
6.0000e+001
```

Therefore, the observer dynamics matrix, $\mathbf{F}$, is calculated as follows:

```
>>F = A22 - L*C*A12 <enter>

F =
   0               1.0000e+000   1.6437e+003
   1.0780e+001     0             1.6987e+004
  -9.8000e-001     0            -6.0000e+001
```

Let us verify that the eigenvalues of $\mathbf{F}$ are at desired locations:

```
>>eig(F) <enter>

ans =
-2.0000e+001+2.0000e+001i
-2.0000e+001-2.0000e+001i
-2.0000e+001
```

which indeed they are. The other observer coefficient matrices, $\mathbf{G}$ and $\mathbf{H}$, are calculated as follows:

```
>>A11 = 0; A21 = [0 0 0]'; B1 = 0; B2 = [0 -1 1]';
  H = B2 - L*C*B1 <enter>

H =
     0
    -1
     1

>>G = F*L + (A21 - L*C*A11)*inv(C) <enter>

G =
    8.1633e+004
    1.0015e+006
   -1.9892e+003
```

A compensator based on the reduced-order observer can be designed by the *separation principle*, in a manner similar to the compensator based on the full-order observer. The control-law defining the reduced-order compensator for a tracking system can be expressed as follows, after substituting Eq. (5.126) into Eq. (5.109):

$$\mathbf{u}(t) = \mathbf{K}[\mathbf{x}_d(t) - \mathbf{x}_o(t)] - \mathbf{K}_d\mathbf{x}_d(t) = (\mathbf{K} - \mathbf{K}_d)\mathbf{x}_d(t) - \mathbf{K}_1\mathbf{x}_1(t) - \mathbf{K}_2\mathbf{x}_{o2}(t) \quad (5.135)$$

where $\mathbf{x}_d(t)$ is the desired state-vector, $\mathbf{K}_d$ is the feedforward gain matrix, and $\mathbf{K}$ is the feedback gain matrix, which can be partitioned into gain matrices that feedback $\mathbf{x}_1(t)$ and $\mathbf{x}_{o2}(t)$, respectively, as $\mathbf{K} = [\mathbf{K}_1; \quad \mathbf{K}_2]$. A schematic diagram of the reduced-order compensator is shown in Figure 5.20.

The estimation error dynamics of the reduced-order compensator is described by the following state-equation, obtained by substituting Eq. (5.132) into Eq. (5.131):

$$\mathbf{e}_{o2}^{(1)}(t) = \mathbf{F}\mathbf{e}_{o2}(t) \quad (5.136)$$

while the state-equation for the tracking error, $\mathbf{e}(t) = \mathbf{x}_d(t) - \mathbf{x}(t)$, is obtained by subtracting Eq. (5.74) from Eq. (5.95), and substituting Eq. (5.135) as follows:

$$\mathbf{e}^{(1)}(t) = \mathbf{A}\mathbf{e}(t) + (\mathbf{A}_d - \mathbf{A} + \mathbf{B}\mathbf{K}_d)\mathbf{x}_d(t) - \mathbf{B}\mathbf{K}[\mathbf{x}_d(t) - \mathbf{x}_o(t)] \quad (5.137)$$

On substituting for $\mathbf{x}_o(t)$ from Eq. (5.126), Eq. (5.137) can be written as follows:

$$\mathbf{e}^{(1)}(t) = (\mathbf{A} - \mathbf{B}\mathbf{K})\mathbf{e}(t) + (\mathbf{A}_d - \mathbf{A} + \mathbf{B}\mathbf{K}_d)\mathbf{x}_d(t) - \mathbf{B}\mathbf{K}_2\mathbf{e}_{o2}(t) \quad (5.138)$$

Hence, the dynamics of the tracking system can be described by Eqs. (5.136) and (5.138). To have the tracking error go to zero in the steady state, irrespective of $\mathbf{x}_d(t)$, we must select the feedforward gain matrix, $\mathbf{K}_d$, such that $(\mathbf{A}_d - \mathbf{A} + \mathbf{B}\mathbf{K}_d)\mathbf{x}_d(t) = \mathbf{0}$, and the feedback gain matrix, $\mathbf{K}$, such that the eigenvalues of $(\mathbf{A} - \mathbf{B}\mathbf{K})$ are in the left-half plane. Since the eigenvalues of $(\mathbf{A} - \mathbf{B}\mathbf{K})$ are the regulator poles, and eigenvalues of $\mathbf{F}$ are

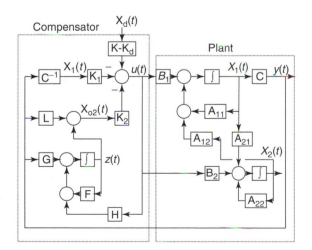

Figure 5.20 Tracking system based on reduced-order compensator

the reduced-order observer poles, it follows from Eqs. (5.136) and (5.138) that the *poles of the tracking system* are of *observer and regulator poles*. (Prove this fact by finding the eigenvalues of the closed-loop system whose state-vector is $[\mathbf{e}(t)^T; \quad \mathbf{e_{o2}}(t)^T]^T$.) According to the separation principle, the design of regulator and observer can be carried out separately by pole-placement. Note from Eqs. (5.136) and (5.138) that the order of the reduced-order tracking system is $(2n - k)$, where k is the number of measurable state-variables. Recall form the previous sub-section that the order of the full-order tracking system was $2n$. Thus, the more state-variables we can measure, the smaller will be the order of the tracking system based on reduced-order observer.

Example 5.20

Let us re-design the tracking system for the inverted pendulum on a moving cart (Example 5.18), using a reduced-order observer. Recall that it is desired to move the cart by 1 m, while not letting the pendulum fall. We have already designed a reduced-order observer for this plant in Example 5.19, using the measurement of the cart's displacement, $x(t)$, such that the observer poles are $s = -20$, $s = -20 \pm 20i$. In Example 5.18, we were able to make $(\mathbf{A_d} - \mathbf{A} + \mathbf{BK_d})\mathbf{x_d}(t) = \mathbf{0}$ with $\mathbf{K_d} = \mathbf{0}$. It remains to select the regulator gain matrix, $\mathbf{K}$, such that the eigenvalues of $(\mathbf{A} - \mathbf{BK})$ are at desired locations in the left-half plane. As in Example 5.18, let us choose the regulator poles to be $s = -0.7853 \pm 3.25328i$ and $s = -0.7853 \pm 0.7853i$. Note that we cannot directly use the regulator gain matrix of Example 5.18, because the state-vector has been *re-defined* in Example 5.19 to be $\mathbf{x}(t) = [x(t); \quad \theta(t); \quad \theta^{(1)}(t); \quad x^{(1)}(t)]^T$, as opposed to $\mathbf{x}(t) = [\theta(t); x(t); \theta^{(1)}(t); x^{(1)}(t)]^T$ of Example 5.18. The new regulator gain matrix would thus be obtained by *switching* the *first* and *second* elements of $\mathbf{K}$ calculated in Example 5.18, or by repeating pole-placement using the re-arranged state coefficient matrices as follows:

```
>>A = [A11 A12; A21 A22]; B = [B1; B2]; <enter>

>>v=[-0.7853+3.25328i -0.7853-3.25328i -0.7853+0.7853i
  -0.7853-0.7853i]'; K=place(A,B,v) <enter>

place: ndigits= 16

K =

-1.4097e+000   -2.7090e+001   -5.1339e+000   -1.9927e+000
```

The partitioning of $\mathbf{K}$ results in $\mathbf{K_1} = -1.4097$ and $\mathbf{K_2} = [-27.090; -5.1339; -1.9927]$. The closed-loop error dynamics matrix, $\mathbf{A_{CL}}$, is the state-dynamics matrix obtained by combining Eqs. (5.136) and (5.138) into a state-equation, with the state-vector, $[\mathbf{e}(t); \quad \mathbf{e_{o2}}(t)]^T$, and is calculated as follows:

```
    >>K2 = K(2:4); ACL = [A-B*K -B*K2; zeros(3,4) F]; <enter>
```

The eigenvalues of $\mathbf{A_{CL}}$ are calculated as follows:

```
>>eig(ACL) <enter>

ans =
-7.8530e-001+3.2533e+000i
-7.8530e-001-3.2533e+000i
-7.8530e-001+7.8530e-001i
-7.8530e-001-7.8530e-001i
-2.0000e+001+2.0000e+001i
-2.0000e+001-2.0000e+001i
-2.0000e+001
```

Note that the closed-loop eigenvalues consist of the regulator and observer poles, as expected. The *closed-loop error response* (i.e. the solution of Eqs. (5.136) and (5.138)) to $\mathbf{x_d}(t) = [1; 0; 0; 0]^T$ is nothing else but the initial response to $[\mathbf{e}(0)^T; \mathbf{e_{o2}}(0)^T]^T = [1; 0; 0; 0; 0; 0; 0]^T$, which is computed as follows:

```
>>sysCL=ss(ACL,zeros(7,1),eye(7),zeros(7,1));
 [y,t,e]= initial(sysCL,[1 zeros(1,6)]); <enter>
```

The estimation error vector, $\mathbf{e_{o2}}(t)$ is *identically zero* for this example, while the tracking errors, i.e. elements of $\mathbf{e}(t)$, are plotted in Figure 5.21. Note in Figure 5.21 that all the error transients decay to zero in about 7 s. The maximum value for the cart's velocity, $x^{(1)}(t)$, is seen to be about 0.9 m/s, while the angular velocity of the pendulum, $\theta^{(1)}(t)$, reaches a maximum value of 0.25 rad/s. The angular displacement of the pendulum, $\theta(t)$, is always less than 0.1 rad ($5.73°$) in magnitude, which is acceptably small for the validity of the linear plant model.

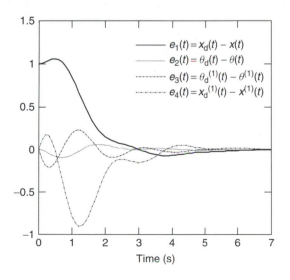

Figure 5.21 Tracking error response of closed-loop system consisting of an inverted pendulum on a moving cart and a reduced-order compensator (Example 5.20)

5.4.3 Noise and robustness issues

If noise is present in the plant, the plant's state-space representation is given by Eqs. (5.26) and (5.27), and the feedback control-law, Eq. (5.109), is modified as follows:

$$\mathbf{u}(t) = \mathbf{K}[\mathbf{x_d}(t) - \mathbf{x_o}(t)] - \mathbf{K_d}\mathbf{x_d}(t) - \mathbf{K_n}\mathbf{x_n}(t) \qquad (5.139)$$

where $\mathbf{x_d}(t)$ and $\mathbf{x_n}(t)$ are the desired state-vector and the noise vector, respectively, and $\mathbf{K_d}$ and $\mathbf{K_n}$ are the feedforward gain matrices. In Eq. (5.139) it is assumed that both $\mathbf{x_d}(t)$ and $\mathbf{x_n}(t)$ can be measured, and thus need not be estimated by the observer. In case $\mathbf{x_d}(t)$ and $\mathbf{x_n}(t)$ are unmeasurable, we have to know the state-space model of the processes by which they are generated, in order to obtain their estimates. While it may be possible to know the dynamics of the desired state-vector, $\mathbf{x_d}(t)$, the noise-vector, $\mathbf{x_n}(t)$, is usually generated by a non-deterministic process whose mathematical model is unknown. In Chapter 7, we will derive observers which include an approximate model for the stochastic processes that generate noise, and design compensators for such plants. In Chapter 7 we will also study the robustness of multivariable control systems with respect to random noise.

SIMULINK can be used to simulate the response of a control system to noise, parameter variations, and nonlinearities, thereby giving a direct information about a system's robustness.

Example 5.21

Let us simulate the inverted-pendulum on a moving cart with the control system designed in Example 5.18 with a full-order compensator, with the addition of *measurement noise* modeled as a *band limited white noise* source block of SIMULINK. *White noise* is a statistical model of a special random process that we will discuss in Chapter 7. The parameter *power* of the white noise block representing the intensity of the noise is selected as 10^{-8}. A SIMULINK block-diagram of the plant with full-order compensator with regulator and observer gains designed in Example 5.18 is shown in Figure 5.22. Note the use of *matrix gain* blocks to synthesize the compensator, and a masked *subsystem* block for the state-space model of the plant. The matrix gain blocks are named B = B1 for the matrix $\mathbf{B}$, C = C1 for the matrix $\mathbf{C}$, L = L1 for the observer gain matrix $\mathbf{L}$, and K = K1 = K2 for the regulator gain matrix, $\mathbf{K}$. The *scope* outputs *thet, x, thdot* and *xdot* are the state-variables $\theta(t)$, $x(t)$, $\theta^{(1)}(t)$, and $x^{(1)}(t)$, respectively, which are *demux-ed* from the state vector of the plant, and are also saved as variables in the MATLAB workspace. The resulting simulation of $\theta(t)$ and $x(t)$ is also shown in Figure 5.22. Note the random fluctuations in both $\theta(t)$ and $x(t)$ about the desired steady-state values of $x_d = 1$ m and $\theta_d = 0$. The maximum magnitude of $\theta(t)$ is limited to 0.1 rad., which is within the range required for linearizing the equations of motion. However, if the intensity of the measurement noise is increased, the $\theta(t)$ oscillations quickly surpass the linear range. The simulation of Figure 5.22 also conforms to our usual experience in trying to balance a stick vertically on a finger.

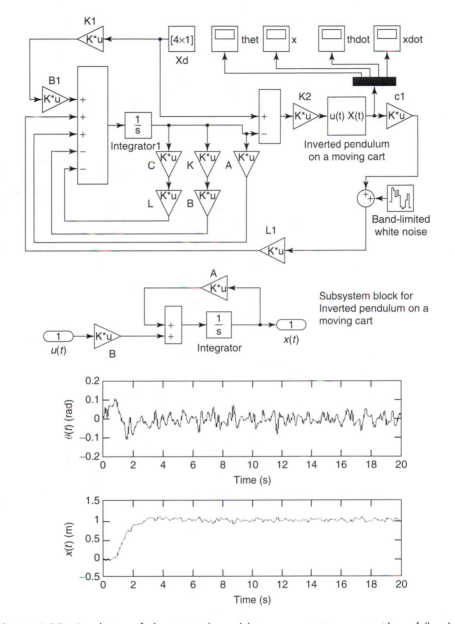

Figure 5.22 Simulation of the inverted pendulum on a moving cart with a full-order compensator and measurement noise with SIMULINK block-diagram

Exercises

5.1. Check the controllability of the plants with the following state-coefficient matrices:

(a)
$$\mathbf{A} = \begin{bmatrix} 0 & 1 \\ 1 & -2 \end{bmatrix}; \quad \mathbf{B} = \begin{bmatrix} 0 & 0 \\ -1 & 1 \end{bmatrix} \tag{5.140}$$

(b)
$$A = \begin{bmatrix} 0 & 1 & 0 \\ 25 & 0 & -1 \\ 0 & 0 & 0 \end{bmatrix}; \quad B = \begin{bmatrix} 0 \\ 1 \\ 5 \end{bmatrix} \qquad (5.141)$$

5.2. As discussed in Section 5.1, an unstable plant is said to be *stabilizable* if all the *uncontrollable* sub-systems have *stable* eigenvalues. Check whether the plants given in Exercise 5.1 are stabilizable.

5.3. If a plant is *stabilizable* (see Exercise 5.2), we can safely ignore the *uncontrollable* sub-systems by removing the rows and columns corresponding to the uncontrollable states from the state coefficient matrices, **A** and **B**. The resulting state-space representation would be controllable, and is called a *minimal realization*. Find the minimal realization of the state coefficient matrices, **A** and **B** for the plants in Exercise 5.1.

5.4. A *distillation column* in a chemical plant has the following state-coefficient matrices:

$$A = \begin{bmatrix} -21 & 0 & 0 & 0 \\ 0.1 & -5 & 0 & 0 \\ 0 & -1.5 & 0 & 0 \\ 0 & -4 & 0 & 0 \end{bmatrix}; \quad B = \begin{bmatrix} 6000 & 0 \\ 0 & 0 \\ 0 & 2.3 \\ 0 & 0.1 \end{bmatrix} \qquad (5.142)$$

(a) Is the plant controllable?

(b) Suppose we would like to control the plant using *only one* input at a time. Is the plant controllable with only the *first* input, i.e. with $B = [6000; 0; 0; 0]^T$? Is the plant controllable with only the *second* input, i.e. with $B = [0; 0; 2.3; 0.1]^T$?

5.5. For the aircraft with lateral dynamics given in Eq. (4.97) in Exercise 4.3:

(a) is the aircraft controllable using both the inputs?

(b) is the aircraft controllable using *only* the aileron input, $\delta_A(t)$?

(c) is the aircraft controllable using *only* the rudder input, $\delta_R(t)$?

5.6. Consider the longitudinal dynamics of a flexible bomber airplane of Example 4.7, with the state-space representation given by Eq. (4.71).

(a) Is the aircraft controllable using both the inputs, $u_1(t)$ and $u_2(t)$?

(b) Is the aircraft controllable using *only* the desired elevator deflection, $u_1(t)$?

(c) Is the aircraft controllable using *only* the desired canard deflection, $u_2(t)$?

5.7. For the aircraft in Exercise 5.5, can you design a full-state feedback regulator which places the closed-loop poles of the aircraft at $s_{1,2} = -1 \pm i$, $s_3 = -15$, $s_4 = -0.8$ using *only one of the inputs* ? If so, which one, and what is the appropriate gain matrix?

5.8. For the aircraft in Exercise 5.6, design a full-state feedback regulator using both the inputs and the MATLAB (CST) command *place*, such that the closed-loop poles are located at $s_{1,2} = -3 \pm 3i$, $s_{3,4} = -1 \pm 2i$, $s_5 = -100$, $s_6 = -75$. Find the maximum overshoots and settling time of the closed-loop initial response if the initial condition vector is $\mathbf{x}(0) = [0; 0.5; 0; 0; 0; 0]^T$.

5.9. For the distillation column of Exercise 5.4, design a full-state feedback regulator to place the closed-loop poles at $s_{1,2} = -0.5 \pm 0.5i$, $s_3 = -5$, $s_4 = -21$.

5.10. Repeat Exercise 5.9 for the closed-loop poles in a Butterworth pattern of radius, $R = 5$. Compare the initial response of the *first state-variable* (i.e. for $\mathbf{C} = [1; 0; 0; 0]$ and $\mathbf{D} = [0; 0]$) of the resulting closed-loop system with that of Exercise 5.9 for initial condition, $\mathbf{x}(0) = [1; 0; 0; 0]^T$. Which of the two (present and that of Exercise 5.9) regulators requires the larger control input magnitudes for this initial condition?

5.11. Consider the turbo-generator of Example 3.14, with the state-space representation given by Eq. (3.117).

(a) Is the plant controllable using both the inputs, $u_1(t)$ and $u_2(t)$?

(b) Is the plant controllable using *only* the input, $u_1(t)$?

(c) Is the plant controllable using *only* the input, $u_2(t)$?

(d) Design a full-state feedback regulator for the plant using only the input, $u_1(t)$, such that the closed-loop eigenvalues are at $s_{1,2} = -2.5 \pm 2.5i$, $s_{3,4} = -1 \pm i$, $s_5 = -10$, $s_6 = -15$.

(e) Repeat part (d) using only the input, $u_2(t)$.

(f) Repeat part (d) using both the inputs, $u_1(t)$ and $u_2(t)$, and the MATLAB (CST) command *place* for designing the multi-input regulator.

(g) Re-design the regulators in parts (d)–(f), such that the maximum overshoot and settling time for the output, $y_1(t)$, are less than 0.3 units and 6 seconds, respectively, if the initial condition vector is $\mathbf{x}(0) = [0.1; 0; 0; 0; 0; 0]^T$.

(h) Re-design the regulators in parts (d)–(f), such that the closed-loop poles are in a Butterworth pattern of radius, $R = 10$, and compare the closed-loop initial responses and input magnitudes with those of part (g).

5.12. Check the observability of the plants with the following state coefficient matrices:

(a) $\mathbf{A} = \begin{bmatrix} -1 & 0 & 0 \\ 0.3 & -0.1 & 0.05 \\ 1 & 0 & 0 \end{bmatrix}$; $\mathbf{C} = [0 \quad 0 \quad -2]$ $\qquad$ (5.143)

(b) $\mathbf{A} = \begin{bmatrix} 0 & 0.1 & -100 & 4 \\ -250 & -7 & 3 & 50 \\ 0 & 0 & -3.3 & 0.06 \\ 2 & 0 & 0 & 0.25 \end{bmatrix}$; $\mathbf{C} = [0 \quad 0 \quad 1 \quad 0]$ $\qquad$ (5.144)

(c) $\mathbf{A} = \begin{bmatrix} 0 & 0 \\ 0 & -1 \end{bmatrix}$; $\mathbf{C} = \begin{bmatrix} 1 & 0 \\ -2 & 0 \end{bmatrix}$ $\qquad$ (5.145)

5.13. An unstable plant is said to be *detectable* if all the *unobservable* sub-systems have *stable* eigenvalues. Check whether the plants given in Exercise 5.12 are detectable.

5.14. If a plant is *detectable* (see Exercise 5.13), we can safely ignore the *unobservable* sub-systems by removing the rows and columns corresponding to the unobservable states from the state coefficient matrices, **A** and **B**. The resulting state-space representation would be observable, and is called a *minimal realization*. Find the minimal realization of the state coefficient matrices, **A** and **B** for the plants in Exercise 5.12.

5.15. For the distillation column of Exercise 5.4, the matrices **C** and **D** are as follows:

$$\mathbf{C} = \begin{bmatrix} 0 & 0 & 1 & 0 \\ 0 & 0 & 0 & 1 \end{bmatrix}; \quad \mathbf{D} = \begin{bmatrix} 0 & 0 \\ 0 & 0 \end{bmatrix} \qquad (5.146)$$

(a) Is the plant observable?

(b) Is the plant observable if *only* the *first output* was measured, i.e. $\mathbf{C} = [0; 0; 1; 0]$, $\mathbf{D} = 0$?

(c) Is the plant observable if *only* the *second output* was measured, i.e. $\mathbf{C} = [0; 0; 0; 1]$, $\mathbf{D} = 0$?

5.16. For the aircraft with lateral dynamics given in Eq. (4.97) in Exercise 4.3:

(a) is the aircraft observable with the output vector, $\mathbf{y}(t) = [p(t); r(t)]^T$?

(b) is the aircraft observable with the output vector, $\mathbf{y}(t) = [p(t); \phi(t)]^T$?

(c) is the aircraft observable with the bank-angle, $\phi(t)$, being the only measured output?

(d) is the aircraft observable with the sideslip-angle, $\beta(t)$, being the only measured output?

(e) design a full-order observer for the aircraft using *only one* of the state-variables as the output, such that the observer poles are placed at $s_{1,2} = -2 \pm 2i$, $s_3 = -16$, $s_4 = -2$.

(f) design a full-order compensator based on the regulator of Exercise 5.7, and the observer designed in part (e). Calculate and plot the initial response, $\phi(t)$, of the compensated system if the initial condition is $\mathbf{x}(0) = [0.5; 0; 0; 0]^T$.

5.17. Consider the longitudinal dynamics of a flexible bomber airplane of Example 4.7.

(a) Is the aircraft observable using both the outputs, $y_1(t)$ and $y_2(t)$?

(b) Is the aircraft observable with *only* the normal acceleration, $y_1(t)$, as the measured output?

(c) Is the aircraft observable with *only* the pitch-rate, $y_2(t)$, as the measured output?

(d) Design a full-order observer for the aircraft using *only* the normal acceleration output, $y_1(t)$, such that the observer poles are placed at $s_{1,2} = -4 \pm 4i$, $s_{3,4} = -3 \pm 3i$, $s_5 = -100$, $s_6 = -75$.

(e) Design a full-order compensator for the aircraft with the regulator of Exercise 5.8 and the observer of part (d). Compute the initial response of the compensated system with the initial condition given by $\mathbf{x}(0) = [0; 0.5; 0; 0; 0; 0]^T$, and compare with that obtained for the regulated system in Exercise 5.8. Also compare the required inputs for the regulated and compensated systems.

5.18. For the distillation column of Exercise 5.15, design a two-output, full-order observer using the MATLAB (CST) command *place* such that the observer poles are placed at $s_{1,2} = -2 \pm 2i$, $s_3 = -5$, $s_4 = -21$. With the resulting observer and the regulator designed in Exercise 5.9, design a full-order compensator and find the initial response of the compensated system for the initial condition $\mathbf{x}(0) = [1; 0; 0; 0]^T$. What are the control inputs required to produce the compensated initial response?

5.19. Consider the turbo-generator of Example 3.14, with the state-space representation given by Eq. (3.117).

(a) Is the plant observable with both the inputs, $y_1(t)$ and $y_2(t)$?

(b) Is the plant observable with *only* the output, $y_1(t)$?

(c) Is the plant observable with *only* the output, $y_2(t)$?

(d) Design a full-order observer for the plant using only the output, $y_1(t)$, such that the observer poles are placed at $s_{1,2} = -3.5 \pm 3.5i$, $s_{3,4} = -5 \pm 5i$, $s_5 = -10$, $s_6 = -15$.

(e) Repeat part (d) using only the output, $y_2(t)$.

(f) Repeat part (d) using both the outputs, $y_1(t)$ and $y_2(t)$, and the MATLAB (CST) command *place* for designing the two-output full-order observer.

(g) Re-design the observers in parts (d)–(f), and combine them with the corresponding regulators designed in Exercise 5.11(g) to form compensators, such that the maximum overshoot and settling time for the compensated initial response, $y_1(t)$, are less than 0.3 units and 6 seconds, respectively, if the initial condition vector is $\mathbf{x}(0) = [0.1; 0; 0; 0; 0; 0]^T$. How do the input magnitudes compare with the required inputs of the corresponding regulators in Exercise 5.11(g)?

5.20. Design a reduced-order observer for the aircraft of Exercise 5.16 using the bank-angle, $\phi(t)$, as the only output, such that the observer poles are in a Butterworth pattern of radius, $R = 16$, and combine it with the regulator of Exercise 5.7, to form a reduced-order compensator. Compare the initial response, $\phi(t)$, and the required inputs of the reduced-order compensated system to that of the full-order compensator in Exercise 5.16 (f) with the initial condition $\mathbf{x}(0) = [0.5; 0; 0; 0]^T$.

5.21. Design a reduced-order observer for the aircraft of Exercise 5.17 with normal acceleration, $y_1(t)$, as the only output, such that the observer poles are in a Butterworth pattern of radius, $R = 100$, and combine it with the regulator of Exercise 5.8, to form a reduced-order compensator. Compare the initial response and the required inputs of the reduced-order compensated system to that of the full-order compensator in Exercise 5.17(e) with the initial condition $\mathbf{x}(0) = [0; 0.5; 0; 0; 0; 0]^T$.

5.22. For the distillation column of Exercise 5.15, design a two-output, reduced-order observer using the MATLAB (CST) command *place* such that the observer poles are placed in Butterworth pattern of radius, $R = 21$. With the resulting observer and the regulator designed in Exercise 5.9, form a reduced-order compensator and compare the initial response and required inputs of the compensated system for the initial condition $\mathbf{x}(0) = [1; 0; 0; 0]^T$, with the corresponding values obtained in Exercise 5.18.

5.23. Using SIMULINK, simulate the tracking system for the inverted pendulum on a moving cart with the reduced-order compensator designed in Example 5.20, including a measurement white noise of intensity (i.e. *power* parameter in the *band-limited white noise* block) when the desired plant state is $\mathbf{x}_d = [0; 1; 0; 0]^T$.

5.24. Using SIMULINK, test the *robustness* of the full-order compensator designed (using linear plant model) in Example 5.18 in controlling the plant model described by the *nonlinear* state-equations of Eqs. (3.17) and (3.18) when the desired plant state is $\mathbf{x}_d = [0; 1; 0; 0]^T$. (Hint: replace the *subsystem* block in Figure 5.22 with a *function* M-file for the nonlinear plant dynamics.)

References

1. Kreyszig, E. *Advanced Engineering Mathematics*. John Wiley & Sons, New York, 1972.
2. Friedland, B. *Control System Design – An Introduction to State-Space Methods*. McGraw-Hill International Edition, Singapore, 1987.
3. Kautsky, J. and Nichols, N.K. *Robust Eigenstructure Assignment in State Feedback Control*. Numerical Analysis Report NA/2/83, School of Mathematical Sciences, Flinders University, Australia, 1983.

6

Linear Optimal Control

6.1 The Optimal Control Problem

After designing control systems by pole-placement in Chapter 5, we naturally ask why we should need to go any further. Recall that in Chapter 5 we were faced with an *overabundance* of design parameters for multi-input, multi-output systems. For such systems, we did not quite know how to determine all the design parameters, because only a limited number of them could be found from the closed-loop pole locations. The MATLAB M-file *place.m* imposes *additional conditions* (apart from closed-loop pole locations) to determine the design parameters for multi-input regulators, or multi-output observers; thus the design obtained by *place.m* cannot be regarded as pole-placement alone. *Optimal control* provides an alternative design strategy by which *all* the control design parameters can be determined even for multi-input, multi-output systems. Also in Chapter 5, we did not know *a priori* which pole locations would produce the desired performance; hence, some trial and error with pole locations was required before a satisfactory performance could be achieved. *Optimal control* allows us to directly formulate the performance objectives of a control system (provided we know how to do so). More importantly – apart from the above advantages – *optimal control* produces the *best possible* control system for a given set of performance objectives. What do we mean by the adjective *optimal*? The answer lies in the fact that there are many ways of doing a particular thing, but only one way which requires the *least effort*, which implies the least expenditure of *energy* (or money). For example, we can hire the most expensive lawyer in town to deal with our inconsiderate neighbor, or we can directly talk to the neighbor to achieve the desired result. Similarly, a control system can be designed to meet the desired performance objectives with the *smallest control energy*, i.e. the energy associated with generating the control inputs. Such a control system which *minimizes the cost* associated with generating control inputs is called an *optimal control system*. In contrast to the pole-placement approach, where the desired performance is *indirectly* achieved through the location of closed-loop poles, the optimal control system *directly addresses* the desired performance objectives, while minimizing the control energy. This is done by formulating an *objective function* which must be *minimized* in the design process. However, one must know how the performance objectives can be precisely translated into the objective function, which usually requires some experience with a given system.

If we define a system's *transient energy* as the total energy of the system when it is undergoing the transient response, then a successful control system must have a transient energy which quickly decays to zero. The maximum value of the transient energy indicates the maximum overshoot, while the time taken by the transient energy to decay to zero indicates the settling time. By including the transient energy in the *objective function*, we can specify the values of the acceptable maximum overshoot and settling time. Similarly, the *control energy* must also be a part of the objective function that is to be minimized. It is clear that the total control energy and total transient energy can be found by integrating the control energy and transient energy, respectively, with respect to time. Therefore, the objective function for the optimal control problem must be a *time integral* of the sum of transient energy and control energy expressed as functions of time.

6.1.1 The general optimal control formulation for regulators

Consider a linear plant described by the following state-equation:

$$\mathbf{x}^{(1)}(t) = \mathbf{A}(t)\mathbf{x}(t) + \mathbf{B}(t)\mathbf{u}(t) \tag{6.1}$$

Note that we have deliberately chosen a *time-varying* plant in Eq. (6.1), because the optimal control problem is generally formulated for time-varying systems. For simplicity, suppose we would like to design a full-state feedback regulator for the plant described by Eq. (6.1) such that the control input vector is given by

$$\mathbf{u}(t) = -\mathbf{K}(t)\mathbf{x}(t) \tag{6.2}$$

The control law given by Eq. (6.2) is linear. Since the plant is also linear, the closed-loop control system would be linear. The *control energy* can be expressed as $\mathbf{u}^{\mathrm{T}}(t)\mathbf{R}(t)\mathbf{u}(t)$, where $\mathbf{R}(t)$ is a *square, symmetric matrix* called the *control cost matrix*. Such an expression for control energy is called a *quadratic form*, because the *scalar* function, $\mathbf{u}^{\mathrm{T}}(t)\mathbf{R}(t)\mathbf{u}(t)$, contains quadratic functions of the elements of $\mathbf{u}(t)$. Similarly, the *transient energy* can also be expressed in a quadratic form as $\mathbf{x}^{\mathrm{T}}(t)\mathbf{Q}(t)\mathbf{x}(t)$, where $\mathbf{Q}(t)$ is a *square, symmetric matrix* called the *state weighting matrix*. The *objective function* can then be written as follows:

$$J(t, t_f) = \int_t^{t_f} [\mathbf{x}^{\mathrm{T}}(\tau)\mathbf{Q}(\tau)\mathbf{x}(\tau) + \mathbf{u}^{\mathrm{T}}(\tau)\mathbf{R}(\tau)\mathbf{u}(\tau)] \, d\tau \tag{6.3}$$

where t and t_f are the *initial* and *final* times, respectively, for the control to be exercised, i.e. the control *begins* at $\tau = t$ and *ends* at $\tau = t_f$, where τ is the variable of integration. The optimal control problem consists of solving for the feedback gain matrix, $\mathbf{K}(t)$, such that the scalar objective function, $J(t, t_f)$, given by Eq. (6.3) is *minimized*. However, the minimization must be carried out in such a manner that the state-vector, $\mathbf{x}(t)$, is the solution of the plant's state-equation (Eq. (6.1)). Equation (6.1) is called a *constraint* (because in its absence, $\mathbf{x}(t)$ would be free to assume *any* value), and the resulting minimization is said to be a *constrained minimization*. Hence, we are looking for a regulator gain matrix, $\mathbf{K}(t)$, which *minimizes* $J(t, t_f)$ *subject to the constraint* given by Eq. (6.1). Note that the transient term, $\mathbf{x}^{\mathrm{T}}(\tau)\mathbf{Q}(\tau)\mathbf{x}(\tau)$, in the objective function implies that a *departure* of the system's state, $\mathbf{x}(\tau)$, from the *final desired state*, $\mathbf{x}(t_f) = \mathbf{0}$, is to be minimized. In other words, the design objective is to bring $\mathbf{x}(\tau)$ to a *constant value of zero* at final time, $\tau = t_f$. If the final desired state is *non-zero*, the objective function can be modified appropriately, as we will see later.

By substituting Eq. (6.2) into Eq. (6.1), the closed-loop state-equation can be written as follows:

$$\mathbf{x}^{(1)}(t) = [\mathbf{A}(t) - \mathbf{B}(t)\mathbf{K}(t)]\mathbf{x}(t) = \mathbf{A}_{\mathbf{CL}}(t)\mathbf{x}(t) \qquad (6.4)$$

where $\mathbf{A}_{\mathbf{CL}}(t) = [\mathbf{A}(t) - \mathbf{B}(t)\mathbf{K}(t)]$, the closed-loop state-dynamics matrix. The solution to Eq. (6.4) can be written as follows:

$$\mathbf{x}(t) = \Phi_{\mathbf{CL}}(t, t_0)\mathbf{x}(t_0) \qquad (6.5)$$

where $\Phi_{\mathbf{CL}}(t, t_0)$ is the *state-transition matrix* of the time-varying closed-loop system represented by Eq. (6.4). Since the system is time-varying, $\Phi_{\mathbf{CL}}(t, t_0)$, is *not* the matrix exponential of $\mathbf{A}_{\mathbf{CL}}(t - t_0)$, but is related in *some other way* (which we do not know) to $\mathbf{A}_{\mathbf{CL}}(t)$. Equation (6.5) indicates that the state at any time, $\mathbf{x}(t)$, can be obtained by post-multiplying the state at some initial time, $\mathbf{x}(t_0)$, with $\Phi_{\mathbf{CL}}(t, t_0)$. On substituting Eq. (6.5) into Eq. (6.3), we get the following expression for the objective function:

$$J(t, t_f) = \int_t^{t_f} \mathbf{x}^{\mathbf{T}}(t)\Phi_{\mathbf{CL}}^{\mathbf{T}}(\tau, t)[\mathbf{Q}(\tau) + \mathbf{K}^{\mathbf{T}}(\tau)\mathbf{R}(\tau)\mathbf{K}(\tau)]\Phi_{\mathbf{CL}}(\tau, t)\mathbf{x}(t) \, d\tau \qquad (6.6)$$

or, taking the *initial state-vector*, $\mathbf{x}(t)$, outside the integral sign, we can write

$$J(t, t_f) = \mathbf{x}^{\mathbf{T}}(t)\mathbf{M}(t, t_f)\mathbf{x}(t) \qquad (6.7)$$

where

$$\mathbf{M}(t, t_f) = \int_t^{t_f} \Phi_{\mathbf{CL}}^{\mathbf{T}}(\tau, t)[\mathbf{Q}(\tau) + \mathbf{K}^{\mathbf{T}}(\tau)\mathbf{R}(\tau)\mathbf{K}(\tau)]\Phi_{\mathbf{CL}}(\tau, t) \, d\tau \qquad (6.8)$$

Equation (6.7) shows that the objective function is a *quadratic function* of the initial state, $\mathbf{x}(t)$. Hence, the linear optimal regulator problem posed by Eqs. (6.1)–(6.3) is also called the *linear, quadratic regulator* (LQR) problem. You can easily show from Eq. (6.8) that $\mathbf{M}(t, t_f)$ is a symmetric matrix, i.e. $\mathbf{M}^{\mathbf{T}}(t, t_f) = \mathbf{M}(t, t_f)$, because both $\mathbf{Q}(t)$ and $\mathbf{R}(t)$ are symmetric. On substituting Eq. (6.5) into Eq. (6.6), we can write the objective function as follows:

$$J(t, t_f) = \int_t^{t_f} \mathbf{x}^{\mathbf{T}}(\tau)[\mathbf{Q}(\tau) + \mathbf{K}^{\mathbf{T}}(\tau)\mathbf{R}(\tau)\mathbf{K}(\tau)]\mathbf{x}(\tau) \, d\tau \qquad (6.9)$$

On differentiating Eq. (6.9) *partially* with respect to the lower limit of integration, t, according to the *Leibniz rule* (see a textbook on integral calculus, such as that by Kreyszig [1]), we get the following:

$$\partial J(t, t_f)/\partial t = -\mathbf{x}^{\mathbf{T}}(t)[\mathbf{Q}(t) + \mathbf{K}^{\mathbf{T}}(t)\mathbf{R}(t)\mathbf{K}(t)]\mathbf{x}(t) \qquad (6.10)$$

where ∂ denotes *partial differentiation*. Also, partial differentiation of Eq. (6.7) with respect to t results in the following:

$$\partial J(t, t_f)/\partial t = [\mathbf{x}^{(1)}(t)]^T \mathbf{M}(t, t_f)\mathbf{x}(t) + \mathbf{x}^{\mathbf{T}}(t)[\partial \mathbf{M}(t, t_f)/\partial t]\mathbf{x}(t) + \mathbf{x}^{\mathbf{T}}(t)\mathbf{M}(t, t_f)\mathbf{x}^{(1)}(t) \qquad (6.11)$$

On substituting $\mathbf{x}^{(1)}(t) = \mathbf{A}_{\mathrm{CL}}(t)\mathbf{x}(t)$ from Eq. (6.4) into Eq. (6.11), we can write

$$\partial J(t, t_f)/\partial t = \mathbf{x}^{\mathrm{T}}(t)[\mathbf{A}_{\mathrm{CL}}^{\mathrm{T}}(t)\mathbf{M}(t, t_f) + \partial \mathbf{M}(t, t_f)/\partial t + \mathbf{M}(t, t_f)\mathbf{A}_{\mathrm{CL}}(t)]\mathbf{x}(t) \qquad (6.12)$$

Equations (6.10) and (6.12) are *quadratic forms* for the *same* scalar function, $\partial J(t, t_f)/\partial t$ in terms of the *initial state*, $\mathbf{x}(t)$. Equating Eqs. (6.10) and (6.12), we get the following *matrix differential equation* to be satisfied by $\mathbf{M}(t, t_f)$:

$$-[\mathbf{Q}(t) + \mathbf{K}^{\mathrm{T}}(t)\mathbf{R}(t)\mathbf{K}(t)] = \mathbf{A}_{\mathrm{CL}}^{\mathrm{T}}(t)\mathbf{M}(t, t_f) + \partial \mathbf{M}(t, t_f)/\partial t + \mathbf{M}(t, t_f)\mathbf{A}_{\mathrm{CL}}(t) \quad (6.13)$$

or

$$-\partial \mathbf{M}(t, t_f)/\partial t = \mathbf{A}_{\mathrm{CL}}^{\mathrm{T}}(t)\mathbf{M}(t, t_f) + \mathbf{M}(t, t_f)\mathbf{A}_{\mathrm{CL}}(t) + [\mathbf{Q}(t) + \mathbf{K}^{\mathrm{T}}(t)\mathbf{R}(t)\mathbf{K}(t)] \quad (6.14)$$

Equation (6.14) is a first order, *matrix partial differential equation* in terms of the initial time, t, whose solution $\mathbf{M}(t, t_f)$ is given by Eq. (6.8). However, since we do not know the state transition matrix, $\mathbf{\Phi}_{\mathrm{CL}}(\tau, t)$, of the general time-varying, closed-loop system, Eq. (6.8) is useless to us for determining $\mathbf{M}(t, t_f)$. Hence, the *only way* to find the unknown matrix $\mathbf{M}(t, t_f)$ is by solving the *matrix differential equation*, Eq. (6.14). We need *only one initial condition* to solve the *first order* matrix differential equation, Eq. (6.14). The simplest initial condition can be obtained by putting $t = t_f$ in Eq. (6.8), resulting in

$$\mathbf{M}(t_f, t_f) = \mathbf{0} \qquad (6.15)$$

The linear optimal control problem is thus posed as finding the *optimal regulator gain matrix*, $\mathbf{K}(t)$, such that the solution, $\mathbf{M}(t, t_f)$, to Eq. (6.14) (and hence the objective function, $J(t, t_f)$) is *minimized*, subject to the *initial condition*, Eq. (6.15). The choice of the matrices $\mathbf{Q}(t)$ and $\mathbf{R}(t)$ is left to the designer. However, as we will see below, these two matrices specifying performance objectives and control effort, cannot be arbitrary, but must obey certain conditions.

6.1.2 Optimal regulator gain matrix and the Riccati equation

Let us denote the *optimal feedback gain matrix* that *minimizes* $\mathbf{M}(t, t_f)$ by $\mathbf{K_o}(t)$. The *minimum* value of $\mathbf{M}(t, t_f)$ which results from the optimal gain matrix, $\mathbf{K_o}(t)$, is denoted by $\mathbf{M_o}(t, t_f)$, and the minimum value of the objective function is denoted by $J_o(t, t_f)$. For simplicity of notation, let us drop the functional arguments for the time being, and denote $\mathbf{M}(t, t_f)$ by $\mathbf{M}$, $J(t, t_f)$ by J, etc. Then, according to Eq. (6.7), the minimum value of the objective function is the following:

$$J_o = \mathbf{x}^T(t)\mathbf{M_o}\mathbf{x}(t) \qquad (6.16)$$

Since J_o is the minimum value of J for any initial state, $\mathbf{x}(t)$, we can write $J_o \leq J$, or

$$\mathbf{x}^T(t)\mathbf{M_o}\mathbf{x}(t) \leq \mathbf{x}^T(t)\mathbf{M}\mathbf{x}(t) \qquad (6.17)$$

If we express $\mathbf{M}$ as follows:

$$\mathbf{M} = \mathbf{M_o} + \mathbf{m} \tag{6.18}$$

and substitute Eq. (6.18) into Eq. (6.17), the following condition must be satisfied:

$$\mathbf{x}^T(t)\mathbf{M_o}\mathbf{x}(t) \leq \mathbf{x}^T(t)\mathbf{M_o}\mathbf{x}(t) + \mathbf{x}^T(t)\mathbf{m}\mathbf{x}(t) \tag{6.19}$$

or

$$\mathbf{x}^T(t)\mathbf{m}\mathbf{x}(t) \geq 0 \tag{6.20}$$

A matrix, $\mathbf{m}$, which satisfies Eq. (6.20), is called a *positive semi-definite* matrix. Since $\mathbf{x}(t)$ is an *arbitrary* initial state-vector, you can show that according to Eq. (6.20), *all eigenvalues of* $\mathbf{m}$ must be *greater than or equal to zero*.

It now remains to derive an expression for the optimum regulator gain matrix, $\mathbf{K_o}(t)$, such that $\mathbf{M}$ is minimized. If $\mathbf{M_o}$ is the minimum value of $\mathbf{M}$, then $\mathbf{M_o}$ must satisfy Eq. (6.14) when $\mathbf{K}(t) = \mathbf{K_o}(t)$, i.e.

$$-\partial \mathbf{M_o}/\partial t = \mathbf{A}_{CL}^T(t)\mathbf{M_o} + \mathbf{M_o}\mathbf{A}_{CL}(t) + [\mathbf{Q}(t) + \mathbf{K_o^T}(t)\mathbf{R}(t)\mathbf{K_o}(t)] \tag{6.21}$$

Let us express the gain matrix, $\mathbf{K}(t)$, in terms of the optimal gain matrix, $\mathbf{K_o}(t)$, as follows:

$$\mathbf{K}(t) = \mathbf{K_o}(t) + \mathbf{k}(t) \tag{6.22}$$

On substituting Eqs. (6.18) and (6.21) into Eq. (6.14), we can write

$$-\partial(\mathbf{M_o} + \mathbf{m})/\partial t = \mathbf{A}_{CL}^T(t)(\mathbf{M_o} + \mathbf{m}) + (\mathbf{M_o} + \mathbf{m})\mathbf{A}_{CL}(t)$$
$$+ [\mathbf{Q}(t) + \{\mathbf{K_o}(t) + \mathbf{k}(t)\}^T \mathbf{R}(t)\{\mathbf{K_o}(t) + \mathbf{k}(t)\}] \tag{6.23}$$

On subtracting Eq. (6.21) from Eq. (6.23), we get

$$-\partial \mathbf{m}/\partial t = \mathbf{A}_{CL}^T(t)\mathbf{m} + \mathbf{m}\mathbf{A}_{CL}(t) + \mathbf{S} \tag{6.24}$$

where

$$\mathbf{S} = [\mathbf{K_o^T}(t)\mathbf{R}(t) - \mathbf{M_o}\mathbf{B}(t)]\mathbf{k}(t) + \mathbf{k}^T(t)[\mathbf{R}(t)\mathbf{K_o}(t) - \mathbf{B}^T(t)\mathbf{M_o}] + \mathbf{k}^T(t)\mathbf{R}(t)\mathbf{k}(t) \tag{6.25}$$

Comparing Eq. (6.24) with Eq. (6.14), we find that the two equations are of the *same form*, with the term $[\mathbf{Q}(t) + \mathbf{K}^T(t)\mathbf{R}(t)\mathbf{K}(t)]$ in Eq. (6.14) replaced by $\mathbf{S}$ in Eq. (6.24). Since the non-optimal matrix, $\mathbf{M}$, in Eq. (6.14) satisfies Eq. (6.8), it must be true that $\mathbf{m}$ satisfies the following equation:

$$\mathbf{m}(t, t_f) = \int_t^{t_f} \mathbf{\Phi}_{CL}^T(\tau, t)\mathbf{S}(\tau, t_f)\mathbf{\Phi}_{CL}(\tau, t)\, d\tau \tag{6.26}$$

Recall from Eq. (6.20) that $\mathbf{m}$ must be positive semi-definite. However, Eq. (6.26) requires that for $\mathbf{m}$ to be positive semi-definite, the matrix $\mathbf{S}$ given by Eq. (6.25) must be positive semi-definite. Looking at Eq. (6.25), we find that $\mathbf{S}$ can be positive semi-definite if and only if the *linear terms* in Eq. (6.25) are zeros, i.e. which implies

$$\mathbf{K_o^T}(t)\mathbf{R}(t) - \mathbf{M_o}\mathbf{B}(t) = \mathbf{0} \qquad (6.27)$$

or the optimal feedback gain matrix is given by

$$\mathbf{K_o}(t) = \mathbf{R}^{-1}(t)\mathbf{B^T}(t)\mathbf{M_o} \qquad (6.28)$$

Substituting Eq. (6.28) into Eq. (6.21), we get the following differential equation to be satisfied by the optimal matrix, $\mathbf{M_o}$:

$$-\partial \mathbf{M_o}/\partial t = \mathbf{A^T}(t)\mathbf{M_o} + \mathbf{M_o}\mathbf{A}(t) - \mathbf{M_o}\mathbf{B}(t)\mathbf{R}^{-1}(t)\mathbf{B^T}(t)\mathbf{M_o} + \mathbf{Q}(t) \qquad (6.29)$$

Equation (6.29) has a special name: the *matrix Riccati equation*. The matrix Riccati equation is special because it's solution, $\mathbf{M_o}$, substituted into Eq. (6.28), gives us the optimal feedback gain matrix, $\mathbf{K_o}(t)$. Exact solutions to the Riccati equation are rare, and in most cases a numerical solution procedure is required. Note that Riccati equation is a first order, nonlinear differential equation, and can be solved by numerical methods similar to those discussed in Chapter 4 for solving the nonlinear state-equations, such as the *Runge–Kutta* method, or other more convenient methods (such as the one we will discuss in Section 6.5). However, in contrast to the state-equation, the solution is a matrix rather than a vector, and the solution procedure has to *march backwards in time*, since the *initial condition* for Riccati equation is specified (Eq. (6.15)) at the *final time*, $t = t_f$, as follows:

$$\mathbf{M_o}(t_f, t_f) = \mathbf{0} \qquad (6.30)$$

For this reason, the condition given by Eq. (6.30) is called the *terminal condition* rather than initial condition. Note that the solution to Eq. (6.29) is $\mathbf{M_o}(t, t_f)$ where $t < t_f$. Let us defer the solution to the matrix Riccati equation until Section 6.5.

In summary, the optimal control procedure using full-state feedback consists of specifying an objective function by suitably selecting the performance and control cost weighting matrices, $\mathbf{Q}(t)$ and $\mathbf{R}(t)$, and solving the Riccati equation subject to the terminal condition, in order to determine the full-state feedback matrix, $\mathbf{K_o}(t)$. In most cases, rather than solving the general time-varying optimal control problem, certain simplifications can be made which result in an easier problem, as seen in the following sections.

6.2 Infinite-Time Linear Optimal Regulator Design

A large number of control problems are such that the control interval, $(t_f - t)$, is *infinite*. If we are interested in a specific *steady-state* behavior of the control system, we are interested in the response, $\mathbf{x}(t)$, when $t_f \rightarrow \infty$, and hence the control interval is *infinite*. The approximation of an infinite control interval results in a simplification in the optimal control problem, as we will see below. For infinite final time, the quadratic objective function can be expressed as follows:

$$J_\infty(t) = \int_t^\infty [\mathbf{x}^T(\tau)\mathbf{Q}(\tau)\mathbf{x}(\tau) + \mathbf{u}^T(\tau)\mathbf{R}(\tau)\mathbf{u}(\tau)]\, d\tau \tag{6.31}$$

where $J_\infty(t)$ indicates the objective function of the infinite final time (or steady-state) optimal control problem. For the infinite final time, the *backward time integration* of the matrix Riccati equation (Eq. (6.29)), beginning from $\mathbf{M_o}(\infty, \infty) = \mathbf{0}$, would result in a solution, $\mathbf{M_o}(t, \infty)$, which is *either* a *constant*, or *does not converge* to any limit. If the numerical solution to the Riccati equation converges to a constant value, then $\partial \mathbf{M_o}/\partial t = \mathbf{0}$, and the Riccati equation becomes

$$0 = \mathbf{A}^T(t)\mathbf{M_o} + \mathbf{M_o}\mathbf{A}(t) - \mathbf{M_o}\mathbf{B}(t)\mathbf{R}^{-1}(t)\mathbf{B}^T(t)\mathbf{M_o} + \mathbf{Q}(t) \tag{6.32}$$

Note that Eq. (6.32) is no longer a differential equation, but an *algebraic equation*. Hence, Eq. (6.32) is called the *algebraic Riccati equation*. The feedback gain matrix is given by Eq. (6.28), in which $\mathbf{M_o}$ is the (constant) solution to the algebraic Riccati equation. It is (relatively) much easier to solve Eq. (6.32) rather that Eq. (6.29). However, a solution to the algebraic Riccati equation *may not* always exist.

What are the conditions for the existence of the positive semi-definite solution to the algebraic Riccati equation? This question is best answered in a textbook devoted to optimal control, such as that by Bryson and Ho [2], and involves precise mathematical conditions, such as *stabilizability, detectability*, etc., for the existence of solution. Here, it suffices to say that for all practical purposes, if *either* the plant is *asymptotically stable, or* the plant is *controllable* and *observable* with the output, $\mathbf{y}(t) = \mathbf{C}(t)\mathbf{x}(t)$, where $\mathbf{C}^T(t)\mathbf{C}(t) = \mathbf{Q}(t)$, and $\mathbf{R}(t)$ is a symmetric, *positive definite* matrix, then there is a *unique, positive definite* solution, $\mathbf{M_o}$, to the algebraic Riccati equation. Note that $\mathbf{C}^T(t)\mathbf{C}(t) = \mathbf{Q}(t)$ implies that $\mathbf{Q}(t)$ must be a *symmetric* and *positive semi-definite* matrix. Furthermore, the requirement that the control cost matrix, $\mathbf{R}(t)$, must be *symmetric* and *positive definite* (i.e. *all eigenvalues* of $\mathbf{R}(t)$ must be *positive* real numbers) for the solution, $\mathbf{M_o}$ to be *positive definite* is clear from Eq. (6.25), which implies that $\mathbf{S}$ (and hence $\mathbf{m}$) will be positive definite only if $\mathbf{R}(t)$ is positive definite. Note that these are *sufficient* (but not *necessary*) conditions for the existence of a unique solution to the algebraic Riccati equation, i.e. there may be plants that *do not* satisfy these conditions, and yet there *may exist* a unique, positive definite solution for such plants. A less restrictive set of sufficient conditions for the existence of a unique, positive definite solution to the algebraic Riccati equation is that the plant must be *stabilizable* and *detectable* with the output, $\mathbf{y}(t) = \mathbf{C}(t)\mathbf{x}(t)$, where $\mathbf{C}^T(t)\mathbf{C}(t) = \mathbf{Q}(t)$, and $\mathbf{R}(t)$ is a symmetric, *positive definite* matrix (see Bryson and Ho [2] for details).

While Eq. (6.32) has been derived for linear optimal control of time-varying plants, its usual application is to *time-invariant* plants, for which the algebraic Riccati equation is written as follows:

$$0 = \mathbf{A}^T\mathbf{M_o} + \mathbf{M_o}\mathbf{A} - \mathbf{M_o}\mathbf{B}\mathbf{R}^{-1}\mathbf{B}^T\mathbf{M_o} + \mathbf{Q} \tag{6.33}$$

In Eq. (6.33), all the matrices are *constant* matrices. MATLAB contains a solver for the algebraic Riccati equation for time-invariant plants in the M-file named *are.m*. The command *are* is used as follows:

```
>>x = are(a,b,c) <enter>
```

where $\mathbf{a} = \mathbf{A}$, $\mathbf{b} = \mathbf{BR}^{-1}\mathbf{B}^{\mathrm{T}}$, $\mathbf{c} = \mathbf{Q}$, in Eq. (6.33), and the returned solution is $\mathbf{x} = \mathbf{M_o}$. For the existence of a unique, positive definite solution to Eq. (6.33), the sufficient conditions remains the same, i.e. the plant with coefficient matrices $\mathbf{A}$, $\mathbf{B}$ must be controllable, $\mathbf{Q}$ must be symmetric and positive semi-definite, and $\mathbf{R}$ must be symmetric and positive definite. Another MATLAB function, *ric*, computes the error in solving the algebraic Riccati equation. Alternatively, MATLAB's Control System Toolbox (CST) provides the functions *lqr* and *lqr2* for the solution of the linear optimal control problem with a quadratic objective function, using two different numerical schemes. The command *lqr* (or *lqr2*) is used as follows:

```
>>[Ko,Mo,E]= lqr(A,B,Q,R) <enter>
```

where $\mathbf{A}$, $\mathbf{B}$, $\mathbf{Q}$, $\mathbf{R}$ are the same as in Eq. (6.33), $\mathbf{Mo} = \mathbf{M_o}$, the returned solution of Eq. (6.33), $\mathbf{Ko} = \mathbf{K_o} = \mathbf{R}^{-1}\mathbf{B}^{\mathrm{T}}\mathbf{M_o}$ the returned optimal regulator gain matrix, and $\mathbf{E}$ is the vector containing the closed-loop eigenvalues (i.e. the eigenvalues of $\mathbf{A_{CL}} = \mathbf{A} - \mathbf{BK_o}$). The command *lqr* (or *lqr2*) is more convenient to use, since it directly works with the plant's coefficient matrices and the weighting matrices. Let us consider a few examples of linear optimal control of time-invariant plants, based upon the solution of the algebraic Riccati equation. (For time-varying plants, the optimal feedback gain matrix can be determined by solving the algebraic Riccati equation at each instant of time, t, using either *lqr* or *lqr2* in a time-marching procedure.)

Example 6.1

Consider the longitudinal motion of a flexible bomber aircraft of Example 4.7. The sixth order, two input system is described by the linear, time-invariant, state-space representation given by Eq. (4.71). The inputs are the *desired elevator deflection* (rad.), $u_1(t)$, and the *desired canard deflection* (rad.), $u_2(t)$, while the outputs are the *normal acceleration* (m/s^2), $y_1(t)$, and the *pitch-rate* (rad./s), $y_2(t)$. Let us design an optimal regulator which would produce a maximum overshoot of less than ± 2 m/s^2 in the normal-acceleration and less than ± 0.03 rad/s in pitch-rate, and a settling time less than 5 s, while requiring elevator and canard deflections *not exceeding* ± 0.1 rad. (5.73°), if the initial condition is 0.1 rad/s perturbation in the pitch-rate, i.e. $\mathbf{x}(0) = [0; 0.1; 0; 0; 0; 0]^T$.

What $\mathbf{Q}$ and $\mathbf{R}$ matrices should we choose for this problem? Note that $\mathbf{Q}$ is a square matrix of size (6×6) and $\mathbf{R}$ is a square matrix of size (2×2). Examining the plant model given by Eq. (4.71), we find that while the normal acceleration, $y_1(t)$, depends upon all the six state variables, the pitch-rate, $y_2(t)$, is equal to the second state-variable. Since we have to enforce the maximum overshoot limits on $y_1(t)$ and $y_2(t)$, we must, therefore, impose certain limits on the maximum overshoots of all the state variables, which is done by selecting an appropriate state weighting matrix, $\mathbf{Q}$. Similarly, the maximum overshoot limits on the two input variables, $u_1(t)$ and $u_2(t)$, must be specified through the control cost matrix, $\mathbf{R}$. The settling time would be determined by both $\mathbf{Q}$ and $\mathbf{R}$. *A priori*, we do not quite know what values of $\mathbf{Q}$ and $\mathbf{R}$ will produce the desired objectives. Hence, some trial and error is required in selecting the appropriate $\mathbf{Q}$ and $\mathbf{R}$. Let us begin by selecting both $\mathbf{Q}$ and $\mathbf{R}$ as *identity matrices*. By doing so, we are specifying that all the six state variables and the two control inputs are *equally important* in the objective function, i.e. it is equally important to bring all the state variables and the control inputs to zero, while minimizing their overshoots. Note that the existence of a unique, positive definite solution to the algebraic Riccati equation will be guaranteed if $\mathbf{Q}$ and $\mathbf{R}$ are positive semi-definite and positive definite, respectively, and the plant is controllable. Let us test whether the plant is controllable as follows:

```
>>rank(ctrb(A,B)) <enter>

ans=
  6
```

Hence, the plant is controllable. By choosing $\mathbf{Q} = \mathbf{I}$, and $\mathbf{R} = \mathbf{I}$, we are ensuring that both are positive definite. Therefore, all the sufficient conditions for the existence of an optimal solution are satisfied. For solving the algebraic Riccati equation, let us use the MATLAB command *lqr* as follows:

```
>>[Ko,Mo,E]=lqr(A,B,eye(6),eye(2)) <enter>

Ko=
 3.3571e+000  -4.2509e-001  -6.2538e-001  -7.3441e-001  2.8190e+000  1.5765e+000
 3.8181e+000   1.0274e+000  -5.4727e-001  -6.8075e-001  2.1020e+000  1.8500e+000


Mo=
 1.7429e+000   2.8673e-001   1.1059e-002  -1.4159e-002  4.4761e-002   3.8181e-002
 2.8673e-001   4.1486e-001   1.0094e-002  -2.1528e-003  -5.6679e-003  1.0274e-002
 1.1059e-002   1.0094e-002   1.0053e+000   4.4217e-003  -8.3383e-003  -5.4727e-003
-1.4159e-002  -2.1528e-003   4.4217e-003   4.9047e-003  -9.7921e-003  -6.8075e-003
 4.4761e-002  -5.6679e-003  -8.3383e-003  -9.7921e-003  3.7586e-002   2.1020e-002
 3.8181e-002   1.0274e-002  -5.4727e-003  -6.8075e-003  2.1020e-002   1.8500e-002

E =
-2.2149e+002+2.0338e+002i
-2.2149e+002-2.0338e+002i
-1.2561e+002
-1.8483e+000+1.3383e+000i
-1.8483e+000-1.3383e+000i
-1.0011e+000
```

To see whether this design is acceptable, we calculate the initial response of the closed-loop system as follows:

```
>>sys1=ss(A-B*Ko,zeros(6,2),C,zeros(2,2));<enter>

>>[Y1,t1,X1]=initial(sys1,[0.1 zeros(1,5)]'); u1=-Ko*X1'; <enter>
```

Let us try another design with $\mathbf{Q} = 0.01\mathbf{I}$, and $\mathbf{R} = \mathbf{I}$. As compared with the previous design, we are now specifying that it is 100 times *more important* to minimize the total control energy than minimizing the total transient energy. The new regulator gain matrix is determined by re-solving the algebraic Riccati equation with $\mathbf{Q} = 0.01\mathbf{I}$ and $\mathbf{R} = \mathbf{I}$ as follows:

```
>>[Ko,Mo,E] = lqr(A,B,0.01*eye(6),eye(2)) <enter>

Ko=
 1.0780e+000  -1.6677e-001  -4.6948e-002  -7.5618e-002  5.9823e-001  3.5302e-001
 1.3785e+000  3.4502e-001   -1.3144e-002  -6.5260e-002  4.7069e-001  3.0941e-001

Mo=
 4.1913e-001   1.2057e-001   9.2728e-003   -2.2727e-003  1.4373e-002   1.3785e-002
 1.2057e-001   1.0336e-001   6.1906e-003   -3.9125e-004  -2.2236e-003  3.4502e-003
 9.2728e-003   6.1906e-003   1.0649e-002   9.7083e-005   -6.2597e-004  -1.3144e-004
-2.2727e-003   -3.9125e-004  9.7083e-005   1.7764e-004   -1.0082e-003  -6.5260e-004
 1.4373e-002   -2.2236e-003  -6.2597e-004  -1.0082e-003  7.9764e-003   4.7069e-003
 1.3785e-002   3.4502e-003   -1.3144e-004  -6.5260e-004  4.7069e-003   3.0941e-003

E =
-9.1803e+001
-7.8748e+001+5.0625e+001i
-7.8748e+001-5.0625e+001i
-1.1602e+000+1.7328e+000i
-1.1602e+000-1.7328e+000i
-1.0560e+000
```

The closed-loop state-space model, closed-loop initial response and the required inputs are calculated as follows:

```
>>sys2=ss(A-B*Ko,zeros(6,2),C,zeros(2,2)); <enter>

>>[Y2,t2,X2] = initial(sys2,[0.1 zeros(1,5)]'); u2=-Ko*X2'; <enter>
```

Note that the closed-loop eigenvalues (contained in the returned matrix $\mathbf{E}$) of the first design are *further inside* the left-half plane than those of the second design, which indicates that the first design would have a *smaller* settling time, and a *larger* input requirement when compared to the second design. The resulting outputs, $y_1(t)$ and $y_2(t)$, for the two regulator designs are compared with the plant's initial response to the same initial condition in Figure 6.1.

The plant's oscillating initial response is seen in Figure 6.1 to have maximum overshoots of -20 m/s^2 and -0.06 rad/s, for $y_1(t)$ and $y_2(t)$, respectively, and a settling time exceeding 5 s (actually about 10 s). Note in Figure 6.1 that while the first design ($\mathbf{Q} = \mathbf{I}, \mathbf{R} = \mathbf{I}$) produces the closed-loop initial response of $y_2(t)$, $u_1(t)$, and $u_2(t)$ within acceptable limits, the response of $y_1(t)$ displays a maximum overshoot of 10 m/s^2 (beginning at -15 m/s^2 at $t = 0$, and shooting to -25 m/s^2), which is unacceptable. The settling time of the first design is about 3 s, while that of the second design ($\mathbf{Q} = 0.01\mathbf{I}, \mathbf{R} = \mathbf{I}$) is slightly less than 5 s. The second design produces a maximum overshoot of $y_1(t)$ less than 2 m/s^2 and that of $y_2(t)$ about -0.025 rad/s, which is acceptable. The required control inputs, $u_1(t)$ and $u_2(t)$, for the two designs are plotted in Figure 6.2. While the first design requires a maximum

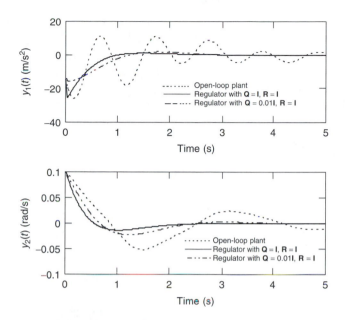

Figure 6.1 Open and closed-loop initial response of the regulated flexible bomber aircraft, for two optimal regulator designs

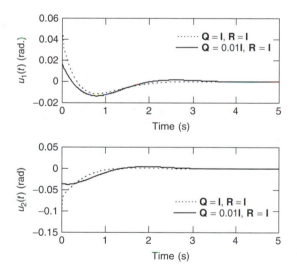

Figure 6.2 Required initial response control inputs of the regulated flexible bomber aircraft, for two optimal regulator designs

value of elevator deflection, $u_1(t)$, about 0.045 rad., the second design is seen to require a maximum value of $u_1(t)$ less than 0.02 rad. Similarly, the canard deflection, $u_2(t)$, for the second design has a smaller maximum value (-0.04 rad) than that of the first design (-0.1 rad.). Hence, the second design fulfills all the design objectives.

In Example 6.1, we find that the total transient energy is *more sensitive* to the *settling time*, than the maximum overshoot. Recall from Chapter 5 that if we try to reduce the settling time, we have to accept an increase in the maximum overshoot. Conversely, to reduce the maximum overshoot of $y_1(t)$, which depends upon all the state variables, we must allow an increase in the settling time, which is achieved in the second design by reducing the importance of minimizing the transient energy by hundred-fold, as compared to the first design. Let us now see what effect a measurement noise will have on the closed-loop initial response. We take the second regulator design (i.e. $\mathbf{Q} = 0.01\mathbf{I}$, $\mathbf{R} = \mathbf{I}$) and simulate the initial response assuming a random error (i.e. measurement noise) in feeding back the pitch-rate (the second state-variable of the plant). The simulation is carried out using SIMULINK block-diagram shown in Figure 6.3, where the measurement noise is simulated by the *band-limited white noise* block with a *power* parameter of 10^{-4}. Note the

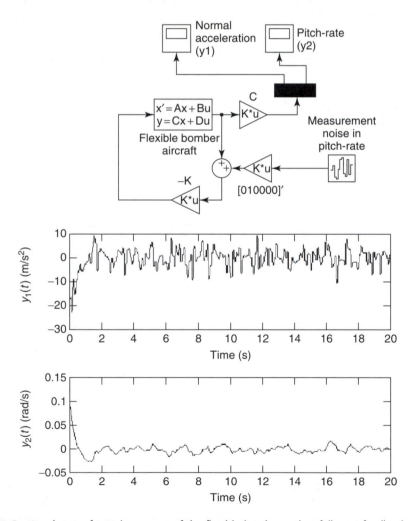

Figure 6.3 Simulation of initial response of the flexible bomber with a full-state feedback regulator and measurement noise in the pitch-rate channel

manner in which the noise is added to the feedback loop through the *matrix gain* block. The simulated initial response is also shown in Figure 6.3. Note the random fluctuations in both normal acceleration, $y_1(t)$, and pitch-rate, $y_2(t)$. The aircraft crew are likely to have a rough ride due to large sustained fluctuations (± 10 m/s^2) in normal acceleration, $y_1(t)$, resulting from the small measurement noise! The feedback loop results in an amplification of the measurement noise. If the elements of the feedback gain matrix, **K**, corresponding to pitch-rate are reduced in magnitude then the noise due to pitch-rate feedback will be alleviated. Alternatively, pitch-rate (or any other state-variable that is noisy) can be removed from state feedback, with the use of an observer based compensator that feeds back only selected state-variables (see Chapter 5).

Example 6.2

Let us design an optimal regulator for the flexible, rotating spacecraft shown in Figure 6.4. The spacecraft consists of a *rigid hub* and four *flexible appendages*, each having a *tip mass*, with three *torque* inputs, $u_1(t)$, $u_2(t)$, $u_3(t)$, and three *angular rotation* outputs in rad., $y_1(t)$, $y_2(t)$, $y_3(t)$. Due to the flexibility of the appendages, the spacecraft is a *distributed parameter* system (see Chapter 1). However, it is approximated by a *lumped parameter*, linear, time-invariant state-space representation using a *finite-element model* [3]. The order of the spacecraft can be reduced to 26 for accuracy in a desired frequency range [4]. The 26th order state-vector, $\mathbf{x}(t)$, of the spacecraft consists of the angular displacement, $y_1(t)$, and *angular velocity* of the rigid hub, combined with individual *transverse* (i.e. *perpendicular* to the appendage) *displacements* and *transverse velocities* of three points on each appendage. The state-coefficient matrices of the spacecraft are given as follows:

$$\mathbf{A} = \begin{bmatrix} \mathbf{0} & \mathbf{I} \\ -\mathbf{M}^{-1}\mathbf{K} & \mathbf{0} \end{bmatrix} \quad \mathbf{B} = \begin{bmatrix} \mathbf{0} \\ \mathbf{M}^{-1}\mathbf{d} \end{bmatrix} \quad \mathbf{C} = [\,\mathbf{d}^{\mathrm{T}}; \quad \mathbf{0}\,]; \quad \mathbf{D} = \mathbf{0} \quad (6.34)$$

where **M, K**, and **d** are the *mass, stiffness,* and *control influence* matrices, given in Appendix C.

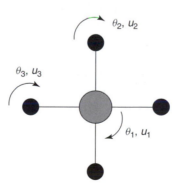

Figure 6.4 A rotating, flexible spacecraft with three inputs, (u_1, u_2, u_3), and three outputs $(\theta_1, \theta_2, \theta_3)$

The eigenvalues of the spacecraft are the following:

```
>>damp(A) <enter>

Eigenvalue                    Damping         Freq.(rad/sec)
9.4299e-013+7.6163e+003i      -6.1230e-017    7.6163e+003
9.4299e-013-7.6163e+003i      -6.1230e-017    7.6163e+003
8.1712e-013+4.7565e+004i      -6.1230e-017    4.7565e+004
8.1712e-013-4.7565e+004i      -6.1230e-017    4.7565e+004
8.0539e-013+2.5366e+003i      -2.8327e-016    2.5366e+003
8.0539e-013-2.5366e+003i      -2.8327e-016    2.5366e+003
7.4867e-013+4.7588e+004i      -6.1230e-017    4.7588e+004
7.4867e-013-4.7588e+004i      -6.1230e-017    4.7588e+004
7.1276e-013+2.5982e+004i      -6.1230e-017    2.5982e+004
7.1276e-013-2.5982e+004i      -6.1230e-017    2.5982e+004
5.2054e-013+1.4871e+004i      -6.1230e-017    1.4871e+004
5.2054e-013-1.4871e+004i      -6.1230e-017    1.4871e+004
4.8110e-013+2.0986e+002i      -2.2817e-015    2.0986e+002
4.8110e-013-2.0986e+002i      -2.2817e-015    2.0986e+002
4.4812e-013+2.6009e+004i      -6.1230e-017    2.6009e+004
4.4812e-013-2.6009e+004i      -6.1230e-017    2.6009e+004
3.1387e-013+7.5783e+003i      -6.1230e-017    7.5783e+003
3.1387e-013-7.5783e+003i      -6.1230e-017    7.5783e+003
2.4454e-013+3.7952e+002i      -7.2736e-016    3.7952e+002
2.4454e-013-3.7952e+002i      -7.2736e-016    3.7952e+002
         0                    -1.0000e+000    0
         0                    -1.0000e+000    0
-9.9504e-013+2.4715e+003i      3.8286e-016    2.4715e+003
-9.9504e-013-2.4715e+003i      3.8286e-016    2.4715e+003
-1.1766e-012+1.4892e+004i      1.6081e-016    1.4892e+004
-1.1766e-012-1.4892e+004i      1.6081e-016    1.4892e+004
```

Clearly, the spacecraft is *unstable* due to a pair of zero eigenvalues (we can ignore the negligible, positive real parts of some eigenvalues, and assume that those real parts are zeros). The natural frequencies of the spacecraft range from 0 to 47 588 rad/s. The nonzero natural frequencies denote structural vibration of the spacecraft. The control objective is to design a controller which stabilizes the spacecraft, and brings the transient response to zero within 5 s, with zero maximum overshoot, while requiring input torques not exceeding 0.1 N-m, when the spacecraft is initially perturbed by a hub rotation of 0.01 rad. due to the movement of astronauts. The initial condition corresponding to the initial perturbation caused by the astronauts' movement is $\mathbf{x}(0) = [0.01;\ \text{zeros}(1,25)]^T$. Let us see whether the spacecraft is controllable:

```
>>rank(ctrb(A,B)) <enter>

ans=

6
```

Since the rank of the controllability test matrix is *less than* 26, the order of the plant, it follows that the spacecraft is *uncontrollable*. The *uncontrollable* modes are the structural vibration modes, while the *unstable* mode is the rigid-body rotation with zero natural frequency. Hence, the spacecraft is *stabilizable* and an optimal regulator can be designed for the spacecraft, since stabilizability of the plant is a sufficient condition for the existence of a unique, positive definite solution to the algebraic Riccati equation. Let us select $\mathbf{Q} = 200\mathbf{I}$, and $\mathbf{R} = \mathbf{I}$, noting that the size of $\mathbf{Q}$ is (26×26) while that of $\mathbf{R}$ is (3×3), and solve the Riccati equation using *lqr* as follows:

```
>>[Ko,Mo,E] = lqr(A,B,200*eye(26),eye(3)); <enter>
```

A positive definite solution to the algebraic Riccati equation *exists* for the present choice of $\mathbf{Q}$ and $\mathbf{R}$, *even though the plant is uncontrollable*. Due to the size of the plant, we avoid printing the solution, $\mathbf{Mo}$, and the optimal feedback gain matrix, $\mathbf{Ko}$, here, but the closed-loop eigenvalues, $\mathbf{E}$, are the following:

```
E =
-1.7321e+003+4.7553e+004i
-1.7321e+003-4.7553e+004i
-1.7502e+003+4.7529e+004i
-1.7502e+003-4.7529e+004i
-1.8970e+003+2.5943e+004i
-1.8970e+003-2.5943e+004i
-1.8991e+003+2.5916e+004i
-1.8991e+003-2.5916e+004i
-1.8081e+003+1.4569e+004i
-1.8081e+003-1.4569e+004i
-1.8147e+003+1.4550e+004i
-1.8147e+003-1.4550e+004i
-7.3743e+002+7.6536e+003i
-7.3743e+002-7.6536e+003i
-7.3328e+002+7.6142e+003i
-7.3328e+002-7.6142e+003i
-2.6794e+002+2.5348e+003i
-2.6794e+002-2.5348e+003i
-2.5808e+002+2.4698e+003i
-2.5808e+002-2.4698e+003i
-3.9190e+001+3.7744e+002i
-3.9190e+001-3.7744e+002i
-1.1482e+000+4.3165e-001i
-1.1482e+000-4.3165e-001i
-1.8066e+001+2.0911e+002i
-1.8066e+001-2.0911e+002i
```

All the closed-loop eigenvalues (contained in the vector $\mathbf{E}$) have negative real-parts, indicating that the closed-loop system is asymptotically stable, which is a bonus! Let us check whether the performance objectives are met by this design by calculating the closed-loop initial response as follows:

```
>>sysCL=ss(A-B*Ko,zeros(26,3),C,D); <enter>

>>[y,t,X] = initial(sysCL, [0.01 zeros(1, 25)]'); u = -Ko*X'; <enter>
```

The closed-loop outputs, $y_1(t)$, $y_2(t)$, and $y_3(t)$, and the required torque inputs, $u_1(t)$, $u_2(t)$, and $u_3(t)$, are plotted in Figure 6.5. We see from Figure 6.5 that all the three outputs settle to zero in 5 s, with zero overshoot, and that the input torque magnitudes are smaller than 0.1 N-m, as desired. Therefore, our design is successful in meeting all the performance objectives and input effort limits.

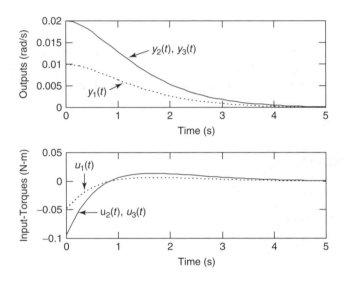

Figure 6.5 Closed-loop initial response and required inputs of the regulated flexible spacecraft for optimal regulator designed with **Q** = 200**I** and **R** = **I**

Examples 6.1 and 6.2 illustrate two practical problems, one with a controllable plant, and the other with an uncontrollable plant, which are successfully solved using linear optimal control. Note the ease with which the two multi-input problems are solved when compared with the pole-placement approach (which, as you will recall, would have resulted in far too many design parameters than can be fixed by pole-placement). We will now consider the application of optimal control to more complicated problems.

6.3 Optimal Control of Tracking Systems

Consider a linear, time-varying plant with state-equation given by Eq. (6.1). It is required to design a tracking system for this plant if the desired state-vector, $\mathbf{x_d}(t)$, is the solution of the following equation:

$$\mathbf{x_d}^{(1)}(t) = \mathbf{A_d}(t)\mathbf{x_d}(t) \tag{6.35}$$

Recall from Chapter 5 that the desired state dynamics is described by a homogeneous state-equation, because $\mathbf{x_d}(t)$ it is unaffected by the input, $\mathbf{u}(t)$. Subtracting Eq. (6.1) from Eq. (6.35), we get the following state-equation for the tracking-error, $\mathbf{e}(t) = \mathbf{x_d}(t) - \mathbf{x}(t)$:

$$\mathbf{e}^{(1)}(t) = \mathbf{A}(t)\mathbf{e}(t) + [\mathbf{A_d}(t) - \mathbf{A}(t)]\mathbf{x_d}(t) - \mathbf{B}(t)\mathbf{u}(t) \qquad (6.36)$$

The control objective is to find the control input, $\mathbf{u}(t)$, such that the tracking-error, $\mathbf{e}(t)$, is brought to zero in the steady-state. To achieve this objective by optimal control, we have to first define the objective function to be minimized. Note that, as opposed to the regulator problem in which the input $\mathbf{u}(t) = -\mathbf{K}(t)\mathbf{x}(t)$, now the control input will also depend linearly on the desired state-vector, $\mathbf{x_d}(t)$. If we express Eqs. (6.1) and (6.35) by a *combined system* of which the state-vector is $\mathbf{x_c}(t) = [\mathbf{e}(t)^T; \mathbf{x_d}(t)^T]^T$, then the control input must be given by the following linear control-law:

$$\mathbf{u}(t) = -\mathbf{K_c}(t)\mathbf{x_c}(t) = -\mathbf{K_c}(t)[\mathbf{e}(t)^T; \mathbf{x_d}(t)^T]^T \qquad (6.37)$$

where $\mathbf{K_c}(t)$ is the *combined feedback gain* matrix. Note that Eqs. (6.35) and (6.36) can be expressed as the following *combined state-equation*:

$$\mathbf{x_c}^{(1)}(t) = \mathbf{A_c}(t)\mathbf{x_c}(t) + \mathbf{B_c}(t)\mathbf{u}(t) \qquad (6.38)$$

where

$$\mathbf{A_c}(t) = \begin{bmatrix} \mathbf{A}(t) & [\mathbf{A_d}(t) - \mathbf{A}(t)] \\ \mathbf{0} & \mathbf{A_d}(t) \end{bmatrix}; \quad \mathbf{B_c}(t) = \begin{bmatrix} -\mathbf{B}(t) \\ \mathbf{0} \end{bmatrix} \qquad (6.39)$$

Since Eqs. (6.1) and (6.2) are now replaced by Eqs. (6.38) and (6.37), respectively, the objective function for the *combined system* can be expressed as an extension of Eq. (6.3) as follows:

$$J(t, t_f) = \int_t^{t_f} [\mathbf{x_c}^T(\tau)\mathbf{Q_c}(\tau)\mathbf{x_c}(\tau) + \mathbf{u}^T(\tau)\mathbf{R}(\tau)\mathbf{u}(\tau)] \, d\tau \qquad (6.40)$$

Note that although we desire that the tracking error, $\mathbf{e}(t) = \mathbf{x_d}(t) - \mathbf{x}(t)$, be reduced to zero in the steady-state (i.e. when $t \to \infty$), we *cannot* pose the tracking system design as an optimal control problem with *infinite* control interval (i.e. $t_f = \infty$). The reason is that the desired state-vector, $\mathbf{x_d}(t)$ (hence $\mathbf{x_c}(t)$), *may not* go to zero in the steady-state, and thus a *non-zero* control input, $\mathbf{u}(t)$, may be required in the steady-state. Also, note that the combined system described by Eq. (6.38) is *uncontrollable*, because the desired state dynamics given by Eq. (6.35) is unaffected by the input, $\mathbf{u}(t)$. Therefore, the combined system's optimal control problem, represented by Eqs. (6.37)–(6.40) is *not guaranteed* to have a unique, positive definite solution. Hence, to have a guaranteed unique, positive definite solution to the optimal control problem, let us *exclude* the uncontrollable desired state-vector from the objective function, by choosing the *combined state-weighting matrix* as follows:

$$\mathbf{Q_c}(t) = \begin{bmatrix} \mathbf{Q}(t) & \mathbf{0} \\ \mathbf{0} & \mathbf{0} \end{bmatrix} \qquad (6.41)$$

which results in the following objective function:

$$J(t, t_f) = \int_t^{t_f} [\mathbf{e}^T(\tau)\mathbf{Q}(\tau)\mathbf{e}(\tau) + \mathbf{u}^T(\tau)\mathbf{R}(\tau)\mathbf{u}(\tau)] \, d\tau \tag{6.42}$$

which is the same as Eq. (6.3), with the crucial difference that $\mathbf{u}(t)$ in Eq. (6.42) is given by Eq. (6.37), rather than Eq. (6.2). By choosing $\mathbf{Q}(t)$ and $\mathbf{R}(t)$ to be positive semi-definite and positive definite, respectively, we satisfy the remaining sufficient conditions for the existence of a unique, positive definite solution to the optimal control problem. Note that the *optimal feedback gain matrix*, $\mathbf{K}_{oc}(t)$, is given by the following extension of Eq. (6.28):

$$\mathbf{K}_{oc}(t) = \mathbf{R}^{-1}(t)\mathbf{B}_c^T(t)\mathbf{M}_{oc} \tag{6.43}$$

where $\mathbf{M}_{oc}$ is the solution to the following Riccati equation:

$$-\partial\mathbf{M}_{oc}/\partial t = \mathbf{A}_c^T(t)\mathbf{M}_{oc} + \mathbf{M}_{oc}\mathbf{A}_c(t) - \mathbf{M}_{oc}\mathbf{B}_c(t)\mathbf{R}^{-1}(t)\mathbf{B}_c^T(t)\mathbf{M}_{oc} + \mathbf{Q}_c(t) \tag{6.44}$$

subject to the terminal condition, $\mathbf{M}_{oc}(t_f, t_f) = \mathbf{0}$. Since $\mathbf{M}_{oc}$ is symmetric (see Section 6.1), it can be expressed as

$$\mathbf{M}_{oc} = \begin{bmatrix} \mathbf{M}_{o1} & \mathbf{M}_{o2} \\ \mathbf{M}_{o2}^T & \mathbf{M}_{o3} \end{bmatrix} \tag{6.45}$$

where $\mathbf{M}_{o1}$ and $\mathbf{M}_{o2}$ correspond to the plant and the desired state dynamics, respectively. Substituting Eqs. (6.45) and (6.39) into Eq. (6.43), we can express the optimal feedback gain matrix as follows:

$$\mathbf{K}_{oc}(t) = -[\mathbf{R}^{-1}(t)\mathbf{B}^T(t)\mathbf{M}_{o1}; \quad \mathbf{R}^{-1}(t)\mathbf{B}^T(t)\mathbf{M}_{o2}] \tag{6.46}$$

and the optimal control input is thus obtained by substituting Eq. (6.46) into Eq. (6.37) as follows:

$$\mathbf{u}(t) = \mathbf{R}^{-1}(t)\mathbf{B}^T(t)\mathbf{M}_{o1}\mathbf{e}(t) + \mathbf{R}^{-1}(t)\mathbf{B}^T(t)\mathbf{M}_{o2}\mathbf{x}_d(t) \tag{6.47}$$

Note that Eq. (6.47) does not require the sub-matrix, $\mathbf{M}_{o3}$. The individual matrix differential equations to be solved for $\mathbf{M}_{o1}$ and $\mathbf{M}_{o2}$ can be obtained by substituting Eqs. (6.39) and (6.45) into Eq. (6.44) as follows:

$$-\partial\mathbf{M}_{o1}/\partial t = \mathbf{A}^T(t)\mathbf{M}_{o1} + \mathbf{M}_{o1}\mathbf{A}(t) - \mathbf{M}_{o1}\mathbf{B}(t)\mathbf{R}^{-1}(t)\mathbf{B}^T(t)\mathbf{M}_{o1} + \mathbf{Q}(t) \tag{6.48}$$

$$-\partial\mathbf{M}_{o2}/\partial t = \mathbf{M}_{o2}\mathbf{A}_d(t) + \mathbf{M}_{o1}[\mathbf{A}_d(t) - \mathbf{A}(t)]$$
$$+ [\mathbf{A}^T(t) - \mathbf{M}_{o1}\mathbf{B}(t)\mathbf{R}^{-1}(t)\mathbf{B}^T(t)]\mathbf{M}_{o2} \tag{6.49}$$

Note that Eq. (6.48) is identical to the *matrix Riccati equation*, Eq. (6.29), which can be solved *independently* of Eq. (6.49), without taking into account the desired state dynamics. Once the optimal matrix, $\mathbf{M}_{o1}$, is obtained from the solution of Eq. (6.48), it can be substituted into Eq. (6.49), which can then be solved for $\mathbf{M}_{o2}$. Equation (6.49) is a *linear, matrix differential equation*, and can be written as follows:

$$-\partial\mathbf{M}_{o2}/\partial t = \mathbf{M}_{o2}\mathbf{A}_d(t) + \mathbf{M}_{o1}[\mathbf{A}_d(t) - \mathbf{A}(t)] + \mathbf{A}_{CL}^T(t)\mathbf{M}_{o2} \tag{6.50}$$

where $\mathbf{A_{CL}}(t) = \mathbf{A}(t) - \mathbf{B}(t)\mathbf{R}^{-1}(t)\mathbf{B}^{\mathrm{T}}(t)\mathbf{M_{o1}}$, the closed-loop state-dynamics matrix. The solution of the optimal tracking system thus requires the solution of the linear differential equation, Eq. (6.50), in addition to the solution of the optimal regulator problem given by Eq. (6.48). The solutions of Eqs. (6.48) and (6.49) are subject to the terminal condition, $\mathbf{M_{oc}}(t_f, t_f) = \mathbf{0}$, which results in $\mathbf{M_{o1}}(t_f, t_f) = \mathbf{0}$, and $\mathbf{M_{o2}}(t_f, t_f) = \mathbf{0}$.

Often, it is required to track a *constant* desired state-vector, $\mathbf{x_d}(t) = \mathbf{x_d^c}$, which implies $\mathbf{A_d}(t) = \mathbf{0}$. Then the matrices $\mathbf{M_{o1}}$ and $\mathbf{M_{o2}}$ are both constants in the *steady-state* (i.e. $t_f \to \infty$), and are the solutions of the following *steady-state* equations (obtained by setting $\partial \mathbf{M_{o1}}/\partial t = \partial \mathbf{M_{o2}}/\partial t = \mathbf{0}$ and $\mathbf{A_d}(t) = \mathbf{0}$ in Eqs. (6.48) and (6.49)):

$$0 = \mathbf{A}^{\mathrm{T}}(t)\mathbf{M_{o1}} + \mathbf{M_{o1}}\mathbf{A}(t) - \mathbf{M_{o1}}\mathbf{B}(t)\mathbf{R}^{-1}(t)\mathbf{B}^{\mathrm{T}}(t)\mathbf{M_{o1}} + \mathbf{Q}(t) \qquad (6.51)$$

$$0 = -\mathbf{M_{o1}}\mathbf{A}(t) + \mathbf{A_{CL}^{\mathrm{T}}}(t)\mathbf{M_{o2}} \qquad (6.52)$$

We immediately recognize Eq. (6.51) as the *algebraic Riccati equation* (Eq. (6.32)) of the *steady-state, optimal regulator* problem. From Eq. (6.52), we must have

$$\mathbf{M_{o2}} = [\mathbf{A_{CL}^{\mathrm{T}}}(t)]^{-1}\mathbf{M_{o1}}\mathbf{A}(t) \qquad (6.53)$$

where $\mathbf{M_{o1}}$ is the solution to the algebraic Riccati equation, Eq. (6.51). Note, however, that even though $\mathbf{M_{o1}}$ and $\mathbf{M_{o2}}$ are finite constants in the steady-state, the matrix $\mathbf{M_{oc}}$ is *not* a finite constant in the steady-state, because as $\mathbf{x}(t_f)$ tends to a constant desired state in the limit $t_f \to \infty$, the objective function (Eq. (6.42)) becomes *infinite*, hence a steady-state solution to Eq. (6.44) does not exist. The only way $\mathbf{M_{oc}}$ can *not* be a finite constant (when both $\mathbf{M_{o1}}$ and $\mathbf{M_{o2}}$ are finite constants) is when $\mathbf{M_{o3}}$ (the discarded matrix in Eq. (6.45)) is *not* a finite constant in the steady-state.

Substituting Eq. (6.53) into Eq. (6.47), we get the following input for the *constant* desired state vector, $\mathbf{x_d^c}$:

$$\mathbf{u}(t) = \mathbf{R}^{-1}(t)\mathbf{B}^{\mathrm{T}}(t)\mathbf{M_{o1}}\mathbf{e}(t) + \mathbf{R}^{-1}(t)\mathbf{B}^{\mathrm{T}}(t)[\mathbf{A_{CL}^{\mathrm{T}}}(t)]^{-1}\mathbf{M_{o1}}\mathbf{A}(t)\mathbf{x_d^c} \qquad (6.54)$$

Substituting Eq. (6.54) into Eq. (6.36), we get the following closed-loop tracking error state-equation with $\mathbf{A_d}(t) = \mathbf{0}$ and $\mathbf{x_d}(t) = \mathbf{x_d^c}$:

$$\mathbf{e}^{(1)}(t) = \mathbf{A_{CL}}(t)\mathbf{e}(t) - [\mathbf{A}(t) + \mathbf{B}(t)\mathbf{R}^{-1}(t)\mathbf{B}^{\mathrm{T}}(t)\{\mathbf{A_{CL}^{\mathrm{T}}}(t)\}^{-1}\mathbf{M_{o1}}\mathbf{A}(t)]\mathbf{x_d^c} \qquad (6.55)$$

From Eq. (6.55), it is clear that the tracking error can go to *zero* in the steady-state (i.e. as $t \to \infty$) for *any non-zero, constant* desired state, $\mathbf{x_d^c}$, if $\mathbf{A_{CL}}(t)$ is *asymptotically stable* and

$$\mathbf{A}(t) + \mathbf{B}(t)\mathbf{R}^{-1}(t)\mathbf{B}^{\mathrm{T}}(t)\{\mathbf{A_{CL}^{\mathrm{T}}}(t)\}^{-1}\mathbf{M_{o1}}\mathbf{A}(t) = \mathbf{0} \qquad (6.56)$$

or

$$\mathbf{M_{o1}}\mathbf{A}(t) = -\left[\mathbf{B}(t)\mathbf{R}^{-1}(t)\mathbf{B}^{\mathrm{T}}(t)\{\mathbf{A_{CL}^{\mathrm{T}}}(t)\}^{-1}\right]^{-1}\mathbf{A}(t) = -\mathbf{A_{CL}^{\mathrm{T}}}(t)\{\mathbf{B}(t)\mathbf{R}^{-1}(t)\mathbf{B}^{\mathrm{T}}(t)\}^{-1}\mathbf{A}(t) \qquad (6.57)$$

Equation (6.57) can be expanded to give the following equation to be satisfied by $\mathbf{M_{o1}}$ for the steady-state tracking error to be zero:

$$\mathbf{M_{o1}}\mathbf{A}(t) = -\mathbf{A}^T(t)\{\mathbf{B}(t)\mathbf{R}^{-1}(t)\mathbf{B}^T(t)\}^{-1}\mathbf{A}(t) + \mathbf{M_{o1}}\mathbf{A}(t) \tag{6.58}$$

which implies that $\mathbf{A}^T(t)\{\mathbf{B}(t)\mathbf{R}^{-1}(t)\mathbf{B}^T(t)\}^{-1}\mathbf{A}(t) = \mathbf{0}$, or $\{\mathbf{B}(t)\mathbf{R}^{-1}(t)\mathbf{B}^T(t)\}^{-1} = \mathbf{0}$. Clearly, this is an *impossible* requirement, because it implies that $\mathbf{R}(t) = \mathbf{0}$. Hence, we *cannot* have an optimal tracking system in which the tracking error, $\mathbf{e}(t)$, goes to zero in the steady-state for *any* constant desired state-vector, $\mathbf{x_d^c}$. As in Chapter 5, the best we can do is to have $\mathbf{e}(t)$ going to zero for *some* values of $\mathbf{x_d^c}$. However, if we want this to happen *while satisfying* the optimality condition for $\mathbf{M_{o2}}$ given by Eq. (6.53), we will be left with the requirement that $[\mathbf{A}(t) + \mathbf{B}(t)\mathbf{R}^{-1}(t)\mathbf{B}^T(t)\{\mathbf{A_{CL}^T}(t)\}^{-1}\mathbf{M_{o1}}\mathbf{A}(t)]\mathbf{x_d^c} = \mathbf{0}$ for some non-zero $\mathbf{x_d^c}$, resulting in $\{\mathbf{B}(t)\mathbf{R}^{-1}(t)\mathbf{B}^T(t)\}^{-1}\mathbf{A}(t)\mathbf{x_d^c} = \mathbf{0}$, which implies that $\{\mathbf{B}(t)\mathbf{R}^{-1}(t)\mathbf{B}^T(t)\}^{-1}$ must be a *singular matrix* – again, an impossible requirement. Therefore, the only possible way we can ensure that the tracking error goes to zero for some desired state is by *dropping the optimality condition* on $\mathbf{M_{o2}}$ given by Eq. (6.53). Then we can write the input vector as follows:

$$\mathbf{u}(t) = \mathbf{R}^{-1}(t)\mathbf{B}^T(t)\mathbf{M_{o1}}\mathbf{e}(t) - \mathbf{K_d}(t)\mathbf{x_d^c} \tag{6.59}$$

where $\mathbf{K_d}(t)$ is the (non-optimal) feedforward gain matrix which would make $\mathbf{e}(t)$ zero in the steady-state for some values of $\mathbf{x_d^c}$. Substituting Eq. (6.59) into Eq. (6.36) we get the following state-equation for the tracking error:

$$\mathbf{e}^{(1)}(t) = \mathbf{A_{CL}}(t)\mathbf{e}(t) - [\mathbf{A}(t) - \mathbf{B}(t)\mathbf{K_d}(t)]\mathbf{x_d^c} \tag{6.60}$$

Equation (6.60) implies that for a zero steady-state tracking error, $\mathbf{K_d}(t)$ must be selected such that $[\mathbf{A}(t) - \mathbf{B}(t)\mathbf{K_d}(t)]\mathbf{x_d^c} = \mathbf{0}$. The closed-loop state-dynamics matrix, $\mathbf{A_{CL}}$, in Eq. (6.60) is an optimal matrix given by $\mathbf{A_{CL}}(t) = \mathbf{A}(t) - \mathbf{B}(t)\mathbf{R}^{-1}(t)\mathbf{B}^T(t)\mathbf{M_{o1}}$, where $\mathbf{M_{o1}}$ is the solution to the algebraic Riccati equation, Eq. (6.51). Hence, the design of a tracking system *does not* end with finding a unique, positive definite solution, $\mathbf{M_{o1}}$, to the algebraic Riccati equation (which would make $\mathbf{A_{CL}}(t)$ asymptotically stable); we should also find a (non-optimal) feedforward gain matrix, $\mathbf{K_d}(t)$, such that $[\mathbf{A}(t) - \mathbf{B}(t)\mathbf{K_d}(t)]\mathbf{x_d^c} = \mathbf{0}$ for some values of the constant desired state-vector, $\mathbf{x_d^c}$. Note that if the plant has *as many inputs* as there are state variables, then $\mathbf{B}(t)$ is a square matrix, and it would be possible to make $\mathbf{e}(t)$ zero in the steady-state for *any arbitrary* $\mathbf{x_d^c}$, by choosing $\mathbf{K_d}(t) = \mathbf{B}^{-1}(t)\mathbf{A}(t)$ (provided $\mathbf{B}(t)$ is non-singular, i.e. the plant is *controllable*.)

Example 6.3

Consider the amplifier-motor of Example 3.7, with the numerical values given as $J = 1$ kg.m^2, $R = 1000$ ohms, $L = 100$ henry, $a = 0.3$ kg.m^2/s^2/Ampere, and $K_A = 10$. Recall from Example 3.7 that the state-vector of the amplifier-motor is $\mathbf{x}(t) = [\theta(t); \theta^{(1)}(t); i(t)]^T$, where $\theta(t)$ is the angular position of the load on the motor, and $i(t)$ is the current supplied to the motor. The input vector is $\mathbf{u}(t) = [v(t); T_L(t)]^T$, where $v(t)$ is the input voltage to the amplifier and $T_L(t)$ is the torque applied by the load on the motor. It is desired to design a tracking system such that the load on the motor moves from an initial angular position, $\theta(0) = 0$, to desired angular position $\theta_d(t) = 0.1$ rad. in about six seconds, and comes to rest at the desired position. The maximum angular velocity of the load, $\theta^{(1)}(t)$, should not exceed 0.05 rad/s. After the load comes to rest at the desired position, the current supplied to the motor should be zero. The desired state-vector is thus $\mathbf{x_d^c} = [0.1; 0; 0]^T$. The plant's state coefficient matrices are the following:

$$\mathbf{A} = \begin{bmatrix} 0 & 1 & 0 \\ 0 & -0.01 & 0.3 \\ 0 & -0.003 & -10 \end{bmatrix}; \quad \mathbf{B} = \begin{bmatrix} 0 & 0 \\ 0 & -1 \\ 0.1 & 0 \end{bmatrix}$$

$$\mathbf{C} = \begin{bmatrix} 1 & 0 & 0 \\ 0 & 1 & 0 \end{bmatrix}; \quad \mathbf{D} = \begin{bmatrix} 0 & 0 \\ 0 & 0 \end{bmatrix} \tag{6.61}$$

The eigenvalues of the linear, time-invariant plant are calculated as follows:

```
>>damp(A) <enter>

Eigenvalue  Damping  Freq. (rad/sec)
    0       -1.0000   0
-0.0101      1.0000   0.0101
-9.9999      1.0000   9.9999
```

The plant is stable, with an eigenvalue at the origin. Since the plant is time-invariant, the controller gain matrices must be constants. Let us first find the optimal feedback gain matrix, $\mathbf{K_{o1}} = -\mathbf{R}^{-1}\mathbf{B}^T\mathbf{M_{o1}}$ by choosing $\mathbf{Q} = \mathbf{I}$ and $\mathbf{R} = \mathbf{I}$, and solving the algebraic Riccati equation as follows:

```
>>[Ko1,Mo1,E] = lqr(A,B,eye(3),eye(2)) <enter>

Ko1 =
   0.0025   0.0046   0.0051
  -1.0000  -1.7220  -0.0462

Mo1 =
   1.7321   1.0000   0.0254
   1.0000   1.7220   0.0462
   0.0254   0.0462   0.0513

E =
 -10.0004
  -0.8660 + 0.5000i
  -0.8660 - 0.5000i
```

The closed-loop eigenvalues are all in the left-half plane, as desired for asymptotic stability of the tracking error dynamics. Next, we calculate the feedforward gain matrix, $\mathbf{K_d}$, which will make the steady-state tracking error zero for the specified constant desired state, $\mathbf{x_d^c}$. This is done by selecting $\mathbf{K_d}$ such that $\mathbf{A}\mathbf{x_d^c} = \mathbf{B}\mathbf{K_d}\mathbf{x_d^c}$ as follows:

$$\mathbf{K_d} = \begin{bmatrix} K_{d1} & K_{d2} & K_{d3} \\ K_{d4} & K_{d5} & K_{d6} \end{bmatrix} \tag{6.62}$$

$$\mathbf{BK_d x_d^c} = \begin{bmatrix} 0 \\ -0.1 K_{d4} \\ 0.01 K_{d1} \end{bmatrix} \tag{6.63}$$

Therefore, $\mathbf{Ax_d^c} = \mathbf{BK_d x_d^c}$ implies that $K_{d1} = 0$ and $K_{d4} = 0$. We can also choose the remaining elements of $\mathbf{K_d}$ as zeros, and still satisfy $\mathbf{Ax_d^c} = \mathbf{BK_d x_d^c}$. Hence, $\mathbf{K_d} = \mathbf{0}$, and the control input is given by $\mathbf{u}(t) = \mathbf{K_{o1} e}(t)$.

Let us obtain the tracking error response of the system to the initial tracking error, $\mathbf{e}(0) = \mathbf{x_d^c} - \mathbf{x}(0) = \mathbf{x_d^c}$ (since $\mathbf{x}(0) = \mathbf{0}$) as follows:

```
>>t=0:0.05:6; sysCL=ss(A-B*Ko1, zeros(3,2),C,D); [y,t,e]
  = initial(sysCL, [0.1 0 0]'); <enter>
```

Then, the state-vector, $\mathbf{x}(t)$, of the closed-loop system can be calculated using $\mathbf{x}(t) = \mathbf{x_d^c} - \mathbf{e}(t)$ as follows:

```
>>n=size(t, 1); for i=1:n; Xd(i,:)=[0.1 0 0]; end; X=Xd-e; <enter>
```

while the input vector, $\mathbf{u}(t)$, is calculated as

```
   >>u = Ko1*e'; <enter>
```

The calculated state variables, $\theta(t)$, $\theta^{(1)}(t)$, $i(t)$, the input voltage, $v(t)$, and the loading torque, $T_L(t)$, are plotted in Figure 6.6. Note that all the state variables reach their desired values in about 6 s, with a maximum overshoot in angular

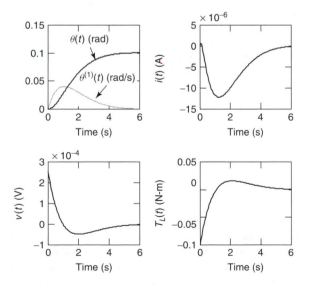

Figure 6.6 Closed-loop initial response of the tracking system for amplifier-motor, with constant desired state, $\theta(t) = 0.1$ rad, $\theta^{(1)}(t) = 0$ rad, and $i(t) = 0$ amperes

velocity, $\theta^{(1)}(t)$, of about 0.04 rad/s, and the maximum overshoot in current, $i(t)$, of -12.5×10^{-6} A. The maximum input voltage, $v(t)$, is 2.5×10^{-4} V and the maximum loading torque is -0.1 N-m.

Using SIMULINK, let us now investigate the effect of an uncertainty in the amplifier gain, K_A, on the tracking system. A SIMULINK block-diagram of the closed-loop tracking system is shown in Figure 6.7. The uncertainty in the amplifier gain, ΔK_A, affects only the third state-variable, $i(t)$, and is incorporated into the

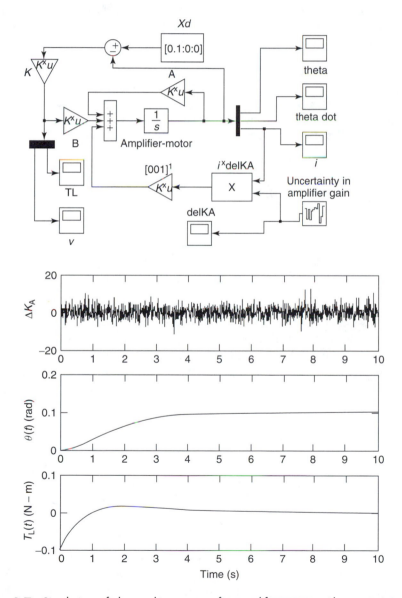

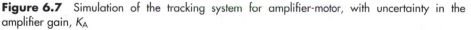

Figure 6.7 Simulation of the tracking system for amplifier-motor, with uncertainty in the amplifier gain, K_A

plant dynamics model using the *band-limited white noise* block output, ΔK_A, multiplied with the vector $[0; 0; i(t)]^T$, and added to the summing junction, which results in the plant dynamics being represented as $\mathbf{x}^{(1)}(t) = (\mathbf{A} + \Delta\mathbf{A})\mathbf{x}(t) + \mathbf{B}\mathbf{u}(t)$, where

$$\Delta\mathbf{A} = \begin{bmatrix} 0 & 0 & 0 \\ 0 & 0 & 0 \\ 0 & 0 & \Delta K_A \end{bmatrix} \tag{6.64}$$

The simulated values of ΔK_A, $\theta(t)$, and $T_L(t)$ are shown in Figure 6.7. Note that despite a random variation in K_A between ± 10, the tracking system's performance is unaffected. This signifies a design which is quite robust to variations in K_A.

Example 6.4

For a particular set of flight conditions, the lateral dynamics of an aircraft are described by a linear, time-invariant state-space representation with the following coefficient matrices:

$$\mathbf{A} = \begin{bmatrix} -9.75 & 0 & -9.75 & 0 \\ 0 & -0.8 & 8 & 0 \\ 0 & -1 & -0.8 & 0 \\ 1 & 0 & 0 & 0 \end{bmatrix}; \quad \mathbf{B} = \begin{bmatrix} 20 & 2.77 \\ 0 & -3 \\ 0 & 0 \\ 0 & 0 \end{bmatrix} \tag{6.65}$$

The state-vector consists of the *roll-rate*, $p(t)$, *yaw-rate*, $r(t)$, *side-slip angle*, $\beta(t)$, and *bank angle*, $\phi(t)$, and is written as $\mathbf{x}(t) = [p(t); r(t); \beta(t); \phi(t)]^T$. The input vector consists of the *aileron deflection angle*, $\delta_A(t)$, and *rudder deflection angle*, $\delta_R(t)$, i.e. $\mathbf{u}(t) = [\delta_A(t); \delta_R(t)]^T$. It is desired to execute a *steady turn* with a constant yaw-rate, $r_d(t) = 0.05$ rad/s, a constant bank angle, $\phi_d(t) = 0.02$ rad, and zero roll-rate and sideslip angle, $p_d(t) = \phi_d(t) = 0$. The desired state-vector is thus $\mathbf{x_d^c} = [0; 0.05; 0; 0.02]^T$. The desired state must be reached in about two seconds, with a maximum roll-rate, $p(t)$, less than 0.1 rad/s and the control inputs ($\delta_A(t)$ and $\delta_R(t)$) not exceeding 0.3 rad. Let us first select a feedforward gain matrix, $\mathbf{K_d}$, which satisfies $\mathbf{A}\mathbf{x_d^c} = \mathbf{B}\mathbf{K_d}\mathbf{x_d^c}$ as follows:

$$\mathbf{K_d} = \begin{bmatrix} K_{d1} & K_{d2} & K_{d3} & K_{d4} \\ K_{d5} & K_{d6} & K_{d7} & K_{d8} \end{bmatrix} \tag{6.66}$$

$$\mathbf{B}\mathbf{K_d}\mathbf{x_d^c} = \begin{bmatrix} 20(0.05K_{d2} + 0.02K_{d4}) + 2.77(0.05K_{d6} + 0.02K_{d8}) \\ -3(0.05K_{d6} + 0.1K_{d8}) \\ 0 \\ 0 \end{bmatrix}; \quad \mathbf{A}\mathbf{x_d^c} = \begin{bmatrix} 0 \\ 0 \\ 0 \\ 0 \end{bmatrix}$$
$$\tag{6.67}$$

Equation (6.67) indicates that, to make $\mathbf{A}\mathbf{x_d^c} = \mathbf{B}\mathbf{K_d}\mathbf{x_d^c}$, we must have $0.05K_{d6} + 0.02K_{d8} = 0$ and $0.05K_{d2} + 0.02K_{d4} = 0$, which can be satisfied by selecting $\mathbf{K_d} = \mathbf{0}$. It now remains for us to calculate the optimal feedback gain matrix, $\mathbf{K_{o1}}$, by solving the algebraic Riccati equation. Note that the plant is stable with the following eigenvalues:

```
>>damp(A) <enter>
```

```
Eigenvalue      Damping   Freq. (rad/sec)
     0            -1.0000         0
-0.8000 + 2.8284i  0.2722      2.9394
-0.8000 - 2.8284i  0.2722      2.9394
-9.7500            1.0000      9.7500
```

Let us select $\mathbf{R} = \mathbf{I}$. After experimenting with several values of $\mathbf{Q}$, we select the following which satisfies the desired transient response and input limitations:

$$\mathbf{Q} = \begin{bmatrix} 20 & 0 & 0 & 0 \\ 0 & 10 & 0 & 0 \\ 0 & 0 & 1 & 0 \\ 0 & 0 & 0 & 200 \end{bmatrix} \tag{6.68}$$

With this combination of $\mathbf{Q}$ and $\mathbf{R}$, the algebraic Riccati equation is solved as follows:

```
>>[Ko1,Mo1,E] = lqr(A,B,Q,R) <enter>
Ko1 =
   4.1351     0.3322    -0.2935    14.0420
   0.5229    -2.7651    -1.1575     1.6797

Mo1 =
   0.2068     0.0166    -0.0147     0.7021
   0.0166     0.9370     0.3723     0.0884
  -0.0147     0.3723     3.6355    -0.0908
   0.7021     0.0884    -0.0908    65.7890

E =
  -90.7740
   -8.4890
   -3.1456
   -1.3872
```

The initial response of the tracking error, $\mathbf{e}(t)$, to $\mathbf{x}(0) = \mathbf{0}$ is calculated as follows:

```
>>sysCL=ss(A-B*Ko1,zeros(4,2),eye(4),zeros(4,2)); <enter>
```

```
>>t=0:0.001:2; [y,t,e]=initial(sysCL,[0 0.05 0 0.02]',t); <enter>
```

and the state-vector and control input vector are then calculated as

```
>>n = size(t,1); for i=1:n; Xd(i,:) = [0 0.05 0 0.02]; end; X = Xd-e;
  u = -Ko1*e'; <enter>
```

The calculated state-vector, $\mathbf{x}(t)$, and input vector, $\mathbf{u}(t)$, are plotted in Figures 6.8 and 6.9, respectively. Note that all the transients settle to their desired values in about two seconds, with a maximum overshoot in roll-rate, $p(t)$, of less than 0.06 rad/s, and a small maximum side-slip angle, $\beta(t)$. The aileron and rudder deflections are seen in Figure 6.9 to be less than 0.3 rad. as desired.

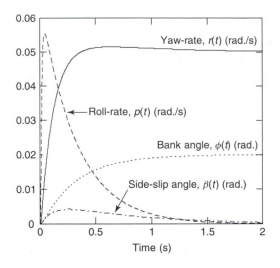

Figure 6.8 Response of the tracking system for aircraft lateral dynamics for a desired steady turn with turn-rate, $r(t) = 0.05$ rad. and bank angle, $\phi(t) = 0.02$ rad

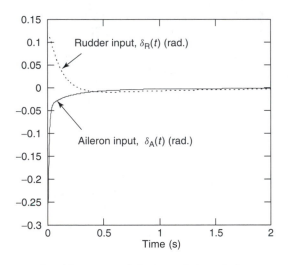

Figure 6.9 Aileron and rudder inputs of the aircraft lateral dynamics tracking system for achieving a desired steady turn with turn-rate, $r(t) = 0.05$ rad. and bank angle, $\phi(t) = 0.02$ rad

6.4 Output Weighted Linear Optimal Control

Many times it is the output, $\mathbf{y}(t)$, rather than the state-vector, $\mathbf{x}(t)$, which is included in the objective function for minimization. The reason for this may be either a lack of physical understanding of some state variables, which makes it difficult to assign weightage to them, or that the desired performance objectives are better specified in terms of the measured output; remember that it is the *output* of the system (rather than the *state*) which indicates the performance to an observer (either a person or a mathematical device discussed in Chapter 5). When the output is used to define the performance, the objective function can be expressed as follows:

$$J(t, t_f) = \int_t^{t_f} [\mathbf{y}^T(\tau)\mathbf{Q}(\tau)\mathbf{y}(\tau) + \mathbf{u}^T(\tau)\mathbf{R}(\tau)\mathbf{u}(\tau)] \, d\tau \qquad (6.69)$$

where $\mathbf{Q}(t)$ is now the *output weighting matrix*. Substituting the output equation given by

$$\mathbf{y}(t) = \mathbf{C}(t)\mathbf{x}(t) + \mathbf{D}(t)\mathbf{u}(t) \qquad (6.70)$$

into Eq. (6.69), the objective function becomes the following:

$$
\begin{aligned}
J(t, t_f) = \int_t^{t_f} & [\mathbf{x}^T(\tau)\mathbf{C}^T(\tau)\mathbf{Q}(\tau)\mathbf{C}(\tau)\mathbf{x}(\tau) + \mathbf{x}^T(\tau)\mathbf{C}^T(\tau)\mathbf{Q}(\tau)\mathbf{D}(\tau)\mathbf{u}(\tau) \\
& + \mathbf{u}^T(\tau)\mathbf{D}^T(\tau)\mathbf{Q}(\tau)\mathbf{C}(\tau)\mathbf{x}(\tau) + \mathbf{u}^T(\tau)\{\mathbf{R}(\tau) + \mathbf{D}^T(\tau)\mathbf{Q}(\tau)\mathbf{C}(\tau)\}\mathbf{u}(\tau)] \, d\tau
\end{aligned}
\qquad (6.71)
$$

or

$$
\begin{aligned}
J(t, t_f) = \int_t^{t_f} & [\mathbf{x}^T(\tau)\mathbf{Q_G}(\tau)\mathbf{x}(\tau) + \mathbf{x}^T(\tau)\mathbf{S}(\tau)\mathbf{u}(\tau) + \mathbf{u}^T(\tau)\mathbf{S}^T(\tau)\mathbf{x}(\tau) \\
& + \mathbf{u}^T(\tau)\mathbf{R_G}(\tau)\mathbf{u}(\tau)] \, d\tau
\end{aligned}
\qquad (6.72)
$$

where $\mathbf{Q_G}(\tau) = \mathbf{C}^T(\tau)\mathbf{Q}(\tau)\mathbf{C}(\tau)$, $\mathbf{S}(\tau) = \mathbf{C}^T(\tau)\mathbf{Q}(\tau)\mathbf{D}(\tau)$, and $\mathbf{R_G}(\tau) = \mathbf{R}(\tau) + \mathbf{D}^T(\tau)\mathbf{Q}(\tau)\mathbf{C}(\tau)$. You can show, using steps similar to Sections 6.1 and 6.2, that the optimal regulator gain matrix, $\mathbf{K_o}(t)$, which minimizes $J(t, t_f)$ given by Eq. (6.72) can be expressed as

$$
\begin{aligned}
\mathbf{K_o}(t) = \mathbf{R_G}^{-1}(t)[\mathbf{B}^T(t)\mathbf{M_o} + \mathbf{S}^T(t)] = & \{\mathbf{R}(t) + \mathbf{D}^T(t)\mathbf{Q}(t)\mathbf{C}(t)\}^{-1}[\mathbf{B}^T(t)\mathbf{M_o} \\
& + \mathbf{D}^T(\tau)\mathbf{Q}(\tau)\mathbf{C}(\tau)]
\end{aligned}
\qquad (6.73)
$$

where $\mathbf{M_o}(t, t_f)$ is the solution to the following matrix Riccati equation:

$$
\begin{aligned}
-\partial \mathbf{M_o}/\partial t = & \mathbf{A_G}^T(t)\mathbf{M_o} + \mathbf{M_o}\mathbf{A_G}(t) - \mathbf{M_o}\mathbf{B}(t)\mathbf{R_G}^{-1}(t)\mathbf{B}^T(t)\mathbf{M_o} \\
& + [\mathbf{Q_G}(t) - \mathbf{S}(t)\mathbf{R_G}^{-1}(t)\mathbf{S}^T(t)]
\end{aligned}
\qquad (6.74)
$$

where $\mathbf{A_G}(t) = \mathbf{A}(t) - \mathbf{B}(t)\mathbf{R_G}^{-1}(t)\mathbf{S}^T(t)$. Equation (6.74) can be solved numerically in a manner similar to that for the solution to Eq. (6.29), with the terminal condition $\mathbf{M_o}(t_f, t_f) = \mathbf{0}$.

The steady-state optimal control problem (i.e. when $t_f \to \infty$) results in the following algebraic Riccati equation:

$$0 = \mathbf{A_G}^T(t)\mathbf{M_o} + \mathbf{M_o}\mathbf{A_G}(t) - \mathbf{M_o}\mathbf{B}(t)\mathbf{R_G}^{-1}(t)\mathbf{B}^T(t)\mathbf{M_o} + [\mathbf{Q_G}(t) - \mathbf{S}(t)\mathbf{R_G}^{-1}(t)\mathbf{S}^T(t)] \qquad (6.75)$$

The sufficient conditions for the existence of a unique, positive definite solution, $\mathbf{M_o}$, to Eq. (6.75) are similar to those for Eq. (6.32), i.e. the system – whose state coefficient matrices are $\mathbf{A_G}(t)$ and $\mathbf{B}(t)$ – is *controllable*, $[\mathbf{Q_G}(t) - \mathbf{S}(t)\mathbf{R_G}^{-1}(t)\mathbf{S}^T(t)]$ is a *positive semi-definite* matrix, and $\mathbf{R}_G(t)$ is a *positive definite* matrix. However, note that for a plant with state-equation given by Eq. (6.1), these conditions are *more restrictive* than those for the existence of a unique, positive definite solution to Eq. (6.32).

Solution to the algebraic Riccati equation, Eq. (6.75), can be obtained using the MATLAB function *are* or CST function *lqr*, by appropriately specifying the coefficient matrices $\mathbf{A}_G(t)$ and $\mathbf{B}(t)$, and the weighting matrices $[\mathbf{Q}_G(t) - \mathbf{S}(t)\mathbf{R}_G^{-1}(t)\mathbf{S}^T(t)]$ and $\mathbf{R}_G(t)$ at each instant of time, t. However, MATLAB (CST) provides the function *lqry* for solving the output weighted linear, quadratic optimal control problem, which only needs the plant coefficient matrices $\mathbf{A}(t)$, $\mathbf{B}(t)$, $\mathbf{C}(t)$, and $\mathbf{D}(t)$, and the output and control weighting matrices, $\mathbf{Q}(t)$ and $\mathbf{R}(t)$, respectively, as follows:

```
>>[K,Mo,E] = lqry(sys,Q,R) <enter>
```

where *sys* is the state-space LTI object of the plant. The *lqry* command is thus easier to use than either *are* or *lqr* for solving the output weighted problem.

Example 6.5

Let us design an optimal regulator for the flexible bomber aircraft (Examples 4.7, 6.1) using output weighting. Recall that the two outputs of this sixth-order, time-invariant plant are the normal acceleration in m/s², $y_1(t)$, and the pitch-rate in rad./s, $y_2(t)$. Our performance objectives remain the same as in Example 6.1, i.e. a maximum overshoot of less than ± 2 m/s² in the normal-acceleration and less than ± 0.03 rad/s in pitch-rate, and a settling time less than 5 s, while requiring elevator and canard deflections (the two inputs) *not exceeding* ± 0.1 rad. (5.73°), if the initial condition is 0.1 rad/s perturbation in the pitch-rate ($\mathbf{x}(0) = [0; 0.1; 0; 0; 0; 0]^T$). After some trial and error, we find that the following weighting matrices satisfy the performance requirements:

$$\mathbf{Q}(t) = \begin{bmatrix} 0.0001 & 0 \\ 0 & 1 \end{bmatrix}; \quad \mathbf{R}(t) = \begin{bmatrix} 1 & 0 \\ 0 & 1 \end{bmatrix} \qquad (6.76)$$

which result in the following optimal gain matrix, $\mathbf{K_o}(t)$, the solution to algebraic Riccati equation, $\mathbf{M_o}$, and the closed-loop eigenvalue vector, $\mathbf{E}$:

```
>>sys=ss(A,B,C,D);[Ko,Mo,E] = lqry(sys,[0.0001 0; 0 1],eye(2)) <enter>

Ko=
 1.2525e+001 -2.3615e-002 3.3557e-001 -3.7672e-003 1.0426e+001 4.4768e+000
 6.6601e+000 3.3449e-001 1.8238e-001 -1.1227e-002 5.9690e+000 2.6042e+000

Mo=
 1.1557e+000  4.6221e-001 2.1004e-002  7.2281e-003  1.6700e-001  6.6601e-002
 4.6221e-001  1.1715e+000 1.3859e-002  1.2980e-002 -3.1487e-004  3.3449e-003
 2.1004e-002  1.3859e-002 7.3455e-004  1.9823e-004  4.4742e-003  1.8238e-003
 7.2281e-003  1.2980e-002 1.9823e-004  1.9469e-004 -5.0229e-005 -1.1227e-004
 1.6700e-001 -3.1487e-004 4.4742e-003 -5.0229e-005  1.3902e-001  5.9690e-002
 6.6601e-002  3.3449e-003 1.8238e-003 -1.1227e-004  5.9690e-002  2.6042e-002
```

```
E =
 -1.1200e+003
 -9.4197e+001
 -1.5385e+000+ 2.2006e+000i
 -1.5385e+000- 2.2006e+000i
 -1.0037e+000+ 6.2579e-001i
 -1.0037e+000- 6.2579e-001i
```

The closed-loop initial response and required control inputs are calculated as follows:

```
>>sysCL=ss(A-B*Ko,zeros(6,2),C,D); [y, t, X]=initial(sysCL, [0 0.1 0 0 0 0]');
  u = -Ko*X'; <enter>
```

Figure 6.10 shows the plots of the calculated outputs and inputs. Note that all the performance objectives are met, and the maximum control values are -0.035 rad. for both desired elevator and canard deflections, $u_1(t)$ and $u_2(t)$, respectively.

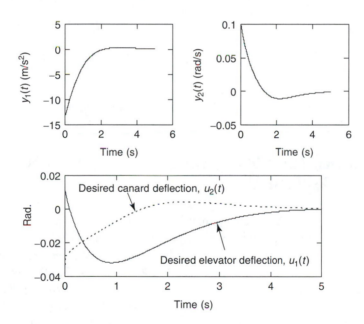

Figure 6.10 Closed-loop initial response and required inputs for the flexible bomber aircraft, with optimal regulator designed using output weighting

Let us simulate the closed-loop initial response of the present design with the regulator designed in Example 6.1 using state-weighting. We use the same SIMULINK block-diagram as shown in Figure 6.3, with the measurement noise modeled by the *band-limited white noise* block of power 10^{-4}. The resulting simulated response is shown in Figure 6.11. Note that the fluctuations in both $y_1(t)$ and $y_2(t)$ are about an *order of magnitude* smaller than those observed in Figure 6.3, which indicates that a better *robustness* with respect to measurement noise has been achieved using output-weighted optimal control.

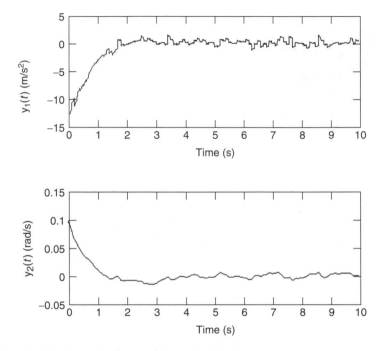

Figure 6.11 Simulated initial response of the flexible bomber with an output-weighted optimal regulator including measurement noise in the pitch channel (SIMULINK block-diagram of Figure 6.3)

6.5 Terminal Time Weighting: Solving the Matrix Riccati Equation

Thus far, we have considered the optimal control problems in which the closed-loop system is brought to a desired state by minimizing the deviation of the *transient response* from the desired state. However, sometimes it is also important to *minimize* the deviation of the system from the desired state at a *specified final time*, t_f. In other words, it is not only important to reach the final state, but to reach there in a *specified final time*, t_f, called the *terminal time*. By assigning a weightage to the state of the system at the terminal time, $\mathbf{x}(t_f)$, in the objective function to be minimized, we can specify the relative importance of minimizing the deviation from the final state. The objective function with terminal time weighting can be expressed as follows:

$$J(t, t_f) = \int_t^{t_f} [\mathbf{x}^T(\tau)\mathbf{Q}(\tau)\mathbf{x}(\tau) + \mathbf{u}^T(\tau)\mathbf{R}(\tau)\mathbf{u}(\tau)]\, d\tau + \mathbf{x}^T(t_f)\mathbf{V}\mathbf{x}(t_f) \qquad (6.77)$$

where $\mathbf{V}$ is the constant *terminal time weighting matrix*. For minimizing the objective function given by Eq. (6.77) for the optimal regulator problem (i.e. when $\mathbf{u}(t) = -\mathbf{K_o}(t)\mathbf{x}(t)$), it can be shown using steps similar to those in Section 6.1, that the optimal feedback gain matrix is given by

$$\mathbf{K_o}(t) = \mathbf{R}^{-1}(t)\mathbf{B}^T(t)\mathbf{M_o} \qquad (6.78)$$

where $\mathbf{M_o}$ is the solution to the matrix Riccati equation, Eq. (6.29), repeated here as follows:

$$-\partial \mathbf{M_o}/\partial t = \mathbf{A}^{\mathrm{T}}(t)\mathbf{M_o} + \mathbf{M_o}\mathbf{A}(t) - \mathbf{M_o}\mathbf{B}(t)\mathbf{R}^{-1}(t)\mathbf{B}^{\mathrm{T}}(t)\mathbf{M_o} + \mathbf{Q}(t) \qquad (6.79)$$

subject to the following *terminal condition*:

$$\mathbf{M_o}(t_f, t_f) = \mathbf{V} \qquad (6.80)$$

Hence, the terminal time weighted problem differs from the standard problem of Section 6.1 only in the *non-zero* terminal condition for the matrix Riccati equation.

Note that the terminal time weighted optimal control *cannot* be simplified in a manner similar to the infinite time control, which results in an algebraic Riccati equation. In other words, we cannot avoid taking the bull by the horns anymore; we have to solve the *matrix* Riccati equation if we want to solve the terminal time control problem. For simplicity, let us confine ourselves to *time-invariant* plants only, $\mathbf{A}(t) = \mathbf{A}$, $\mathbf{B}(t) = \mathbf{B}$, and $\mathbf{Q}(t) = \mathbf{Q}$, $\mathbf{R}(t) = \mathbf{R}$, where $\mathbf{A}$, $\mathbf{B}$, $\mathbf{Q}$, and $\mathbf{R}$ are constants. Then the matrix Riccati equation becomes

$$-\partial \mathbf{M_o}/\partial t = \mathbf{A}^{\mathrm{T}}\mathbf{M_o} + \mathbf{M_o}\mathbf{A} - \mathbf{M_o}\mathbf{B}\mathbf{R}^{-1}\mathbf{B}^{\mathrm{T}}\mathbf{M_o} + \mathbf{Q} \qquad (6.81)$$

which is to be solved with the terminal condition given by Eq. (6.80). We remarked in Section 6.1 on the difficulty of solving the matrix Riccati equation, owing to its *nonlinear* nature. However, if we look carefully at Eq. (6.81), we find that by expressing $\mathbf{M_o}$ in the following manner

$$\mathbf{M_o}(t, t_f) = \mathbf{E}(t, t_f)\mathbf{F}^{-1}(t, t_f) \qquad (6.82)$$

where $\mathbf{E}(t, t_f)$ and $\mathbf{F}(t, t_f)$ are two unknown matrices, we can write Eq. (6.81), as the following set of coupled, *linear* differential equations:

$$\begin{bmatrix} \partial \mathbf{F}(t, t_f)/\partial t \\ \partial \mathbf{E}(t, t_f)/\partial t \end{bmatrix} = \begin{bmatrix} \mathbf{A} & -\mathbf{B}\mathbf{R}^{-1}\mathbf{B}^{\mathrm{T}} \\ -\mathbf{Q} & -\mathbf{A}^{\mathrm{T}} \end{bmatrix} \begin{bmatrix} \mathbf{F}(t, t_f) \\ \mathbf{E}(t, t_f) \end{bmatrix} \qquad (6.83)$$

If we define a matrix $\mathbf{H}$, called the *Hamiltonian,* as follows

$$\mathbf{H} = \begin{bmatrix} \mathbf{A} & -\mathbf{B}\mathbf{R}^{-1}\mathbf{B}^{\mathrm{T}} \\ -\mathbf{Q} & -\mathbf{A}^{\mathrm{T}} \end{bmatrix} \qquad (6.84)$$

then we can write the solution to the homogeneous set of first order, *matrix* differential equation, Eq. (6.83), as an extension of the solution given by Eq. (4.16) to the homogeneous, *vector* state-equation. Hence, we can write

$$\begin{bmatrix} \mathbf{F}(t, t_f) \\ \mathbf{E}(t, t_f) \end{bmatrix} = \exp\{\mathbf{H}(t - t_f)\} \begin{bmatrix} \mathbf{F}(t_f, t_f) \\ \mathbf{E}(t_f, t_f) \end{bmatrix} \qquad (6.85)$$

where $\exp\{\mathbf{H}(t - t_f)\}$ is a *matrix exponential* which can be calculated by the methods of Chapter 4. To satisfy the terminal condition, Eq. (6.80), we can choose $\mathbf{E}(t_f, t_f) = \mathbf{V}$, and $\mathbf{F}(t_f, t_f) = \mathbf{I}$. Thus, for a given time, t, the matrices $\mathbf{E}(t, t_f)$ and $\mathbf{F}(t, t_f)$ can be calculated using Eq. (6.85), where the matrix exponential, $\exp\{\mathbf{H}(t - t_f)\}$, is calculated by an appropriate algorithm, such as *expm*, *expm1*, *expm2*, or *expm3* of MATLAB. However, care must be taken to select $\mathbf{Q}$ and $\mathbf{R}$ matrices, such that the calculated matrix $\mathbf{F}(t, t_f)$ is *non-singular*, otherwise $\mathbf{M_o}$ cannot be computed using Eq. (6.82). A MATLAB M-file called *matricc.m*, which solves the matrix Riccati equation for linear, time-invariant systems using Eq. (6.85) is given in Table 6.1, and can be called as follows:

```
>>Mo = matricc(A,B,Q,R,V,t,tf,dt) <enter>
```

where $\mathbf{A}$, $\mathbf{B}$, $\mathbf{Q}$, $\mathbf{R}$, $\mathbf{V}$, t, and $tf\,(t_f)$ are the same as in Eq. (6.77), dt is the time-step, and $\mathbf{Mo}(\mathbf{M_o})$ is the returned solution of the matrix Riccati equation, Eq. (6.81).

For solving the terminal time weighted optimal control problem, it is clear that while the matrix Riccati equation is solved by marching *backwards* in time using Eq. (6.85) – starting with the *terminal condition*, Eq. (6.82) – the state-equation of the closed-loop system must be solved by marching *forwards* in time using the methods of Chapter 4, starting with the *initial condition*, $\mathbf{x}(t_0) = \mathbf{x_0}$. Hence, the solution requires marching in both forward and backward direction (in time), beginning from the conditions specified at the two time boundaries, i.e. the terminal time, t_f, and initial time, t_0. For this reason, the general optimal control problem is referred to as *two-point boundary-value problem*. We can write a computer program combining the solution to the matrix Riccati equation and the solution to the closed-loop state-equation. Such a program is given in Table 6.2 as an M-file called *tpbvlti.m*, which solves the two-point boundary-value

Table 6.1 Listing of the M-file *matricc.m* for the solution of the terminal-time weighted, linear optimal control problem

matricc.m

```
function M=matricc(A,B,Q,R,V,t,tf);
% Program for solving the Matrix Riccati equation resulting
% from the terminal-time weighted optimal control problem for
% linear, time-invariant systems.
% A = State dynamics matrix of the plant
% B = Input coefficient matrix of the plant
% Q = State weighting matrix
% R = Control weighting matrix
% V = Terminal state weighting matrix
% t = present time
% tf = terminal time
% M = returned solution of the Matrix Riccati equation
% Copyright(c)2000 by Ashish Tewari
%
% Construct the Hamiltonian matrix:-
H=[A -B*inv(R)*B';-Q -A'];
% Solve the Matrix Riccati equation using the matrix exponential:-
n=size(A,1);
FE=expm2(H*(t-tf))*[eye(n);V];
F=FE(1:n,:);E=FE(n+1:2*n,:);
M=E*inv(F);
```

Table 6.2 Listing of the M-file *tpbvlti.m* for solving the two-point boundary value problem associated with the solution of the closed-loop state-equations with a terminal-time weighted, linear time-invariant, optimal regulator

---
tpbvlti.m

```
function [u,X,t] = tpbvlti(A,B,Q,R,V,t0,tf,X0)
% Time-marching solution of the two-point boundary-value problem
% resulting from the terminal-time weighted, optimal regulator
% for a linear, time-invariant system.
% A= State dynamics matrix of the plant
% B= Input coefficient matrix of the plant
% Q= State weighting matrix
% R= Control weighting matrix
% V= Terminal state weighting matrix
% t0= initial time; tf= terminal time (tf-t0 should be small for
% convergence)
% X0= initial state vector; t= time vector
% u=matrix with the ith input stored in the ith column, and jth row
% corresponding to the jth time point
% X= returned matrix with the ith state variable stored in the ith column,
% and jth row corresponding to the jth time point
% copyright(c)2000 by Ashish Tewari
[w,z]=damp(A);
mw=max(w);
if mw==0;
dt=(tf-t0)/20;
else
dt=1/max(w);
end
t=t0:dt:tf;
n=size(t,2);
% initial condition:-
X(1,:)=X0';
% solution of the matrix Riccati equation for t=t0:-
M=matricc(A,B,Q,R,V,t0,tf);
% calculation of input vector for t=t0:-
u(1,:)=-X(1,:)*M*B*inv(R);
% beginning of the time-loop:-
for i=1:n-1
% solution of the matrix Riccati equation:-
M=matricc(A,B,Q,R,V,t(i),tf);
% calculation of the closed-loop state-dynamics matrix
Ac=A-B*inv(R)*B'*M;
% conversion of system from continuous-time to digital
[ad,bd]=c2d(Ac,B,dt);
% solution of digitized closed-loop state-equations
X(i+1,:)=X(i,:)*ad';
% updating the input vector for time t=t(i):-
u(i+1,:)=-X(i+1,:)*M*B*inv(R);
end
```

problem for a linear, *time-invariant* plant using *matricc.m* for the solution of matrix Riccati equation. This M-file can be used as follows:

```
>>[u,X,t] = tpbvlti(A,B,Q,R,V,t0,tf,X0) <enter>
```

where **A**, **B**, **Q**, **R**, **V**, t, and $tf(t_f)$ have their usual meanings, $t0$ is the initial time at which the initial condition vector, **x0**, is specified, and **u**, **x**, and **t** contain the returned input, state and time vectors, respectively. However, the usage of this M-file is restricted to small value of the interval $t_f - t_0$. For larger time intervals, a time-marching procedure of Chapter 4 could be used, employing $t_f - t_0$ as the time step.

Example 6.6

Terminal time weighting is very common in problems where two objects are desired to be brought together in a specified final time, such as missile guidance to a target, a rendezvous (or docking) of two spacecraft, or a rendezvous of an aircraft and a tanker aircraft for refuelling. When posing the optimal guidance strategy for such problems, usually the state variables are the relative distance between the two objects, $x_1(t)$, and the relative velocity (also called the *closure velocity*), $x_2(t)$. At some specified final time, t_f, it is desired that the relative distance becomes zero, i.e. $x_1(t_f) = 0$. Whereas in a rendezvous problem, it is also desired that the final closure velocity also becomes zero, i.e. $x_2(t_f) = 0$, such a condition is not imposed on guiding a missile to its target. If $x_1(t_f) \neq 0$, a *miss* is said to occur, and the *miss distance*, $x_1(t_f)$, is a measure of the success (or failure) of either the rendezvous or the missile intercept. The linear, time-invariant state-equation for a missile guidance or a rendezvous problem can be written in terms of the state variables, $x_1(t)$ and $x_2(t)$, and single input, $u(t)$ – which is the *normal acceleration* provided to the object – with the following state coefficient matrices:

$$\mathbf{A} = \begin{bmatrix} 0 & 1 \\ 0 & 0 \end{bmatrix}; \quad \mathbf{B} = \begin{bmatrix} 0 \\ 1 \end{bmatrix} \tag{6.86}$$

When we do not care how the missile (or spacecraft) moves before the intercept (or rendezvous) occurs, we may not wish to assign any weightage to the transient response by choosing $\mathbf{Q} = \mathbf{0}$, and $\mathbf{R} = 1$. In such cases, the matrix Riccati equation can be solved *analytically*, and you may refer to Bryson and Ho [2] for the resulting exact solutions. However, it is generally desired that the plant adheres to certain limitations in its transient response, while proceeding from the initial time to the final time, hence **Q** is appropriately chosen to be non-zero. Let us consider a spacecraft docking (rendezvous) problem. The initial relative distance and closure velocity between the two spacecraft are 100 m and −10 m/s, respectively. It is desired to complete the rendezvous in 10 seconds, with the closure velocity never exceeding a magnitude of 50 m/s, while requiring a maximum normal acceleration input magnitude of less than 50 m/s². The docking will be considered successful if the final magnitudes of relative distance, $x_1(t_f)$, and relative velocity, $x_2(t_f)$, are less than 0.15 m and 0.5 m/s, respectively. Since both relative distance,

$x_1(t)$, and closure velocity, $x_2(t)$, are required to be minimized at the terminal time, the terminal time weighting matrix, $\mathbf{V}$, should assign weightage to both the state variables. The following choice of $\mathbf{Q}$, $\mathbf{R}$, and $\mathbf{V}$ is seen to meet the performance requirements:

$$\mathbf{Q} = \begin{bmatrix} 1 & 0 \\ 0 & 0 \end{bmatrix}; \quad \mathbf{R} = 3; \quad \mathbf{V} = \begin{bmatrix} 10 & 0 \\ 0 & 0 \end{bmatrix} \tag{6.87}$$

and the two-point boundary-value problem is solved using *tpbvlti.m* as follows:

```
>>[u,X,t] = tpbvlti(A,B,[1 0; 0 0],3,[10 0; 0 0],0,10,
  [100 -10]'); <enter>
```

The calculated state variables, $x_1(t)$ and $x_2(t)$, and the input, $u(t)$, are plotted in Figure 6.12. Note that the performance objectives are met quite successfully. The calculated values of $x_1(t_f)$ and $x_2(t_f)$ are not *exactly* zeros, but -0.1438 m and 0.3610 m/s, respectively; however, these are small enough to be acceptable.

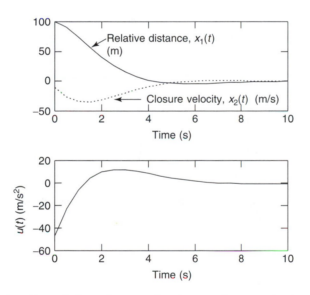

Figure 6.12 Closed-loop docking of spacecraft with terminal time weighted, optimal regulator

Example 6.7

Suppose it is desired to bring both normal acceleration, $y_1(t)$, and the pitch-rate, $y_2(t)$, of the flexible bomber aircraft (Example 6.5) to zero in *exactly* 0.2 seconds, after encountering a perturbation in the pitch-rate of 0.1 rad/s, *regardless* of the maximum overshoot and settling time of the transient response. Since we do not care what happens to the transient response, we can choose $\mathbf{Q} = \mathbf{0}$. Then the choice of

R and **V** matrices that achieves the desired performance is $\mathbf{R} = 10\mathbf{I}$, and $\mathbf{V} = 907\mathbf{I}$, and the closed-loop response is calculated using *tpbvlti* as follows:

```
>>[u,X,t] = tpbvlti(A,B,zeros(6),10*eye(2),907*eye(6),0,0.2,
  [0 0.1 0 0 0 0]'); y = C*X'; <enter>
```

The resulting outputs $y_1(t)$ and $y_2(t)$, and the required inputs, $u_1(t)$ and $u_2(t)$, are plotted in Figure 8.13. Note that both $y_1(t)$ and $y_2(t)$ are minimized in 0.2 seconds, as desired. The maximum input magnitudes do not exceed ± 0.3 rad.($17°$). This design has been carried out after experimenting with various values of v, where $\mathbf{V} = v\mathbf{I}$, and settling with the one ($v = 907$) that gives the minimum magnitudes of $y_1(t_f)$ and $y_2(t_f)$, which are -0.009 m/s^2 and -1.4×10^{-4} rad., respectively.

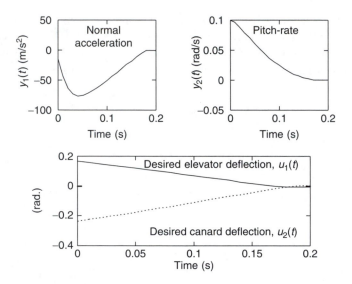

Figure 6.13 Closed-loop response of the flexible bomber aircraft with terminal-time weighted, optimal regulator ($t_f = 0.2$ s)

Exercises

6.1. Design an optimal, full-state feedback regulator for the distillation column whose state-space representation is given in Exercises 5.4 and 5.15, using $\mathbf{Q} = 1 \times 10^{-4}\mathbf{I}$ and $\mathbf{R} = \mathbf{I}$. Determine the initial response of the regulated system to the initial condition, $\mathbf{x}(0) = [1; 0; 0; 0]^T$. Calculate the inputs required for the closed-loop initial response.

6.2. Re-design the optimal regulator for the distillation column in Exercise 6.1, choosing **Q** and **R** such that the settling time of the closed-loop initial response to $\mathbf{x}(0) = [1; 0; 0; 0]^T$

is less than 10 seconds, and the maximum overshoot magnitudes of both the outputs are less than 1×10^{-3} units. Calculate the inputs required for the closed-loop initial response.

6.3. For the aircraft lateral dynamics given in Exercise 4.3, design a two-input optimal regulator with $\mathbf{Q} = \mathbf{I}$ and $\mathbf{R} = \mathbf{I}$. Calculate and plot the initial response, $p(t)$, of the regulated system if the initial condition is $\mathbf{x}(0) = [p(0); r(0); \beta(0); \phi(0)]^T = [0.5; 0; 0; 0]^T$. What are the settling time and the maximum overshoot of the closed-loop initial response? What are the largest input magnitudes required for the closed-loop initial response?

6.4. For the turbo-generator of Example 3.14, with the state-space representation given by Eq. (3.117), design a two-input optimal regulator with $\mathbf{Q} = \mathbf{I}$ and $\mathbf{R} = \mathbf{I}$. Calculate and plot the initial response of the regulated system if the initial condition is $\mathbf{x}(0) = [0.1; 0; 0; 0; 0; 0]^T$. What are the settling time and the maximum overshoots of the closed-loop initial response? What are the largest control input magnitudes required for the closed-loop initial response?

6.5. Re-design the regulator in Exercise 6.4 using output-weighted optimal control such that the maximum overshoot magnitudes of both the outputs in the initial response to $\mathbf{x}(0) = [0.1; 0; 0; 0; 0; 0]^T$ is less than 0.035 units, with a settling time less than 0.6 second, and the required input magnitudes of $u_1(t)$ and $u_2(t)$ should not exceed 0.05 units and 0.4 units, respectively.

6.6. Re-design the regulator for the distillation column to the specifications of Exercise 6.2 using output weighted optimal control.

6.7. Repeat Exercise 6.3 using output weighted optimal control with $\mathbf{Q} = \mathbf{I}$ and $\mathbf{R} = \mathbf{I}$. Compare the initial response and input magnitudes of the new regulator with that designed in Exercise 6.3.

6.8. It is required that the bank-angle, $\phi(t)$, of the aircraft in Exercise 6.3 must track a desired bank-angle given by $\phi_d(t) = 0.1$ rad. Design an optimal tracking system to achieve the desired bank angle in less than 5 seconds with both the input magnitudes less than 0.1 rad., and plot the closed-loop tracking error, $\phi_d(t) - \phi(t)$, if the initial condition of the airplane is zero, i.e. $\mathbf{x}(0) = \mathbf{0}$. Use SIMULINK to investigate the robustness of the tracking system to a random measurement noise in feeding back the roll-rate, $p(t)$.

6.9. Can you design a tracking system for the aircraft in Exercise 6.3 to track a desired constant roll-rate, $p_d(t)$?

6.10. Can you design a tracking system for the turbo-generator of Exercise 6.4 such that a desired state vector is $\mathbf{x}_d(t) = [10; 0; 0; 0; 0; 0]^T$?

6.11. The angular motion of a tank-gun turret [5] is described by the following state-coefficient matrices:

$$
\mathbf{A} = \begin{bmatrix}
0 & 1 & 0 & 0 & 0 & 0 & 0 & 0 \\
0 & 0 & 1 & 0 & 0 & 0 & 0 & 0 \\
0 & 1071 & -46 & -1071 & 0 & 0 & 0 & 0 \\
0 & 0 & 4.7 & -94.3 & 0 & 0 & 0 & 0 \\
0 & 0 & 0 & 0 & 0 & 1 & 0 & 0 \\
0 & 0 & 0 & 0 & 0 & 0 & 1 & 0 \\
0 & 0 & 0 & 0 & 0 & 947 & -17.3 & -947 \\
0 & 0 & 0 & 0 & 0 & 0 & -7.5 & -101
\end{bmatrix}
$$

$$
\mathbf{B} = \begin{bmatrix}
0 & 0 \\
0 & 0 \\
0 & 0 \\
94.3 & 0 \\
0 & 0 \\
0 & 0 \\
0 & 0 \\
0 & 94.3
\end{bmatrix}
\tag{6.88}
$$

The state-vector is given by $\mathbf{x}(t) = [x_1(t); x_2(t); \ldots; x_8(t)]^T$, where $x_1(t)$ is the *turret azimuth angle*, $x_2(t)$ is the *turret azimuth angular rate*, $x_3(t)$ is the *turret azimuth angular acceleration*, $x_4(t)$ is the *azimuth control hydraulic servo-valve displacement*, $x_5(t)$ is the *turret elevation angle*, $x_6(t)$ is the *turret elevation angular rate*, $x_7(t)$ is the *turret elevation angular acceleration*, and $x_8(t)$ is the *elevation control hydraulic servo-valve displacement*. The input vector, $\mathbf{u}(t) = [u_1(t); u_2(t)]^T$, consists of the input to the *azimuth control servo-valve*, $u_1(t)$, and the input to the *elevation control servo-valve*, $u_2(t)$. Design an optimal tracking system for achieving a constant desired state-vector, $\mathbf{x_d}^c = [1.57; 0; 0; 0; 0.5; 0; 0; 0]^T$, in less than seven seconds, if the initial condition is zero, i.e. $\mathbf{x}(0) = \mathbf{0}$, with the control inputs not exceeding five units.

Plot the initial response and the required control inputs. Using SIMULINK, study the robustness of the tracking system with respect to the following:

(a) measurement noise in the turret azimuth angular acceleration, $x_3(t)$, channel.

(b) measurement noise in the turret elevation angular acceleration, $x_7(t)$.

(c) saturation limits on the two servo-valves, $u_1(t)$ and $u_2(t)$.

6.12. Design a terminal-time weighted optimal regulator for the tank-gun turret if it is desired to move the turret to a zero final state in exactly 0.2 seconds, beginning with an initial condition of $\mathbf{x}(0) = [0.05; 0; 0; 0; 0.01; 0; 0; 0]^T$. Plot the azimuth and elevation angles of the turret, $x_1(t)$, and $x_2(t)$, respectively, and the required control inputs. What are the maximum control input magnitudes?

6.13. For the turbo-generator of Exercise 6.4, design a terminal-time weighted optimal regulator such that the system is brought to a zero final state in exactly 1 second, beginning with the initial condition $\mathbf{x}(0) = [10; 0; 0; 0; 0; 0]^T$. Plot the outputs and the required control inputs.

6.14. For the amplifier-motor of Example 6.3, design a terminal-time weighted optimal regulator such that the load angle, $\theta(t)$, is brought to zero in exactly 0.5 seconds, beginning with the initial condition $\mathbf{x}(0) = [0.1; 0; 0]^T$. Plot the outputs and the required control inputs. What is the maximum overshoot of the angular velocity, $\theta^{(1)}(t)$, and what is the final value of $\theta^{(1)}(t)$?

References

1. Kreyszig, E. *Advanced Engineering Mathematics*. John Wiley & Sons, New York, 1972.
2. Bryson, A.E. and Ho, Y.C. *Applied Optimal Control*. Hemisphere, New York, 1975.
3. Junkins, J.L. and Kim, Y. *Introduction to Dynamics and Control of Flexible Structures*. AIAA Education Series, American Institute of Aeronautics and Astronautics, Washington, DC, 1993.
4. Tewari, A. Robust model reduction of a flexible spacecraft. *J. of Guidance, Control, and Dynamics*, Vol. 21, No.5, 1998, 809–812.
5. Loh, N.K. Cheok, K.C. and Beck, R.R. Modern control design for gun-turret control system. Paper 1/5, *Southcon/85 Convention Record*, Atlanta, GA, 1983.

7

Kalman Filters

7.1 Stochastic Systems

We defined *deterministic systems* in Chapter 1 as those systems whose governing physical laws are such that if the state of the system at some time (i.e. the *initial condition*) is specified, then one can precisely predict the state at a later time. Most (if not all) natural processes are *non-deterministic* systems (i.e. systems that are *not* deterministic). *Non-deterministic* systems can be divided into two categories: *stochastic* and *random* systems. A *stochastic* (also called *probabilistic*) system has such governing physical laws that even if the initial conditions are known precisely, it is impossible to determine the system's state at a later time. In other words, based upon the *stochastic* governing laws and the initial conditions, one could only determine the *probability* of a state, rather than the state itself. When we flip a perfect coin, we do not know if head or tail will come up; we only know that both the possibilities have an equal *probability* of 50%. The disturbances encountered by many physical systems – such as atmospheric turbulence and disturbance due to an uneven ground – are produced by *stochastic* systems. A *random* system is one which has no apparent governing physical laws. While it is a human endeavour to ascribe physical laws to observed natural phenomena, some natural phenomena are so complex that it is impossible to pin down the physical laws obeyed by them. The human brain presently appears to be a *random* system. Environmental temperature and rainfall are outputs of a random system. It is very difficult to practically distinguish between *random* and *stochastic* systems. Also, frequently we are unable to practically distinguish between a non-deterministic (stochastic or random) system, and a deterministic system whose future state we *cannot* predict based upon an *erroneous measurement* of the initial condition. A *double pendulum* (Figure 1.5) is a classic example of *unpredictable*, deterministic systems. For all practical purposes, we will treat all unpredictable systems – deterministic or non-deterministic – (stochastic or random) as *stochastic* systems, since we have to employ the same statistical methods while studying such systems, regardless of the nature of their physical governing laws. For the same reason, it is a common practice to use the words *random*, *stochastic*, and *unpredictable* interchangeably. While we are unable to predict the state of a random process, we can evolve a strategy to deal with such processes when they affect a control system in the form of noise. Such a strategy has to be based on a branch of mathematics dealing with unpredictable systems, called *statistics*.

Since the initial state, $\mathbf{x}(t_0)$, of a stochastic system is insufficient to determine the future state, $\mathbf{x}(t)$, we have to make an *educated guess* as to what the future state might be, based upon a statistical analysis of many similar systems, and taking the *average* of their future states at a given time, t. For example, if we want to know how a stray dog would behave if offered a bone, we will offer N stray dogs N different bones, and record the state variables of interest, such as intensity of the sound produced by each dog, the forward acceleration, the angular position of the dog's tail, and, perhaps, the intensity of the dog's bite, as functions of time. Suppose $\mathbf{x}_i(t)$ is the recorded state-vector of the ith dog. Then the *mean* state-vector is defined as follows:

$$\mathbf{x_m}(t) = (1/N) \sum_{i=1}^{N} \mathbf{x}_i(t) \tag{7.1}$$

Note that $\mathbf{x_m}(t)$ is the state-vector we would *expect* after studying N similar stochastic systems. Hence, it is also called the *expected value* of the state-vector, $\mathbf{x}(t)$, and denoted by $\mathbf{x_m}(t) = E[\mathbf{x}(t)]$. The *expected value operator*, $E[\cdot]$, has the following properties (which are clear from Eq. (7.1)):

(a) $E[random\ signal] = mean\ of\ the\ random\ signal.$

(b) $E[deterministic\ signal] = deterministic\ signal.$

(c) $E[\mathbf{x}_1(t) + \mathbf{x}_2(t)] = E[\mathbf{x}_1(t)] + E[\mathbf{x}_2(t)].$

(d) $E[\mathbf{Cx}(t)] = \mathbf{C}E[\mathbf{x}(t)]$; $\mathbf{C} = constant\ matrix.$

(e) $E[\mathbf{x}(t)\mathbf{C}] = E[\mathbf{x}(t)]\mathbf{C}$; $\mathbf{C} = constant\ matrix.$

(f) $E[\mathbf{x}(t)\mathbf{y}(t)] = E[\mathbf{x}(t)]\mathbf{y}(t)$; $\mathbf{x}(t) = random\ signal$; $\mathbf{y}(t) = deterministic\ signal.$

(g) $E[\mathbf{y}(t)\mathbf{x}(t)] = \mathbf{y}(t)E[\mathbf{x}(t)]$; $\mathbf{x}(t) = random\ signal$; $\mathbf{y}(t) = deterministic\ signal.$

We can define another statistical quantity, namely a *correlation matrix* of the state-vector as follows:

$$\mathbf{R_x}(t, \tau) = (1/N) \sum_{i=1}^{N} \mathbf{x}_i(t)\mathbf{x}_i^T(\tau) \tag{7.2}$$

The correlation matrix, $\mathbf{R_x}(t, \cdot\tau)$, is a measure of a statistical property called *correlation* among the different state variables, as well as between the same state variable at two different times. For two scalar variables, $x_1(t)$ and $x_2(t)$, if the expected value of $x_1(t)x_2(\tau)$ is zero, i.e. $E[x_1(t)x_2(\tau)] = 0$, where τ is different from t, then $x_1(t)$ and $x_2(t)$ are said to be *uncorrelated*. Comparing Eqs. (7.1) and (7.2), it is clear that the correlation matrix is the expected value of the matrix $\mathbf{x}_i(t)\mathbf{x}_i^T(\tau)$, or $\mathbf{R_x}(t, \tau) = E[\mathbf{x}_i(t)\mathbf{x}_i^T(\tau)]$. When $t = \tau$, the correlation matrix, $\mathbf{R}_x(t, t) = E[\mathbf{x}_i(t)\mathbf{x}_i^T(t)]$, is called the *covariance matrix*. It is obvious that the covariance matrix, $\mathbf{R}_x(t, t)$, is symmetric. If $\mathbf{R}_x(t, \tau)$ is a *diagonal* matrix, it implies that all the state variables are *uncorrelated*, i.e. $E[x_i(t)x_j(\tau)] = 0$, where $i \neq j$. You are referred to a textbook on probability and statistics, such as that by Papoulis [1], for further details on expected values, correlation and covariance matrices.

There are special stochastic systems, called *stationary systems*, for which all the statistical properties, such as the mean value, $\mathbf{x_m}(t)$, and correlation matrix, $\mathbf{R_x}(t, \tau)$, *do not* change with a *translation* in time, i.e. when time, t, is replaced by $(t + \theta)$. Hence, for a *stationary system*, $\mathbf{x_m}(t + \theta) = \mathbf{x_m}(t) = $ constant, and $\mathbf{R_x}(t + \theta, \tau + \theta) = \mathbf{R_x}(t, \tau)$ for *all* values of θ. Expressing $\mathbf{R_x}(t, \tau) = \mathbf{R_x}(t, t + \alpha)$ where $\tau = t + \alpha$, we can show that for a stationary system, $\mathbf{R_x}(t - \alpha, t) = \mathbf{R_x}(t, t + \alpha)$, which implies that for a stationary system, the correlation matrix is only a function of the *time-shift*, α, i.e. $\mathbf{R_x}(t, t + \alpha) = \mathbf{R_x}(\alpha)$. Many stochastic systems of interest are assumed to be stationary, which greatly simplifies the statistical analysis of such systems.

The expected value, $\mathbf{x_m}(t)$, and the correlation matrix, $\mathbf{R_x}(t, \tau)$, are examples of *ensemble statistical properties*, i.e. properties of an *ensemble* (or *group*) of N samples. Clearly, the accuracy by which the expected value, $\mathbf{x_m}(t)$, approximates the actual state-vector, $\mathbf{x}(t)$, depends upon the number of samples, N. If N is increased, the accuracy is improved. For a *random* system, an *infinite* number of samples are required for predicting the state-vector, i.e. $N = \infty$. However, we can usually obtain good accuracy with a *finite* (but large) number of samples. Of course, the samples must be taken in as many different situations as possible. For example, if we confined our sample of stray dogs to our own neighborhood, the accuracy of ensemble properties would suffer. Instead, we should pick the dogs from many different parts of the town, and repeat our experiment at various times of the day, month, and year. However, as illustrated by the stray dog example, one has to go to great lengths merely to collect sufficient data for arriving at an accurate ensemble average. Finding an ensemble average in some cases may even be impossible, such as trying to calculate the expected value of annual rainfall in London – which would require constructing N Londons and taking the ensemble average of the annual rainfall recorded in all the Londons! However, we can measure annual rainfall in London for *many years*, and take the *time average* by dividing the total rainfall by the number of years. Taking a time average is entirely different from taking the ensemble average, especially if the system is *non-stationary*. However, there is a sub-class of stationary systems, called *ergodic* systems, for which a *time average* is the same as an ensemble average. For those stationary systems that are *not* ergodic, it is inaccurate to substitute a time average for the ensemble average, but we still do so routinely because there is no other alternative (there is only one London in the world). Hence, we will substitute time averaged statistics for ensemble statistics of *all* stationary systems. For a stationary system, by taking the time average, we can evaluate the *mean*, $\mathbf{x_m}$, and the *correlation matrix*, $\mathbf{R_x}(\tau)$, over a *large* time period, $T \to \infty$, as follows:

$$\mathbf{x_m} = \lim_{T \to \infty}(1/T) \int_{-T/2}^{T/2} \mathbf{x}(t)\, dt \tag{7.3}$$

$$\mathbf{R_x}(\tau) = \lim_{T \to \infty}(1/T) \int_{-T/2}^{T/2} \mathbf{x}(t)\mathbf{x}^T(t + \tau)\, dt \tag{7.4}$$

Note that since the system is stationary, the mean value, $\mathbf{x_m}$, is a constant, and the *correlation matrix*, $\mathbf{R_x}(\tau)$, is only a function of the *time-shift*, τ. For frequency domain analysis of stochastic systems, it is useful to define a *power spectral density matrix*, $\mathbf{S_x}(\omega)$, as the Fourier transform of the correlation matrix, $\mathbf{R_x}(\tau)$, given by

$$\mathbf{S_x}(\omega) = \int_{-\infty}^{\infty} \mathbf{R_x}(\tau)e^{-i\omega\tau}\, d\tau \tag{7.5}$$

where ω is the frequency of excitation (i.e. the frequency of an oscillatory input applied to the stochastic system). The power spectral density matrix, $\mathbf{S_x}(\omega)$, is a measure of how the *power* of a random signal, $\mathbf{x}(t)$, varies with frequency, ω. The Fourier transform of the random signal, $\mathbf{x}(t)$, is given by

$$X(i\omega) = \int_{-\infty}^{\infty} \mathbf{x}(t)e^{-i\omega t}dt \tag{7.6}$$

It can be shown from Eqs. (7.4)–(7.6) that

$$\mathbf{S_x}(\omega) = X(i\omega)X^T(-i\omega) \tag{7.7}$$

The correlation matrix, $\mathbf{R_x}(\tau)$, can be obtained by calculating the inverse Fourier transform of the power spectral density matrix, $\mathbf{S_x}(\omega)$, as follows:

$$\mathbf{R_x}(\tau) = (1/2\pi) \int_{-\infty}^{\infty} \mathbf{S_x}(\omega)e^{i\omega \tau}d\omega \tag{7.8}$$

Comparing Eqs. (7.4) and (7.8), we find that at $\tau = 0$, the correlation matrix becomes the *covariance matrix*, given by

$$\mathbf{R_x}(0) = \lim_{T\to\infty}(1/T) \int_{-T/2}^{T/2} \mathbf{x}(t)\mathbf{x}^T(t)dt = (1/2\pi) \int_{-\infty}^{\infty} \mathbf{S_x}(\omega)e^{i\omega \tau}d\omega \tag{7.9}$$

or

$$\int_{-\infty}^{\infty} \mathbf{S_x}(\omega)e^{i\omega \tau}d\omega = 2\pi \mathbf{x_{ms}} \tag{7.10}$$

where $\mathbf{x_{ms}}$, called the *mean-square value of* $\mathbf{x}(t)$, is the following:

$$\mathbf{x_{ms}} = \lim_{T\to\infty}(1/T) \int_{-T/2}^{T/2} \mathbf{x}(t)\mathbf{x}^T(t)\, dt \tag{7.11}$$

Usually, the state-vector, $\mathbf{x}(t)$, is available at *discrete* time points, rather than as a *continuous* function of time. The *discrete* Fourier transform of a *discrete* time state-vector, $\mathbf{x}(j\Delta t)$, where Δt is the time step size and $j = 1, 2, \ldots, N$, is calculated by the following expression:

$$X(k\Delta\omega) = \sum_{j=1}^{N}\mathbf{x}(j\Delta t)e^{-2\pi(j-1)(k-1)/N} \tag{7.12}$$

where $\Delta\omega = 2\pi/(N\Delta t)$ is the frequency step size. Similarly, the *inverse* discrete Fourier transform can be calculated as follows:

$$\mathbf{x}(j\Delta t) = (1/N) \sum_{j=1}^{N} X(k\Delta\omega)e^{2\pi(j-1)(k-1)/N} \tag{7.13}$$

When the discrete Fourier transform is used to calculate the *power spectral density*, the result must be divided by N, the number of frequency points in $X(k\Delta\omega)$ as follows:

$$\mathbf{S_x}(k\Delta\omega) = X(ik\Delta\omega)X^T(-ik\Delta\omega)/N \tag{7.14}$$

The discrete Fourier transform of a signal, $x(t)$, can be calculated by using the MATLAB command for discrete Fourier transform, *fft*, as follows:

```
>>X = fft(x,n) <enter>
```

where **x** is the returned n-point *discrete* Fourier transform of **x**. If n is not specified, an N-point Fourier transform is calculated, where N is the number of discrete time points in the matrix **x**. **x** has as many rows as there are time points, and as many columns as there are variables. Thus, each column of **x** is the Fourier transform of the corresponding column of **x**. The inverse discrete Fourier transform is similarly calculated using the MATLAB command $ifft$ as follows:

```
>>x = ifft(X,n) <enter>
```

Example 7.1

Consider a scalar random signal, $x(t)$, which is generated using the MATLAB command *randn* as follows:

```
>>t=0:0.1:10; randn('seed',0); x=randn(size(t)); <enter>
```

The command *randn* generates a random number, according to a special random process with a *zero* mean value, called *normal(or Gaussian) probability distribution* [1]. The 'seed' of the random number generator, *randn*, is set to zero to initialize the generating process to the value when MATLAB is started. The random signal, $x(t)$, is generated in time steps of 0.1 seconds, for 10 seconds – a total of 101 time points. The full-order discrete Fourier transform of $x(t)$ is calculated as follows:

```
>>X=fft(x); <enter>
```

The discrete frequencies, ω, at which the Fourier transform of $x(t)$ is calculated are calculated and stored in vector w as follows:

```
>>w = (0: length(X)-1)'*2*pi/(0.1*length(X)); <enter>
```

The power spectral density, $S_x(\omega)$, of the discrete random signal, $x(t)$, is calculated as follows:

```
>>S = X.*conj(X)/length(X); <enter>
```

The correlation function, $R_x(\tau)$ is calculated by taking the inverse Fourier transform of $S(\omega)$ with the help of the MATLAB command *ifft* as follows:

```
>>Rx = ifft(S); <enter>
```

The scalar plots $x(t)$, $|X(\omega)|$, $S_x(\omega)$, and $R_X(t)$ are shown in Figure 7.1. Note the special shape of the $S_x(\omega)$ plot – it is symmetrical about the *mid-point frequency* $\omega = 51\Delta\omega = 31.1$ rad/s. This is a characteristic of the normal distribution.

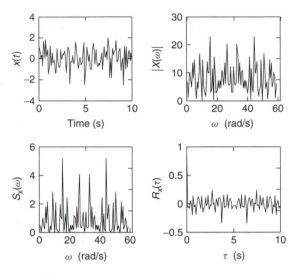

Figure 7.1 Random signal, $x(t)$, its Fourier transform magnitude, $|X(\omega)|$, power spectral density, $S_x(\omega)$, and correlation function, $R_x(\tau)$

The mean value of $x(t)$ can be obtained using the MATLAB command *mean* as follows:

```
>>mean(x) <enter>
ans =
  0.0708
```

which is very close to zero as expected (the mean value of $x(t)$ will be *exactly* zero only if the number of sample points, i.e. the length of $x(t)$, is *infinity*). The *covariance*, $R_x(0)$ can be read from the plot of $R_x(\tau)$ to be approximately 0.9, or obtained more precisely using the MATLAB command *cov* as follows:

```
>>cov(x) <enter>
ans =
  0.9006
```

The *mean-square* of $x(t)$, x_{ms}, can be calculated as follows:

```
>>xms = mean(x.*x) <enter>
ans =
  0.8967
```

Note that x_{ms} is equal to the *mean value* of the power spectral density $S_x(\omega)$ plot, calculated as follows:

```
>>mean(S) <enter>
ans =
  0.8967
```

You can repeat Example 7.1 using a different MATLAB random number generator called *rand*, which follows another stochastic process called *uniform probability distribution*. Definition and discussion of *probability distributions* of random processes is beyond the scope of this book, but can be found in a textbook on probability, such as that by Popoulis [1].

7.2 Filtering of Random Signals

When random signals are passed through a deterministic system, their statistical properties are modified. A deterministic system to which random signals are input, so that the output is a random signal with desired statistical properties is called a *filter*. Filters can be linear or nonlinear, time-invariant or time varying. However, for simplicity we will usually consider linear, time-invariant filters. Linear, time-invariant filters are commonly employed in control systems to reduce the effect of measurement noise on the control system. In such systems, the output is usually a superposition of a deterministic signal and a random measurement noise. Often, the measurement noise has a predominantly *high-frequency* content (i.e. its power spectral density has *more peaks* at *higher* frequencies). To filter-out the high-frequency noise, a special filter called a *low-pass* filter is employed, which blocks *all* signal frequencies *above* a specified *cut-off frequency*, ω_0. The output of a low-pass filter thus contains only lower frequencies, i.e. $\omega < \omega_0$, which implies a *smoothening* of the input signal fed to the filter. Sometimes, it is desired to block both high- and low-frequency contents of a noisy signal. This is achieved by passing the signal through a *band-pass* filter, which allows only a specified band of frequencies, $\omega_1 < \omega < \omega_2$, to pass through as the output signal. Similarly, a *high-pass* filter blocks all frequencies *below* a specified *cut-off frequency*, ω_0, and has an output containing only the *higher* frequencies, i.e. $\omega > \omega_0$.

Since it is impossible for a filter to *perfectly block* the undesirable signals, it is desired that the magnitude of signal above or below a given frequency decays rapidly with frequency. Such a decay of signal magnitude with frequency is called *attenuation*, or *roll-off*. The output of a filter not only has a frequency content different from the input signal, but also certain other characteristics of the filter, such as a phase-shift or a change in magnitude. In other words, the signal passing through a filter is also *distorted* by the filter, which is undesirable. It is inevitable that a filter would produce an output signal based upon its characteristics, described by the transfer-function, frequency response, impulse response, or a state-space representation of the filter. However, a filter can be designed to achieve a desired set of performance objectives – such as cut-off frequencies, desired *attenuation (roll-off)* of signals above or below the cut-off frequencies, etc., with the *minimum* possible *distortion* of the signals passing through the filter. Usually, it is observed that a greater attenuation of noise also leads to a greater distortion of the filtered signal. It is beyond the scope of this book to discuss the many different approaches followed in filter design, and you may refer to Parks and Burrus [2] for details. It suffices here to state that the numerator and denominator polynomials of the filter's transfer function, or coefficient matrices of the filter's state-space model, can be selected by a design process to achieve the conflicting requirements of maximum noise attenuation and minimum signal distortion.

Example 7.2

Consider a single-input, single-output filter with the following transfer function:

$$G(s) = \omega_0/(s + \omega_0) \tag{7.15}$$

This is the simplest possible low-pass filter with cut-off frequency, ω_0. Let us pass the following signal – which is a deterministic system's output $(\sin(10t))$ corrupted by a normally distributed random noise – through this low-pass filter:

$$u(t) = \sin(10t) + 0.2^*randn(t) \tag{7.16}$$

where *randn(t)* denotes the random noise generated by the MATLAB random number generator with a normal distribution (see Example 7.1). The random input signal is generated as follows:

```
>> t=0:0.001:1;randn('seed',0);u=sin(10*t)+0.2*randn(size(t)); <enter>
```

To block the high-frequency random noise, a cut-off frequency $\omega_0 = 10$ rad/s is initially selected for the low-pass filter, and the Bode plot of the filter is obtained as follows:

```
>> n = 10; d = [1 10]; sys=tf(n,d); [mag, phase,w] = bode(sys); <enter>
```

The filter is re-designed with cut-off frequency values, $\omega_0 = 40$ and 100 rad/s. The Bode plots of the low-pass filter with cut-off frequency values $\omega_0 = 10$, 40, and 100 rad/s, are plotted in Figure 7.2. Note that the magnitude of the filter decays with frequency at $\omega \geq \omega_0$, providing *noise attenuation* for frequencies above ω_0. The filter does not *totally* block the frequencies above the cut-off frequency, but decreases the magnitude (*attenuates*) almost linearly with the logarithmic frequency. The ideal *roll-off* (i.e. slope of decreasing magnitude with frequency) for filtering noise is 20 dB per 10 divisions of logarithmic frequency scale in rad/s (called *20 dB per decade*). The smaller the value of the cut-off frequency, the earlier would noise attenuation begin, and smaller will be the noise content of the filtered signal. It can be seen in the magnitude plot of Figure 7.2 that the filter with $\omega_0 = 10$ rad/s achieves a roll-off of 20 dB per decade at frequencies above 20 rad/s, while the other two filters achieve the same roll-off at much higher frequencies. The phase plot in Figure 7.2 shows that the low-pass filter *decreases* the phase of the input signal. Hence, a first order low-pass filter is also said to act as a *phase-lag* device. However, the phase of filter with $\omega_0 = 10$ and 40 rad/s begins to change at frequencies below 10 rad/s, which implies that the deterministic part of the random input signal, $\sin(10t)$, would be appreciably *distorted* (i.e. changed in wave-form) by these two filters, which is undesirable. The phase of the filter with $\omega_0 = 100$ rad/s is relatively unchanged until 10 rad/s, indicating little distortion of the deterministic signal.

A simulation of the filter's output is obtained using Control System Toolbox (CST) function *lsim* as follows:

```
>>[y,t,X]=lsim(sys,u,t); <enter>
```

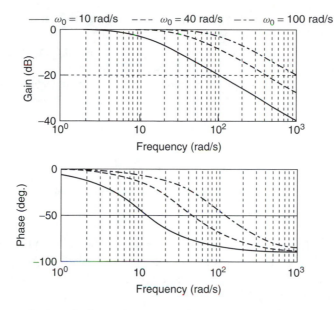

Figure 7.2 Bode plots of a first-order, low-pass filter with cut-off frequency, $\omega_0 = 10$, 40, and 100 rad/s

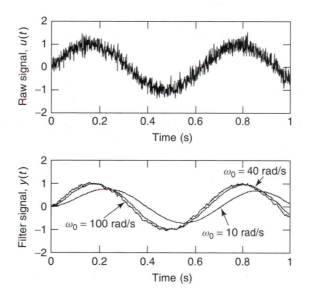

Figure 7.3 Raw signal and filtered signals after passing through a first-order, low-pass filter with cut-off frequency, 10, 40, and 100 rad/s, respectively

The input, $u(t)$, (i.e. raw signal) and the output, $y(t)$ (i.e. the filtered signal) for $\omega_0 = 10$, 40, and 100 rad/s are plotted in Figure 7.3. Note in Figure 7.3 that the filtered signals are smoother than the raw signal, but have an appreciable *distortion* in the wave-form, compared to the desired noise-free signal $\sin(10t)$). Among the three filter designs, for the filter with $\omega_0 = 10$ rad/s the distortion is maximum, but the filtered signal is the smoothest. For the filter with $\omega_0 = 100$ rad/s, the waveform distortion is minimum, but the filtered signal is the roughest indicating that the filter has allowed a lot of noise to pass through. An intermediate value of the cutoff frequency, $\omega_0 = 40$ rad/s, provides a good compromise between the conflicting requirements of smaller signal distortion and greater noise attenuation.

Example 7.3

Let us consider filtering a random signal which consists of a linear superposition of a deterministic signal, $\sin(20t) + \sin(50t)$, with a random noise given by

$$u(t) = \sin(20t) + \sin(50t) + 0.5^*randn(t) \tag{7.17}$$

where *randn(t)* denotes the random noise generated by the MATLAB's normal distribution random number generator. To filter the noise with least possible distortion, a sophisticated low-pass filter, called *elliptic filter*, is used with the following state-space representation:

$$\mathbf{A} = \begin{bmatrix} -153.34 & -94.989 & 0 & 0 \\ 94.989 & 0 & 0 & 0 \\ -153.34 & 104\,940 & -62.975 & -138.33 \\ 0 & 0 & 138.33 & 0 \end{bmatrix}$$

$$\mathbf{B} = \begin{bmatrix} 120 \\ 0 \\ 120 \\ 0 \end{bmatrix}$$

$$\mathbf{C} = [-1.2726 \times 10^{-5} \quad 8.7089 \times 10^{-3} \quad -5.2265 \times 10^{-6} \quad 1.0191 \times 10^{-3}]$$

$$\mathbf{D} = 9.9592 \times 10^{-6}$$

The Bode plot of the fourth order elliptic filter, shown in Figure 7.4, is obtained as follows:

```
>>sfilt=ss(A,B,C,D); bode(sfilt) <enter>
```

Figure 7.4 shows a *cut-off frequency* of 120 rad/s, a roll-off of about 100 dB per decade between the cut-off frequency and 1000 rad/s, and *ripples* in the magnitude for frequency greater than 1000 rad/s. The *passband*, i.e. the band of frequencies which the filter lets pass ($\omega \leq 120$ rad/s), is seen to be *flat* in magnitude, which implies a negligible magnitude distortion of the deterministic part of the filtered signal. The *stopband*, i.e. the band of frequencies which the filter is supposed to block ($\omega > 120$ rad/s), is seen to have ripples at -100 dB magnitude. As this magnitude is very small, there is expected to be a negligible influence of the ripples on the filtered signal. The phase plot shows a gradual $180°$ phase change in the passband, and rapid phase changes in the stopband above 1000 rad/s.

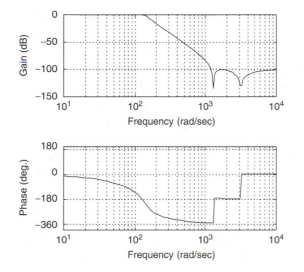

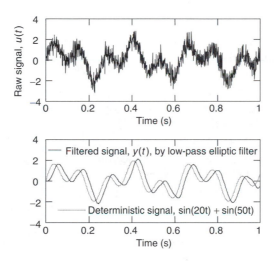

Figure 7.4 Bode plot of a fourth-order, low-pass elliptic filter with cut-off frequency 120 rad/s (Example 7.3)

Figure 7.5 Raw random plus deterministic signal, and filtered signal after passing through a fourth-order, low-pass elliptic filter with cut-off frequency 120 rad/s

The filter output is simulated using MATLAB command *lsim* as follows:

```
>>[y,t,X] = lsim(sfilt,u,t); <enter>
```

Figure 7.5 shows time-plots of the raw signal, $u(t)$, and the filtered signal, $y(t)$, compared with the desired deterministic output, $\sin(20t) + \sin(50t)$. The filtered signal is seen to be smooth, and with only a phase-lag as the signal distortion, when compared with the desired deterministic signal.

7.3 White Noise, and White Noise Filters

In the previous section, we saw how random signals can be generated using a random number generator, such as the MATLAB command *randn*, which generates random numbers with a *normal (or Gaussian) probability distribution*. If we could theoretically use a random number generator to generate an *infinite set of random numbers* with *zero mean value* and a *normal probability distribution*, we would have a *stationary* random signal with a *constant* power spectral density [1]. Such a theoretical random signal with zero mean and a constant power spectral density is called a *stationary white noise*. A *white noise vector*, $\mathbf{w}(t)$, has each element as a white noise, and can be regarded as the state-vector of a stochastic system called a *white noise process*. For such a process, we can write the mean value, $\mathbf{w_m} = 0$, and the power spectral density matrix, $\mathbf{S}(\omega) = \mathbf{W}$, where $\mathbf{W}$ is a *constant* matrix. Since the power spectral density of a white noise is a constant, it follows from Eq. (7.9) that the *covariance matrix* of the white noise is a matrix with all elements as *infinities*, i.e. $\mathbf{R_w}(0) = \infty$. Also, from Eq. (7.8) we can write the following expression for the *correlation matrix* of the white noise process (recall that the *inverse* Laplace (or Fourier) transform of a constant is the constant multiplied by the *unit impulse function*, $\delta(t)$):

$$\mathbf{R_w}(\tau) = \mathbf{W}\delta(\tau) \tag{7.18}$$

Equation (7.18) re-iterates that the covariance of white noise is infinite, i.e. $\mathbf{R_w}(0) = \infty$. Note that the correlation matrix of white noise is *zero*, i.e. $\mathbf{R_w}(\tau) = \mathbf{0}$, for $\tau \neq 0$, which implies that the white noise is *uncorrelated in time* (there is absolutely no correlation between $\mathbf{w}(t)$ and $\mathbf{w}(t + \tau)$). Hence, white noise can be regarded as *perfectly* random. However, a physical process with constant power spectral density is unknown; all physical processes have a power spectrum that tends to zero as $\omega \rightarrow \infty$. All known physical processes have a finite *bandwidth*, i.e. range of frequencies at which the process can be *excited* (denoted by peaks in the power spectrum). The white noise, in contrast, has an *infinite* bandwidth. Hence, the white noise process appears to be a figment of our imagination. However, it is a useful figment of imagination, as we have seen in Chapters 5 and 6 that white noise can be used to approximate random disturbances while simulating the response of control systems.

Let us see what happens if a linear, time-invariant system is placed in the path of a white noise. Such a linear system, into which the white noise is input, would be called a *white noise filter*. Since the input is a white noise, from the previous section we expect that the output of the filter would be a random signal. What are the statistical properties of such a random signal? We can write the following expression for the output, $\mathbf{y}(t)$, of a linear system with transfer matrix, $\mathbf{G}(s)$, with a white noise input, $\mathbf{w}(t)$, using the matrix form of the *superposition integral* of Eq. (2.120), as follows:

$$\mathbf{y}(t) = \int_{-\infty}^{t} \mathbf{g}(t - \tau)\mathbf{w}(\tau)\,d\tau \tag{7.19}$$

where $\mathbf{g}(t)$ is the *impulse response matrix* of the filter related to the transfer matrix, $\mathbf{G}(s)$, by the inverse Laplace transform as follows:

$$\mathbf{g}(t) = \int_0^\infty \mathbf{G}(s)e^{st}\,ds \qquad (7.20)$$

The lower limit of integration in Eq. (7.19) can be changed to zero if the white noise, $\mathbf{w}(t)$, starts acting on the system at $t = 0$, and the system is *causal*, i.e. it starts producing the output *only after* receiving the input, but *not before that* (most physical systems are causal). Since the input white noise is a stationary process (with mean, $\mathbf{w_m} = 0$), and the filter is time-invariant, we expect that the output, $\mathbf{y}(t)$, should be the output of a stationary process, and we can determine its mean value by taking time average as follows:

$$\mathbf{y_m} = \lim_{T\to\infty}(1/T)\int_{-T/2}^{T/2}\mathbf{y}(t)\,dt = \lim_{T\to\infty}(1/T)\int_{-T/2}^{T/2}\left[\int_{-\infty}^t \mathbf{g}(t-\tau)\mathbf{w}(\tau)\,d\tau\right]dt$$

$$= \int_{-\infty}^t \mathbf{g}(t-\tau)\left[\lim_{T\to\infty}(1/T)\int_{-T/2}^{T/2}\mathbf{w}(t)\,dt\right]d\tau = \int_{-\infty}^t \mathbf{g}(t-\tau)\mathbf{w_m}d\tau = 0 \qquad (7.21)$$

Hence, the mean of the output of a linear system for a white noise input is zero. The correlation matrix of the output signal, $\mathbf{R_y}(\tau)$ can be calculated as follows, using the properties of the expected value operator, $E[\cdot]$:

$$\mathbf{R_y}(\tau) = E[\mathbf{y}(t)\mathbf{y}^\mathrm{T}(t+\tau)] = E\left[\int_{-\infty}^t \mathbf{g}(t-\alpha)\mathbf{w}(\alpha)\,d\alpha \cdot \int_{-\infty}^{t+\tau}\mathbf{w}^\mathrm{T}(\beta)\mathbf{g}^\mathrm{T}(t+\tau-\beta)\,d\beta\right]$$

$$= E\left[\int_{-\infty}^t\int_{-\infty}^{t+\tau}\mathbf{g}(t-\alpha)\mathbf{w}(\alpha)\mathbf{w}^\mathrm{T}(\beta)\mathbf{g}^\mathrm{T}(t+\tau-\beta)\,d\alpha\,d\beta\right]$$

$$= \int_{-\infty}^t\int_{-\infty}^{t+\tau}\mathbf{g}(t-\alpha)E[\mathbf{w}(\alpha)\mathbf{w}^\mathrm{T}(\beta)]\mathbf{g}^\mathrm{T}(t+\tau-\beta)\,d\alpha\,d\beta \qquad (7.22)$$

Using the fact that $E[\mathbf{w}(\alpha)\mathbf{w}^\mathrm{T}(\beta)] = \mathbf{R_w}(\alpha-\beta) = \mathbf{W}\delta(\beta-\alpha)$, we can write

$$\mathbf{R_y}(\tau) = \int_{-\infty}^t\int_{-\infty}^{t+\tau}\mathbf{g}(t-\alpha)\mathbf{W}\delta(\beta-\alpha)\mathbf{g}^\mathrm{T}(t+\tau-\beta),\,d\alpha\,d\beta$$

$$= \int_{-\infty}^t \mathbf{g}(t-\alpha)\mathbf{W}\mathbf{g}^\mathrm{T}(t+\tau-\alpha)\,d\alpha = \int_{-\infty}^{t+\tau}\mathbf{g}(\lambda)\mathbf{W}\mathbf{g}^\mathrm{T}(\lambda+\tau)\,d\lambda \qquad (7.23)$$

Since $\mathbf{y}(t)$ is the output of a stationary process, we can replace the upper limit of integration in Eq. (7.23) by ∞, provided $\mathbf{y}(t)$ reaches a *steady-state* as $t \to \infty$, i.e. the filter is *asymptotically stable*. Therefore, the correlation matrix of the output of an asymptotically stable system to the white noise can be written as follows:

$$\mathbf{R_y}(\tau) = \int_{-\infty}^\infty \mathbf{g}(\lambda)\mathbf{W}\mathbf{g}^\mathrm{T}(\lambda+\tau)\,d\lambda \qquad (7.24)$$

The power spectral density matrix, $\mathbf{S}_y(\omega)$, for the output, $\mathbf{y}(t)$, can then be obtained by taking the Fourier transform of the correlation matrix, $\mathbf{R}_y(\tau)$, as follows:

$$\mathbf{S}_y(\omega) = \int_{-\infty}^{\infty} \left[\int_{-\infty}^{\infty} \mathbf{g}(\lambda) \mathbf{W} \mathbf{g}^T(\lambda + \tau) \, d\lambda \right] e^{-i\omega\tau} \, d\tau \tag{7.25}$$

Inverting the order of integration in Eq. (7.25), we can write

$$\mathbf{S}_y(\omega) = \int_{-\infty}^{\infty} \mathbf{g}(\lambda) \mathbf{W} \left[\int_{-\infty}^{\infty} \mathbf{g}^T(\lambda + \tau) e^{-i\omega\tau} \, d\tau \right] d\lambda \tag{7.26}$$

The inner integral in Eq. (7.26) can be expressed as follows:

$$\int_{-\infty}^{\infty} \mathbf{g}^T(\lambda + \tau) e^{-i\omega\tau} \, d\tau = \int_{-\infty}^{\infty} \mathbf{g}^T(\xi) e^{-i\omega(\xi-\lambda)} \, d\xi = e^{i\omega\lambda} \mathbf{G}^T(i\omega) \tag{7.27}$$

where $\mathbf{G}(i\omega)$ is the *frequency response matrix* of the filter, i.e. $\mathbf{G}(i\omega) = \mathbf{G}(s = i\omega)$. Substituting Eq. (7.27) into Eq. (7.26), we get the following expression for the power spectral density matrix:

$$\mathbf{S}_y(\omega) = \left[\int_{-\infty}^{\infty} \mathbf{g}(\lambda) e^{i\omega\lambda} \, d\lambda \right] \mathbf{W} \mathbf{G}^T(i\omega) = \mathbf{G}(-i\omega) \mathbf{W} \mathbf{G}^T(i\omega) \tag{7.28}$$

Equation (7.28) is an important result for the output of the white noise filter, and shows that the filtered white noise *does not* have a constant power spectral density, but one which depends upon the frequency response, $\mathbf{G}(i\omega)$, of the filter.

Example 7.4

Let us determine the power spectral density of white noise after passing through the first order low-pass filter of Eq. (7.15). The frequency response of this single-input, single-output filter is the following:

$$G(i\omega) = \omega_0/(i\omega + \omega_0) \tag{7.29}$$

Using Eq. (7.28), we can write the power spectral density as follows:

$$S(\omega) = [\omega_0/(-i\omega + \omega_0)]W[\omega_0/(i\omega + \omega_0)] = W\omega_0^2/(\omega^2 + \omega_0^2) \tag{7.30}$$

Equation (7.30) indicates that the spectrum of the white noise passing through this filter is no longer flat, but begins to decay at $\omega = \omega_0$. For this reason, the filtered white noise is called *colored noise* whose 'color' is indicated by the frequency, ω_0. The correlation function is obtained by taking the inverse Fourier transform of Eq. (7.30) as follows:

$$R_y(\tau) = (1/2\pi) \int [W\omega_0^2/(\omega^2 + \omega_0^2)] e^{i\omega\tau} \, d\omega$$

$$= (W\omega_0/2) \exp(-\omega_0|\tau|) \quad (\tau > 0) \tag{7.31}$$

Since for a stationary process, $R_y(-\tau) = R_y(\tau)$, we can write a general expression for $R_y(\tau)$ valid for all values of τ as follows:

$$R_y(\tau) = (W\omega_0/2)\exp(-\omega_0|\tau|) \tag{7.32}$$

Finally, the covariance matrix of the filtered white noise, $R_y(0)$, is given by

$$R_y(0) = (W\omega_0/2) \tag{7.33}$$

Example 7.5

Atmospheric turbulence is a random process, which has been studied extensively. A semi-empirical model of *vertical gust velocity*, $x(t)$, caused by atmospheric turbulence is the *Dryden spectrum* with the following expression for the power spectral density:

$$S_x(\omega) = a^2T[1 + 3(\omega T)^2]/[1 + (\omega T)^2]^2 \tag{7.34}$$

where a and T are constants. The correlation function, $R_x(\tau)$, of the Dryden spectrum can be calculated by taking the inverse Fourier transform of $S_x(\omega)$ according to Eq. (7.8), either analytically or using MATLAB. It can be shown that the analytical expression of $R_x(\tau)$ is the following:

$$R_x(\tau) = a^2(1 - 0.5|\tau|/T)e^{-|\tau|/T} \tag{7.35}$$

It is customary to express stochastic systems as filters of white noise. What is the *transfer function* of a filter through which white noise must be passed to get the filtered output as the Dryden turbulence? We can employ Eq. (7.28) for the relationship between the power spectral density, $S_x(\omega)$, and the filter transfer function, $G(s)$, which can be written for the scalar case as follows:

$$S_x(\omega) = G(-i\omega)WG(i\omega) = WG(-i\omega)G(i\omega) \tag{7.36}$$

Comparing Eqs. (7.34) and (7.36), we see that a *factorization* of the power spectral density, $S_x(\omega)$, given by

$$S_x(\omega) = a^2T[1 - \sqrt{3}T(i\omega)][1 + \sqrt{3}T(i\omega)]/\{[1 - T(i\omega)]^2[1 + T(i\omega)]^2\}$$
$$= WG(-i\omega)G(i\omega) \tag{7.37}$$

leads to the following possibilities for $G(s)$:

$$G(s) = a(T/W)^{1/2}(1 \pm \sqrt{3}Ts)/(1 \pm Ts)^2 \tag{7.38}$$

Since we want a stable filter (i.e. all poles of $G(s)$ in the left-half plane), we should have a '$+$' sign in the denominator of $G(s)$ in Eq. (7.38). Also, we do not want a *zero* in the right-half plane, because it leads to an undesirable phase plot of $G(i\omega)$, called *non-minimum phase*. A stable transfer function with all zeros in the left-half plane is called *minimum phase*. Hence, the following choice of $G(s)$ would yield a stable and minimum phase filter:

$$G(s) = a(T/W)^{1/2}(1 + \sqrt{3}Ts)/(1 + Ts)^2 \qquad (7.39)$$

Such a method of obtaining the filter's transfer function through a factorization of the power spectral density is called *spectral factorization*, and is commonly employed in deriving white noise filter representations of stochastic systems. For $W = 1$ and $a = 1$, we can obtain a state-space representation of the filter with $T = 1$ s using MATLAB as follows:

```
>> T=1; n = sqrt(T)*[T*sqrt(3) 1]; d = conv([T 1],[T 1]); sysfilt = tf
   (n,d) <enter>

Transfer function:
 1.732 s+1
 --------
s^2+2 s+1
```

A simulated response of the filter to white noise, representing the vertical gust velocity, $x(t)$, due to Dryden turbulence, can be obtained using MATLAB as follows:

```
>>t=0:0.001:1; u = randn(size(t)); [x,t,X] = lsim(sysfilt,u,t);
  <enter>
```

Figure 7.6 shows the Dryden turbulence power spectrum, $S(\omega)/a^2$ plotted against ωT, the Bode plots of the white noise filter transfer function, $G(s)$, for $T = 0.1$,

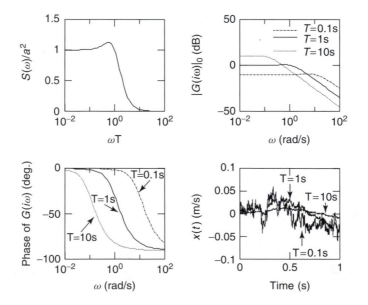

Figure 7.6 Power spectrum, white noise filter Bode plots, and simulated gust vertical velocity of the Dryden turbulence model

1, and 10 s, and simulated vertical gust velocity, $x(t)$, of the filtered white noise with $T = 0.1$, 1, and 10 s. Note that the maximum value of $S(\omega)$ is $1.125a^2$ which occurs near $\omega T = 1$. Note from the Bode plots of the filter with various T values, that the white noise filter for Dryden turbulence acts as a *low-pass filter* with a cut-off frequency, $1/T$. Consequently, as T increases the filtered output (representing turbulence) becomes smoother, because more high-frequency noise is blocked by the filter.

7.4 The Kalman Filter

In the previous section we saw how we can represent stochastic systems by passing white noise through linear systems. Such a representation of stochastic systems is useful for dealing with a plant which we cannot model accurately using only a deterministic model, because of the presence of modeling uncertainties (called *process noise*) and *measurement noise*. A noisy plant is thus a stochastic system, which is modeled by passing white noise through an appropriate linear system. Consider such a plant with the following linear, time-varying state-space representation:

$$\mathbf{x}^{(1)}(t) = \mathbf{A}(t)\mathbf{x}(t) + \mathbf{B}(t)\mathbf{u}(t) + \mathbf{F}(t)\mathbf{v}(t) \tag{7.40}$$

$$\mathbf{y}(t) = \mathbf{C}(t)\mathbf{x}(t) + \mathbf{D}(t)\mathbf{u}(t) + \mathbf{z}(t) \tag{7.41}$$

where $\mathbf{v}(t)$ is the *process noise vector* which may arise due to modeling errors such as neglecting nonlinear or higher-frequency dynamics, and $\mathbf{z}(t)$ is the *measurement noise vector*. By assuming $\mathbf{v}(t)$ and $\mathbf{z}(t)$ to be *white noises*, we will only be extending the methodology of the previous section for a description of the stochastic plant. However, since now we are dealing with a time-varying system as the plant, our definition of white noise has to be modified. For a time-varying stochastic system, the output is a *non-stationary* random signal. Hence, the random noises, $\mathbf{v}(t)$ and $\mathbf{z}(t)$, could in general be *non-stationary white noises*. A *non-stationary white noise* can be obtained by passing the stationary white noise through an *amplifier* with a time-varying gain. The *correlation matrices* of non-stationary white noises, $\mathbf{v}(t)$ and $\mathbf{z}(t)$, can be expressed as follows:

$$\mathbf{R_v}(t, \tau) = \mathbf{V}(t)\delta(t - \tau) \tag{7.42}$$

$$\mathbf{R_z}(t, \tau) = \mathbf{Z}(t)\delta(t - \tau) \tag{7.43}$$

where $\mathbf{V}(t)$ and $\mathbf{Z}(t)$ are the time-varying power spectral density matrices of $\mathbf{v}(t)$ and $\mathbf{z}(t)$, respectively. Note that Eqs. (7.42) and (7.43) yield *infinite* covariance matrices, $\mathbf{R_v}(t, t)$ and $\mathbf{R_z}(t, t)$, respectively, which can be regarded as a characteristic of white noise – stationary or non-stationary.

For designing a control system based on the stochastic plant, we cannot rely on full-state feedback, because we cannot predict the state-vector, $\mathbf{x}(t)$, of the stochastic plant. Therefore, an *observer* is required for estimating the state-vector, based upon a measurement of the output, $\mathbf{y}(t)$, given by Eq. (7.41) and a known input, $\mathbf{u}(t)$. Using the pole-placement methods of Chapter 5 we can come up with an observer that has poles at desired locations. However, such an observer would not take into account the power spectra of the process and measurement noise. Also, note the difficulty encountered in Chapter 5 for designing observers for multi-input, multi-output plants, which limits the pole-placement approach of observer design largely to single-output plants. To take into account the fact that the measured output, $\mathbf{y}(t)$, and state-vector of the plant, $\mathbf{x}(t)$, are random vectors, we need an observer that estimates the state-vector based upon *statistical* (rather than deterministic) description of the *vector* output and plant state. Such an observer is the *Kalman Filter*. Rather than being an ordinary observer of Chapter 5, the *Kalman filter* is an *optimal* observer, which *minimizes* a statistical measure of the *estimation error*, $\mathbf{e_o}(t) = \mathbf{x}(t) - \mathbf{x_o}(t)$, where $\mathbf{x_o}(t)$ is the *estimated state-vector*. The state-equation of the Kalman filter is that of a time-varying observer (similar to Eq. (5.103)), and can be written as follows:

$$\mathbf{x_o^{(1)}}(t) = \mathbf{A}(t)\mathbf{x_o}(t) + \mathbf{B}(t)\mathbf{u}(t) + \mathbf{L}(t)[\mathbf{y}(t) - \mathbf{C}(t)\mathbf{x_o}(t) - \mathbf{D}(t)\mathbf{u}(t)] \qquad (7.44)$$

where $\mathbf{L}(t)$ is the *gain matrix* of the Kalman filter (also called the *optimal observer gain matrix*). Being an *optimal* observer, the Kalman filter is a counterpart of the optimal regulator of Chapter 6. However, while the optimal regulator minimizes an objective function based on transient and steady-state response and control effort, the Kalman filter minimizes the *covariance of the estimation error*, $\mathbf{R_e}(t, t) = E[\mathbf{e_o}(t)\mathbf{e_o^T}(t)]$. Why is it useful to minimize the covariance of estimation error? Recall that the state-vector, $\mathbf{x}(t)$, is a random vector. The estimated state, $\mathbf{x_o}(t)$, is based on the measurement of the output, $\mathbf{y}(t)$, for a *finite* time, say T, where $t \geq T$. However, a true statistical average (or *mean*) of $\mathbf{x}(t)$ would require measuring the output for an infinite time (i.e. taking *infinite* number of samples), and then finding the expected value of $\mathbf{x}(t)$. Hence, the *best estimate* that the Kalman filter could obtain for $\mathbf{x}(t)$ is not the *true mean*, but a *conditional mean*, $\mathbf{x_m}(t)$, based on only a *finite* time record of the output, $\mathbf{y}(t)$, for $T \leq t$, written as follows:

$$\mathbf{x_m}(t) = E[\mathbf{x}(t) : \mathbf{y}(T), T \leq t] \qquad (7.45)$$

There may be a *deviation* of the estimated state-vector, $\mathbf{x_o}(t)$, from the conditional mean, $\mathbf{x_m}(t)$, and we can write the estimated state-vector as follows:

$$\mathbf{x_o}(t) = \mathbf{x_m}(t) + \Delta\mathbf{x}(t) \qquad (7.46)$$

where $\Delta\mathbf{x}(t)$ is the deviation from the conditional mean. The *conditional covariance matrix* (i.e. the covariance matrix based on a finite record of the output) of the estimation error is given by

$$\mathbf{R_e}(t, t) = E[\mathbf{e_o}(t)\mathbf{e_o^T}(t) : \mathbf{y}(T), T \leq t] = E[\{\mathbf{x}(t) - \mathbf{x_o}(t)\}\{\mathbf{x^T}(t) - \mathbf{x_o^T}(t)\} : \mathbf{y}(T), T \leq t] \qquad (7.47)$$

Equation (7.47) can be simplified using Eq. (7.45) as follows:

$$\mathbf{R_e}(t, t) = E[\mathbf{x}(t)\mathbf{x^T}(t)] - \mathbf{x_o}(t)\mathbf{x_m^T}(t) - \mathbf{x_o^T}(t)\mathbf{x_m}(t) + \mathbf{x_o}(t)\mathbf{x_o^T}(t) \qquad (7.48)$$

Finally, substituting Eq. (7.46) into Eq. (7.48) and simplifying, we get

$$\mathbf{R_e}(t, t) = E[\mathbf{x}(t)\mathbf{x^T}(t)] - \mathbf{x_m}(t)\mathbf{x_m^T}(t) + \Delta\mathbf{x}(t)\Delta\mathbf{x^T}(t) \qquad (7.49)$$

From Eq. (7.49) it is clear that the best estimate of state-vector, implying $\Delta \mathbf{x}(t) = 0$ (i.e. $\mathbf{x_o}(t) = \mathbf{x_m}(t)$), would result in a *minimization* of the conditional covariance matrix, $\mathbf{R_e}(t, t)$. In other words, minimization of $\mathbf{R_e}(t, t)$ yields the optimal (i.e. the best) observer, which is the Kalman filter.

Let us derive the expression for the gain matrix of the Kalman filter, $\mathbf{L}(t)$, which minimizes $\mathbf{R_e}(t, t)$, i.e. which makes the estimated state-vector equal to the conditional mean vector. The *optimal* estimation error is thus $\mathbf{e_o}(t) = \mathbf{x}(t) - \mathbf{x_m}(t)$. Subtracting Eq. (7.44) from Eq. (7.40) and substituting Eq. (7.41), we can write the following state-equation for the optimal estimation error:

$$\mathbf{e_o}^{(1)}(t) = [\mathbf{A}(t) - \mathbf{L}(t)\mathbf{C}(t)]\mathbf{e_o}(t) + \mathbf{F}(t)\mathbf{v}(t) - \mathbf{L}(t)\mathbf{z}(t) \tag{7.50}$$

Note that since $\mathbf{v}(t)$ and $\mathbf{z}(t)$ are white noises, the following vector is also a (non-stationary) white noise

$$\mathbf{w}(t) = \mathbf{F}(t)\mathbf{v}(t) - \mathbf{L}(t)\mathbf{z}(t) \tag{7.51}$$

To find the covariance of the estimation error, we must somehow find an expression for the solution of Eq. (7.50), which is a linear, time-varying system excited by a non-stationary white noise, $\mathbf{w}(t)$. Let us write Eqs. (7.50) and (7.51) as follows:

$$\mathbf{e_o}^{(1)}(t) = \mathbf{A_o}(t)\mathbf{e_o}(t) + \mathbf{w}(t) \tag{7.52}$$

where $\mathbf{A_o}(t) = [\mathbf{A}(t) - \mathbf{L}(t)\mathbf{C}(t)]$. The solution to Eq. (7.52) for a given initial condition, $\mathbf{e_o}(t_0)$, can be expressed as follows:

$$\mathbf{e_o}(t) = \Phi(t, t_0)\mathbf{e_o}(t_0) + \int_{t_0}^{t} \Phi(t, \lambda)\mathbf{w}(\lambda)\, d\lambda \tag{7.53}$$

where $\Phi(t, t_0)$ is the *state-transition matrix* of the time-varying estimation error state-equation, Eq. (7.52). Then the conditional covariance of estimation error can be written as follows (dropping the notation $\mathbf{y}(T)$, $T \leq t$, from the expected value for convenience):

$$\mathbf{R_e}(t, t) = E[\mathbf{e_o}(t)\mathbf{e_o}^{\mathbf{T}}(t)] = E\left[\Phi(t, t_0)\mathbf{e_o}(t_0)\mathbf{e_o}^{\mathbf{T}}(t_0)\Phi^{\mathbf{T}}(t, t_0) \right.$$
$$\left. +\mathbf{e_o}(t_0)\int_{t_0}^{t} \Phi(t, \lambda)\mathbf{w}(\lambda)\, d\lambda + \left\{\int_{t_0}^{t} \Phi(t, \lambda)\mathbf{w}(\lambda)\, d\lambda\right\} \mathbf{e_o}(t_0) \right.$$
$$\left. + \int_{t_0}^{t}\int_{t_0}^{t} \Phi(t, \lambda)\mathbf{w}(\lambda)\mathbf{w}^{\mathbf{T}}(\xi)\Phi^{\mathbf{T}}(t, \xi)\, d\lambda d\xi \right] \tag{7.54}$$

or, using the properties of the expected value operator, we can write

$$\mathbf{R_e}(t, t) = \Phi(t, t_0)E[\mathbf{e_o}(t_0)\mathbf{e_o}^{\mathbf{T}}(t_0)]\Phi^{\mathbf{T}}(t, t_0) + \mathbf{e_o}(t_0)\int_{t_0}^{t} \Phi(t, \lambda)E[\mathbf{w}(\lambda)]\, d\lambda$$
$$+ \left\{\int_{t_0}^{t} \Phi(t, \lambda)E[\mathbf{w}(\lambda)]\, d\lambda\right\} \mathbf{e_o}(t_0)$$
$$+ \int_{t_0}^{t}\int_{t_0}^{t} \Phi(t, \lambda)E[\mathbf{w}(\lambda)\mathbf{w}^{\mathbf{T}}(\xi)]\Phi^{\mathbf{T}}(t, \xi)\, d\lambda d\xi \tag{7.55}$$

Since the expected value of white noise is zero, i.e. $E[\mathbf{w}(t)] = \mathbf{0}$, and the *correlation matrix* of white noise is given by

$$E[\mathbf{w}(\lambda)\mathbf{w}^{\mathbf{T}}(\xi)] = \mathbf{W}(\lambda)\delta(\lambda - \xi) \tag{7.56}$$

we can simplify Eq. (7.55) as follows:

$$\mathbf{R_e}(t, t) = \Phi(t, t_0)E[\mathbf{e_o}(t_0)\mathbf{e_o^T}(t_0)]\Phi^{\mathbf{T}}(t, t_0) + \int_{t_0}^{t} \Phi(t, \lambda)\mathbf{W}(\lambda)\Phi^{\mathbf{T}}(t, \lambda)\, d\lambda \tag{7.57}$$

If the initial estimation error, $\mathbf{e_o}(t_0)$, is also a random vector, we can write the initial conditional covariance matrix as follows:

$$\mathbf{R_e}(t_0, t_0) = E[\mathbf{e_o}(t_0)\mathbf{e_o^T}(t_0)] \tag{7.58}$$

Substituting Eq. (7.58) into Eq. (7.57), we can write

$$\mathbf{R_e}(t, t) = \Phi(t, t_0)\mathbf{R_e}(t_0, t_0)\Phi^{\mathbf{T}}(t, t_0) + \int_{t_0}^{t} \Phi(t, \lambda)\mathbf{W}(\lambda)\Phi^{\mathbf{T}}(t, \lambda)\, d\lambda \tag{7.59}$$

Equation (7.59) describes how the optimal estimation error covariance evolves with time. However, the state-transition matrix for the time varying system, $\Phi(t, t_0)$, is an *unknown quantity*. Fortunately, we have already encountered an integral similar to that in Eq. (7.59) while deriving the *optimal regulator gain* in Chapter 6. An equivalent integral for the *optimal control problem* is given in Eq. (6.8) for $\mathbf{M}(t, t_f)$. Comparing Eqs. (7.59) and (6.8) we find that $\Phi_{\mathbf{CL}}^{\mathbf{T}}(\tau, t)$ in Eq. (6.8) is replaced by $\Phi(t, \lambda)$ in Eq. (7.59), where τ and λ are the variables of integration in Eqs. (6.8) and (7.59), respectively. Furthermore, the matrix $[\mathbf{Q}(\tau) + \mathbf{K}^{\mathbf{T}}(\tau)\mathbf{R}(\tau)\mathbf{K}(\tau)]$ in Eq. (6.8) is replaced by $\mathbf{W}(\lambda)$ in Eq. (7.59). Also, the direction of integration in time is *opposite* in Eqs. (6.8) $(t \to t_f)$ and (7.59) $(t_0 \to t)$. Thus, taking a cue from the similarity (and differences) between Eqs. (6.8) and (7.59), we can write a *differential equation* for $\mathbf{R_e}(t, t)$ similar to that for $\mathbf{M}(t, t_f)$, Eq. (6.14), as follows:

$$d\mathbf{R_e}(t, t)/dt = \mathbf{A_o}(t)\mathbf{R_e}(t, t) + \mathbf{R_e}(t, t)\mathbf{A_o^T}(t) + \mathbf{W}(t) \tag{7.60}$$

Note that Eq. (7.60) is an *ordinary differential equation*, rather than a *partial differential equation* (Eq. (6.14)). Also, due to the fact that time progresses in a *forward direction* $(t_0 \to t)$ in Eq. (7.60), rather than in a *backward direction* $(t_f \to t)$ in Eq. (6.14), the negative sign on the left-hand side of Eq. (6.14) is replaced by a positive sign on the left-hand side of Eq. (7.60). Equation (7.60) is called the *covariance equation* for the Kalman filter, and must be solved with the initial condition given by Eq. (7.58). Note that we *do not* need to know the state-transition matrix, $\Phi(t, t_0)$, for solving for the optimal covariance matrix. Equation (7.60) is the counterpart of the matrix Riccati equation for the Kalman filter.

Substituting Eqs. (7.51), (7.42) and (7.43) into Eq. (7.56), and *assuming* that the two white noise signals, $\mathbf{v}(t)$ and $\mathbf{z}(t)$, are *uncorrelated with each other*, i.e. $E[\mathbf{v}(t)\mathbf{z}^{\mathbf{T}}(\tau)] = E[\mathbf{z}(t)\mathbf{v}^{\mathbf{T}}(\tau)] = \mathbf{0}$, we can write the following expression relating $\mathbf{W}(t)$ to the spectral densities of the two white noise signals, $\mathbf{V}(t)$ and $\mathbf{Z}(t)$, as follows:

$$\mathbf{W}(t) = \mathbf{F}(t)\mathbf{V}(t)\mathbf{F}^\mathbf{T}(t) + \mathbf{L}(t)\mathbf{Z}(t)\mathbf{L}^\mathbf{T}(t) \tag{7.61}$$

Substituting Eq. (7.61) into Eq. (7.60) and substituting $\mathbf{A_o}(t) = [\mathbf{A}(t) - \mathbf{L}(t)\mathbf{C}(t)]$, we can express the covariance equation as follows:

$$d\mathbf{R_e}(t, t)/dt = [\mathbf{A}(t) - \mathbf{L}(t)\mathbf{C}(t)]\mathbf{R_e}(t, t) + \mathbf{R_e}(t, t))[\mathbf{A}(t) - \mathbf{L}(t)\mathbf{C}(t)]^\mathbf{T}$$

$$+ \mathbf{F}(t)\mathbf{V}(t)\mathbf{F}^\mathbf{T}(t) + \mathbf{L}(t)\mathbf{Z}(t)\mathbf{L}^\mathbf{T}(t) \tag{7.62}$$

Comparing Eq. (7.62) with Eq. (6.21) and using the steps similar to those of Section 6.1.2, we can write the *optimal Kalman filter gain*, $\mathbf{L^o}(t)$, as follows:

$$\mathbf{L^o}(t) = \mathbf{R_e^o}(t, t)\mathbf{C}^\mathbf{T}(t)\mathbf{Z}^{-1}(t) \tag{7.63}$$

where $\mathbf{R_e^o}(t, t)$ is the *optimal covariance matrix* satisfying the following *matrix Riccati equation*:

$$d\mathbf{R_e^o}(t, t)/dt = \mathbf{A}(t)\mathbf{R_e^o}(t, t) + \mathbf{R_e^o}(t, t)\mathbf{A}^\mathbf{T}(t)$$

$$- \mathbf{R_e^o}(t, t)\mathbf{C}^\mathbf{T}(t)\mathbf{Z}^{-1}(t)\mathbf{C}(t)\mathbf{R_e^o}(t, t) + \mathbf{F}(t)\mathbf{V}(t)\mathbf{F}^\mathbf{T}(t) \tag{7.64}$$

We can derive a more general matrix Riccati equation for the Kalman filter when the two noise signals are correlated with each other with the following *cross-correlation matrix*:

$$E[\mathbf{v}(t)\mathbf{z}^\mathbf{T}(\tau)] = \Psi(t)\delta(t - \tau) \tag{7.65}$$

where $\Psi(t)$ is the *cross-spectral density matrix* between $\mathbf{v}(t)$ and $\mathbf{z}(t)$. Then the optimal Kalman filter gain can be shown to be given by

$$\mathbf{L^o}(t) = [\mathbf{R_e^o}(t, t)\mathbf{C}^\mathbf{T}(t) + \mathbf{F}(t)\Psi(t)]\mathbf{Z}^{-1}(t) \tag{7.66}$$

where $\mathbf{R_e^o}(t, t)$ is the *optimal covariance matrix* satisfying the following general *matrix Riccati equation*:

$$d\mathbf{R_e^o}(t, t)/dt = \mathbf{A_G}(t)\mathbf{R_e^o}(t, t) + \mathbf{R_e^o}(t, t)\mathbf{A_G^T}(t)$$

$$- \mathbf{R_e^o}(t, t)\mathbf{C}^\mathbf{T}(t)\mathbf{Z}^{-1}(t)\mathbf{C}(t)\mathbf{R_e^o}(t, t) + \mathbf{F}(t)\mathbf{V_G}(t)\mathbf{F}^\mathbf{T}(t) \tag{7.67}$$

with

$$\mathbf{A_G}(t) = \mathbf{A}(t) - \mathbf{F}(t)\Psi(t)\mathbf{Z}^{-1}(t)\mathbf{C}(t) \tag{7.68}$$

$$\mathbf{V_G}(t) = \mathbf{V}(t) - \Psi(t)\mathbf{Z}^{-1}(t)\Psi^\mathbf{T}(t) \tag{7.69}$$

For simplicity of notation, we will use $\mathbf{L}$ to denote the optimal gain matrix of the Kalman filter, rather than $\mathbf{L^o}(t)$.

The appearance of matrix Riccati equation for the Kalman filter problem is not surprising, since the Kalman filter is an optimal observer. Hence, Kalman filter problem is solved quite similarly to the optimal control problem. Usually, we are interested in a *steady Kalman filter*, i.e. the Kalman filter for which the covariance matrix converges to a constant in the limit $t \to \infty$. Such a Kalman filter results naturally when the plant is *time-invariant* and the noise signals are *stationary white noises*. In such a case, the estimation error would also be a stationary white noise with a *constant optimal covariance matrix*, $\mathbf{R_e^o}$. For the steady-state time-varying problem with non-stationary white noise signals, or the time-invariant problem with stationary white noise signals, the following *algebraic Riccati equation* results for the optimal covariance matrix, $\mathbf{R_e^o}$:

$$0 = \mathbf{A_G R_e^o} + \mathbf{R_e^o A_G^T} - \mathbf{R_e^o C^T Z^{-1} C R_e^o} + \mathbf{F V_G F^T} \tag{7.70}$$

where the matrices on the right-hand side are either constant (time-invariant), or steady-state values for the time-varying plant. The sufficient conditions for the existence of a *unique, positive definite* solution to the algebraic Riccati equation (Eq. (7.70)) are the same as those stated in Chapter 6: the system with state-dynamics matrix, $\mathbf{A}$, and observation matrix, $\mathbf{C}$, is *detectable*, and the system with state-dynamics matrix, $\mathbf{A}$, and controls coefficient matrix, $\mathbf{B} = \mathbf{F V}^{1/2}$, is *stabilizable* ($\mathbf{V}^{1/2}$ denotes the *matrix square-root* of $\mathbf{V}$, which satisfies $\mathbf{V}^{1/2}(\mathbf{V}^{1/2})^T = \mathbf{V}$). These sufficient conditions will be met if the system with state-dynamics matrix, $\mathbf{A}$, and observation matrix, $\mathbf{C}$, is *observable*, $\mathbf{V}$ is a *positive semi-definite* matrix, and $\mathbf{Z}$ is a *positive definite matrix*. We can solve the algebraic Riccati equation for steady-state Kalman filter using either the MATLAB command *are*, or more specifically, the specialized Kalman filter commands *lqe* or *lqe2*, which are called as follows:

```
>>[L,P,E] = lqe(A,F,C,V,Z,Psi) <enter>
```

where $\mathbf{A}$, $\mathbf{F}$, $\mathbf{C}$, are the plant's state coefficient matrices, $\mathbf{V}$ is the process noise spectral density matrix, $\mathbf{Z}$ is the measurement noise spectral density matrix, and $\mathbf{Psi} = \Psi$, the cross-spectral density matrix of process and measurement noises. If $\mathbf{Psi}$ is not specified (by having only the first four input arguments in *lqe*), it is assumed that $\Psi = \mathbf{0}$. $\mathbf{L}$ is the returned Kalman filter optimal gain, $\mathbf{P} = \mathbf{R_e^o}$, the returned optimal (conditional) covariance matrix of the estimation error, and $\mathbf{E}$ is a vector containing the eigenvalues of the Kalman filter (i.e. the eigenvalues of $\mathbf{A\text{-}LC}$). The command *lqe2*, is used in a manner similar to *lqe*, but utilizes a more numerically robust algorithm for solving the algebraic Riccati equation than *lqe*. A third MATLAB command, *lqew*, is also available, which solves a special Kalman filter problem in which the output equation is $\mathbf{y}(t) = \mathbf{C}(t)\mathbf{x}(t) + \mathbf{D}(t)\mathbf{u}(t) + \mathbf{G}(t)\mathbf{v}(t) + \mathbf{z}(t)$, where $\mathbf{v}(t)$ is the process noise that affects the output through the coefficient matrix, $\mathbf{G}(t)$, and is *uncorrelated* with the measurement noise, $\mathbf{z}(t)$.

Example 7.6

Let us design a Kalman filter for the fighter aircraft of Example 5.13. It is assumed that only the first two of the three state variables are measured. The state coefficient matrices for the linear, time-invariant model are as follows:

$$\mathbf{A} = \begin{bmatrix} -1.7 & 50 & 260 \\ 0.22 & -1.4 & -32 \\ 0 & 0 & -12 \end{bmatrix}; \quad \mathbf{B} = \begin{bmatrix} -272 \\ 0 \\ 14 \end{bmatrix} 0; \quad \mathbf{F} = \begin{bmatrix} 0.02 & 0.1 \\ -0.0035 & 0.004 \\ 0 & 0 \end{bmatrix}$$

$$\mathbf{C} = \begin{bmatrix} 1 & 0 & 0 \\ 0 & 1 & 0 \end{bmatrix}; \quad \mathbf{D} = 0 \tag{7.71}$$

The third-order plant has a single input, two outputs and two process noise variables. The process noise spectral density matrix for the bomber is assumed to be $V = F^T F$, while that for the measurement noise is, $Z = 10CC^T$. Also, assume that the cross-spectral density of process and measurement noise is zero, i.e. $\Psi = 0$. Then the Kalman filter gain is calculated using the MATLAB command *lqe* as follows:

```
>>A=[-1.7 50 260; 0.22 -1.4 -32;0 0 -12];F=[0.02 0.1;-0.0035 0.004;0 0];
  C=[1 0 0; 0 1 0]; <enter>

>>[L,P,E] = lqe(A,F,C,F'*F,10*C*C') <enter>
L =
         3.5231          0.2445
         0.2445          0.0170
         0               0


P =
        35.2306          2.4450          0
         2.4450          0.1697          0
         0               0               0


E =
 -12.0000
 -4.8700
 -1.7700
```

The Kalman filter is thus *stable* with eigenvalues at $\lambda = -12$, $\lambda = -4.87$, and $\lambda = -1.77$. Let us simulate the Kalman filter estimation error with $v(t)$ and $z(t)$ generated using *randn* as follows:

```
>>randn('seed',0); t=0:0.01:10; v = randn(size(t,2),2); z = randn(size
  (t,2),2); w = F*v'-L*z'; <enter>

>>[e,X] = lsim(A-L*C,eye(3),eye(3),zeros(3,3),w',t); <enter>
```

The simulated white noise, representing an element of $v(t)$ or $z(t)$, and the simulated estimation error, $e_o(t) = [e_{01}(t)e_{02}(t)e_{03}(t)]^T$, are plotted in Figure 7.7.

Note that the third element of the estimation error vector, $e_o(t)$, is identically zero (because the last row of F is zero), while the first two elements are much smoother and with smaller magnitudes than the noise. The first two elements of the estimation error are random variables with zero mean values. How accurate is our simulated estimation error? Not very accurate, because the noise vector we have simulated does not have *exactly* the same power spectral density that we have assumed. In fact, the *simulated* white noise is far from being a perfect white noise, which we can verify by calculating the covariance matrices and mean values of $v(t)$ and $z(t)$, as follows:

```
    >>cov(v) <enter>
    ans =
          1.0621 -0.0307
         -0.0307          1.0145
```

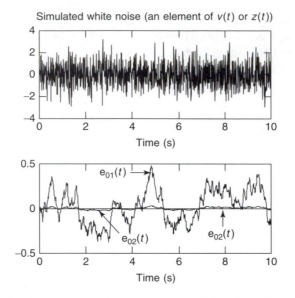

Figure 7.7 Simulated white noise and estimation errors of the Kalman filter for a fighter aircraft

```
>>cov(z) <enter>
ans =
        1.0229          0.0527  -0.0133
        0.0527          0.8992          0.0131
       -0.0133          0.0131          0.9467

>>mean(v) <enter>
ans =
 -0.0139 -0.0047

>>mean(z) <enter>
ans =
 -0.0125 0.0025 -0.0421
```

Recall that a white noise vector must have *infinite* covariance matrix and a *zero* mean vector. The covariance of the *simulated* estimation error is the following:

```
>>cov(e) <enter>
ans =
        0.0347          0.0024          0
        0.0024          0.0002          0
        0               0               0
```

Note that the simulated estimation error's covariance is different from the optimal covariance matrix, **P**, obtained as the solution to the algebraic Riccati equation of the Kalman filter. However, the ratios between the elements of **P** are the same as those between the elements of *cov(e)*. Hence, the matrix *P* must be scaled by a scalar constant to represent the covariance of estimation error. To reduce the difference between the covariance matrices of the simulated and optimal estimation error, we should modify our assumptions of the power spectral densities, or simulate the white noise more accurately using the random number generator, *randn*. Since the actual noise will almost *never* be a white noise, there is no point in spending the time to accurately model white noise better on a computer. Instead, we should *fine tune* the Kalman filter gain by appropriately selecting the spectral densities $\mathbf{V}$, $\mathbf{Z}$, and Ψ. After some trial and error, by selecting $\mathbf{V} = \mathbf{F}^T\mathbf{F}$, $\mathbf{Z} = 0.01\mathbf{CC}^T$, and $\Psi = \mathbf{0}$, we get the Kalman filter gain and optimal estimation error covariance as follows:

```
>>[L,P,E] = lqe(A,F,C,F'*F,0.01*C*C') <enter>

L =
        3.5247          0.2446
        0.2446          0.0170
        0               0

P =
        0.0352          0.0024          0
        0.0024          0.0002          0
        0               0               0

E =
        -12.0000
        -4.8708
        -1.7709
```

and the simulated estimation error is re-calculated as

```
>>w = F*v'-L*z';sysob=ss(A-L*C,eye(3),eye(3),zeros(3,3));[e,t,X] = lsim
  (sysob,w',t); <enter>
```

with the simulated estimation error's covariance matrix given by

```
>>cov(e) <enter>
ans =
        0.0347          0.0024          0
        0.0024          0.0002          0
        0               0               0
```

which is the same as calculated previously, and quite close to the new optimal covariance, $\mathbf{P}$. The new Kalman filter gain and eigenvalues are not changed by much (indicating little change in the estimation error time response), but the scaling of the optimal covariance of the estimation error is now greatly improved. The *mean* estimation error vector, $\mathbf{e}_{om}$, is calculated as follows:

```
>> mean(e) <enter>

ans =
        0.0305          0.0021          0
```

which is quite close to zero vector, as desired. The accuracy of the mean value will, of course, improve by taking more time points in the simulation, and in the limit of *infinite* number of time points, the mean would become *exactly* zero.

The Kalman filter approach provides us with a procedure for designing observers for multivariable plants. Such an observer is guaranteed to be optimal in the presence of white noise signals. However, since white noise is rarely encountered, the power spectral densities used for designing the Kalman filter can be treated as *tuning* parameters to arrive at an observer for multivariable plants that has desirable properties, such as performance and robustness. The linear Kalman filter can also be used to design observers for *nonlinear* plants, by treating nonlinearities as process noise with appropriate power spectral density matrix.

Example 7.7

Let us design a Kalman filter to estimate the states of a double-pendulum (Example 4.12, Figure 1.5). A choice of the state variables for this fourth order system is $x_1(t) = \theta_1(t)$, $x_2(t) = \theta_2(t)$, $x_3(t) = \theta_1^{(1)}(t)$; $x_4(t) = \theta_2^{(1)}(t)$, which results in *nonlinear* state-equations given by Eq. (4.93). The function M-file for evaluating the time derivative of the state-vector, $\mathbf{x}^{(1)}(t)$, is called *doub.m* and is tabulated in Table 4.7. Thus, M-file assumes a known input torque acting on the pendulum given by $u(t) = 0.01\sin(5t)N - m$. It is desired to design a *linear* Kalman filter based on the known input, $u(t)$, and measurement of the angular position and angular velocity of the mass, m_1, i.e. $\mathbf{y}(t) = [\theta_1(t); \theta_1^{(1)}(t)]^T$, with the following *linearized* plant model:

$$x_1^{(1)}(t) = x_3(t)$$
$$x_2^{(1)}(t) = x_4(t)$$
$$x_3^{(1)}(t) = [m_2gx_2(t) - (m_1 + m_2)gx_1(t)]/(m_1L_1) \quad (7.72)$$
$$x_4^{(1)}(t) = -[gx_2(t) + L_1dx_3(t)/dt]/L_2 + u(t)/(m_2L_2^2)$$

which results in the following linear state coefficient matrices for the plant:

$$\mathbf{A} = \begin{bmatrix} 0 & 0 & 1 & 0 \\ 0 & 0 & 0 & 1 \\ -(m_1+m_2)g/(m_1L_1) & m_2g/(m_1L_1) & 0 & 0 \\ (m_1+m_2)g/(m_1L_2) & -g(m_2/m_1+1)/L_2 & 0 & 0 \end{bmatrix};$$

$$\mathbf{B} = \begin{bmatrix} 0 \\ 0 \\ -1/(m_2L_2^2) \end{bmatrix}$$

$$\mathbf{C} = \begin{bmatrix} 1 & 0 & 0 & 0 \\ 0 & 0 & 1 & 0 \end{bmatrix}; \quad \mathbf{D} = \begin{bmatrix} 0 \\ 0 \end{bmatrix} \quad (7.73)$$

The solution of the exact nonlinear state-equations of the plant with zero initial condition, i.e. $\mathbf{x}(0) = \mathbf{0}$, is obtained using the MATLAB Runge–Kutta solver *ode45* as follows:

```
>>[t,X] = ode45(@doub,[0 20],zeros(4,1)); <enter>
```

However, it is assumed that the state-vector, $\mathbf{x}(t)$, solved above, is unavailable for measurement, and only the output, $\mathbf{y}(t)$, can be measured, which is calculated by

```
>>C = [1 0 0 0; 0 0 1 0]; y = C*X'; <enter>
```

The state coefficient matrices, $\mathbf{A}$ and $\mathbf{B}$, of the linearized plant are calculated with $m_1 = 1$ kg, $m_2 = 2$ kg, $L_1 = 1$ m, $L_2 = 2$ m, and $g = 9.8$ m/s^2 (same values as those used in *doub.m* for the nonlinear plant) as follows:

```
>>m1=1;m2=2;L1=1;L2=2;g=9.8; A = [0 0 1 0; 0 0 0 1; -(m1+m2)*g/(m1*L1)
  m2*g/(m1*L1) 0 0; (m1+m2)*g/(m1*L2) -g*(m2/m1+1)/L2 0 0] <enter>

A =
   0             0         1.0000        0
   0             0         0             1.0000
  -29.4000      19.6000    0             0
   14.7000     -14.7000    0             0

>>B = [0 0 0 1/(m2*L2*L2)]' <enter>


B =
      0
      0
      0
   0.1250
```

Since nonlinearities appear as *process noise* for all the state variables of the linear plant model, the process noise coefficient matrix is assumed to be an identity matrix, i.e. $\mathbf{F} = \mathbf{I}$. The linear Kalman filter is to be designed using the spectral densities of process and measurement noise, $\mathbf{V}$, $\mathbf{Z}$, and $\mathbf{\Psi}$ such that the exact state-vector, $\mathbf{x}(t)$, is accurately estimated. After some trial and error, we select $\mathbf{V} = 10^6\mathbf{I}$, $\mathbf{Z} = \mathbf{CC^T}$, and $\mathbf{\Psi} = 0$, and calculate the Kalman filter gain as follows:

```
>>[L,P,E]=lqe(A,eye(4),C,1e6*eye(4),C*C') <enter>

L =
   9.9989e+002  -1.3899e+001
   1.5303e+001   1.0256e+003
  -1.3899e+001   1.0202e+003
   2.1698e+001   4.8473e+002
```

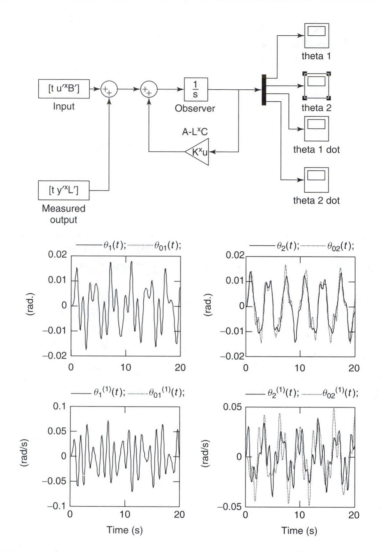

Figure 7.8 Actual (simulated) state-vector, $\mathbf{x}(t) = [\theta_1(t); \theta_2(t); \theta_1^{(1)}(t); \theta_2^{(1)}(t)]^T$, and estimated state-vector, $\mathbf{x_o}(t) = [\theta_{o1}(t); \theta_{o2}(t); \theta_{o1}^{(1)}(t); \theta_{o2}^{(1)}(t)]^T$, with a linear Kalman filter for the nonlinear dynamics of a double pendulum

```
P  =
  9.9989e+002  1.5303e+001  -1.3899e+001  2.1698e+001
  1.5303e+001  5.3371e+004   1.0256e+003  2.6027e+004
 -1.3899e+001  1.0256e+003   1.0202e+003  4.8473e+002
  2.1698e+001  2.6027e+004   4.8473e+002  1.2818e+006

E  =
 -1.0000e+003+  1.5202e+001i
 -1.0000e+003-  1.5202e+001i
 -1.8785e+001
 -1.3039e+000
```

Finally, we calculate the estimated state-vector, $\mathbf{x_o}(t)$, by solving the Kalman filter state-equation (Eq. (7.44) with the known input, $u(t)$, and the measured output vector, $\mathbf{y}(t)$, using the SIMULINK block-diagram shown in Figure 7.8. The input, $u(t)$, is calculated in MATLAB work-space as follows, at the *same* time points previously used for generating the state-vector, $\mathbf{x}(t)$, with *ode45*:

```
>>u = 0.01*sin(5*t'); <enter>
```

The *actual* (i.e. generated by solving nonlinear equations through *ode45*) and estimated (SIMULINK) state variables are compared in Figure 7.8. Note that the state variables $\theta_1(t)$ and $\theta_1^{(1)}(t)$ are almost exactly estimated, as expected, because these state variables are directly measured. The estimation errors for $\theta_2(t)$ and $\theta_2^{(1)}(t)$ are appreciable, but reasonable, since we are trying to estimate a nonlinear plant by a linear Kalman filter.

7.5 Optimal (Linear, Quadratic, Gaussian) Compensators

In Chapter 5, we had used the *separation principle* to separately design a regulator and an observer using pole-placement, and put them together to form a *compensator* for the plant whose state-vector was unmeasurable. In Chapter 6, we presented *optimal control* techniques for designing linear regulators for multi-input plants that minimized a quadratic objective function, which included transient, terminal, and control penalties. In the present chapter, we have introduced the Kalman filter, which is an *optimal* observer for multi-output plants in the presence of process and measurement noise, modeled as white noises. Therefore, using a separation principle similar to that of Chapter 5, we can combine the optimal regulator of Chapter 6 with the optimal observer (the Kalman filter), and end up with an *optimal compensator* for multivariable plants. Since the optimal compensator is based upon a linear plant, a quadratic objective function, and an assumption of white noise that has a *normal*, or *Gaussian*, probability distribution, the optimal compensator is popularly called the *Linear, Quadratic, Gaussian* (or LQG) compensator. In short, the optimal compensator design process is the following:

(a) Design an *optimal* regulator for a linear plant assuming *full-state feedback* (i.e. assuming all the state variables are available for measurement) and a quadratic objective function (such as that given by Eq. (6.3)). The regulator is designed to generate a control input, $\mathbf{u}(t)$, based upon the measured state-vector, $\mathbf{x}(t)$.

(b) Design a Kalman filter for the plant assuming a known control input, $\mathbf{u}(t)$, a measured output, $\mathbf{y}(t)$, and white noises, $\mathbf{v}(t)$ and $\mathbf{z}(t)$, with known power spectral densities. The Kalman filter is designed to provide an *optimal estimate* of the state-vector, $\mathbf{x_o}(t)$.

(c) Combine the separately designed optimal regulator and Kalman filter into an optimal compensator, which generates the input vector, $\mathbf{u}(t)$, based upon the estimated state-vector, $\mathbf{x_o}(t)$, rather than the actual state-vector, $\mathbf{x}(t)$, and the measured output vector, $\mathbf{y}(t)$.

Since the optimal regulator and Kalman filter are designed separately, they can be selected to have desirable properties that are independent of one another. The closed-loop eigenvalues consist of the regulator eigenvalues and the Kalman filter eigenvalues, as seen in Chapter 5. The block diagram and state-equations for the closed-loop system with optimal compensator are the same as those for the pole-placement compensator designed in Chapter 5, except that now the plant contains process and measurement noise. The closed-loop system's performance can be obtained as desired by suitably selecting the optimal regulator's weighting matrices, $\mathbf{Q}$ and $\mathbf{R}$, and the Kalman filter's spectral noise densities, $\mathbf{V}$, $\mathbf{Z}$, and Ψ. Hence, the matrices $\mathbf{Q}$, $\mathbf{R}$, $\mathbf{V}$, $\mathbf{Z}$, and Ψ are the *design parameters* for the closed-loop system with an optimal compensator.

A state-space realization of the optimal compensator for regulating a noisy plant with state-space representation of Eqs. (7.40) and (7.41) is given by the following state and output equations:

$$\mathbf{x_o}^{(1)}(t) = (\mathbf{A} - \mathbf{BK} - \mathbf{LC} + \mathbf{LDK})\mathbf{x_o}(t) + \mathbf{L}\mathbf{y}(t) \qquad (7.74)$$

$$\mathbf{u}(t) = -\mathbf{Kx_o}(t) \qquad (7.75)$$

where $\mathbf{K}$ and $\mathbf{L}$ are the optimal regulator and Kalman filter gain matrices, respectively. For a corresponding optimal tracking system, the state and output equations derived in Section 5.4.1 should be used with the understanding that $\mathbf{K}$ is the optimal feedback gain matrix and $\mathbf{L}$ is the optimal Kalman filter gain matrix.

Using MATLAB (CST), we can construct a state-space model of the regulating closed-loop system, *sysCL*, as follows:

```
>>sysp=ss(A,B,C,D); sysc=ss(A-B*K-L*C+L*D*K,L,K,zeros(size(D'))); <enter>

>>sysCL=feedback(sysp,sysc) <enter>
```

where *sysp* is the state-space model of the plant, and *sysc* is the state-space model of the LQG compensator. The resulting closed-loop system's state-vector is $\mathbf{x_{CL}}(t) = [\mathbf{x}^T(t); \mathbf{x_o}^T(t)]^T$. Alternatively, MATLAB (CST) provides a readymade command *reg* to construct a state-space model of the optimal compensator, given a state-space model of the plant, *sysp*, the optimal regulator feedback gain matrix, $\mathbf{K}$, and the Kalman filter gain matrix, $\mathbf{L}$. This command is used as follows:

```
>>sysc= reg(sysp,K,L) <enter>
```

where *sysc* is the state-space model of the compensator. To find the state-space representation of the closed-loop system, *sysCL*, the command *reg* should be followed by the command *feedback* as shown above.

Example 7.8

Let us design an optimal compensator for the flexible bomber aircraft (Examples 4.7, 6.1, 6.5, 6.7), with the process noise coefficient matrix, $\mathbf{F} = \mathbf{B}$. Recall that the sixth order, two input system is described by a linear, time-invariant, state-space representation given by Eq. (4.71). The inputs are the desired elevator deflection (rad.), $u_1(t)$, and the desired canard deflection (rad.), $u_2(t)$, while the outputs are the normal acceleration (m/s^2), $y_1(t)$, and the pitch-rate (rad./s), $y_2(t)$. In Example 6.1, we designed an optimal regulator for this plant to achieve a maximum overshoot of less than ± 2 m/s^2 in the normal-acceleration, and less than ± 0.03 rad/s in pitch-rate, and a settling time less than 5 s, while requiring elevator and canard deflections *not exceeding* ± 0.1 rad. (5.73°), if the initial condition is 0.1 rad/s perturbation in the pitch-rate, i.e. $\mathbf{x}(0) = [0; 0.1; 0; 0; 0; 0]^T$. This was achieved with $\mathbf{Q} = 0.01\mathbf{I}$ and $\mathbf{R} = \mathbf{I}$, resulting in the following optimal feedback gain matrix:

$$\mathbf{K} = \begin{bmatrix} 1.0780 & -0.16677 & -0.046948 & -0.075618 & 0.59823 & 0.35302 \\ 1.3785 & 0.34502 & -0.013144 & -0.065260 & 0.47069 & 0.30941 \end{bmatrix},$$

(7.76)

and the following eigenvalues of the regulator (i.e. eigenvalues of $\mathbf{A\text{-}BK}$):

```
>>eig(A-B*K) <enter>
ans =
-7.8748e+001+ 5.0625e+001i
-7.8748e+001- 5.0625e+001i
-9.1803e+001
-1.1602e+000+ 1.7328e+000i
-1.1602e+000- 1.7328e+000i
-1.0560e+000
```

Note that the dominant regulator pole is at $s = -1.056$, which determines the *speed of response* of the full-state feedback control system. Since the closed-loop eigenvalues of the compensated system are the eigenvalues of the regulator and the eigenvalues of the Kalman filter, if we wish to achieve the same performance in the compensated system as the full-state feedback system, ideally we must select a Kalman filter such that the Kalman filter eigenvalues *do not dominate* the closed-loop system, i.e. they *should not* be closer to the imaginary axis than the regulator eigenvalues. As the Kalman filter does not require a control input, its eigenvalues can be pushed deeper into the left-half plane without causing concern of large required control effort (as in the case of the regulator). In other words, the Kalman filter can have *faster dynamics* than the regulator, which is achieved *free of cost*. However, as we will see in the present example, it is not always possible to push *all* the Kalman filter poles deeper into the left-half plane than the regulator poles by varying the noise spectral densities of the Kalman filter. In such cases, a judicious choice of Kalman filter spectral densities would yield the *best recovery of the full-state feedback dynamics*.

To design the Kalman filter to *recover* the full-state feedback performance of Example 6.1, the process noise spectral density matrix for the bomber is selected after some trial and error to be $\mathbf{V} = 0.0007\mathbf{B}^T\mathbf{B}$, while the spectral density matrix of the measurement noise is $\mathbf{Z} = \mathbf{CC}^T$. We also assume that the process and measurement noises are uncorrelated, i.e. $\Psi = \mathbf{0}$. Then the Kalman filter gain is calculated using the MATLAB command *lqe* as follows:

```
>>[L,P,E] = lqe(A,B,C,0.0007*B'*B,C*C') <enter>
L =
-1.8370e-005  5.1824e-001
 9.8532e-004  7.0924e+000


-1.5055e-001 -2.6107e+001
-3.1291e-002 -6.7694e+001
-3.0779e-002 -1.0403e+001
-3.3715e-002  1.7954e-001

P =
 1.2307e+000 5.2093e-001 -3.9401e+001 5.5461e+001 -1.8815e-001 2.7655e-002
 5.2093e-001 6.9482e+000 -4.0617e+000 -6.3112e+001 -5.8958e+000 5.1163e+000
-3.9401e+001 -4.0617e+000 2.8063e+004 4.7549e+004 -2.7418e+002 -2.3069e+002
 5.5461e+001 -6.3112e+001 4.7549e+004 6.0281e+005 -1.1015e+003 -1.7972e+003
-1.8815e-001 -5.8958e+000 -2.7418e+002 -1.1015e+003 1.2093e+002 -2.4702e+001
 2.7655e-002 5.1163e+000 -2.3069e+002 -1.7972e+003 -2.4702e+001 3.2603e+002

E =
-1.4700e+002
-8.9112e+001
-7.9327e+000
-1.9548e+000+ 4.1212e+000i
-1.9548e+000- 4.1212e+000i
-5.6453e-001
```

Note that the Kalman filter's dominant pole is at $s = -0.56453$, which is closer
to the imaginary axis than the dominant regulator pole. Thus, the two closed-loop

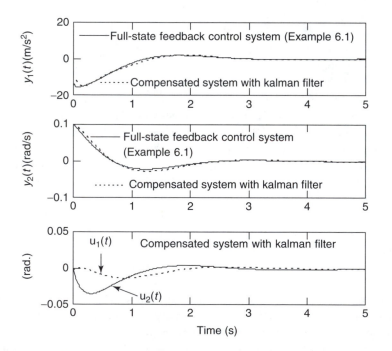

Figure 7.9 Closed-loop system's initial response and control inputs for the flexible bomber with optimal (LQG) compensator

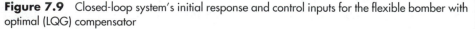

poles *closest to the imaginary axis* are the dominant pole the Kalman filter and that of the optimal regulator, and the closed-loop response's speed and overshoots are largely determined by these two poles. We can construct a state-space model of the closed-loop system as follows:

```
>>sysp=ss(A,B,C,D); sysc=ss(A-B*K-L*C+L*D*K,L,K,zeros(size (D')));
  <enter>

>>sysCL=feedback(sysp,sysc) <enter>
```

You may confirm that the closed-loop eigenvalues are indeed the eigenvalues of the regulator and the Kalman filter using *damp(sysCL)*. The initial response of the closed-loop system to $\mathbf{x}(0) = [0; 0.1; 0; 0; 0; 0]^T$ and the required control input vector are calculated as follows:

```
>>[y,t,X]=initial(sysCL,[0 0.1 zeros(1,10)]'); u=-K* X(:,7:12)'; <enter>
```

The closed-loop responses, $y_1(t)$ and $y_2(t)$, are compared to their corresponding values for the full-state feedback control system (Example 6.1) in Figure 7.9. Figure 7.9 also contains the plots of the required control inputs, $u_1(t)$ and $u_2(t)$. Note that the speed of the response, indicated by the settling time of less than five seconds, maximum overshoots, and control input magnitudes are all quite similar to those of the full-state feedback system. This is the best recovery of the full-state feedback performance obtained by varying the Kalman filter's process noise spectral density scaling parameter, ρ, where $\mathbf{V} = \rho \mathbf{B}^T \mathbf{B}$.

In Example 7.8, we saw how the full-state feedback control system's performance can be *recovered* by properly designing a Kalman filter to estimate the state-vector in the compensated system. In other words, the Kalman filter part of the optimal (LQG) compensator can be designed to yield approximately the same performance as that of the full-state feedback regulator. Now let us examine the robustness of the designed LQG compensated closed-loop system with respect to measurement noise in comparison with the full-state feedback system of Example 6.1. Such a comparison is valid, because both control systems use the same feedback gain matrix, $\mathbf{K}$. A SIMULINK block-diagram and the simulated initial response of the compensated closed-loop system are shown in Figure 7.10. The same white noise intensity (*power parameter of 10^{-8}*) in the pitch-rate channel is used as in Example 6.1. Comparing Figure 7.10 with Figure 6.3, we observe that the normal acceleration and pitch-rate fluctuations are about half the magnitudes and smaller in frequency than those seen in Figure 6.3. This indicates that the LQG compensated system is more robust with respect to measurement noise than the full-state feedback system of the same regulator gain matrix, $\mathbf{K}$.

We have thus far confined our attention to white noise model of disturbances. How robust would an LQG compensator be to *actual* parameter variations and noise which is not white? This is a crucial question, which is answered in the next section.

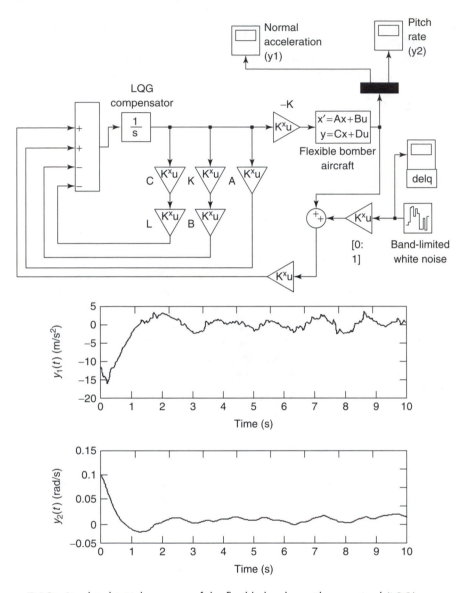

Figure 7.10 Simulated initial response of the flexible bomber with an optimal (LQG) compensator and measurement noise in the pitch-rate channel

7.6 Robust Multivariable LQG Control: Loop Transfer Recovery

In Chapter 2, we observed that in principle, the closed-loop systems are *less sensitive* (or *more robust*) to variations in the mathematical model of the plant (called *process noise and measurement noise* in Section 7.5), when compared to the corresponding open-loop systems. It was also observed that the *robustness* of a single-input, single-output feedback

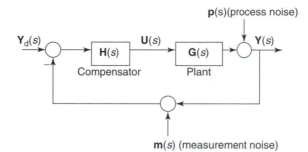

Figure 7.11 A multivariable feedback control system with compensator transfer matrix, $\mathbf{H}(s)$, plant transfer matrix, $\mathbf{G}(s)$, process noise, $\mathbf{p}(s)$, and measurement noise, $\mathbf{m}(s)$

control system is related to the *return difference*, $1 + G(s)H(s)$, where $G(s)$ and $H(s)$ are the transfer functions of the plant and controller, respectively. The larger the return difference of the feedback loop, the greater will be the robustness when compared to the corresponding open-loop system.

Consider a linear, time-invariant, multivariable feedback control system with the block-diagram shown in Figure 7.11. The control system consists of a feedback controller with transfer matrix, $\mathbf{H}(s)$, and a plant with transfer matrix, $\mathbf{G}(s)$, with desired output, $\mathbf{Y}_d(s)$. The *process noise*, $\mathbf{p}(s)$, and *measurement noise*, $\mathbf{m}(s)$, are present in the control system as shown. Using Figure 7.11, the output, $\mathbf{Y}(s)$, can be expressed as follows:

$$\mathbf{Y}(s) = [\mathbf{I} + \mathbf{G}(s)\mathbf{H}(s)]^{-1}\mathbf{p}(s) - [\mathbf{I} - \{\mathbf{I} + \mathbf{G}(s)\mathbf{H}(s)\}^{-1}]\mathbf{m}(s)$$
$$+ \{\mathbf{I} - [\mathbf{I} + \mathbf{G}(s)\mathbf{H}(s)]^{-1}\}\mathbf{Y}_d(s) \qquad (7.77)$$

while the control input, $\mathbf{U}(s)$, can be expressed as

$$\mathbf{U}(s) = [\mathbf{I} + \mathbf{H}(s)\mathbf{G}(s)]^{-1}\mathbf{H}(s)\mathbf{Y}_d(s) - [\mathbf{I} + \mathbf{H}(s)\mathbf{G}(s)]^{-1}\mathbf{H}(s)\mathbf{p}(s)$$
$$- [\mathbf{I} + \mathbf{H}(s)\mathbf{G}(s)]^{-1}\mathbf{H}(s)\mathbf{m}(s) \qquad (7.78)$$

From Eqs. (7.77) and (7.78), it is clear that the *sensitivity* of the *output* with respect to process and measurement noise depends upon the matrix $[\mathbf{I} + \mathbf{G}(s)\mathbf{H}(s)]^{-1}$, while the sensitivity of the input to process and measurement noise depends upon the matrix $[\mathbf{I} + \mathbf{H}(s)\mathbf{G}(s)]^{-1}$. The larger the elements of these two matrices, the larger will be the sensitivity of the output and input to process and measurement noise. Since *robustness* is inversely proportional to *sensitivity*, we can extend the analogy to multivariable systems by saying that the *robustness of the output* is measured by the matrix $[\mathbf{I} + \mathbf{G}(s)\mathbf{H}(s)]$, called the *return-difference matrix at the output*, and the *robustness of the input* is measured by the matrix $[\mathbf{I} + \mathbf{H}(s)\mathbf{G}(s)]$, called the *return difference matrix at the plant input*. Therefore, for multivariable control-systems, there are *two* return difference matrices to be considered: the return difference at the output, $[\mathbf{I} + \mathbf{G}(s)\mathbf{H}(s)]$, and that at the plant's input, $[\mathbf{I} + \mathbf{H}(s)\mathbf{G}(s)]$. Alternatively, we can define the *return ratio matrices at the plant's output and input*, as $\mathbf{G}(s)\mathbf{H}(s)$ and $\mathbf{H}(s)\mathbf{G}(s)$, respectively, and measure robustness properties in terms of the *return ratios* rather than the *return differences*.

Continuing our analogy with single-input, single-output systems, we would like to assign a *scalar measure* to robustness, rather than deal with the two return-difference (or return ratio) matrices. We can define a *matrix norm* (introduced in Chapter 4) to assign a scalar measure to a matrix. For a complex matrix, **M**, with n rows and m columns, one such norm is the *spectral* (or *Hilbert*) norm, given by

$$\|\mathbf{M}\|_s = \sigma_{\max} \tag{7.79}$$

where $\sigma_{\max}$ is the *positive square-root of the maximum eigenvalue* of the matrix $\mathbf{M}^{\mathbf{H}}\mathbf{M}$ if $n \geq m$, or of the matrix $\mathbf{M}\mathbf{M}^{\mathbf{H}}$ if $n \leq m$. Here $\mathbf{M}^{\mathbf{H}}$ denotes the *Hermitian of* **M** defined as the *transpose of the complex conjugate of* **M**. In MATLAB, the Hermitian of a *complex* matrix, **M**, is calculated by M', i.e. the same command as used for evaluating the *transpose* of *real* matrices. All positive square-roots of the eigenvalues of $\mathbf{M}^{\mathbf{H}}\mathbf{M}$ if $n \geq m$ (or $\mathbf{M}\mathbf{M}^{\mathbf{H}}$ if $n \leq m$) are called the *singular values of* **M**, and are denoted by σ_k, $k = 1, 2, .., n$, where n is the size of **M**. The largest among σ_k is denoted by $\sigma_{\max}$, and the smallest among σ_k is denoted by $\sigma_{\min}$. If **M** varies with frequency, then each singular value also varies with frequency.

A useful algorithm for calculating singular values of a complex matrix, **M**, of n rows and m columns is the *singular value decomposition*, which expresses **M** as follows:

$$\mathbf{M} = \mathbf{U}\mathbf{S}\mathbf{V}^{\mathbf{H}} \tag{7.80}$$

where **U** and **V** are complex matrices with the property $\mathbf{U}^{\mathbf{H}}\mathbf{U} = \mathbf{I}$ and $\mathbf{V}^{\mathbf{H}}\mathbf{V} = \mathbf{I}$, and **S** is a *real matrix* containing the singular values of **M** as the *diagonal elements of a square sub-matrix* of size $(n \times n)$ or $(m \times m)$, whichever is *smaller*. MATLAB also provides the function *svd* for computing the singular values by singular value decomposition, and is used as follows:

```
>>[U,S,V] = svd(M) <enter>
```

Example 7.9

Find the singular values of the following matrix:

$$\mathbf{M} = \begin{bmatrix} 1+i & 2+2i & 3+i \\ 1-i & 2-2i & 3-5i \\ 5 & 4+i & 1+2i \\ 7-i & 2 & 3+3i \end{bmatrix} \tag{7.81}$$

Using the MATLAB command *svd*, we get the matrices **U**, **S**, **V** of the singular value decomposition (Eq. (7.80)) as follows:

```
>>[U,S,V] = svd(M) <enter>

U =
 0.2945+0.0604i   0.4008-0.2157i  -0.0626-0.2319i  -0.5112-0.6192i
 0.0836-0.4294i  -0.0220-0.7052i   0.2569+0.1456i   0.4269-0.2036i
 0.5256+0.0431i  -0.3392-0.0455i  -0.4622-0.5346i   0.3230+0.0309i
 0.6620-0.0425i  -0.3236+0.2707i   0.5302+0.2728i  -0.1615-0.0154i
```

```
S =
  12.1277   0        0
       0    5.5662   0
       0    0        2.2218
       0    0        0


V =
  0.6739              -0.6043            0.4251
  0.4292-0.1318i      -0.0562-0.3573i   -0.7604-0.2990i
  0.4793-0.3385i       0.6930-0.1540i    0.2254+0.3176i
```

Hence, the singular values of $\mathbf{M}$ are the diagonal elements of the (3×3) sub-matrix of $\mathbf{S}$, i.e. $\sigma_1(\mathbf{M}) = 2.2218$, $\sigma_2(\mathbf{M}) = 5.5662$, and $\sigma_3(\mathbf{M}) = 2.1277$, with the *largest* singular value, $\sigma_{\max}(\mathbf{M}) = 12.1277$ and the *smallest* singular value, $\sigma_{\min}(\mathbf{M}) = 2.2218$. Alternatively, we can directly use their definition to calculate the singular values as follows:

```
>>sigma= sqrt(eig(M'*M)) <enter>

sigma =
  12.1277+0.0000i
   5.5662+0.0000i
   2.2218+0.0000i
```

The singular values help us analyze the properties of a multivariable feedback (called a *multi-loop*) system in a manner quite similar to a single-input, single-output feedback (called a *single-loop*) system. For analyzing robustness, we can treat the *largest* and *smallest* singular values of a return difference (or return ratio) matrix as providing the *upper* and *lower* bounds on the *scalar* return difference (or return ratio) of an *equivalent* single-loop system. For example, to maximize robustness with respect to the process noise, it is clear from Eq. (7.77) that we should *minimize* the singular values of the *sensitivity matrix*, $[\mathbf{I} + \mathbf{G}(s)\mathbf{H}(s)]^{-1}$, which implies *minimizing* the *largest* singular value, $\sigma_{\max}[\{\mathbf{I} + \mathbf{G}(s)\mathbf{H}(s)\}^{-1}]$, or *maximizing* the singular values of the return difference matrix at the output, i.e. *maximizing* $\sigma_{\min}[\mathbf{I} + \mathbf{G}(s)\mathbf{H}(s)]$. The latter requirement is equivalent to *maximizing* the *smallest* singular value of the return ratio at the output, $\sigma_{\min}[\mathbf{G}(s)\mathbf{H}(s)]$. Similarly, *minimizing* the sensitivity to the measurement noise requires *minimizing* the *largest* singular value of the matrix $[\mathbf{I} - \{\mathbf{I} + \mathbf{G}(s)\mathbf{H}(s)\}^{-1}]$, which is equivalent to *minimizing* the *largest* singular value of the return ratio at the output, $\sigma_{\max}[\mathbf{G}(s)\mathbf{H}(s)]$. On the other hand, tracking a desired output requires from Eq. (7.77) that the *sensitivity* to $\mathbf{Y}_d(s)$ be *maximized*, which requires *maximizing* $\sigma_{\min}[\mathbf{I} - \{\mathbf{I} + \mathbf{G}(s)\mathbf{H}(s)\}^{-1}]$, which is equivalent to maximizing $\sigma_{\min}[\mathbf{G}(s)\mathbf{H}(s)]$. Also, it is clear from Eq. (7.77) and the relationship $\mathbf{U}(s) = \mathbf{H}(s)\mathbf{Y}(s)$ that *optimal control* (i.e. minimization of control input magnitudes) requires *minimization* of $\sigma_{\max}[\mathbf{H}(s)\{\mathbf{I} + \mathbf{G}(s)\mathbf{H}(s)\}^{-1}]$, or alternatively, a minimization of $\sigma_{\max}[\mathbf{H}(s)]$. In summary, the following conditions on the singular values of the return ratio at output result from robustness, optimal control and tracking requirements:

(a) For *robustness* with respect to the *process noise*, $\sigma_{\min}[\mathbf{G}(s)\mathbf{H}(s)]$ should be *maximized*.

(b) For *robustness* with respect to the *measurement noise*, $\sigma_{\max}[\mathbf{G}(s)\mathbf{H}(s)]$ should be *minimized*.

(c) For *optimal control*, $\sigma_{\max}[\mathbf{H}(s)]$ should be *minimized*.

(d) For *tracking* a changing desired output, $\sigma_{\min}[\mathbf{G}(s)\mathbf{H}(s)]$ should be *maximized*.

Clearly, the second requirement conflicts with the first and the fourth. Also, since $\sigma_{\max}[\mathbf{G}(s)\mathbf{H}(s)] \leq \sigma_{\max}[\mathbf{G}(s)]\sigma_{\max}[\mathbf{H}(s)]$ (a property of scalar norms), the third requirement is in conflict with the first and the fourth. However, since measurement noise usually has a predominantly high-frequency content (i.e. more peaks in the power spectrum at high frequencies), we achieve a compromise by minimizing $\sigma_{\max}[\mathbf{G}(s)\mathbf{H}(s)]$ (and $\sigma_{\max}[\mathbf{H}(s)]$) at *high frequencies*, and maximizing $\sigma_{\min}[\mathbf{G}(s)\mathbf{H}(s)]$ at *low frequencies*. In this manner, good robustness properties, optimal control, and tracking system performance can be obtained throughout a given frequency range.

The singular values of the return difference matrix at the output in the frequency domain, $\sigma[\mathbf{I} + \mathbf{G}(i\omega)\mathbf{H}(i\omega)]$, can be used to estimate the gain and phase margins (see Chapter 2) of a multivariable system. One can make singular value plots against frequency, ω, in a similar manner as the Bode gain plots. The way in which multivariable gain and phase margins are defined with respect to the singular values is as follows: take the *smallest* singular value, $\sigma_{\min}$, of all the singular values of the return difference matrix at the output, and find a real constant, a, such that $\sigma_{\min}[\mathbf{I} + \mathbf{G}(i\omega)] \geq a$ for all frequencies, ω, in the frequency range of interest. Then the gain and phase margins can be defined as follows:

$$\text{Gain margin} = 1/(1 \pm a) \tag{7.82}$$

$$\text{Phase margin} = \pm 2\sin^{-1}(a/2) \tag{7.83}$$

Example 7.10

Figure 7.12 shows the singular values of the rotating flexible spacecraft of Example 6.2. We observe that the smallest singular value reaches a minimum of -70 dB in the frequency range $0.01-10,000$ rad/s at frequency 0.01 rad/s. Hence, $a = 10^{-(70/20)} = 3.16 \times 10^{-4}$. Therefore, the gain margins are $1/(1 + a) = 0.9997$ and $1/(1 - a) = 1.0003$, and phase margins are $\pm 0.0181°$ (which are quite small!). These margins are quite conservative, because they allow for *simultaneous* gain and phase variations of all the controller transfer functions. The present analysis indicates that the control system for the spacecraft cannot tolerate an appreciable variation in the phase of the return difference matrix before its eigenvalue cross into the right-half s-plane. However, the spacecraft is already *unstable* due to double eigenvalues at $s = 0$ (see Example 6.2), which the classical measures of gain and phase margins *do not* indicate (recall from Chapter 2 that gain and phase margins only indicate poles crossing over into right-half s-plane). Hence, gain and phase margins have limited utility for indicating closed-loop robustness.

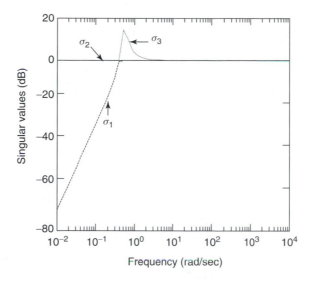

Figure 7.12 Singular value plot of the return difference matrix $[\mathbf{I} + \mathbf{G}(s)\mathbf{H}(s)]$ of the rotating spacecraft

For designing a control system in which the input vector is the *least sensitive* to process and measurement noise, we can derive appropriate conditions on the singular values of the return ratio at the plant's *input*, $\mathbf{H}(s)\mathbf{G}(s)$, by considering Eq. (7.78) as follows:

(a) For *robustness* with respect to the *process noise*, $\sigma_{\min}[\mathbf{H}(s)\mathbf{G}(s)]$ should be *maximized*.

(b) For *robustness* with respect to the *measurement noise*, $\sigma_{\min}[\mathbf{H}(s)\mathbf{G}(s)]$ should be *maximized*.

Thus, there is no conflict in achieving the robustness of the plant's input to process and measurement noise. However, Eq. (7.78) indicates that for tracking a changing desired output, $\sigma_{\max}[\mathbf{H}(s)\mathbf{G}(s)]$ should be *minimum*, which *conflicts with the robustness requirement*. This conflict can again be resolved by selecting different frequency ranges for maximizing and minimizing the singular values.

The adjustment of the singular values of return ratio matrices to achieve desired closed-loop robustness and performance is called *loop shaping*. This term is derived from single-loop systems where scalar return ratios of a loop are to be adjusted. For *compensated* systems based on an observer (i.e. the Kalman filter), generally there is a *loss* of robustness, when compared to *full-state feedback* control systems. To *recover* the robustness properties associated with full-state feedback, the Kalman filter must be designed such that the sensitivity of the *plant's input* to process and measurement noise is minimized. As seen above, this requires that the smallest singular value of the return ratio at plant's input, $\sigma_{\min}[\mathbf{H}(s)\mathbf{G}(s)]$, should be maximized. Theoretically, this maximum value of $\sigma_{\min}[\mathbf{H}(s)\mathbf{G}(s)]$ should be *equal* to that of the return ratio at the plant's input with full-state feedback. Such a process of designing a Kalman filter based compensator to recover the robustness of full-state feedback is called *loop transfer recovery*

(LTR). Optimal compensators designed with loop transfer recovery are called LQG/LTR compensators. The loop transfer recovery can either be conducted at the plant's the input, as described below, or at the plant's output. The design of optimal (LQG) compensators for loop transfer recovery at the plant's input can be stated as follows:

1. Design a full-state feedback optimal regulator by selecting $\mathbf{Q}$ and $\mathbf{R}$ matrices such that the desired performance objectives are met, and the singular values of the return ratio at the plant's input are maximized. With full-state feedback, the return ratio at the plant's input is $\mathbf{H}(s)\mathbf{G}(s) = -\mathbf{K}(s\mathbf{I} - \mathbf{A})^{-1}\mathbf{B}$, where $\mathbf{A}$, $\mathbf{B}$ are the plant's state coefficient matrices and $\mathbf{K}$ is the full-state feedback regulator gain matrix.

2. Design a Kalman filter by selecting the noise coefficient matrix, $\mathbf{F}$, and the white noise spectral densities, $\mathbf{V}$, $\mathbf{Z}$, and Ψ, such that the singular values of the return ratio at the plant's input, $\mathbf{H}(s)\mathbf{G}(s)$, approach the corresponding singular values with full-state feedback. Hence, $\mathbf{F}$, $\mathbf{V}$, $\mathbf{Z}$, and Ψ are treated as *design parameters* of the Kalman filter to achieve full-state feedback return ratio at the plant's input, rather than *actual* parameters of process and measurement (white) noises.

The compensated system's return ratio matrix at plant's input can be obtained by taking the Laplace transform of Eqs. (7.74) and (7.75), and combining the results as follows:

$$\mathbf{U}(s) = -\mathbf{K}(s\mathbf{I} - \mathbf{A_c})^{-1}\mathbf{LY}(s) \qquad (7.84)$$

where $\mathbf{A_c} = (\mathbf{A} - \mathbf{BK} - \mathbf{LC} + \mathbf{LDK})$ is the compensator's state-dynamics matrix, $\mathbf{L}$ is the Kalman filter's gain matrix, and $\mathbf{Y}(s)$ is the plant's output, written as follows:

$$\mathbf{Y}(s) = [\mathbf{C}(s\mathbf{I} - \mathbf{A})^{-1}\mathbf{B} + \mathbf{D}]\mathbf{U}(s) \qquad (7.85)$$

Substituting Eq. (7.85) into Eq. (7.84), we get the following expression for $\mathbf{U}(s)$:

$$\mathbf{U}(s) = -\mathbf{K}(s\mathbf{I} - \mathbf{A_c})^{-1}\mathbf{L}[\mathbf{C}(s\mathbf{I} - \mathbf{A})^{-1}\mathbf{B} + \mathbf{D}]\mathbf{U}(s) \qquad (7.86)$$

Note that the return ratio matrix at the plant's input is the matrix by which $\mathbf{U}(s)$ gets pre-multiplied in passing around the feedback loop and returning to itself, i.e. $\mathbf{U}(s) = \mathbf{H}(s)\mathbf{G}(s)$ in Figure 7.11 if all other *inputs* to the control system, $\mathbf{Y_d}(s)$, $\mathbf{p}(s)$, $\mathbf{m}(s)$, are zero. Hence, the return ratio at the compensated plant's input is

$$\mathbf{H}(s)\mathbf{G}(s) = -\mathbf{K}(s\mathbf{I} - \mathbf{A_c})^{-1}\mathbf{L}[\mathbf{C}(s\mathbf{I} - \mathbf{A})^{-1}\mathbf{B} + \mathbf{D}] \qquad (7.87)$$

There is no unique set of Kalman filter design parameters $\mathbf{F}$, $\mathbf{V}$, $\mathbf{Z}$, and Ψ to achieve loop transfer recovery. Specifically, if the plant is *square* (i.e. it has *equal* number of outputs and inputs) and *minimum-phase* (i.e. the plant's transfer matrix has *no zeros* in the right-half plane), then by selecting the noise coefficient matrix of the plant as $\mathbf{F} = \mathbf{B}$, the cross-spectral density as $\Psi = \mathbf{0}$, the measurement noise spectral density as $\mathbf{Z} = \mathbf{I}$, and the process noise spectral density as $\mathbf{V} = \mathbf{V_o} + \rho\mathbf{I}$, where ρ is a scaling parameter, it can be shown from the Kalman filter equations that in the limit $\rho \to \infty$, the compensated system's return ratio, given by Eq. (7.87), converges to $-\mathbf{K}(s\mathbf{I} - \mathbf{A})^{-1}\mathbf{B}$, the return ratio of the full-state feedback system at the plant input. In most cases, *better* loop transfer recovery can be obtained by choosing $\mathbf{Z} = \mathbf{CC}^{\mathrm{T}}$ and $\mathbf{V} = \rho \mathbf{B}^{\mathrm{T}}\mathbf{B}$, and making ρ large. However, making ρ extremely large reduces the *roll-off* of the closed-loop transfer function at high frequencies, which is undesirable. Hence, instead of making ρ very large to achieve perfect loop transfer recovery at all frequencies, we should choose a value of ρ which is sufficiently large to *approximately* recover the return ratio over a given range of frequencies.

MATLAB's *Robust Control Toolbox* [3] provides the command *sigma* to calculate the singular values of a transfer matrix, $\mathbf{G}(s = i\omega)$, as a function of frequency, ω, as follows:

```
>>[sv,w] = sigma(sys) <enter>
```

where *sys* is an LTI object of $\mathbf{G}(s) = \mathbf{C}(s\mathbf{I} - \mathbf{A})^{-1}\mathbf{B} + \mathbf{D}$, and **sv** and **w** contain the returned singular values and frequency points, respectively. The user can specify the set of frequencies at which the singular values of $\mathbf{G}(s)$ are to be computed by including the frequency vector, **w**, as an additional input argument of the command as follows:

```
>>[sv,w] = sigma(sys,w) <enter>
```

Also, *sigma* can calculate the singular values of some commonly encountered functions of $\mathbf{G}(i\omega)$ by using the command as follows:

```
>>[sv,w] = sigma(sys,w,type) <enter>
```

where $type = 1$, 2, or 3 specify that singular values of $\mathbf{G}^{-1}(i\omega)$, $\mathbf{I} + \mathbf{G}(i\omega)$, or $\mathbf{I} + \mathbf{G}^{-1}(i\omega)$, respectively, are to be calculated. (Of course, *sigma* requires that $\mathbf{G}(i\omega)$ should be a square matrix.) Hence, *sigma* is a versatile command, and can be easily used to compute singular values of return ratios, or return difference matrices. If you do not have the *Robust Control Toolbox*, you can write your own M-file for calculating the singular value spectrum using the MATLAB functions *svd* or *eig* as discussed above.

Example 7.11

Re-consider the flexible bomber airplane of Example 7.8, where we designed an optimal compensator using an optimal regulator with $\mathbf{Q} = 0.01\mathbf{I}$, $\mathbf{R} = \mathbf{I}$, and a Kalman filter with $\mathbf{F} = \mathbf{B}$, $\mathbf{V} = 0.0007\mathbf{B}^{\mathrm{T}}\mathbf{B}$, $\mathbf{Z} = \mathbf{CC}^{\mathrm{T}}$, and $\Psi = \mathbf{0}$, to *recover* the *performance* of the full-state feedback regulator. Let us now see how *robust* such a compensator is by studying the return ratio at the plant's input. Recall that the return ratio at the plant's input is $\mathbf{H}(s)\mathbf{G}(s)$, which is the transfer matrix of a hypothetical system formed by placing the plant, $\mathbf{G}(s)$, in series with the compensator, $\mathbf{H}(s)$ (the plant *is followed* by the compensator). Hence, we can find a state-space representation of $\mathbf{H}(s)\mathbf{G}(s)$ in terms of the state-space model, *sysHG*, with plant model, *sysp*, and compensator model, *sysc*, as follows:

```
>>sysp=ss(A,B,C,D);sysc=ss(A-B*K-L*C+L*D*K,L,-K,zeros(size(K,1)));
  <enter>

>> sysHG = series(sysp,sysc); <enter>
```

The singular value spectrum of the return ratio at the plant's input for frequency range 10^{-2}–10^4 rad/s is calculated as follows:

```
>>w = logspace(-2,4); [sv,w] = sigma(sysHG,w); <enter>
```

while the singular value spectrum of the return ratio at plant's input of the full-state feedback system is calculated in the same frequency range by

```
>>sysfs=ss(A,B,-K,zeros(size(K,1))); [sv1,w1] = sigma(sysfs,w);
  <enter>
```

The two sets of singular values, **sv** and **sv1**, are compared in Figure 7.13, which is plotted using the following command:

```
>>semilogx(w,20*log10(sv),':',w1,20*log10(sv1)) <enter>
```

Note that there is a large difference between the smallest singular values of the compensated and full-state feedback systems, indicating that the compensated system is *much less* robust than the full-state feedback system. For recovering the full-state feedback robustness at the plant's input, we re-design the Kalman filter using $\mathbf{V} = \rho \mathbf{B}^{\mathrm{T}} \mathbf{B}$, $\mathbf{Z} = \mathbf{C}\mathbf{C}^{\mathrm{T}}$, and $\Psi = \mathbf{0}$, where ρ is a scaling parameter for the process noise spectral density. As ρ is increased, say, from 10 to 10^8, the return ratio of the compensated plant approaches that of the full-state feedback system *over a larger range of frequencies*, as seen in the singular value plots of Figure 7.14. For $\rho = 10$, the smallest singular value of the compensated system's return ratio becomes equal to that of the full-state feedback system in the frequency range 1–100 rad/s, while for $\rho = 10^8$ the range of frequencies (or bandwidth) over which

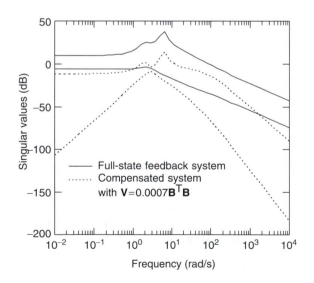

Figure 7.13 Singular values of the return ratio matrix, $\mathbf{H}(s)\mathbf{G}(s)$, at the plant's input of compensated system and full-state feedback system for the flexible bomber

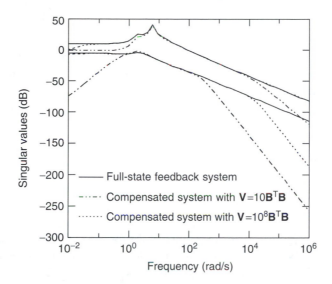

Figure 7.14 Comparison of singular values of return ratio matrix at the plant's input of full-state feedback system and compensated system for loop transfer recovery with process noise spectral density, $\mathbf{V} = 10\mathbf{B}^T\mathbf{B}$ and $\mathbf{V} = 10^8\mathbf{B}^T\mathbf{B}$ (flexible bomber)

the loop transfer recovery occurs increases to 10^{-2}–10^4 rad/s. At frequencies higher than the loop transfer recovery bandwidth the return ratio is seen to roll-off at more than 50dB/decade. Such a roll-off would also be present in the singular values of the closed-loop transfer matrix, which is good for rejection of noise at high frequencies. However, within the loop transfer recovery bandwidth, the roll-off is only about 20dB/decade. Therefore, the larger the bandwidth for loop transfer recovery, the smaller would be the range of frequencies over which high noise attenuation is provided by the compensator. Hence, the loop transfer recovery bandwidth must not be chosen to be too large; otherwise high frequency noise (usually the measurement noise) would get unnecessarily amplified by smaller roll-off provided within the LTR bandwidth.

Note that the Kalman filter designed in Example 7.8 with $\mathbf{V} = 0.0007\mathbf{B}^T\mathbf{B}$ recovers the *performance* of the full-state feedback system (with a loss of robustness), whereas the re-designed Kalman filter with $\mathbf{V} = 10^8\mathbf{B}^T\mathbf{B}$ recovers the *robustness* of the full-state feedback system over a bandwidth of 10^6 rad/s (with an expected loss of performance). By choosing a large value of the process noise spectral density for loop transfer recovery, a pair of Kalman filter poles comes very close to the imaginary axis and becomes the dominant pole configuration, thereby playing havoc with the performance. Hence, there is a *contradiction* in recovering *both* performance and robustness with the same Kalman filter, and a compromise must be made between the two.

It is interesting to note that Example 7.11 has a *non-minimum phase* plant, with a plant zero at s = 2.5034×10^{-7}. The loop transfer recovery is *not guaranteed* for non-minimum phase plants, because it is pointed out above that $\mathbf{H}(s)\mathbf{G}(s)$ converges to $-\mathbf{K}(s\mathbf{I} - \mathbf{A})^{-1}\mathbf{B}$ in the limit of *infinite* process noise spectral density, provided that the plant is *square* and *minimum phase*. The reason why LQG/LTR compensators generally cannot be designed for non-minimum phase plants is that perfect loop transfer recovery requires placing *some poles* of the Kalman filter at the *zeros* of the plant [4]. If the plant is *non-minimum phase*, it implies that the Kalman filter for perfect loop transfer recovery should be *unstable*. However, if the right-half plane zeros of a non-minimum phase plant are very close to the imaginary axis (as in Example 7.11), the frequency associated with it lies *outside* the selected bandwidth for loop transfer recovery, and hence loop transfer recovery in a given bandwidth is still possible, as seen above.

Example 7.12

Let us design an optimal LQG/LTR compensator for the flexible, rotating spacecraft of Example 6.2. The spacecraft consists of a rigid hub and four flexible appendages, each having a tip mass, with three torque inputs in N-m, $u_1(t)$, $u_2(t)$, $u_3(t)$, and three angular rotation outputs in rad., $y_1(t)$, $y_2(t)$, $y_3(t)$. A linear, time-invariant state-space representation of the 26th order spacecraft was given in Example 6.2, where it was observed that the spacecraft is *unstable* and *uncontrollable*. The natural frequencies of the spacecraft, including structural vibration frequencies, range from 0–$47\,588$ rad/s. The *uncontrollable* modes are the structural vibration modes, while the *unstable* mode is the rigid-body rotation with zero natural frequency. Hence, the spacecraft is *stabilizable* and an optimal regulator with $\mathbf{Q} = 200\mathbf{I}$, and $\mathbf{R} = \mathbf{I}$ was designed in Example 6.2 to stabilize the spacecraft, with a settling time of 5 s, with zero maximum overshoots, while requiring input torques not exceeding 0.1 N-m, when the spacecraft is initially perturbed by a hub rotation of 0.01 rad. (i.e. the initial condition is $\mathbf{x}(0) = [0.01;\ \text{zeros}(1, 25)]^T$).

We would like to combine the optimal regulator already designed in Example 6.2 with a Kalman filter that recovers the return ratio at the plant's input in the frequency range 0–$50\,000$ rad/s (approximately the bandwidth of the plant). To do so, we select $\mathbf{F} = \mathbf{B}$, $\mathbf{Z} = \mathbf{CC}^{\mathbf{T}}$, $\Psi = \mathbf{0}$, and $\mathbf{V} = \rho\mathbf{B}^{\mathbf{T}}\mathbf{B}$, where ρ is a scaling parameter. By comparing the singular values of $\mathbf{H}(s)\mathbf{G}(s)$ with those of $-\mathbf{K}(s\mathbf{I} - \mathbf{A})^{-1}\mathbf{B}$, for various values of ρ, we find that loop transfer recovery occurs in the desired bandwidth for $\rho = 10^{22}$, for which the Kalman filter gain, covariance, and eigenvalues are obtained as follows (only eigenvalues are shown below):

```
>>[L,P,E] = lqe(A,B,C,1e12*B'*B,C*C') <enter>

E =
-1.0977e+008+ 1.0977e+008i
-1.0977e+008- 1.0977e+008i
-1.0977e+008+ 1.0977e+008i
-1.0977e+008- 1.0977e+008i
-8.6042e+004+ 8.6074e+004i
-8.6042e+004- 8.6074e+004i
-4.6032e+000+ 4.5876e+004i
-4.6032e+000- 4.5876e+004i
```

```
-8.7040e-001+ 2.3193e+004i
-8.7040e-001- 2.3193e+004i
-3.4073e-001+ 1.0077e+004i
-3.4073e-001- 1.0077e+004i
-3.5758e-002+ 3.7121e+003i
-3.5758e-002- 3.7121e+003i
-1.0975e-002+ 4.0386e+002i
-1.0975e-002- 4.0386e+002i
-1.2767e-004+ 4.5875e+004i
-1.2767e-004- 4.5875e+004i
-9.3867e-005+ 3.7121e+003i
-9.3867e-005- 3.7121e+003i
-7.4172e-005+ 1.0077e+004i
-7.4172e-005- 1.0077e+004i
-5.8966e-005+ 2.3193e+004i
-5.8966e-005- 2.3193e+004i
-3.8564e-005+ 4.0385e+002i
-3.8564e-005- 4.0385e+002i
```

Note that the Kalman filter is stable, but has some eigenvalues very close to the imaginary axis, which is likely to degrade the performance of the closed-loop system. The singular values of the return ratio at the plant's input, for the compensated system are calculated as follows, with **K** obtained in Example 6.2:

```
>>sysp=ss(A,B,C,D);sysc=ss(A-B*K-L*C+L*D*K,L,-K,zeros(size(K,1)));
  <enter>

>> sysHG = series(sysp,sysc); <enter>

>>w = logspace(-2,6); [sv,w] = sigma(sysHG,w); <enter>
```

The singular values of the full-state feedback return ratio at the plant's input are obtained as follows:

```
>>sysfs=ss(A,B,-K,zeros(size(K,1))); [sv1,w1] = sigma(sysfs,w); <enter>
```

The two sets of singular values (full-state feedback and compensated system) are compared in Figure 7.15. Note that the smallest singular value of the return ratio is recovered in the range 0–50 000 rad/s, as desired, while the other two singular values are recovered in a larger frequency range. The plant has some zeros very close to the imaginary axis (largest real part is 3.3×10^{-8}) in the *right-half plane*, hence, the plant is *non-minimum phase* and perfect loop transfer recovery is not guaranteed. Also, due to this nature of the plant, the function *lqe* may yield inaccurate results.

Owing to the presence of some Kalman filter poles very close to the imaginary axis, the closed-loop system would have unacceptable performance. Hence, for improving the closed-loop performance, the loop transfer recovery bandwidth must be reduced. A good compromise between performance and robustness is achieved by choosing $\rho = 10^6$. Figure 7.16 shows the spectrum of the smallest singular value of

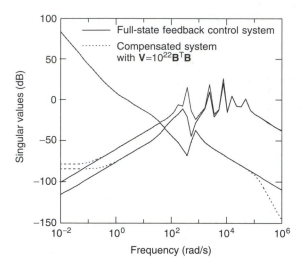

Figure 7.15 Singular values of the return ratio matrix at the plant's input of the full-state feedback system compared with those of the compensated plant with $\mathbf{V} = 10^{22}\mathbf{B}^{\mathsf{T}}\mathbf{B}$ for the flexible, rotating spacecraft

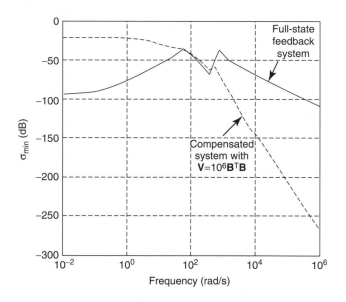

Figure 7.16 Smallest singular value of $\mathbf{H}(s)\mathbf{G}(s)$ of the optimal (LQG) compensated system with $\mathbf{V} = 10^{6}\mathbf{B}^{\mathsf{T}}\mathbf{B}$ for the flexible rotating spacecraft, compared with that of the full-state feedback system showing loop-transfer recovery in the frequency range 5–500 rad/s

$\mathbf{H}(s)\mathbf{G}(s)$ which indicates a loop transfer recovery bandwidth of 50–500 rad/s. The performance of the closed-loop system is determined using a SIMULINK simulation of the initial response with $\mathbf{x}(0) = [0.01; \ zeros(1, 25)]^{T})$ and measurement noise in the hub-rotation angle, $y_1(t)$, as shown in Figure 7.17. A closed-loop settling time

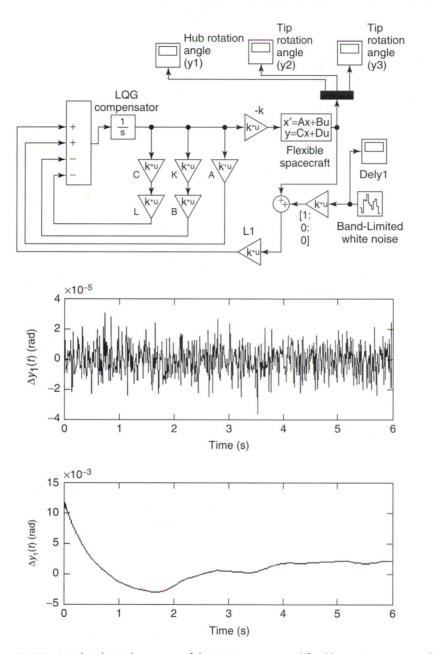

Figure 7.17 Simulated initial response of the LQG compensated flexible rotating spacecraft's hub-rotation angle, $y_1(t)$, with measurement noise, $\Delta y_1(t)$, in the hub-rotation channel

of five seconds, with maximum overshoots less than 1% and a steady-state error of 0.002 rad, is observed when the measurement error in $y_1(t)$ is $\pm 3.5 \times 10^{-5}$ rad. Comparing with the full-state feedback response with zero noise seen in Figure 6.5, the performance is fairly robust with respect to measurement noise.

Exercises

7.1 Repeat Example 7.1 using the MATLAB's random number generator, *rand*, which is a random process with *uniform probability distribution* [1]. What significant differences, if any, do you observe from the normally distributed random process illustrated in Example 7.1?

7.2. The following deterministic signal, $\sin(2t)$, corrupted by a *uniformly distributed* random noise is input to a low-pass filter whose transfer function is given by Eq. (7.15):

$$u(t) = \sin(2t) + 0.5^*rand(t) \tag{7.88}$$

where *rand* is the MATLAB's uniformly distributed random number generator (see Exercise 7.1). Determine the filter's cut-off frequency, ω_o, such that the noise is attenuated without too much distortion of the output, $y(t)$, compared to the noise-free signal, $\sin(2t)$. Generate the input signal, $u(t)$, from $t = 0$ to $t = 10$ seconds, using a time step size of 0.01 second, and simulate the filtered output signal, $y(t)$. Draw a Bode plot of the filter.

7.3. Simulate the output to the following noisy signal input to the elliptic filter described in Example 7.3:

$$u(t) = \sin(10t) + \sin(100t) + 0.5^*rand(t) \tag{7.89}$$

How does the filtered signal compare with the desired deterministic signal, $\sin(10t) + \sin(100t)$? What possible changes are required in the filter's Bode plot such that the signal distortion and noise attenuation are both improved?

7.4. For a linear system with a state-space representation given by Eqs. (7.40) and (7.41), and the process white noise, $\mathbf{v}(t)$, as the only input (i.e. $\mathbf{u}(t) = \mathbf{z}(t) = \mathbf{0}$), derive the differential equation to be satisfied by the *covariance matrix of the output*, $\mathbf{R_y}(t, t)$. (Hint: use steps similar to those employed in deriving Eq. (7.64) for the covariance matrix of the estimation error $\mathbf{R_e}(t, t)$.)

7.5. An interesting *non-stationary* random process is the *Wiener process*, $\mathbf{w}_i(t)$, defined as the *time integral* of the white noise, $\mathbf{w}(t)$, such that

$$\mathbf{w}_i^{(1)}(t) = \mathbf{w}(t) \tag{7.90}$$

Consider the Wiener process as the output, $\mathbf{y}(t) = \mathbf{w}_i(t)$, of a linear system, into which white noise, $\mathbf{w}(t)$, is input. Using the covariance equation derived in Exercise 7.4, show that the covariance matrix of the Wiener process, $\mathbf{R_{wi}}(t)$, grows linearly with time.

7.6. Design a Kalman filter for the distillation column whose state-space representation is given in Exercises 5.4 and 5.15, using $\mathbf{F} = \mathbf{B}$, $\mathbf{Z} = \mathbf{CC^T}$, $\Psi = \mathbf{0}$, and $\mathbf{V} = 0.01\mathbf{B^T B}$. Where are the poles of the Kalman filter in the s-plane?

7.7. Design a Kalman filter for the aircraft lateral dynamics given by Eq. (4.97) in Exercise 4.3 using $\mathbf{F} = \mathbf{I}$, $\mathbf{Z} = \mathbf{CC^T}$, $\Psi = \mathbf{0}$, and $\mathbf{V} = 0.001\mathbf{F^TF}$. Where are the poles of the Kalman filter in the s-plane?

7.8. Design a Kalman filter for the turbo-generator of Example 3.14, using $\mathbf{F} = \mathbf{B}$, $\mathbf{Z} = \mathbf{CC^T}$, $\Psi = \mathbf{0}$, and $\mathbf{V} = 0.0001\mathbf{B^TB}$. Where are the poles of the Kalman filter in the s-plane?

7.9. Design a linear Kalman filter to estimate the *nonlinear* state of wing-rock model described in Example 4.13, and programmed in the M-file *wrock.m* which is listed in Table 4.8. Plot the elements of the estimation error vector as functions of time, if the initial condition is $\mathbf{x}(0) = [0.2; 0; 0; 0; 0]^T$. Which state variables have the largest estimation error, and what are the magnitudes of the largest estimation errors?

7.10. Re-design a Kalman filter for the distillation column (Exercise 7.6), such that the LQG optimal compensator formed by including the optimal regulator deigned in Exercise 6.2 has loop-transfer recovery at the plant's input in the frequency range 0.01–10 rad/s. What are the largest control input magnitudes required for the resulting LQG/LTR compensator if the initial condition is $\mathbf{x}(0) = [1; 0; 0; 0]^T$?

7.11. Re-design a Kalman filter for the aircraft lateral dynamics (Exercise 7.7), such that the LQG optimal compensator formed by including the optimal regulator deigned in Exercise 6.3 has loop-transfer recovery at the plant's input in the frequency range 0.1–100 rad/s. What are the largest control input magnitudes required for the resulting LQG/LTR compensator if the initial condition is $\mathbf{x}(0) = [0.5; 0; 0; 0]^T$? Repeat the simulation of the closed-loop initial response using a measurement noise of ±0.01 rad/s in the roll-rate, $p(t)$, channel.

7.12. Re-design a Kalman filter for the turbo-generator (Exercise 7.8), such that the LQG optimal compensator formed by including the optimal regulator designed in Exercise 6.5 has loop-transfer recovery at the plant's input in the frequency range 0–10 rad/s. What are the largest control input magnitudes required for the resulting LQG/LTR compensator if the initial condition is $\mathbf{x}(0) = [0.1; 0; 0; 0; 0; 0]^T$?

References

1. Papoulis, A. *Probability, Random Variables and Stochastic Processes*. McGraw-Hill, New York, 1984.
2. Parks, T.W. and Burrus, C.S. *Digital Filter Design*. Wiley, New York, 1987.
3. *Robust Control Toolbox for use with MATLAB, User's Guide*. MathWorks, Inc., Natick, MA, 2000.
4. Maciejowski, J.M. *Multivariable Feedback Design*. Addison-Wesley, New York, 1989.

8

Digital Control Systems

8.1 What are Digital Systems?

So far, we have confined our attention to *continuous time* (or *analog*) systems, i.e. systems whose inputs and outputs are *continuous* functions of time. However, we often have to deal with systems whose output signals are *discontinuous* in time. Such systems are called *discrete time* (or *digital*) systems. Where are digital systems encountered? Any continuous time signal that is processed (or *sampled*) by either a human being, or by a machine – mechanical or electrical – becomes a digital signal by the very process of *sampling*, because a human being or a machine takes some *non-zero* time, albeit small, to *evaluate* or *record* the value of the signal at any given time. Over this non-zero time interval, the current value of the signal is *held* and analyzed, before the value of the signal at the *next* time interval is processed. Hence, by sampling a continuous time signal, we essentially *convert* it into a digital signal, and the *sampling process* is a digital system. Another type of digital system is the one which produces discontinuous outputs due to the intrinsic nature of its mechanism. Watches and clocks of any type, tick and tock in a *discontinuous* manner (i.e. the needle or electronic display moves in *jerks* rather than in a perfectly smooth manner) while trying to simulate a *continuous* passage of time. The jerks or discontinuities in the motion of a time-piece can be minimized by choosing an appropriate mechanism. However, the *time interval* between two successive ticks, no matter how small, is non-zero for an atomic clock as well as for a grandfather clock. Hence, we have to deal with inherently digital systems, as well as digital systems obtained by sampling continuous time signals. However, if the *sampling interval* of the output of a continuous system, or the discontinuity in time of the output of an inherently digital system, is *small* in comparison to the *time constants* (i.e. inverse of natural frequencies) of the system, then such a signal, even though digital by nature, can be *approximated* to be a continuous signal, and we can apply all the methods developed in the previous chapters to study and design a control system that produces such an output. Interestingly, the *solution* of the state-equations for an analog system requires its *approximation* by a corresponding *digital* system, as we saw in Chapter 4!

Most mechanical and electrical control systems before the advent of the digital computer had previously been approximated as continuous time (or *analog*) systems. Nowadays, control systems are invariably based on the *digital computer*, which is an *inherently* digital system, because it receives and produces *electrical impulses*, rather than continuous electrical signals, on which *binary arithmetic* is based. There are several advantages in having a digital computer as a part of the control system (usually the feedback controller), namely the ease and versatility of implementing complex control laws merely by programming the computer, relative insensitivity to noise, and low cost. Hence, modern control systems employ digital computers in almost all applications, such as automobiles, pacemakers for heart, aircraft and spacecraft, electronic appliances, personal computers, robotic manipulators, chemical and nuclear reactors and industrial processes. Such applications are usually approximated as an analog plant controlled by a digital computer. However, by having a digital computer in the control system for controlling an analog plant, we introduce an additional characteristic to the control system – the *sampling rate*, i.e. the *inverse* of the non-zero time interval required by the digital computer to process the analog output at each instant of time. The *sampling rate* can drastically modify the behavior of a control system – such as stability, performance, and robustness – when compared to an analog control system. In this chapter, we study how to take into account the sampling rate of digital control systems while studying stability, performance, and robustness, and how such control systems can be designed.

A system into which a continuous time signal is input and from which a digital signal is output (perhaps due to processing by a digital computer) is called an *analog-to-digital converter* (or *A/D converter*). A system which converts digital inputs to analog outputs is called a *digital-to-analog* (or D/A) converter. Since a digital computer processes only digital input signals and produces digital outputs, while using a digital computer in a feedback control system for controlling an analog plant, we must have *both* A/D and D/A converters, as shown in Figure 8.1. However, whereas D/A conversion is a simple and continuous time process (where the analog output is a weighted sum of a set of input impulses), the A/D conversion is a digital process consisting of *sampling* and *holding* the analog input (as pointed out above), and producing an impulse as the output.

We discussed in Chapter 4 how a good approximation of an analog system by a digital system can be obtained if the discrete time step used to solve the digitized state-equations is appropriately small. Conversely, the *sampling rate* at which an analog signal is sampled is crucial for the successful implementation of a digital control system. The *sampling rate* must not be smaller than the *bandwidth* (i.e. the range of frequencies in the power spectrum) of the analog signal, otherwise a *distorted* digital output will be produced. Thus, we may regard the A/D converter as a *low-pass filter* (see Example 7.2) with sampling rate as the cut-off frequency. To reduce signal distortion, the cut-off frequency (sampling rate) should be appreciably larger than the bandwidth of the signal we wish to pass through the filter. Generally, the sampling rate should be at least *twice the bandwidth*

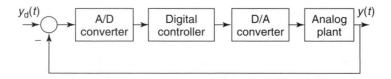

Figure 8.1 A block diagram of a negative feedback control system with an analog plant and a digital controller

(in Hertz) of the analog signal. This minimum value of the sampling rate is known as the *Nyquist sampling rate*.

It is clear from the above discussion that a successful implementation of a digital control system requires a *mathematical model* for the A/D converter. Since A/D converter is based on *two* distinct processes – *sampling* and *holding* – such a model must include separate models for both of these processes. As with continuous time systems, we can obtain the mathematical model for A/D converter in either *frequency domain* (using transform methods and transfer function), or in the *time domain* (using a state-space representation). We begin with the frequency domain modeling and analysis of single-input, single-output digital systems.

8.2 A/D Conversion and the z-Transform

The simplest model for the *sampling process* of the A/D converter is a *switch* which repeatedly closes for a *very short duration*, t_w, after every T seconds, where $1/T$ is the *sampling rate* of the analog input, $f(t)$. The output of such a switch would consist of *series* of *pulses* separated by T seconds. The *width* of each pulse is the duration, t_w, for which the switch remains closed. If t_w is very small in comparison to T, we can assume that $f(t)$ *remains constant* during each pulse (i.e. the pulses are rectangular). Thus, the *height* of the kth pulse is the value of the analog input $f(t)$ at $t = kT$, i.e. $f(kT)$. If t_w is very small, the kth pulse can be approximated by a *unit impulse*, $\delta(t - kT)$, scaled by the area $f(kT)t_w$. Thus, we can use Eq. (2.35) for approximating $f(t)$ by a series of impulses (as shown in Figure 2.16) with $\tau = kT$ and $\Delta\tau = t_w$, and write the following expression for the sampled signal, $f_{tw}^*(t)$:

$$f_{tw}^*(t) = \sum_{k=0}^{\infty} f(kT)t_w\delta(t - kT) = t_w \sum_{k=0}^{\infty} f(kT)\delta(t - kT) \qquad (8.1)$$

Equation (8.1) denotes the fact that the sampled signal, $f_{tw}^*(t)$, is obtained by sampling $f(t)$ at the sampling rate, $1/T$, with pulses of duration t_w. The *ideal sampler* is regarded as the sampler which produces a series of impulses, $f^*(t)$, weighted by the input value, $f(kT)$ as follows:

$$f^*(t) = \sum_{k=0}^{\infty} f(kT)\delta(t - kT) \qquad (8.2)$$

Clearly, the *ideal sampler* output, $f^*(t)$, does not depend upon t_w, which is regarded as a *characteristic* of the *real sampler* described by Eq. (8.1). The ideal sampler is thus a real sampler with $t_w = 1$ second.

Since the sampling process gives a non-zero value of $f^*(t)$ *only* for the duration for which the switch remains closed, there are repeated *gaps* in $f^*(t)$ of approximately T seconds, in which $f^*(t)$ is zero. The *holding* process is an *interpolation* of the sampled input, $f^*(t)$, in each time interval, T, so that the gaps are filled. The simplest holding process is the *zero-order hold* (z.o.h.), which holds the input constant over each time interval, T (i.e. applies a zero-order interpolation to $f^*(t)$). As a result,

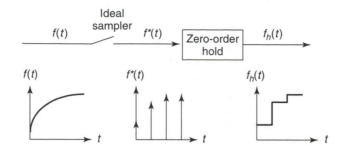

Figure 8.2 Schematic diagram and input, sampled, and output signals of an analog-to-digital (A/D) converter with an ideal sampler and a zero-order hold

the held input, $f_h(t)$, has a staircase time plot. A block diagram of the *ideal sampling* and *holding* processes with a zero-order hold, and their input and output signals, are shown in Figure 8.2.

Since the input to the z.o.h. is a series of impulses, $f^*(t)$, while the output, $f_h(t)$, is a series of steps with amplitude $f(kT)$, it follows that the *impulse response*, $h(t)$, of the z.o.h. must be a *step* that starts at $t = 0$ and ends at $t = T$, and is represented as follows:

$$h(t) = u_s(t) - u_s(t - T) \tag{8.3}$$

or, taking the Laplace transform of Eq. (8.2), the transfer function of the z.o.h. is given by

$$G(s) = (1 - e^{-Ts})/s \tag{8.4}$$

Note that we have used a special property of the Laplace transform in Eq. (8.3) called the *time-shift property*, which is denoted by $\mathcal{L}[y(t - T)] = e^{-Ts}\mathcal{L}[y(t)]$. Similarly, taking the Laplace transform of Eq. (8.2), we can write the Laplace transform of the ideally sampled signal, $F^*(s)$, as follows:

$$F^*(s) = \sum_{k=0}^{\infty} f(kT)e^{-kTs} \tag{8.5}$$

If we define a variable z such that $z = e^{Ts}$, we can write Eq. (8.5) as follows:

$$F(z) = \sum_{k=0}^{\infty} f(kT)z^{-k} \tag{8.6}$$

In Eq. (8.6), $F(z)$ is called the *z-transform* of $f(kT)$, and is denoted by $z\{f(kT)\}$. The z-transform is more useful in studying digital systems than the Laplace transform, because the former incorporates the sampling interval, T, which is a characteristic of digital systems. The expressions for digital transfer functions in terms of the z-transform are easier to manipulate, as they are free from the time-shift factor, e^{-Ts}, of the Laplace transform. In a manner similar to the Laplace transform, we can derive z-transforms of some frequently encountered functions.

Example 8.1

Let us derive the z-transform of $f(kT) = u_s(kT)$, the unit step function. Using the definition of the z-transform, Eq. (8.6), we can write

$$F(z) = \sum_{k=0}^{\infty} u_s(kT)z^{-k} = \sum_{k=0}^{\infty} z^{-k} = 1 + z^{-1} + z^{-2} + z^{-3} + \cdots$$

$$= 1/(1 - z^{-1}) \tag{8.7}$$

or

$$F(z) = z/(z - 1) \tag{8.8}$$

Thus, the z-transform of the unit step function, $u_s(kT)$, is $z/(z - 1)$. (Note that we have used the *binomial series expansion* in Eq. (8.7), given by $(1 - x)^{-1} = 1 + x + x^2 + x^3 + \ldots$).

Example 8.2

Let us derive the z-transform of the function $f(kT) = e^{-akT}$, where $t \geq 0$. Using the definition of the z-transform, Eq. (8.6), we can write

$$F(z) = \sum_{k=0}^{\infty} e^{-akT} z^{-k} = \sum_{k=0}^{\infty} (ze^{aT})^{-k} = 1 + (ze^{aT})^{-1} + (ze^{aT})^{-2} + \cdots$$

$$= 1/[1 - (ze^{aT})^{-1}] = z/(z - ze^{-aT}) \tag{8.9}$$

Thus, $z\{e^{-akT}\} = z/(z - ze^{-aT})$.

The z-transforms of other commonly used functions can be similarly obtained by manipulating series expressions involving z, and are listed in Table 8.1.

Some important properties of the z-transform are listed below, and may be verified by using the definition of the z-transform (Eq. (8.6)):

(a) Linearity:

$$z\{af(kT)\} = az\{f(kT)\} \tag{8.10}$$

$$z\{f_1(kT) + f_2(kT)\} = z\{f_1(kT)\} + z\{f_2(kT)\} \tag{8.11}$$

(b) Scaling in the z-plane:

$$z\{e^{-akT} f(kT)\} = F(e^{aT} z) \tag{8.12}$$

(c) Translation in time:

$$z\{f(kT + T)\} = zF(z) - zf(0^-) \tag{8.13}$$

where $f(0^-)$ is the *initial value* of $f(kT)$ for $k = 0$. Note that if $f(kT)$ has a *jump* at $k = 0$ (such as $f(kT) = u_s(kT)$), then $f(0^-)$ is *understood* to be the value of $f(kT)$

Table 8.1 Some commonly encountered z-transforms

Discrete Time Function, $f(kT)$	z-transform, $F(z)$	Laplace Transform of Equivalent Analog Function, $F(s)$
$u_s(kT)$	$z/(z-1)$	$1/s$
kT	$Tz/(z-1)^2$	$1/s^2$
e^{-akT}	$z/(z-e^{-aT})$	$1/(s+a)$
$1-e^{-akT}$	$z(1-e^{-aT})/[(z-1)(z-e^{-aT})]$	$a/[s(s+a)]$
kTe^{-akT}	$Tze^{-aT}/(z-e^{-aT})^2$	$1/(s+a)^2$
$kT-(1-e^{-akT})/a$	$Tz/(z-1)^2 - z(1-e^{-aT})/[a(z-1)(z-e^{-aT})]$	$a/[s^2(s+a)]$
$\sin(akT)$	$z\sin(aT)/[z^2-2z\cos(aT)+1]$	$a/(s^2+a^2)$
$\cos(akT)$	$z[z-\cos(aT)]/[z^2-2z\cos(aT)+1]$	$s/(s^2+a^2)$
$e^{-akT}\sin(bkT)$	$ze^{-aT}\sin(bT)/[z^2-2ze^{-aT}\cos(bT)-e^{-2aT}]$	$b/[(s+a)^2+b^2]$
$e^{-akT}\cos(bkT)$	$[z^2-ze^{-aT}\cos(bT)]/[z^2-2ze^{-aT}\cos(bT)-e^{-2aT}]$	$(s+a)/[(s+a)^2+b^2]$
$(kT)^n$	$\lim_{a\to 0}(-1)^n d^n/da^n[z/(z-e^{-aT})]$	$n!/s^{n+1}$

before the jump. Thus, for $f(kT)=u_s(kT)$, $f(0^-)=0$. A *negative* translation in time is given by $z\{f(kT-T)\}=z^{-1}F(z)+zf(0^-)$.

(d) Differentiation with z:

$$z\{kTf(kT)\}=-TzdF(z)/dz=-TzF^{(1)}(z) \qquad (8.14)$$

(e) Initial value theorem:

$$f(0^+)=\lim_{z\to\infty}F(z) \qquad (8.15)$$

Equation (8.15) holds if and only if the said limit exists. Note that if $f(kT)$ has a *jump* at $k=0$ (such as $f(kT)=u_s(kT)$), then $f(0^+)$ is *understood* to be the value of $f(kT)$ *after* the jump. Thus, for $f(kT)=u_s(kT)$, $f(0^+)=1$.

(f) Final value theorem:

$$f(\infty)=\lim_{z\to 1}(1-z^{-1})F(z) \qquad (8.16)$$

Equation (8.16) holds if and only if the said limit exists. Using Table 8.1, and the properties of the z-transform, we can evaluate z-transforms of rather complicated functions.

Example 8.3

Let us find the z-transform of $f(kT)=10e^{-akT}\sin(bkT-2T)$. From Table 8.1, we know that

$$z\{\sin(bkT)\}=z\sin(bT)/[z^2-2z\cos(bT)+1] \qquad (8.17)$$

Then, using the linearity property of the z-transform given by Eq. (8.10), we can write

$$z\{10\sin(bkT)\}=10z\sin(bT)/[z^2-2z\cos(bT)+1] \qquad (8.18)$$

Furthermore, the scaling in the z-plane given by Eq. (8.12) yields

$$z\{10e^{-akT}\sin(bkT)\} = 10ze^{aT}\sin(bT)/[z^2e^{2aT} - 2ze^{aT}\cos(bT) + 1] \quad (8.19)$$

Finally, the translation in time given by Eq. (8.13) yields (noting that $10e^0\sin(0) = 0$)

$$z\{10e^{-akT}\sin(bkT - 2T)\} = 10z^{-1}e^{aT}\sin(bT)/[z^2e^{2aT} - 2ze^{aT}\cos(bT) + 1]$$

$$(8.20)$$

8.3 Pulse Transfer Functions of Single-Input, Single-Output Systems

In a manner similar to the transfer function in Laplace domain for a single-input, single-output analog system, we can define a *pulse transfer function* for a single-input, single-output digital system in the *z-domain* as the z-transform of the *output* signal *divided* by the z-transform of the *input* signal. Finding the pulse transfer functions of digital systems in the z-domain is a useful application of the z-transform. However, before finding the pulse transfer function of a digital system, we must distinguish between an *inherently digital system*, and an analog system *rendered digital* due to the process of *data sampling*, as shown in Figure 8.3. It is clear from Figure 8.3 that while an inherently digital system would always produce a *discrete time* output, an analog system would produce a *continuous time output*, even though the input is a digitally sampled signal (recall the continuous time outputs to impulse inputs of analog systems calculated in Chapters 2 and 4).

Finding an *inherently digital* system's pulse transfer function is straightforward, because both input and output are discrete time signals with the same interval, T. On the other hand, taking the z-transform of the transfer function, $G(s)$, of a *sampled-data analog* system – in which the input is *discrete* with a sampling interval, T, while the output is a *continuous* time signal – is quite problematic. However, if we assume that we are *only* interested in finding the output of a sampled-data analog system at the *same*

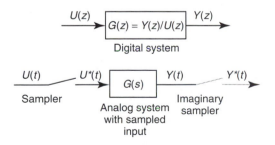

Figure 8.3 Block diagrams of an inherently digital system and a sampled-data analog system rendered digital by an *imaginary sampler* at the output

discrete time points at which the input is sampled, the difficulty in finding the transfer function in the z-domain can be removed. By making such an assumption, we would be disregarding the continuous nature of the output of a sampled-data analog system, and treat the system as if it's output is sampled by an *imaginary sampler*. By having such an imaginary sampler at the output, the sampled-data analog system effectively becomes an inherently digital system. The ideally sampled input of an analog system is given by Eq. (8.2) as follows:

$$u^*(t) = \sum_{n=0}^{\infty} u(nT)\delta(t - nT) \tag{8.21}$$

The continuous time output, $y(t)$, is given by the *discretized convolution integral* of Eq. (2.119) as follows:

$$y(t) = \sum_{n=0}^{\infty} u(nT)g(t - nT) \tag{8.22}$$

where $g(t - nT)$ is the *impulse response* of the analog system to a unit impulse applied at time $t = nT$. The digitized output, $y^*(t)$, resulting from sampling of $y(t)$ by an imaginary sampler shown in Figure 8.3, with the same sampling interval, T, as that of the input, $u^*(t)$, can be written as follows:

$$y^*(t) = y(kT) = \sum_{n=0}^{\infty} u(nT)g(kT - nT) \tag{8.23}$$

Then, the z-transform of the (imaginary) sampled output, $y^*(t) = y(kT)$, can be written as

$$Y(z) = z\{y(kT)\} = \sum_{k=0}^{\infty} \sum_{n=0}^{\infty} u(nT)g(kT - nT)z^{-k} \tag{8.24}$$

or, expressing the double summation as a product of two summations,

$$Y(z) = \sum_{k=0}^{\infty} g(kT - nT)z^{-k+n} \sum_{n=0}^{\infty} u(nT)z^{-n} \tag{8.25}$$

Since $g(kT - nT) = 0$ for $k < n$, we can begin the first summation in Eq. (8.25) at $k = n$ instead of $k = 0$, and defining $m = k - n$, we can write

$$Y(z) = \sum_{m=0}^{\infty} g(mT)z^{-m} \sum_{n=0}^{\infty} u(nT)z^{-n} \tag{8.26}$$

or, recognizing the second summation of Eq. (8.26) as the z-transform of the input, $U(z)$,

$$Y(z) = \left[\sum_{m=0}^{\infty} g(mT)z^{-m} \right] U(z) \tag{8.27}$$

Comparing Eq. (8.27) with the definition of the pulse transfer function, $G(z) = Y(z)/U(z)$, we can write the pulse transfer function of a sampled-data analog system as follows:

$$G(z) = \sum_{m=0}^{\infty} g(mT)z^{-m} \tag{8.28}$$

Equation (8.28) indicates that the pulse transfer function of a sampled-data analog system is the z-transform of the *digitized* impulse response, $g(mT)$. This is an important result, which shows that although the output, $y(t)$, of a sampled-data analog system is continuous in time, by having an imaginary sampler at the output (Figure 8.3), the analog system essentially becomes a digital system with the pulse transfer function, $G(z)$. Since the impulse response, $g(t)$, of the analog system is the inverse Laplace transform of its transfer function, $G(s)$, we can directly obtain $G(z)$ from $G(s)$ by using Table 8.1, and denote $G(z)$ by the convenient notation, $G(z) = z\{G(s)\}$.

Example 8.4

Let us find the pulse transfer function of the z.o.h. The transfer function of z.o.h is given by Eq. (8.4) as $G(s) = (1 - e^{-Ts})/s$. We can express $G(s)$ as $G(s) = 1/s - e^{-Ts}/s$, and write

$$G(z) = z\{1/s\} - z\{e^{-Ts}/s\} \tag{8.29}$$

From Table 8.1, we note that

$$z\{1/s\} = z/(z-1) \tag{8.30}$$

Furthermore, note that e^{-Ts}/s is the Laplace transform of $u_s(t - T)$. Therefore, using the time-shift property of the z-transform given by Eq. (8.13), we can write

$$z\{e^{-Ts}/s\} = z\{u_s(kT - T)\} = z^{-1} \cdot z\{u_s(kT)\} = 1/(z-1) \tag{8.31}$$

Substituting Eqs. (8.30) and (8.31) into Eq. (8.29), we can write the pulse-transfer function of the z.o.h as follows:

$$G(z) = z/(z-1) - 1/(z-1) = (z-1)/(z-1) = 1 \tag{8.32}$$

Thus, the pulse transfer function of the z.o.h is unity.

Example 8.5

Let us find the pulse transfer function of a sampled-data analog system obtained by placing the z.o.h in *series* with an analog system with transfer function $G_1(s) = s/(s^2 + 1)$. The overall transfer function of the resulting sampled-data analog system is written as

$$G(s) = G_1(s)(1 - e^{-Ts})/s = (1 - e^{-Ts})/(s^2 + 1) \tag{8.33}$$

For finding the pulse transfer function, $G(z)$, we first express $G(s)$ as follows:

$$G(s) = 1/(s^2 + 1) - e^{-Ts}/(s^2 + 1) \qquad (8.34)$$

Note that $1/(s^2+1)$ is the Laplace transform of $\sin(kT)$. Furthermore, $e^{-Ts}/(s^2+1)$ is the Laplace transform of $\sin(kT - T)$. Hence, using the time-shift property of the z-transform and Eq. (8.34), we can write

$$G(z) = z\{\sin(kT)\} - z^{-1} \cdot z\{\sin(kT)\} = (1 - z^{-1}) \cdot z\{\sin(kT)\} \qquad (8.35)$$

or, using Table 8.1,

$$G(z) = (1 - z^{-1})z \sin(T)/[z^2 - 2z \cos(T) + 1]$$
$$= (z - 1) \sin(T)/[z^2 - 2z \cos(T) + 1] \qquad (8.36)$$

Examples 8.4 and 8.5 indicate that we can obtain the pulse transfer function of a *general* sampled-data analog system consisting of the z.o.h in series with a system with *proper* transfer function, $G_1(s)$, by expressing the overall transfer function as $G(s) = (1 - e^{-Ts})[G_1(s)/s]$. Then, using the time-shift property of the z-transform, and noting that $s_1(t) = \mathcal{L}^{-1}[G_1(s)/s]$ is the *step response* of the system, $G_1(s)$, with $s(0^-) = 0$, we can write the pulse transfer function, $G(z)$, as follows:

$$G(z) = (1 - z^{-1}) \cdot z\{G_1(s)/s\} = (1 - z^{-1}) \cdot z\{s_1(kT)\} \qquad (8.37)$$

In Chapter 2, we learnt how to calculate the step response, $s_1(t)$, of a proper transfer function, $G_1(s)$, using the *partial fraction expansion* of $G_1(s)/s$, either by hand, or by the M-file *stepresp.m*. Note that in the M-file *stepresp.m*, the step response, $s_1(t)$, is essentially computed at *discrete* time points, with time interval, $dt = T$. Hence, *stepresp.m* gives us the *digitized step response*, $s_1(kT)$, which can be thought of as the result of sampling $s_1(t)$ by an imaginary sampler (Figure 8.3) with a specified sampling interval, T. If only we can compute the z-transform of a discrete time signal, $s_1(kT)$, by an appropriate M-file, we will have all the necessary tools to compute the pulse transfer function of a general sampled-data system. For the moment, let us confine ourselves to hand calculation of the pulse transfer function.

Example 8.6

Let us find the pulse transfer function of the sampled-data analog system consisting of an A/D converter with z.o.h in *closed-loop* with an analog plant of transfer function, $G_1(s)$, as shown in Figure 8.4(a), if $G_1(s) = (s + 1)/[(s + 2)(s + 3)]$.

To find the pulse transfer function of the closed-loop sampled-data analog system, $G(z) = Y(z)/Y_d(z)$, we have to express the block diagram of the system in such a way that both $y(t)$ and $y_d(t)$ are digital signals. This is done in Figure 8.4(b) by moving the sampler at the input of z.o.h to outside the feedback loop before

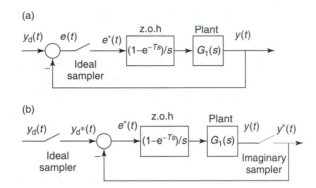

Figure 8.4 Two equivalent block diagrams of closed-loop sampled-data analog system with zero-order hold and analog plant, $G_1(s)$

the summing junction, as well as placing an imaginary sampler in the feedback loop. In doing so, we are not changing the essential characteristic of the closed-loop system. Hence, the block-diagrams shown in Figures 8.4(a) and (b) are *equivalent*. The *open-loop* transfer function of the system is given by

$$G_o(s) = (1 - e^{-Ts})G_1(s)/s \tag{8.38}$$

The pulse transfer function of the closed-loop system is given by $G(z) = Y(z)/Y_d(z)$. From the block-diagram of Figure 8.4(b) it is clear that $e^*(t) = y_d^*(t) - y^*(t)$, or, taking the z-transform, $E(z) = Y_d(z) - Y(z)$. Furthermore, the open-loop pulse transfer function is $z\{G_o(s)\} = G_o(z) = Y(z)/E(z)$. Therefore, the closed-loop pulse transfer function is $G(z) = Y(z)/Y_d(z) = G_o(z)/[1 + G_o(z)]$, where $G_o(z)$ is the pulse transfer function of the open-loop system. We can calculate $G_o(z)$ by finding the z-transform of $G_o(s)$ by Eq. (8.37) as follows:

$$G_o(z) = z\{G_o(s)\} = (1 - z^{-1}) \cdot z\{G_1(s)/s\} \tag{8.39}$$

where

$$z\{G_1(s)/s\} = z\{1/(6s) + 1/[2(s + 2)] - 2/[3(s + 3)]\}$$
$$= z/[6(z - 1)] + z/[2(z - e^{-2T})] - 2z/[3(z - e^{-3T})] \tag{8.40}$$

Substituting Eq. (8.40) into Eq. (8.39) and simplifying, we get

$$G_o(z) = [z(1 + 3e^{-2T} - 4e^{-3T}) + e^{-5T} + 3e^{-3T}$$
$$- 4e^{-2T}]/[6(z - e^{-2T})(z - e^{-3T})] \tag{8.41}$$

Finally, the pulse transfer function of the closed-loop system is obtained as follows:

$$G(z) = G_o(z)/[1 + G_o(z)]$$
$$= [z(1 + 3e^{-2T} - 4e^{-3T}) + e^{-5T} + 3e^{-3T} - 4e^{-2T}]/[6(z - e^{-2T})(z - e^{-3T})$$
$$+ z(1 + 3e^{-2T} - 4e^{-3T}) + e^{-5T} + 3e^{-3T} - 4e^{-2T}] \tag{8.42}$$

Examples 8.5 and 8.6 show that finding the pulse transfer functions of sampled-data analog systems requires converting the system into an equivalent system in which both output and input appear to be digital signals by appropriate placement of samplers. In doing so, care must be exercised that the characteristics of the system are unchanged. It is clear that the *pulse transfer function* of a general *sampled-data analog* system consisting of several *analog* sub-systems cannot be handled in the *same* manner as the *transfer function* of an *analog system*. For example, the pulse transfer function, $G(z)$, of a sampled-data analog system consisting of two analog sub-systems with transfer functions $G_1(s)$ and $G_2(s)$ in series *cannot* be written as $G(z) = G_1(z)G_2(z)$, but only as $G(z) = z\{G_1(s)G_2(s)\}$. This is due to the fact that, generally, $z\{G_1(s)G_2(s)\} \neq G_1(z)G_2(z)$. However, if we are dealing with a *digital* system consisting of *digital* sub-systems, the *pulse transfer functions* of the sub-systems are handled in precisely the same manner as the *transfer functions* of *analog* sub-systems systems. For example, the pulse transfer function of a digital system consisting of *two* digital sub-systems with pulse transfer functions $G_1(z)$ and $G_2(z)$ in *series* is merely $G(z) = G_1(z)G_2(z)$. All of this shows us that we must be extremely careful in deriving the pulse transfer functions. It always helps to write down the input-output relationships (such as in Example 8.6) of the various sub-systems as *separate* equations, and then derive the overall input-output relationship from those equations.

8.4 Frequency Response of Single-Input, Single-Output Digital Systems

In a manner similar to an analog system with transfer function, $G(s)$, whose frequency response is the value of $G(s)$ when $s = i\omega$, we can define the *frequency response* of a *digital* system with a *pulse transfer function*, $G(z)$, as the value of $G(z)$ when $z = e^{i\omega T}$, or $G(e^{i\omega T})$. In so doing, we can plot the gain and phase of the frequency response, $G(e^{i\omega T})$, as functions of the frequency, ω, somewhat like the Bode plots of an analog system. However, the *digital Bode plots* crucially depend upon the sampling interval, T. For instance, the gain, $|G(e^{i\omega T})|$, would become *infinite* for some values of ωT. The frequency response of a digital system is related to the steady-state response to a harmonic input, provided the system is stable (i.e. its harmonic response at large time exists and is finite). The sampling rate of a harmonic output is crucial in obtaining the digital system's frequency response. If a high-frequency signal is sampled at a rate smaller than the signal frequency, then a large distortion and *ambiguity* occur in the sampled data. For example, if we sample two very different harmonic signals, say $\sin(0.25\pi t/T)$ and $\sin(1.75\pi t/T)$, with the same sampling interval, T, then the two sampled data would be *identical*. In other words, we lose information about the higher frequency signal by sampling it at a lower rate. This important phenomenon of sampled-data analog systems is known as *aliasing*. To avoid aliasing, the sampling rate must be at least *twice* the signal's *bandwidth*, ω_b (i.e. the *largest* frequency contained in the signal). The minimum acceptable sampling rate is called the *Nyquist frequency*. The Nyquist frequency in rad/s is thus given by *half* the required *sampling rate*, i.e. $2\pi/T = 2\omega_b$, or $\pi/T = \omega_b$, where T is the sampling interval. Hence, the frequency response of a digital system with a given sampling interval, T, is usually calculated only for $\omega \leq \pi/T$.

MATLAB's Control System Toolbox (CST) provides the command *dbode* for computing the frequency response of a digital system, and is used as follows:

```
>>[mag,phase,w] = dbode(num,den,T) <enter>
```

where *num* and *den* are the numerator and denominator polynomials of the system's pulse transfer function, $G(z)$, in decreasing powers of z, T is the sampling interval, *mag* and *phase* are the returned magnitude and phase (degrees), respectively, of the computed frequency response, $G(e^{i\omega T})$, and w is the vector of discrete frequency points (upto the Nyquist frequency) at which the frequency response is computed. The user can optionally specify the desired frequency points as the *fourth* input argument of the command *dbode*. When used without the output arguments on the left-hand side, *dbode* produces the digital Bode plots on the screen.

Example 8.7

Let us find the frequency response of a digital system with pulse transfer function, $G(z) = z/(z^2 + 2)$, and sampling interval, $T = 0.1$ second. To get the magnitude and phase of the frequency response, we can use the command *dbode* as follows:

```
>>num = [1 0];den = [1 0 2];dbode(num,den,0.1) <enter>
```

The resulting digital Bode magnitude and phase plots are shown in Figure 8.5. Note that the Nyquist frequency is $\pi/T = 31.4159$ rad/s, indicated as the highest frequency in Figure 8.5 at which the frequency response is plotted. At the Nyquist frequency, the phase is seen to approach $180°$. There is a peak in the gain plot at $\omega = 15.7$ rad/s at which the gain is 0 dB and the phase is $90°$. A gain of 0 dB

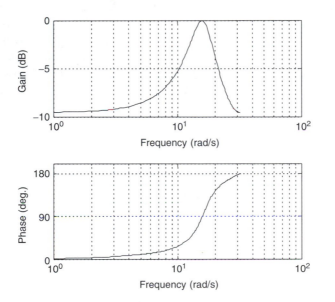

Figure 8.5 Digital Bode plot of the pulse transfer function, $G(z) = z/(z^2 + 2)$ with a sampling interval, $T = 0.1$ second

corresponds to $|G(e^{i\omega T})| = 1$, or $|e^{i\omega T}/(e^{2i\omega T} + 2)| = 1$, which is seen in Figure 8.5 to occur at $\omega T = 1.5708 = \pi/2$. Thus, the value of z at which to 0 dB and the phase is $90°$ is $z = e^{i\pi/2} = i$, and the corresponding value of $G(z) = G(i) = i$. Note that both Nyquist frequency, $\omega = \pi/T$, and the frequency, $\omega = \pi/(2T)$, at which the gain peaks to 0 dB with phase $90°$, will be modified if we change the sampling interval, T.

8.5 Stability of Single-Input, Single-Output Digital Systems

The non-zero sampling interval, T, of digital systems crucially affects their *stability*. In Chapter 2, we saw how the location of *poles* of an analog system (i.e. roots of the denominator polynomial of transfer function, $G(s)$) determined the system's stability. If none of the poles lie in the right-half s-plane, then the system is stable. It would be interesting to find out what is the appropriate location of a *digital* system's *poles* – defined as the roots of the denominator polynomial of the *pulse transfer function*, $G(z)$ – for stability. Let us consider a *mapping* (or *transformation*) of the Laplace domain on to the z-plane. Such a mapping is defined by the transformation, $z = e^{Ts}$. The region of *instability*, namely the right-half s-plane, can be denoted by expressing $s = \alpha + i\omega$, where $\alpha > 0$. The corresponding region in the z-plane can be obtained by $z = e^{(\alpha+i\omega)T} = e^{\alpha T}e^{i\omega T} = e^{\alpha T}[\cos(\omega T) + i\sin(\omega T)]$, where $\alpha > 0$. Note that a *line* with $\alpha = $ constant in the s-plane corresponds to a *circle* $z = e^{\alpha T}[\cos(\omega T) + i\sin(\omega T)]$ in the z-plane of *radius* $e^{\alpha T}$ *centered at* $z = 0$. Thus, a line with $\alpha = $ constant in the right-half s-plane ($\alpha > 0$) corresponds to a circle in the z-plane of *radius larger than unity* ($e^{\alpha T} > 1$). In other words, the right-half s-plane corresponds to the region *outside* a *unit circle* centered at the origin in the z-plane. Therefore, stability of a digital system with pulse transfer function, $G(z)$, is determined by the location of the poles of $G(z)$ with respect to the unit circle in the z-plane. If any pole of $G(z)$ is located *outside* the unit circle, then the system is *unstable*. If all the poles of $G(z)$ lie *inside* the unit circle, then the system is *asymptotically stable*. If some poles lie *on* the unit circle, then the system is *stable*, but *not* asymptotically stable. If poles on the unit circle are *repeated*, then the system is *unstable*.

Example 8.8

Let us analyze the stability of a digital system with pulse transfer function, $G(z) = z/[(z - e^{-T})(z^2 + 1)]$, if the sampling interval is $T = 0.1$ second. The poles of $G(z)$ are given by $z = e^{-T}$, and $z = \pm i$. With $T = 0.1$ s, the poles are $z = e^{-0.1} = 0.9048$, and $z = \pm i$. Since none of the poles of the system lie outside the unit circle centered at $z = 0$, the system is *stable*. However, the poles $z = \pm i$ are *on* the unit circle. Therefore, the system is *not* asymptotically stable. Even if the sampling interval, T, is made very small, the pole at $z = e^{-T}$ will approach – but remain *inside* – the unit circle. Hence, this system is stable for all possible values of the sampling interval, T.

Example 8.9

Let us find the *range* of sampling interval, T, for which the digital system with pulse transfer function, $G(z) = 100/(z^2 - 5e^{-T} + 4)$, is stable. The poles of $G(z)$ are given by the solution of the characteristic equation $z^2 - 5e^{-T} + 4 = 0$, or $z = \pm(5e^{-T} - 4)^{1/2}$. If $T > 0.51$, then $|z| > 1.0$, indicating instability. Hence, the system is stable for $T \leq 0.51$ second.

Example 8.10

A sampled-data closed-loop system shown in Figure 8.4 has a plant transfer function, $G_1(s) = s/(s^2 + 1)$. Let us find the range of sampling interval, T, for which the closed-loop system is stable.

In Example 8.6, we derived the pulse transfer function of the closed-loop system to be the following:

$$G(z) = [z(1 + 3e^{-2T} - 4e^{-3T}) + e^{-5T} + 3e^{-3T} - 4e^{-2T}]/[6(z - e^{-2T})(z - e^{-3T})$$
$$+ z(1 + 3e^{-2T} - 4e^{-3T}) + e^{-5T} + 3e^{-3T} - 4e^{-2T}] \tag{8.43}$$

The closed-loop poles are the roots of the following characteristic polynomial:

$$6(z - e^{-2T})(z - e^{-3T}) + z(1 + 3e^{-2T} - 4e^{-3T}) + e^{-5T}$$
$$+ 3e^{-3T} - 4e^{-2T} = 0 \tag{8.44}$$

or

$$6z^2 + z(1 - 3e^{-2T} - 10e^{-3T}) + 7e^{-5T} + 3e^{-3T} - 4e^{-2T} = 0 \tag{8.45}$$

An easy way of solving Eq. (8.45) is through the MATLAB command *roots* as follows:

```
>>T=0.0001;z=roots([6 1-3*exp(-2*T)-10*exp(-3*T) 7*exp(-5*T)+3*exp(-3
  *T)-4*exp(-2*T)]) <enter>

z =
  0.9998
  0.9996
```

Since the poles of $G(z)$ approach the unit circle only in the limit $T \rightarrow 0$, it is clear that the system is stable for all non-zero values of the sampling interval, T.

In a manner similar to stability analysis of the closed-loop analog, single-input, single-output system shown in Figure 2.32 using the *Nyquist plot* of the open-loop transfer function, $G_o(s) = G(s)H(s)$, we can obtain the Nyquist plots of *closed-loop digital systems* with an *open-loop* pulse transfer function, $G_o(z)$, using the MATLAB (CST) command *dnyquist* as follows:

```
>>dnyquist(num,den,T) <enter>
```

where *num* and *den* are the numerator and denominator polynomials of the system's *open-loop* pulse transfer function, $G_o(z)$, in decreasing powers of z, and T is the sampling interval. The result is a *Nyquist plot* (i.e. *mapping* of the *imaginary axis* of the Laplace domain in the $G_o(z)$ plane) on the screen. The user can supply a vector of desired frequency points, w, at which the Nyquist plot is to be obtained, as the fourth input argument of the command *dnyquist*. The command *dnyquist* obtains the Nyquist plot of the digital system in the $G_o(z)$ plane. Hence, the stability analysis is carried out in *exactly the same* manner as for analog systems, using the Nyquist stability theorem of Chapter 2. According to the Nyquist stability theorem for digital systems, the closed-loop system given by the pulse transfer function, $G_o(z)/[1 + G_o(z)]$, is *stable* if the Nyquist plot of $G_o(z)$ encircles the point -1 *exactly* P times in the *anti-clockwise* direction, where P is the number of *unstable* poles of $G_o(z)$.

Example 8.11

Let us use the digital Nyquist plot to analyze the stability of the closed-loop sampled data analog system shown in Figure 8.4 with the plant, $G_1(s) = 1/[(s(s + 1)]$. The open-loop transfer function of the system is $G_o(s) = (1 - e^{-Ts})/[s^2(s + 1)]$, while the open-loop *pulse transfer function* can be written as follows:

$$G_o(z) = [z(T - 1 + e^{-T}) + 1 - e^{-T} - Te^{-T}]/[(z - 1)(z - e^{-T})] \quad (8.46)$$

For a specific value of the sampling interval, such as $T = 0.1$ second, the Nyquist plot is obtained as follows:

```
>>T=0.1;num = [T-1+exp(-T) 1-exp(-T)-T*exp(-T)],den = conv([1
  -1],[1 -exp(-T)]) <enter>

num =
  0.0048    0.0047
den =
  1.0000   -1.9048    0.9048
```

The poles of $G_o(z)$ are obtained as follows:

```
    >>roots(den) <enter>

    ans =
       1.0000
       0.9048
```

Thus, $G_o(z)$ does not have any poles outside the unit circle, hence $P = 0$. The digital Nyquist plot is obtained in the frequency range $1-10$ rad/s as follows:

```
    >>w=logspace(0,1);dnyquist(num,den,T,w) <enter>
```

The resulting Nyquist plot is shown in Figure 8.6. Note that the Nyquist plot of $G_o(z)$ *does not* encircle the point -1 in the $G_o(z)$ plane. Hence, the closed-loop system with pulse transfer function, $G_o(z)/[1 + G_o(z)]$, is stable for $T = 0.1$

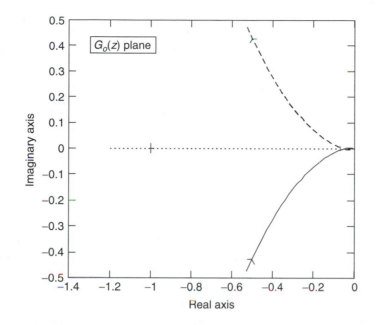

Figure 8.6 Digital Nyquist plot of open-loop pulse transfer function, $G_o(z)$, for the sampled-data closed-loop system of Example 8.10 with sampling interval $T = 0.1$ s

second. We can also analyze the stability of the closed-loop system by checking the closed-loop pole locations as follows:

```
>>sysp=tf(num,den);sysCL=feedback(sysp,1) <enter>

Transfer function:
0.004837z+0.004679
--------------
z^2-1.9z+0.9095


>>[wn,z,p]=damp(sysCL); wn, p <enter>

wn =
 0.9537
 0.9537

p=
 0.9500+0.0838i
 0.9500-0.0838i
```

Note that the returned arguments of the CST command *damp* denote the closed-loop poles, *p*, and the *magnitudes* of the closed-loop poles, *wn*. (*Caution: z* and *wn do not* indicate the *digital* system's damping and natural frequencies, for which you should use the command *ddamp* as shown below.) Since *wn* indicates that both the closed-loop poles are inside the unit circle, the closed-loop system is stable.

In addition to *dbode* and *dnyquist*, MATLAB (CST) provides the command *dnichols* for drawing the digital Nichols plot. You may refer to the User's Guide for Control Systems Toolbox [1] for details about this as well as other digital frequency response commands.

8.6 Performance of Single-Input, Single-Output Digital Systems

Performance of a digital system can be studied by obtaining the time response to specified inputs. In a manner similar to the inverse Laplace transform for analog systems, we can use the definition of the z-transform to find the *inverse z-transform* of the output, $Y(z)$, denoted by $z^{-1}\{Y(z)\}$, to find the discrete time response of a single-input, single-output digital system with pulse transfer function, $G(z)$, and input, $U(z)$. From Eq. (8.6) it is clear that the inverse z-transform, $z^{-1}\{Y(z)\}$, yields the discrete time output, $y(kT)$. The method of finding the inverse z-transform of a function, $F(z)$, is quite similar to that followed in finding the inverse Laplace transform, and involves expressing $F(z)$ as a *partial fraction* expansion in z (there is an alternative method of using *power series expansion* in z, but we will not consider it due to its complexity). However, examining the z-transforms given in Table 8.1, it is clear that z is usually present as a numerator factor in most z-transforms. Hence, to find the inverse z-transform of $F(z)$, it will be useful to express $F(z)$ as the following expansion:

$$F(z) = K_1 z/(z - z_1) + K_2 z/(z - z_2) + \cdots + K_n z/(z - z_n) \qquad (8.47)$$

where $z_1, z_2, \ldots, z_n$ are the poles of $F(z)$ and $K_1, K_2, \ldots, K_n$ are the residues of the following partial fraction expansion:

$$F(z) = K_1/(z - z_1) + K_2/(z - z_2) + \cdots + K_n/(z - z_n) \qquad (8.48)$$

(Note that, for simplicity, we have assumed that all the poles of $F(z)$ are distinct. If some poles are repeated, we can use the more general partial fraction expansion introduced in Chapter 2.) Taking the inverse z-transform of each term on the right-hand side of Eq. (8.47), we can write

$$f(kT) = K_1 z_1^k + K_2 z_2^k + \cdots + K_n z_n^k \qquad (8.49)$$

We can apply the above strategy to find the discrete time output, $y(kT)$, of a digital system, $G(z)$, with input $U(z)$ and zero initial conditions, by finding the inverse z-transform of $Y(z) = G(z)U(z)$.

Example 8.12

Let us find the discrete time response of $G(z) = z/(z^2 - 0.7z + 0.3)$ to a unit step input, $u(kT) = u_s(kT)$. The z-transform of $u_s(kT)$ is $z/(z - 1)$. Hence, $Y(z) = G(z)U(z) = z^2/[(z - 1)(z^2 - 0.7z + 0.3)]$. We begin by finding the partial fraction expansion of $Y(z)/z$ as follows by using the MATLAB command *residue*:

```
>>num = [1 0];den = conv([1 -1],[1 -0.7 0.3]); [K,z,c] = residue(num,
  den)<enter>

K =
   1.6667
  -0.8333+0.0989i
  -0.8333-0.0989i
z =
   1.0000
   0.3500+0.4213i
   0.3500-0.4213i
c =
    []
```

Hence, the partial fraction expansion of $Y(z)/z$ is the following:

$$Y(z)/z = z/[(z-1)(z^2 - 0.7z + 0.3)]$$

$$= (5/3)/(z-1) + (-0.8333 + 0.0989i)/[z - (0.3500 + 0.4213i)]$$

$$+ (-0.8333 - 0.0989i)/[z - (0.3500 - 0.4213i)] \tag{8.50}$$

or

$$Y(z) = (5/3)z/(z-1) + (-0.8333 + 0.0989i)z/[z - (0.3500 + 0.4213i)]$$

$$+ (-0.8333 - 0.0989i)z/[z - (0.3500 - 0.4213i)] \tag{8.51}$$

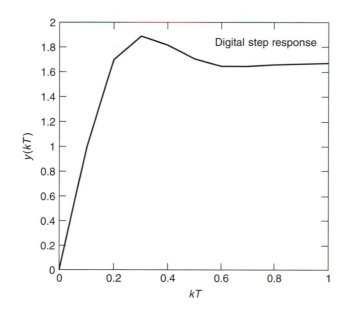

Figure 8.7 Step response of the digital system of Example 8.11 with sampling interval, $T = 0.1$ s

Then, taking the inverse z-transform of each term on the right-hand side of Eq. (8.51), we get the discrete time output at the kth sampling instant as follows:

$$f(kT) = 5/3 + (-0.8333 + 0.0989i)(0.3500 + 0.4213i)^k$$
$$+ (-0.8333 - 0.0989i)(0.3500 - 0.4213i)^k \qquad (8.52)$$

The step response, $f(kT)$, calculated by Eq. (8.52) is plotted in Figure 8.7 for $T = 0.1$ seconds using the following MATLAB command:

```
>>for k=1:11;f(k)=K(1)+K(2)*z(2)^(k-1)+K(3)*z(3)^(k-1);t(k)=0.1*
  (k-1);end;plot(t,f) <enter>
```

Note that the step response has a maximum overshoot of 13% with a steady state value of about 1.67. The peak time and the settling time are defined for the digital response as occurring after so many sampling intervals. The peak time occurs for $k = 3$, while the settling time occurs for $k = 5$. With $T = 0.1$ second, the peak time is 0.3 second, while the settling time is 0.5 second.

MATLAB M-files can be written to compute step and impulse responses of digital systems using the inverse z-transform of partial fraction expansion (such as *impresp.m* and *stepresp.m* for impulse and step responses, respectively, of analog systems in Chapter 2).

The steady state value of the response, $Y(z)$, can be directly obtained using the final-value theorem given by Eq. (8.16), as follows:

$$y(\infty) = \lim_{z \to 1}(1 - z^{-1})Y(z) = \lim_{z \to 1}(1 - z^{-1})G(z)U(z) \qquad (8.53)$$

(Of course, the existence of the steady-state limit in Eq. (8,53) translates into the requirement that all the poles of $G(z)U(z)$ should lie inside the unit circle.) The steady state value of the response can thus be found using Eq. (8.53), without having to compute and plot the response. Let us use Eq. (8.53) to find the steady state value of the *step response* of a digital system with pulse transfer function, $G(z)$, as follows:

$$y(\infty) = \lim_{z \to 1}(1 - z^{-1})Y(z) = \lim_{z \to 1}(1 - z^{-1})G(z)U(z)$$
$$= \lim_{z \to 1}(1 - z^{-1})[z/(z - 1)]G(z) = \lim_{z \to 1} G(z) \qquad (8.54)$$

Thus, the steady state value of the step response of $G(z)$ is given by $\lim_{z \to 1} G(z)$ called the *digital DC gain* of $G(z)$. However, instead of having to calculate the limit in Eq. (8.54) by hand, we can use the MATLAB (CST) command *ddcgain* to find the digital DC gain as follows:

```
>>ddcgain(num,den) <enter>
```

where *num* and *den* are the numerator and denominator polynomials of the pulse transfer function, $G(z)$, in decreasing powers of z. Let us find the steady state value of the step response in Example 8.12 using *ddcgain* as follows:

```
>>num = [1 0];den = [1 -0.7 0.3];ddcgain(num,den) <enter>

ans =
   1.6667
```

Note that we can find the steady state value of response to an *arbitrary* input, $U(z)$, by calculating the digital DC gain of the pulse transfer function $(1 - z^{-1})G(z)U(z)$. Also, an idea about a digital system's response can be obtained from the location of the *dominant* poles with respect to the unit circle in the z-domain. The MATLAB command *ddamp* is quite useful in determining the natural frequencies and damping ratios associated with the poles of a digital system, and is used in a manner similar to the command *damp* for analog systems, with the difference that the sampling interval, T, is specified as the second input argument:

```
>>[mag,wn,z] = ddamp(den,T) <enter>
```

where *den* is the denominator polynomial of the system's pulse transfer function, T is the sampling interval, *mag* is the magnitude of the pole in the z-domain, *wn* is the equivalent natural frequency in the s-plane, and z is the equivalent damping ratio. The command *ddamp* used without the sampling interval as the second input argument gives only the magnitudes of the poles of the system. When used without the output arguments on the left-hand side, the magnitudes, natural frequencies, and damping ratios are printed on the screen. Instead of *den*, you can input the z-plane pole locations as a vector, and get the returned damping ratios and natural frequencies associated with the poles. Let us determine the natural frequencies and damping ratios of the digital system of Example 8.12 when the sampling interval is $T = 0.1$ second as follows:

```
>>ddamp([1 -0.7 0.3],0.1) <enter>

Eigenvalue        Magnitude Equiv. Damping Equiv. FrEq. (rad/sec)
  0.3500+0.4213i  0.5477     0.5657         10.6421
  0.3500-0.4213i  0.5477     0.5657         10.6421
```

A damping ratio of $\zeta = 0.5657$ indicates a well damped response, as seen in Figure 8.7.

For systems of order greater than two, we can identify the dominant poles which are the closest to the unit circle in the z-domain. The performance of such systems is primarily determined by the location of the dominant poles.

8.7 Closed-Loop Compensation Techniques for Single-Input, Single-Output Digital Systems

For improving the performance of digital systems, we can apply closed-loop compensation techniques similar to analog systems discussed in Section 2.12, by placing a compensator with pulse-transfer function, $H(z)$, in closed-loop with a digital plant, $G(z)$, as shown in Figure 8.8. However, we have to be mindful of the dependence of the closed-loop digital system's properties on the sampling interval, T, and the properties of the z-transform, while designing compensators in the z-domain. Let us consider an example of closed-loop digital compensation.

Just as in continuous systems, we can have the ideal *proportional-integral-derivative* (PID) control of digital systems. The digital equivalent of a PID compensator's transfer

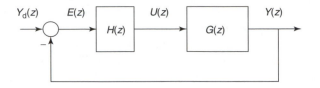

Figure 8.8 A single-input, single-output digital feedback control system with controller pulse transfer function, $H(z)$, and plant pulse transfer function, $G(z)$

function is obtained by combining the *proportional* term with the z-transform of the *derivative* term and the z-transform of the integral term. Note that the time derivative of the error in Figure 8.8, $e^{(1)}(t)$, is approximated by the digital equivalent of a first order *difference* as $e^{(1)}(t) = [e(kT) - e(kT - T)]/T$, whose z-transform is obtained using the time-shift property of z-transform (Eq. (8.13)) as $(1 - z^{-1})E(z)/T$. Thus, the derivative term of PID compensator has the form $K_D(1 - z^{-1})$. The z-transform of the integral term, K_I/s, is $K_I z/(z - 1)$ according to Table 8.1. The PID compensator's pulse transfer is thus given by $H(z) = K_P + K_D(1 - z^{-1}) + K_I z/(z - 1)$, or $H(z) = [(K_P + K_D + K_I)z^2 - (K_P - 2K_D)z + K_D]/(z^2 - z)$.

Example 8.13

Consider a digital system with the following pulse-transfer function and a sampling interval, $T = 0.2$ second:

$$G(z) = 0.0001(z - 1)/(z^2 + 1.97z + 0.99) \qquad (8.55)$$

The poles, natural frequencies and damping ratios of the system are obtained as follows:

```
>>ddamp([1 1.97 0.99],0.2) <enter>

Eigenvalue        Magnitude Equiv. Damping Equiv. FrEq. (rad/sec)
 -0.9850+0.1406i  0.9950    0.0017            14.9990
 -0.9850-0.1406i  0.9950    0.0017            14.9990
```

A damping ratio of $\zeta = 0.0017$ indicates a lightly damped response, which is unacceptable. It is desired to increase the closed-loop damping to $\zeta = 0.7$ while keeping the natural frequency unchanged at $\omega_n = 15$ rad/s, through a *proportional-integral* (PI) compensator in the closed-loop configuration of Figure 8.8. With $H(z) = K_P + K_I z/(z - 1)$, the closed-loop pulse-transfer function is given by

$$Y(z)/Y_d(z) = G(z)H(z)/[1 + G(z)H(z)]$$

$$= 0.0001[K_P(z - 1) + K_I z]/[z^2 + \{1.97 + 0.0001(K_P + K_I)\}z$$

$$+ 0.99 - 0.0001K_P] \qquad (8.56)$$

The closed-loop poles are the roots of $z^2 + \{1.97 + 0.0001(K_P + K_I)\}z + 0.99 - 0.0001K_P = 0$, and must be the same as the desired pole locations $z_{1,2} =$

$e^{sT} = \exp(-\zeta\omega_n T)[\cos\{\omega_n T(1-\zeta^2)^{1/2}\} \pm i\sin\{\omega_n T(1-\zeta^2)^{1/2}\}]$. Plugging $\zeta = 0.7$ and $\omega_n = 15$ rad/s in the desired characteristic polynomial, $z^2 - 2\exp(-\zeta\omega_n T)\cos\{\omega_n T(1-\zeta^2)^{1/2}\}z + \exp(-2\zeta\omega_n T)$, and comparing the coefficients with those of the denominator polynomial Eq. (8.56), we get $K_P = 9750$ and $K_I = -28\,125$. Let us check whether the closed-loop poles are at desired locations as follows:

```
>>Kp=9750;Ki=-28125;sysGH=tf(0.0001*[Kp+Ki-Kp],[1 1.97 0.99]);
  sysCL=feedback(sysGH,1) <enter>
Transfer function:
-1.838z-0.975
---------------
z^2+0.1325z+0.015


>>ddamp([1 0.1325 0.015],0.2) <enter>


Eigenvalue              Magnitude Equiv. Damping Equiv. Freq.(rad/s)
-6.63e-002+1.03e-001i 1.22e-001 7.00e-001       1.50e+001
-6.63e-002-1.03e-001i 1.22e-001 7.00e-001       1.50e+001
```

which confirms that the desired closed-loop damping and natural frequency have been achieved.

The steady-state error of the *stable* closed-loop system shown in Figure 8.8 to a *unit step* desired output can be obtained from the final value theorem (Eq. (8.16)) as follows:

$$e(\infty) = \lim_{z\to 1}(1 - z^{-1})E(z) = \lim_{z\to 1}(1 - z^{-1})Y_d(z)/[1 + G(z)H(z)]$$

$$= \lim_{z\to 1} 1/[1 + G(z)H(z)] \tag{8.57}$$

or the steady state error is the digital DC gain of the pulse transfer function $1/[1 + G(z)H(z)]$. For the system of Example 8.13, the steady-state error is computed as follows:

```
>>num=0.0001*[Kp+Ki -Kp];den=[1 1.97 0.99];a=ddcgain(num,den);
  e=1/(1+a) <enter>

e =
  3.4510
```

which is very large. The closed-loop steady state error to a step desired output can be reduced to zero by a compensator which has *two* poles at $z = 1$, such as $H(z) = K(z+\alpha)/(z-1)^2$, since the pole at $z = 1$ of the PI compensator got canceled by the plant zero at the same location. Recall that a pole at $s = 0$ in the s-plane is equivalent to a pole at $z = 1$ in the z-plane; hence, the *type* of a digital closed-loop system is determined by the number of poles at $z = 1$ in the open-loop pulse transfer function, $G(z)H(z)$. For the system of Example 8.13, you can find the compensator, $H(z) = K(z+\alpha)/(z-1)^2$, as an exercise such that the closed-loop damping and natural frequency are as desired, and the steady-state error to a unit step function is brought to zero.

Example 8.13 illustrates that the closed-loop compensation of digital systems in the z-domain can be carried out similarly to that of analog systems in the s-domain, which was discussed in Section 2.12. You can extend the remaining techniques presented in Section 2.12, namely, *lead compensation*, *lag compensation*, and *lead-lag compensation* for the z-domain design of digital systems.

8.8 State-Space Modeling of Multivariable Digital Systems

Consider a single-input, single-output digital system described by the pulse transfer function, $G(z)$. Instead of having to find the partial fraction expansion of $Y(z) = G(z)U(z)$ for calculating a system's response, an alternative approach would be to express the pulse transfer function, $G(z)$, as a *rational function* in z, given by $G(z) = num(z)/den(z)$, and then obtaining the inverse z-transforms of both sides of the equation, $Y(z)den(z) = U(z)num(z)$. Let us write a general expression for a rational pulse transfer function, $G(z)$, as follows:

$$G(z) = (b_m z^m + b_{m-1} z^{m-1} + \cdots + b_1 z + b_0)/(z^n + a_{n-1} z^{n-1} + \cdots + a_1 z + a_0) \quad (8.58)$$

We can write $Y(z)den(z) = U(z)num(z)$ as follows:

$$Y(z)(z^n + a_{n-1} z^{n-1} + \cdots + a_1 z + a_0) = U(z)(b_m z^m + b_{m-1} z^{m-1} + \cdots + b_1 z + b_0) \quad (8.59)$$

or

$$z^n Y(z) + a_{n-1} z^{n-1} Y(z) + \cdots + a_1 z Y(z) + a_0 Y(z)$$
$$= b_m z^m U(z) + b_{m-1} z^{m-1} U(z) + \cdots + b_1 z U(z) + b_0 U(z) \quad (8.60)$$

Recall the time-shift property of the z-transform given by Eq. (8.13). For a single translation in time, $z\{y(kT + T)\} = zY(z) - zy(0^-)$, where $Y(z)$ is the z-transform of $y(kT)$. Taking another step forward in time, we can similarly write $z\{y(kT + 2T)\} = z\{y(kT + T)\} - zy(T^-) = z^2 Y(z) - z^2 y(0^-) - zy(T^-)$, and for n time steps the corresponding result would be $z\{y(kT + nT)\} = z^n Y(z) - z^{n-1} y(0^-) - z^{n-2} y(T^-) - \cdots - zy\{n - 1)T^-\}$. Similarly, for the input we could write $z\{u(kT + mT)\} = z^m U(z) - z^{m-1} u(0^-) - z^{m-2} u(T^-) - \cdots - zu\{(m - 1)T^-\}$. The values of the output at n previous instants, $y(0^-), y(T^-), \ldots, y\{(n - 1)T^-\}$ are analogous to the *initial conditions* that must be specified for solving an nth order continuous-time differential equation. In addition, the input, $u(kT)$, is assumed to be known at the m previous time instants, $u(0^-), u(T^-), \ldots, u\{(m - 1)T^-\}$. For simplicity, let us assume that both input and output are zero at the n and m previous time instants, respectively, i.e. $y(0^-) = y(T^-) = \cdots = y\{(n - 1)T^-\} = u(0^-) = u(T^-) = \cdots = u\{(m - 1)T^-\} = 0$. Then, it follows that $zY(z) = z\{y(kT + T)\}, \ldots, z^n Y(z) = z\{y(kT + nT)\}$, and $zU(z) = z\{u(kT + T)\}, \ldots, z^m U(z) = z\{u(kT + mT)\}$. Hence, we can easily take the inverse z-transform of Eq. (8.60) and write

$$y(kT + nT) = -a_{n-1} y(kT + nT - T) - \cdots - a_1 y(kT + T) - a_0 y(kT)$$
$$+ b_0 u(kT) + b_1 u(kT + T) + \cdots + b_m u(kT + mT) \quad (8.61)$$

Equation (8.61) is called a *difference equation* (digital equivalent of a *differential equation* for analog systems) of order n. It denotes how the output is *propagated* in time by n sampling instants, given the values of the output at all *previous* instants, as well as the values of the input at all instants up to $t = kT + mT$. Note that for a proper system, $m \leq n$, a simplified notation for Eq. (8.61) is obtained by dropping T from the brackets, and writing the difference equation as follows:

$$y(k + n) = -a_{n-1}y(k + n - 1) - \cdots - a_1 y(k + 1) - a_0 y(k)$$
$$+ b_0 u(k) + b_1 u(k + 1) + \cdots + b_m u(k + m) \qquad (8.62)$$

where $y(k + n)$ is understood to denote $y(kT + nT)$, and so on. Solving the difference equation, Eq. (8.62), for the discrete time output, $y(k)$, is the digital equivalent of solving an nth order differential equation for the output, $y(t)$, of an analog system. The most convenient way of solving an nth order difference equation is by expressing it as a set of n first order difference equations, called the *digital state-equations* (analogous to the state-equations of the continuous time systems).

For simplicity, let us assume that $n = m$ in Eq. (8.62), which can then be written as follows:

$$y(k + n) = -a_{n-1}y(k + n - 1) - \cdots - a_1 y(k + 1) - a_0 y(k)$$
$$+ b_0 u(k) + b_1 u(k + 1) + \cdots + b_n u(k + n) \qquad (8.63)$$

To derive a digital state-space representation for the digital system described by Eq. (8.63), let us draw a *schematic diagram* for the system – shown in Figure 8.9–in a manner similar to Figure 3.4 for an nth order analog system. The drawing of the schematic diagram is made simple by the introduction of a *dummy variable*, $q(k)$, in a manner similar to the analog system of Example 3.6 (since the right-hand side of Eq. (8.63) contains time delays of the input), such that

$$q(k + n) + a_{n-1}q(k + n - 1) + \cdots + a_1 q(k + 1) + a_0 q(k) = u(k) \qquad (8.64)$$

and

$$b_n q(k + n) + b_{n-1}q(k + n - 1) + \cdots + b_1 q(k + 1) + b_0 q(k) = y(k) \qquad (8.65)$$

In Figure 8.9, the *integrator* of Figure 3.4 is replaced by a *delay element* (denoted by a rectangle inscribed with $\mathcal{D}$). While the output, $x(t)$, of an integrator is the time-integral of its input, $x^{(1)}(t)$, the output, $x(k)$, of a delay element is the input, $x(k + 1)$, *delayed* by one sampling interval. Note from Eq. (8.13) that the pulse transfer function of a delay element is $1/z$, while the transfer function of an integrator is $1/s$. Thus, a delay element is the digital equivalent of an integrator. We can arbitrarily select the state variables, $x_1(k), x_2(k), \ldots, x_n(k)$, to be the *outputs* of the delay elements, *numbered from the right*, as shown in Figure 8.9. Therefore, $x_1(k) = q(k)$, $x_2(k) = q(k + 1), \ldots, x_n(k) = q(k + n - 1)$, and the digital state-equations are the following:

$$x_1(k + 1) = x_2(k) \qquad (8.66a)$$

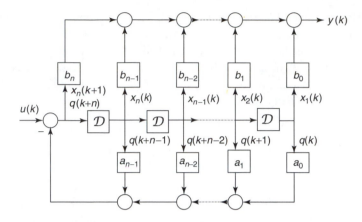

Figure 8.9 Schematic diagram for an nth order digital system

$$x_2(k + 1) = x_3(k) \tag{8.66b}$$

$$\vdots$$

$$x_{n-1}(k + 1) = x_n(k) \tag{8.66c}$$

$$x_n(k + 1) = -[a_{n-1}x_n(k) + a_{n-2}x_{n-1}(k) + \cdots + a_0x_1(k) - u(k)] \tag{8.66d}$$

The output equation is obtained from the summing junction at the top of Figure 8.9 as follows:

$$y(k) = b_0x_1(k) + b_1x_2(k) + \cdots + b_{n-1}x_n(k) + b_nx_n(k + 1) \tag{8.67}$$

and substituting Eq. (8.66d) into Eq. (8.67) the output equation is expressed as follows:

$$y(k) = (b_0 - a_0b_n)x_1(k) + (b_1 - a_1b_n)x_2(k) + \cdots + (b_{n-1} - a_{n-1}b_n)x_n(k) + b_nu(k) \tag{8.68}$$

The matrix form of the digital state-equations is the following:

$$
\begin{bmatrix}
x_1(k+1) \\
x_2(k+1) \\
\vdots \\
x_{n-1}(k+1) \\
x_n(k+1)
\end{bmatrix}
=
\begin{bmatrix}
0 & 1 & 0 & \cdots & 0 \\
0 & 0 & 1 & \cdots & 0 \\
\vdots & \vdots & \vdots & \vdots & \\
0 & 0 & 0 & \cdots & 1 \\
-a_0 & -a_1 & -a_2 & \cdots & -a_{n-1}
\end{bmatrix}
\begin{bmatrix}
x_1(k) \\
x_2(k) \\
\vdots \\
x_{n-1}(k) \\
x_n(k)
\end{bmatrix}
+
\begin{bmatrix}
0 \\
0 \\
\vdots \\
0 \\
1
\end{bmatrix}
u(k)
\tag{8.69}
$$

and the output equation in matrix form is as follows:

$$y(k) = [(b_0 - a_0b_n)(b_1 - a_1b_n) \ldots (b_{n-1} - a_{n-1}b_n)]
\begin{bmatrix}
x_1(k) \\
x_2(k) \\
\vdots \\
x_n(k)
\end{bmatrix}
+ b_nu(k) \tag{8.70}$$

On comparing Eqs. (8.69) and (8.70) with Eqs. (3.59) and (3.60), respectively, we find that the digital system has been expressed in the *controller companion form*. Alternatively, we can obtain the *observer companion form* and the *Jordan canonical form* using the same approach as given in Chapter 3 for analog systems.

Whereas we used a single-input, single-output system to illustrate the digital state-space representation, the main utility of digital state-space representations lie in modeling *multivariable* digital systems. A general multivariable, linear, time-invariant digital system can be expressed by the following state-space representation:

$$\mathbf{x}(k+1) = \mathbf{A_d}\mathbf{x}(k) + \mathbf{B_d}\mathbf{u}(k) \tag{8.71}$$

$$\mathbf{y}(k) = \mathbf{C_d}\mathbf{x}(k) + \mathbf{D_d}\mathbf{u}(k) \tag{8.72}$$

where $\mathbf{A_d}$, $\mathbf{B_d}$, $\mathbf{C_d}$, and $\mathbf{D_d}$ are the constant coefficient matrices, and $\mathbf{x}(k)$ and $\mathbf{u}(k)$ are the digital state-vector and the digital input vector, respectively, at the kth sampling instant. The subscript d on the coefficient matrices is used to indicate a digital state-space representation (as opposed to an analog state-space representation, which is denoted by $\mathbf{A}$, $\mathbf{B}$, $\mathbf{C}$, $\mathbf{D}$). All the properties of the state-space representation as discussed in Chapter 3 for analog systems – namely linear transformation, system characteristics, and block-building – apply for the digital state-space representations, *with the crucial difference* that the s-plane for analog systems is transformed into the z-plane for digital systems. For example, a *multivariable digital state-space* representation given by the coefficient matrices $\mathbf{A_d}$, $\mathbf{B_d}$, $\mathbf{C_d}$, and $\mathbf{D_d}$ has a *pulse transfer matrix*, $\mathbf{G}(z) = \mathbf{C_d}(z\mathbf{I} - \mathbf{A_d})^{-1}\mathbf{B_d} + \mathbf{D_d}$.

Example 8.14

Let us find a state-space representation for the digital system of Example 8.12. The pulse transfer function of the system is $G(z) = z/(z^2 - 0.7z + 0.3)$. We can employ the MATLAB (CST) function *ss* to get a state-space representation of the digital system as follows:

```
>>num = [1 0];den = [1 -0.7 0.3];sys=tf(num,den);sys=ss(sys) <enter>

a =
              x1        x2
       x1     0.7      -0.6
       x2     0.5       0

b =
              u1
       x1     1
       x2     0

c =
              x1        x2
       y1     1         0

d =
              u1
       y1     0
```

The eigenvalues of the digital state-dynamics matrix, $\mathbf{A}_d$, must be the *poles* of the system in the z-domain. We can determine the eigenvalues, natural frequencies, and damping ratios associated with $\mathbf{A_d}$ for a sampling interval, $T = 0.1$ second, by using the MATLAB (CST) function *ddamp* as follows (*ddamp* allows the use of the state-dynamics matrix as an input, instead of the denominator polynomial of $G(z)$):

```
>>[Ad,Bd,Cd,Dd]=ssdata(sys);ddamp(Ad,0.1) <enter>
```

```
Eigenvalue       Magnitude Equiv. Damping Equiv. Freq. (rad/sec)
 0.3500+0.4213i 0.5477     0.5657           10.6421
 0.3500-0.4213i 0.5477     0.5657           10.6421
```

Note that these values are the same as those obtained in Example 8.12. Let us find the *Jordan canonical form* of the digital system using the command *canon* as follows:

```
>>sysc = canon(sys,'modal') <enter>

a =
              x1          x2
    x1       0.35     0.42131
    x2    -0.42131       0.35

b =
            u1
    x1    1.354
    x2   1.1248

c =
            x1          x2
    y1   0.73855            0

d =
            u1
    y1      0
```

Note the appearance of the real and imaginary parts of the poles in z-domain as the elements of the Jordan form state-dynamics matrix.

While obtaining the pulse transfer functions of single-input, single-output, sampled-data analog systems, we employed a sampling and holding of continuous time input signal. This procedure can be extended to multivariable systems, where each input is sampled and held. Such a procedure leads to a *digital approximation* of multivariable analog system, which we studied in Chapter 4. Instead of the pulse transfer functions (or pulse transfer matrices), it makes a better sense to talk of *digital state-space representation* of a sampled-data, multivariable analog system. Chapter 4 presented analog-to-digital conversion of analog state-equations, to obtain their time domain solution on a digital computer. The same techniques of converting an analog system to an equivalent digital system can be employed for other purposes, such as controlling an analog plant by a digital controller. In Chapter 4, the MATLAB (CST) commands *c2d* and *c2dm* were introduced for approximating an analog system by an equivalent digital system. While *c2d* employs a zero-order hold (z.o.h) for holding the sampled signals, the command *c2dm* also allows a *first-order hold* (f.o.h), a *bilinear* (Tustin), or a *prewarp* interpolation for better accuracy when the continuous time input signals are smooth. The command c2dm is used as follows:

```
>>[Ad,Bd,Cd,Dd] = c2dm(A,B,C,D,T,'method') <enter>
```

where **A**, **B**, **C**, **D** are the continuous-time state-space coefficient matrices, **Ad**, **Bd**, **Cd**, **Dd** are the returned digital state-space coefficient matrices, T is the specified sampling interval, *method* is the method of interpolation for the input vector to be specified as *zoh* (for zero-order hold), *foh* (for first-order hold), *tustin* (for Tustin interpolation), or *prewarp* (for Tustin interpolation with *frequency prewarping* – a more accurate interpolation than plain Tustin). While the *first-order hold* was covered in Chapter 4, the *bilinear* (or Tustin) approximation refers to the use of the approximation, $s = \ln(z)/T \approx 2(z-1)/[T(z+1)]$ while mapping from the s-plane to the z-plane. Such an approximation is obtained by expanding $\ln(z) = 2(z-1)/(z+1) + (1/3)[(z-1)/(z+1)]^3 + \cdots$ and retaining only the first term of the series. While the imaginary axis on the s-plane is not *precisely* mapped as a unit circle in the z-plane, as it would be according to $z = e^{Ts}$ (or $s = \ln(z)/T$). A frequency response of the bilinear transformed digital system would be thus *warped* by this approximation. To remove the frequency warping of the bilinear transformation, a *pre-warping* technique is applied, by which the *critical frequencies* of the s-plane transfer function fall precisely on the z-plane at the points where they belong. You may refer to a textbook on digital control [2] for pre-warping techniques applied to bilinear transformation.

Example 8.15

Let us find a digital state-space representation of a sampled-data analog system with the following state coefficient matrices:

$$\mathbf{A} = \begin{bmatrix} -1 & 0 & 0 \\ 0 & -0.5 & 0.7 \\ 0 & -0.7 & -0.5 \end{bmatrix}; \quad \mathbf{B} = \begin{bmatrix} 1 & 0 \\ 0 & -1 \\ 0 & -1 \end{bmatrix}$$

$$\mathbf{C} = \begin{bmatrix} 1 & 0 & 0 \\ 0 & 0 & 1 \end{bmatrix}; \quad \mathbf{D} = \begin{bmatrix} 0 & 0 \\ 0 & -2 \end{bmatrix} \tag{8.73}$$

if the sampling interval is 0.2 second. We begin by a digital conversion using a z.o.h with the command *c2dm* as follows:

```
>>A = [-1 0 0; 0 -0.5 0.7; 0 -0.7 -0.5]; B = [1 0; 0 -1; 0 -1]; C = [1 0
  0; 0 0 1]; <enter>

>>D = [0 0; 0 -2]; [Ad,Bd,Cd,Dd] = c2dm(A,B,C,D,0.1,'zoh')
  <enter>
```

```
Ad =
  0.9048    0         0
       0    0.9489    0.0665
       0   -0.0665    0.9489

Bd =
  0.0952    0
       0   -0.1008
       0   -0.0941

Cd =
     1      0         0
     0      0         1

Dd =
     0      0
     0     -2
```

Let us also find an alternative digital approximation using f.o.h as follows:

```
>>[Ad1,Bd1,Cd1,Dd1] = c2dm(A,B,C,D,0.1,'foh') <enter>

Ad1 =
  0.9048         0         0
       0    0.9489    0.0665
       0   -0.0665    0.9489

Bd1 =
  0.0906         0
       0    -0.1015
       0    -0.0883

Cd1 =
     1    0    0
     0    0    1

Dd1 =
  0.0484         0
       0    -2.0480
```

Note that while the digital state-dynamics matrix, $\mathbf{A_d}$, and the output state coefficient matrix, $\mathbf{C_d}$, are unchanged by changing the hold from z.o.h to f.o.h, the matrices $\mathbf{B_d}$ and $\mathbf{D_d}$ are appreciably modified. This is expected, because the matrices $\mathbf{B_d}$ and $\mathbf{D_d}$ are the coefficient matrices for the input vector, and the hold affects only the sampled, continuous time inputs.

8.9 Solution of Linear Digital State-Equations

A great part of Chapter 4 discussed the solution of the digitized approximation to analog linear, state-equations. Since we have extensively studied the solution of digital state-equations – time varying and time-invariant – in Chapter 4, we shall briefly review the solution of digital, time-invariant state-equations here. The time propagation of a linear, time-invariant scalar first-order difference equation is directly obtained from the difference equation, namely $x(k+1) = ax(k) + bu(k)$, where $u(k)$ is an arbitrary digital input. Similarly, a linear, time-invariant, digital state-equation is a vector difference equation, expressed as $\mathbf{x}(k+1) = \mathbf{A_d}\mathbf{x}(k) + \mathbf{B_d}\mathbf{u}(k)$, where $\mathbf{u}(k)$ is the digital input vector, and $\mathbf{x}(k)$ is the digital state-vector. Since all the quantities on the right-hand side of a first-order, vector difference equation are known, the value of the state-vector at the next sampling instant, $\mathbf{x}(k+1)$, is directly obtained. MATLAB provides a fast algorithm in the command *ltitr* for the time propagation of a linear, time-invariant digital state-equation, and is used as follows:

```
>>X = ltitr(Ad,Bd,U,X0) <enter>
```

where **Ad** and **Bd** are the state coefficient matrices of the digital system, **U** is a matrix containing the values of the input variables in its columns (each row of **U** corresponds to a different time point), **x0** is the initial condition vector, and **x** is the returned matrix containing values of the state-variables in its columns, with each row corresponding to a different time point. If the initial condition is not specified, *ltitr* assumes zero initial conditions.

Example 8.16

Let us solve the digital state-equation obtained in Example 8.16 with z.o.h, for the first 10 sampling instants, if each of the two input variables is a unit step function, and the initial condition is zero, i.e. $\mathbf{x}(0) = \mathbf{0}$. We begin with specifying the inputs with a sampling interval of $T = 0.2$ second as follows:

```
>>t=0:0.2:2; U=[ones(size(t)); ones(size(t))]'; <enter>
```

Then, the command *ltitr* is used to solve the digital state-equation as follows:

```
>>X = ltitr(Ad,Bd,U) <enter>

X =
   0          0          0
   0.0952    -0.1008    -0.0941
   0.1813    -0.2028    -0.1766
   0.2592    -0.3050    -0.2482
   0.3297    -0.4068    -0.3093
   0.3935    -0.5074    -0.3605
   0.4512    -0.6063    -0.4024
   0.5034    -0.7030    -0.4356
   0.5507    -0.7969    -0.4606
   0.5934    -0.8877    -0.4781
   0.6321    -0.9750    -0.4887
```

Let us also find the response of the digital system obtained in Example 8.15 with f.o.h if the initial condition is $\mathbf{x}(0) = [0.1; 1; 1]^T$, and the input vector is

$\mathbf{u}(k) = [\sin(kT); -\sin(kT)]^T$. We begin by specifying the inputs as follows:

```
>>t=0:0.2:2; U=[sin(t); -sin(t)]'; <enter>
```

Then, the command *ltitr* is used with the appropriate initial condition as follows:

```
>>X0 = [0.1 1 1]'; X = ltitr(Ad1,Bd1,U,X0) <enter>

X =
    0.1000   1.0000   1.0000
    0.0905   1.0154   0.8824
    0.0999   1.0424   0.7873
    0.1256   1.0810   0.7121
    0.1648   1.1305   0.6536
    0.2141   1.1890   0.6083
    0.2699   1.2541   0.5724
    0.3286   1.3226   0.5420
    0.3866   1.3911   0.5133
    0.4403   1.4556   0.4828
    0.4866   1.5122   0.4472
```

Note that the digital state-space representation allows us to find the response of multi-variable digital systems to arbitrary inputs with ease. MATLAB (CST) also provides the commands *dstep*, *dinitial*, and *dimpulse* for the computation of step, initial and impulse response, respectively, of a digital system. The commands *dstep*, *dinitial*, and *dimpulse* are used in a manner similar to the *step*, *initial*, and *impulse* for analog systems, with the difference that the sampling interval is an additional input argument.

For the solution to both Eqs. (8.71) and (8.72) to arbitrary inputs and initial conditions, MATLAB provides the command *dlsim* which is used as follows:

```
>>[y,X] = dlsim(Ad,Bd,Cd,Dd,U,X0) <enter>
```

where **Ad**, **Bd**, **Cd** and **Dd** are the state coefficient matrices of the digital system, **U** is a matrix containing the values of the input variables in its columns (each row of **U** corresponds to a different time point), **x0** is the initial condition vector, **x** is the returned matrix containing values of the state-variables in its columns, with each row corresponding to a different time point, and **y** is the returned matrix containing the output variables at different time points. If the initial condition is not specified, *dlsim* assumes zero initial conditions. Note that while for analog systems, the corresponding MATLAB (CST) command, *lsim*, digitized the continuous time state and output equations before solving them, *dlsim* does not need to do so. For single-input, single-output systems, *dlsim* can be used with pulse transfer function instead of state-space representation (*dlsim* computes a state-space representation internally).

Example 8.17

Let us solve the digital state-equation obtained in Example 8.16 with z.o.h, for the first 10 sampling instants, if each of the two input variables is a *normally distributed random function*, and the initial condition is $\mathbf{x}(0) = [0.1; 1; 1]^T$. We begin with specifying the inputs with a sampling interval of $T = 0.2$ second as follows:

```
>>randn('seed',0); t=0:0.2:2; U=[randn(size(t)); randn(size(t)]' <enter>
```

```
U =
    1.1650   -0.7012
    0.6268    1.2460
    0.0751   -0.6390
    0.3516    0.5774
   -0.6965   -0.3600
    1.6961   -0.1356
    0.0591   -1.3493
    1.7971   -1.2704
    0.2641    0.9846
    0.8717   -0.0449
   -1.4462   -0.7989
```

Then, the initial condition is specified and the response is obtained using *dlsim* as follows:

```
>>X0 = [0.1 1 1]'; [y,X] = dlsim(Ad,Bd,Cd,Dd,U,X0) <enter>
```

```
y =
    0.1000    2.4023
    0.2013   -1.5436
    0.2418    1.9883
    0.2260   -0.4849
    0.2379    1.2328
    0.1490    0.7274
    0.2962    3.0788
    0.2737    2.9634
    0.4186   -1.5212
    0.4039    0.3430
    0.4484    1.7718
```

```
X =
0.1000    1.0000    1.0000
0.2013    1.0861    0.9483
0.2418    0.9681    0.7104
0.2260    1.0303    0.6698
0.2379    0.9640    0.5127
0.1490    0.9852    0.4562
0.2962    0.9788    0.3801
0.2737    1.0902    0.4225
0.4186    1.1907    0.4479
0.4039    1.0604    0.2532
0.4484    1.0276    0.1739
```

The solution of digital time-varying and nonlinear state-equations also can be obtained using the techniques presented in Chapter 4.

8.10 Design of Multivariable, Digital Control Systems Using Pole-Placement: Regulators, Observers, and Compensators

In the previous section, we noted how a digital state-space representation is equivalent to an analog state-space representation, with the crucial difference that while the latter's characteristics are analyzed in the s-plane, the former is characterized in the z-plane. Therefore, whereas the techniques of Chapter 5 were employed to design an analog control system by placing its poles in the *s-plane*, a digital control system can be similarly designed by placing its poles at desired locations in the *z-plane*. Of course, if a digital system is *controllable*, its poles can be placed anywhere in the *z-plane* using full-state feedback. It can be shown that the conditions and tests for controllability for a linear, time-invariant digital system are the *same* as those presented in Chapter 5 for the analog systems, with the coefficient pair $(\mathbf{A}, \mathbf{B})$ *replaced* by the pair $(\mathbf{A_d}, \mathbf{B_d})$. Hence, one can directly apply the MATLAB (CST) command *ctrb* $(\mathbf{A_d}, \mathbf{B_d})$ to construct the controllability test matrix. Furthermore, using a full-state feedback, $\mathbf{u}(k) = -\mathbf{Kx}(k)$, results in a closed-loop system with poles at the eigenvalues of the closed-loop state-dynamics matrix, $\mathbf{A_{dc}} = (\mathbf{A_d} - \mathbf{B_d K})$. For single-input systems, one can derive the equivalent expression for the full-state feedback gain matrix, $\mathbf{K}$, as the *Ackermann's pole-placement formula* of Eq. (5.52), namely $\mathbf{K} = (\boldsymbol{\alpha} - \mathbf{a})\mathbf{P'P}^{-1}$, where $\mathbf{P'}$ is the controllability test matrix of the plant in *controller companion form*, $\boldsymbol{\alpha}$ is the vector containing coefficients of the closed-loop characteristic polynomial (except the highest power of z) (i.e. the coefficients of z in $|z\mathbf{I} - \mathbf{A_{dc}}|$), and $\mathbf{a}$ is the vector containing coefficients of the digital plant's characteristic polynomial (except the highest power of z) (i.e. the coefficients of z in $|z\mathbf{I} - \mathbf{A_d}|$). Such a direct equivalence between analog and digital systems allows us to use the MATLAB (CST) commands *place* and *acker* to design full-state feedback regulators for digital systems in the same manner as presented in Chapter 5.

Example 8.18

It is desired to use a digital computer based controller with a sampling interval of 0.2 seconds, to control the inverted pendulum on a moving cart of Example 5.9. We begin by converting the plant to an equivalent digital plant using the command *c2dm* as follows:

```
>>A = [0 0 1 0; 0 0 0 1; 10.78 0 0 0; -0.98 0 0 0];B = [0 0 -1 1]';
  C=[1 0 0 0;0 1 0 0];D=[0; 0]; <enter>


>>[Ad,Bd,Cd,Dd] = c2dm(A,B,C,D,0.2,'zoh') <enter>
```

```
Ad =
        1.2235    0        0.2147    0
       -0.0203    1.0000  -0.0013    0.2000
        2.3143    0        1.2235    0
       -0.2104    0       -0.0203    1.0000


Bd =
       -0.0207
        0.0201
       -0.2147
        0.2013


Cd =
     1  0  0  0
     0  1  0  0

Dd =
   0
   0
```

The plant's pole locations in the z-plane are as follows:

```
>>ddamp(Ad,0.2) <enter>
Eigenvalue   Magnitude   Equiv. Damping   Equiv. Freq. (rad/sec)
   1.9283      1.9283       -1.0000          3.2833
   1.0000      1.0000       -1.0000          0
   1.0000      1.0000       -1.0000          0
   0.5186      0.5186        1.0000          3.2833
```

Note that the plant is *unstable*. Let us choose to place *all* the four closed-loop poles at $z = 0$. Then the full-state feedback gain matrix is obtained by using the command *acker* as follows (we could not have used *place* because the multiplicity of poles to be placed is greater than rank(**B**), i.e. 1):

```
>>P=zeros(1,4); K = acker(Ad,Bd,P) <enter>


K =
-142.7071 -61.5324 -41.5386 -30.7662
```

Let us check the closed-loop pole locations, and calculate the closed-loop natural frequencies and damping ratios as follows:

```
>>ddamp(Ad-Bd*K, 0.2) <enter>

Eigenvalue                 Magnitude  Equiv. Damping Equiv. Freq. (rad/s)
 3.64e-004                 3.64e-004  1.00e+000       3.96e+001
 6.03e-008+3.64e-004i      3.64e-004  9.81e-001       4.04e+001
 6.03e-008-3.64e-004i      3.64e-004  9.81e-001       4.04e+001
-3.64e-004                 3.64e-004  9.30e-001       4.26e+001
```

Note that the closed-loop poles have not been placed with a great precision, but the error is acceptable for our purposes. The closed-loop digital system is now stable,

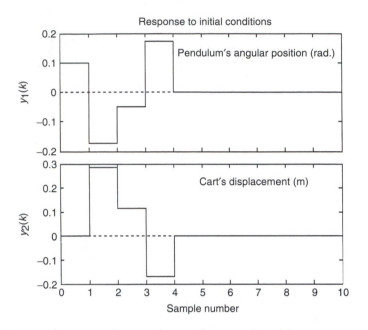

Figure 8.10 Initial response of a control system for inverted pendulum on a moving cart with all closed-loop poles placed at $z = 0$, illustrating the *deadbeat response*

with all damping ratios of $\zeta \approx 1$. Let us see how a closed-loop system with all poles at $z \approx 0$ behaves. Figure 8.10 shows an initial response of the closed-loop system obtained using *dinitial* as follows:

```
>>dinitial(Ad-Bd*K, zeros(4,1),Cd,Dd, [0.1;0;0;0]); <enter>
```

Note the interesting feature of the closed-loop response seen in Figure 8.10: both the outputs settle to a steady-state in *exactly* four time steps. This phenomenon of a sampled data analog system settling to a steady-state in a finite number of sampling intervals is called a *deadbeat response*, and is a property of a stable system with all poles at $z = 0$.

In addition to the full-state feedback regulator problem, all other concepts presented in Chapter 5 for analog control system design – tracking systems, observability, observers, and compensators – can be similarly extended to digital systems with the modification that the time derivative, $dx(t)/dt$, is replaced by the corresponding value of $x(t)$ at the next sampling instant, i.e. $x(k+1)$, and the analog state coefficient matrices, **A**, **B**, **C**, **D**, are replaced by the digital coefficient matrices, **Ad**, **Bd**, **Cd**, **Dd**. For instance, let us consider a full-order observer for a digital system described by Eqs. (8.71) and (8.72). As an extension of the analog observer described by Eq. (5.97), the state-equation for the digital observer can be written as follows:

$$\mathbf{x_o}(k+1) = \mathbf{A_o}\mathbf{x_o}(k) + \mathbf{B_o}\mathbf{u}(k) + \mathbf{L}\mathbf{y}(k) \tag{8.74}$$

where $\mathbf{x_o}(k)$ is the *estimated state-vector*, $\mathbf{u}(k)$ is the input vector, $\mathbf{y}(k)$ is the output vector (all at the kth sampling instant), $\mathbf{A_o}$, $\mathbf{B_o}$ are the *digital* state-dynamics and control coefficient matrices of the observer, and $\mathbf{L}$ is the *digital observer gain matrix*. The matrices $\mathbf{A_o}$, $\mathbf{B_o}$, and $\mathbf{L}$ must be selected in a design process such that the *estimation error*, $\mathbf{e_o}(k) = \mathbf{x}(k) - \mathbf{x_o}(k)$, is brought to zero in the steady state. On subtracting Eq. (8.74) from Eq. (8.71), we get the following *error dynamics* state-equation:

$$\mathbf{e_o}(k+1) = \mathbf{A_o}\mathbf{e_o}(k) + (\mathbf{A_d} - \mathbf{A_o})\mathbf{x}(k) + (\mathbf{B_d} - \mathbf{B_o})\mathbf{u}(k) - \mathbf{L}\mathbf{y}(k) \qquad (8.75)$$

Substitution of Eq. (8.72) into Eq. (8.75) yields

$$\mathbf{e_o}(k+1) = \mathbf{A_o}\mathbf{e_o}(k) + (\mathbf{A_d} - \mathbf{A_o})\mathbf{x}(k) + (\mathbf{B_d} - \mathbf{B_o})\mathbf{u}(k) - \mathbf{L}[\mathbf{C_d}\mathbf{x}(k) + \mathbf{D_d}\mathbf{u}(k)] \qquad (8.76)$$

or

$$\mathbf{e_o}(k+1) = \mathbf{A_o}\mathbf{e_o}(k) + (\mathbf{A_d} - \mathbf{A_o} - \mathbf{L}\mathbf{C_d})\mathbf{x}(k) + (\mathbf{B_d} - \mathbf{B_o} - \mathbf{L}\mathbf{D_d})\mathbf{u}(k) \qquad (8.77)$$

From Eq. (8.77), it is clear that estimation error, $\mathbf{e_o}(k)$, will go to zero in the steady state irrespective of $\mathbf{x}(k)$ and $\mathbf{u}(k)$, if all the *eigenvalues* of $\mathbf{A_o}$ are inside the *unit circle*, and the coefficient matrices of $\mathbf{x}(k)$ and $\mathbf{u}(k)$ are *zeros*, i.e. $(\mathbf{A_d} - \mathbf{A_o} - \mathbf{L}\mathbf{C_d}) = \mathbf{0}$, $(\mathbf{B_d} - \mathbf{B_o} - \mathbf{L}\mathbf{D_d}) = \mathbf{0}$. The latter requirement leads to the following expressions for $\mathbf{A_o}$ and $\mathbf{B_o}$:

$$\mathbf{A_o} = \mathbf{A_d} - \mathbf{L}\mathbf{C_d}; \quad \mathbf{B_o} = \mathbf{B_d} - \mathbf{L}\mathbf{d_d} \qquad (8.78)$$

The error dynamics state-equation is thus the following:

$$\mathbf{e_o}(k+1) = (\mathbf{A_d} - \mathbf{L}\mathbf{C_d})\mathbf{e_o}(k) \qquad (8.79)$$

The observer gain matrix, $\mathbf{L}$, must be selected to place all the eigenvalues of $\mathbf{A_o}$ (which are also the poles of the observer) at desired locations inside the unit circle in the z-plane, which implies that the estimation error dynamics given by Eq. (8.79) is *asymptotically stable* (i.e. $\mathbf{e_o}(k) \to \mathbf{0}$ as $k \to \infty$).

The digital observer described by Eq. (8.74) estimates the state-vector at a given sampling instant, $\mathbf{x_o}(k+1)$, based upon the measurement of the output, $\mathbf{y}(k)$, which is *one sampling instant old*. Such an estimate is likely to be *less accurate* than that based on the current value of the measured output, $\mathbf{y}(k+1)$. Thus, a *more accurate* linear, digital observer is described by the following state-equation:

$$\mathbf{x_o}(k+1) = \mathbf{A_o}\mathbf{x_o}(k) + \mathbf{B_o}\mathbf{u}(k) + \mathbf{L}^*\mathbf{y}(k+1) \qquad (8.80)$$

where $\mathbf{L}^*$ is the new observer gain matrix. The observer described by Eq. (8.80) is referred to as the *current observer*, because it employs the *current* value of the output, $\mathbf{y}(k+1)$. Equation (8.80) yields the following state-equation for the estimation error:

$$\mathbf{e_o}(k+1) = \mathbf{A_o}\mathbf{e_o}(k) + (\mathbf{A_d} - \mathbf{A_o})\mathbf{x}(k) + (\mathbf{B_d} - \mathbf{B_o})\mathbf{u}(k)$$
$$- \mathbf{L}^*[\mathbf{C_d}\mathbf{x}(k+1) + \mathbf{D_d}\mathbf{u}(k+1)] \qquad (8.81)$$

Substituting for $\mathbf{x}(k+1)$ from Eq. (8.71), we get

$$\mathbf{e_o}(k+1) = \mathbf{A_o}\mathbf{e_o}(k) + (\mathbf{A_d} - \mathbf{A_o} - \mathbf{L^*C_dA_d})\mathbf{x}(k) + (\mathbf{B_d} - \mathbf{B_o}$$
$$-\mathbf{L^*C_dB_d})\mathbf{u}(k) - \mathbf{L^*D_d}\mathbf{u}(k+1) \tag{8.82}$$

For $\mathbf{e_o}(k)$ to go to zero in the steady state, the following conditions must be satisfied in addition to $\mathbf{A_o}$ having all eigenvalues inside the unit circle:

$$\mathbf{A_o} = \mathbf{A_d} - \mathbf{L^*C_dA_d} \tag{8.83}$$

$$(\mathbf{B_d} - \mathbf{B_o} - \mathbf{L^*C_dBd})\mathbf{u}(k) = \mathbf{L^*D_d}\mathbf{u}(k+1) \tag{8.84}$$

which result in the following state equations for the estimation error and the observer:

$$\mathbf{e_o}(k+1) = (\mathbf{A_d} - \mathbf{L^*C_dA_d})\mathbf{e_o}(k) \tag{8.85}$$

$$\mathbf{x_o}(k+1) = (\mathbf{A_d} - \mathbf{L^*C_dA_d})\mathbf{x_o}(k) + (\mathbf{B_d} - \mathbf{L^*C_dB_d})\mathbf{u}(k)$$
$$- \mathbf{L^*D_d}\mathbf{u}(k+1) + \mathbf{L^*y}(k+1) \tag{8.86}$$

Equation (8.86) can be alternatively expressed as the following *two* difference equations:

$$\mathbf{x_1}(k+1) = \mathbf{A_d}\mathbf{x_o}(k) + \mathbf{B_d}\mathbf{u}(k) \tag{8.87}$$

$$\mathbf{x_o}(k+1) = \mathbf{x_1}(k+1) + \mathbf{L^*}[\mathbf{y}(k+1) - \mathbf{D_d}\mathbf{u}(k+1) - \mathbf{C_d}\mathbf{x_1}(k+1)] \tag{8.88}$$

where $\mathbf{x_1}(k+1)$ is the *first estimate* of the state (or *predicted state*) at the $(k+1)$th sampling instant based upon the quantities at the *previous* sampling instant, and $\mathbf{x_o}(k+1)$ is the *corrected* – or *final* – state estimate due to the measurement of the output at the current sampling instant, $\mathbf{y}(k+1)$. The *predictor-corrector* observer formulation given by Eqs. (8.87) and (8.88) is sometimes more useful when implemented in a computer program.

For practical purposes, it is not the observer alone that we are interested in, but the implementation of the digital observer in a feedback *digital compensator*. Using a full-order current digital observer in a full-state feedback compensator results in a feedback control law of the form

$$\mathbf{u}(k) = -\mathbf{Kx_o}(k) \tag{8.89}$$

Substituting Eq. (8.89) into Eq. (8.86), we can write the state-equation for the estimated state as follows:

$$\mathbf{x_o}(k+1) = (\mathbf{I} - \mathbf{L^*D_dK})^{-1}(\mathbf{A_d} - \mathbf{L^*C_dA_d} - \mathbf{B_dK} + \mathbf{L^*C_dB_dK})\mathbf{x_o}(k)$$
$$+ (\mathbf{I} - \mathbf{L^*D_dK})^{-1}\mathbf{L^*y}(k+1) \tag{8.90}$$

Note that Eq. (8.90) requires that the matrix $(\mathbf{I} - \mathbf{L^*D_dK})$ must be *non-singular*. Substituting Eq. (8.89) into Eq. (8.71), we get the other state-equation for the closed-loop system as follows:

$$\mathbf{x}(k+1) = \mathbf{A_d}\mathbf{x}(k) - \mathbf{B_dKx_o}(k) \tag{8.91}$$

Equations (8.90) and (8.91) must be solved simultaneously to get the composite state-vector for the closed-loop system, $\mathbf{x_c}(k) = [\mathbf{x^T}(k); \mathbf{x_o^T}(k)]^T$. We can combine Eqs. (8.90) and (8.91) into the following state-equation:

$$\begin{bmatrix} \mathbf{x}(k+1) \\ \mathbf{x_o}(k+1) \end{bmatrix} = \begin{bmatrix} \mathbf{A_d} & -\mathbf{B_d K} \\ \mathbf{0} & (\mathbf{I} - \mathbf{L^* D_d K})^{-1}(\mathbf{A_d} - \mathbf{L^* C_d A_d} - \mathbf{B_d K} + \mathbf{L^* C_d B_d K}) \end{bmatrix}$$
$$\times \begin{bmatrix} \mathbf{x}(k) \\ \mathbf{x_o}(k) \end{bmatrix} + \begin{bmatrix} \mathbf{0} \\ (\mathbf{I} - \mathbf{L^* D_d K})^{-1}\mathbf{L^*} \end{bmatrix} \mathbf{y}(k+1) \qquad (8.92)$$

Substituting $\mathbf{y}(k+1) = \mathbf{C_d x}(k+1) + \mathbf{D_d u}(k+1) = \mathbf{C_d x}(k+1) - \mathbf{D_d K x_o}(k+1)$ into Eq. (8.92), and re-arranging, we can write

$$\begin{bmatrix} \mathbf{x}(k+1) \\ \mathbf{x_o}(k+1) \end{bmatrix} = \begin{bmatrix} \mathbf{A_d} & -\mathbf{B_d K} \\ (\mathbf{I} + \mathbf{F^* D_d K})^{-1}\mathbf{F^* C_d A_d} & (\mathbf{I} + \mathbf{F^* D_d K})^{-1}(\mathbf{I} - \mathbf{L^* D_d K})^{-1}(\mathbf{A_d} - \mathbf{L^* C_d A_d} - \mathbf{B_d K}) \end{bmatrix}$$
$$\times \begin{bmatrix} \mathbf{x}(k) \\ \mathbf{x_o}(k) \end{bmatrix} \qquad (8.93)$$

where $\mathbf{F^*} = (\mathbf{I} - \mathbf{L^* D_d K})^{-1}\mathbf{L^*}$. Note that Eq. (8.93) is a homogeneous digital state-equation representing closed-loop dynamics of the compensated system.

Example 8.19

Let us design a full-order observer based digital compensator with a sampling interval of 0.2 seconds, to control the inverted pendulum on a moving cart of Example 8.18. The cart's displacement, $y_2(t)$, is the only measured output, which results in $\mathbf{C} = [0; 1; 0; 0]$, and $\mathbf{D} = 0$. We begin by converting the plant to an equivalent digital plant with a z.o.h using the command *c2dm* as follows:

```
>>A = [0 0 1 0; 0 0 0 1; 10.78 0 0 0; -0.98 0 0 0]; B = [0 0 -1 1]';
  C = [0 1 0 0]; D = 0; <enter>

>>[Ad,Bd,Cd,Dd] = c2dm(A,B,C,D,0.2,'zoh'); <enter>

Ad =
   1.2235       0   0.2147        0
  -0.0203  1.0000  -0.0013   0.2000
   2.3143       0   1.2235        0
  -0.2104       0  -0.0203   1.0000


Bd =
  -0.0207
   0.0201
  -0.2147
   0.2013

Cd =
   0   1   0   0
```

```
Dd =
  0
```

Note that since $\mathbf{A_o} = \mathbf{A_d} - \mathbf{L^*C_dA_d}$, the observability test matrix is now computed by *ctrb* $(\mathbf{A_d}^T, \mathbf{A_d}^T\mathbf{C_d}^T)$. Let us check whether the digital plant is observable as follows:

```
>>rank(ctrb(Ad',Ad'*Cd')) <enter>

ans =
4
```

Since the rank of the observability test matrix is equal to the order of the plant, the plant is observable. The observer gain matrix is calculated by placing the observer poles well inside the unit circle, such as at $z = \pm 0.7$, and $z = 0.5 \pm 0.5i$. Then the observer gain matrix, $\mathbf{L^*}$, is computed using *place* as follows:

```
>>P=[-0.7 0.7 0.5-0.5i 0.5+0.5i]; Lstar = place(Ad', Ad'*Cd', P)'
  <enter>

place: ndigits= 15
Lstar =
 -38.3080
   1.2450
-102.1822
   6.4364
```

Let us check whether observer poles are placed as desired:

```
>>eig(Ad - Lstar*Cd*Ad) <enter>

ans =
        0.5000+0.5000i
        0.5000-0.5000i
        0.7000
       -0.7000
```

Note that the plant is strictly proper. Hence, the observer's state coefficient matrices are computed as follows:

```
>>Ao = Ad - Lstar*Cd*Ad, Bo = Bd - Lstar*Cd*Bd <enter>

Ao =
    0.4453    38.3080     0.1635     7.6616
    0.0050    -0.2450     0.0003    -0.0490
    0.2385   102.1822     1.0870    20.4364
   -0.0796    -6.4364    -0.0117    -0.2873

Bo =
    0.7480
   -0.0049
    1.8357
    0.0722
```

A full-state feedback regulator was designed for this plant in Example 8.18, to place all the regulated system's poles at $z = 0$. Using this regulator gain matrix, **K**, the digital closed-loop state-dynamics matrix (Eq. (8.93)) is calculated as follows:

```
>>Ac=[Ad -Bd*K; Lstar*Cd*Ad Ao-Bd*K]; <enter>
```

Finally, let us calculate the compensated digital system's initial response to $\mathbf{x_c}(0) = [\mathbf{x}^T(0); \mathbf{x_o}^T(0)]^T = [0.1; 0; 0; 0; 0; 0; 0; 0]^T$ (implying an initial pendulum angle of 0.1 rad.) using the command *dinitial* as follows:

```
>>dinitial(Ac,zeros(8,1),[Cd zeros(1,4)],0,[0.1 zeros(1,7)]') <enter>
```

The resulting digital initial response is plotted in Figure 8.11. The initial response settles to zero in about 25 sampling instants, with a maximum overshoot of 0.6 m. Note that the closed-loop system's performance has deteriorated when compared to the deadbeat response of Figure 8.10, due to the presence of dominant observer poles. Let us see what happens if we place all the observer poles also at $z = 0$, as follows:

```
>>P=zeros(1,4);Lstar = acker(Ad',Ad'*Cd',P)'; <enter>
```

and re-calculate the closed-loop initial response, which is plotted in Figure 8.12. Figure 8.12 shows that the deadbeat response of Figure 8.10 has been largely

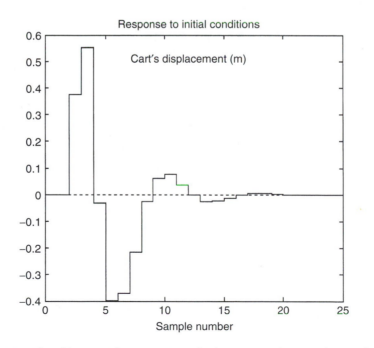

Figure 8.11   Closed-loop initial response (cart's displacement, $y_2(k)$, in m) of inverted pendulum on a moving cart with full-order compensator designed using observer poles at $z = \pm 0.7$, and $z = 0.5 \pm 0.5i$

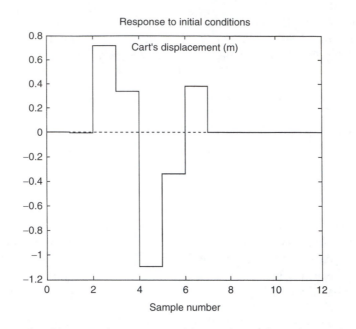

Figure 8.12 Closed-loop initial response (pendulum angle in radian) of inverted pendulum on a moving cart with full-order compensator designed using all observer poles at $z = 0$, and all regulator poles also at $z = 0$. Note the restoration of the *deadbeat response* of Figure 8.10, but with larger overshoots

restored, but with twice the overshoots of Figure 8.10, and with a larger settling time (seven steps) when compared to that of Figure 8.10 (four steps).

As an alternative to the calculations given above, MATLAB (CST) provides the command *dreg* for constructing the state-space representation of the full-order digital compensator of Eq. (8.93), and can be used to derive the full-order, full-state feedback closed-loop digital system as follows:

```
>>[ac,bc,cc,dc] = dreg(Ad,Bd,Cd,Dd,K,Lstar) <enter>
```

where (**ac**, **bc**, **cc**, **dc**) is the digital state-space representation of the digital compensator whose input is the plant output, $\mathbf{y}(k + 1)$, and whose output is the plant input, $\mathbf{u}(k)$. Then the digital compensator is put in a closed-loop with the plant, to obtain the closed-loop state-space representation (**Ac**, **Bc**, **Cc**, **Dc**) as follows:

```
>>sysc=ss(ac,bc,cc,dc); sysp=ss(Ad,Bd,Cd,Dd); <enter>
```

```
>>syss=series(sysc,sysp);sysCL=feedback(syss,eye(size(syss))); <enter>
```

```
>>[Ac,Bc,Cc,Dc] = ssdata(sysCL); <enter>
```

The command *dreg* allows constructing a compensator in which only selected outputs are measured that are specified using an additional input argument, *SENSORS*. Also, some additional known, non-stochastic inputs of the plant (such as desired outputs) can be specified as an additional argument, *KNOWN*, while control inputs (i.e. inputs applied by the compensator to the plant) are specified by an additional argument, *CONTROLS*. The resulting compensator has control feedback commands as outputs and the known inputs and sensors as inputs. The command *dreg* is thus used in its most general form as follows:

```
>>[ac,bc,cc,dc] = dreg(Ad,Bd,Cd,Dd,K,Lstar,SENSORS,KNOWN,CONTROLS); <enter>
```

The greatest utility of *dreg* lies in the digital equivalent of LQG compensators. We will discuss digital optimal control in the next section.

8.11 Linear Optimal Control of Digital Systems

In a manner similar to the linear optimal control of analog systems presented in Chapter 6, we can devise equivalent techniques for linear optimal digital control. The digital optimal control expressions are in terms of *difference equations* rather than *differential equations* implementation presented in Chapter 6. We saw in the previous section how the implementation of a difference equation for a current digital observer resulted in a state-space model (Eq. (8.86)) that was quite different from that of the analog observer of Eq. (5.97). A similar modification is expected in the solution to the digital optimal control problem. Let us begin with the full-state feedback regulator design. An objective function to be minimized for the control of a linear, *time-varying* digital system can be expressed as the digital equivalent of Eq. (6.31) as follows:

$$J(0, N) = \sum_{k=0}^{N} \mathbf{x}^{\mathrm{T}}(k)\mathbf{Q}(k)\mathbf{x}(k) + \mathbf{u}^{\mathrm{T}}(k)\mathbf{R}(k)\mathbf{u}(k) \qquad (8.94)$$

The *quadratic form* of the objective function, $J(0, N)$, given by Eq. (8.94) suffers no loss of generality if we assume both $\mathbf{Q}(k)$ and $\mathbf{R}(k)$ to be *symmetric* matrices. The *optimal* control gain matrix, $\mathbf{K}(k)$, of the control law, $\mathbf{u}(k) = -\mathbf{K}(k)\mathbf{x}(k)$, is to be chosen such that $J(0, N)$ is minimized *with respect to* the control input, $\mathbf{u}(k)$, *subject to* the constraint that the state-vector, $\mathbf{x}(k)$, is the solution of the following time-varying state-equation:

$$\mathbf{x}(k + 1) = \mathbf{A}(k)\mathbf{x}(k) + \mathbf{B}(k)\mathbf{u}(k) \qquad (8.95)$$

Note that we have dropped the subscript d from the coefficient matrices, $\mathbf{A}(k)$ and $\mathbf{B}(k)$, in Eq. (8.95) – indicating a digital system – for simplicity of notation. To derive the optimal regulator gain matrix which minimizes $J(0, N)$, it is more expedient to work with $J(m, N)$ which can be expressed as follows:

$$J(m, N) = J(0, N) - J(0, m - 1) = F(m) + F(m + 1) + \cdots + F(N - 1) + F(N) \qquad (8.96)$$

where $F(k) = \mathbf{x}^{\mathrm{T}}(k)\mathbf{Q}(k)\mathbf{x}(k) + \mathbf{u}^{\mathrm{T}}(k)\mathbf{R}(k)\mathbf{u}(k)$. To Eq. (8.96), we can apply the *principle of optimality* [2], which states that if the regulator, $\mathbf{u}(k) = -\mathbf{K}(k)\mathbf{x}(k)$, is optimal for the interval $0 \leq k \leq N$, then it is *also* optimal over *any* sub-interval, $m \leq k \leq N$, where $0 \leq m \leq N$. The principle of optimality can be applied by first minimizing $J(N, N) = F(N)$, and then finding $F(N - 1)$ to minimize $J(N - 1, N) = F(N - 1) + F(N) = F(N - 1) + J_o(N, N)$, where $J_o(N, N)$ refers to the minimum value of $J(N, N)$, and continuing this process until $J(0, N)$ is minimized. Such a sequential minimization is called *dynamic programming* in optimization parlance. Note that as in the case of analog optimal control (Chapter 6), the digital optimal control solution is obtained by marching *backwards* in time, beginning with the minimization of $J(N, N)$.

The minimum value of $J(N, N) = F(N) = \mathbf{x}^{\mathrm{T}}(N)\mathbf{Q}(N)\mathbf{x}(N) + \mathbf{u}^{\mathrm{T}}(N)\mathbf{R}(N)\mathbf{u}(N)$ with respect to the input, $\mathbf{u}(N)$, is easily obtained to be

$$J_o(N, N) = \mathbf{x}^{\mathrm{T}}(N)\mathbf{Q}(N)\mathbf{x}(N) \tag{8.97}$$

because the state at the final sampling instant, $\mathbf{x}(N)$, is *independent* of the final control input, $\mathbf{u}(N)$, and the minimum value of $\mathbf{u}^{\mathrm{T}}(N)\mathbf{R}(N)\mathbf{u}(N) = 0$, which occurs for the *optimal control input*, $\mathbf{u_o}(N) = \mathbf{0}$. Substituting Eq. (8.95) into Eq. (8.97) for $k = N - 1$, we get

$$J_o(N, N) = [\mathbf{A}(N - 1)\mathbf{x}(N - 1) + \mathbf{B}(N - 1)\mathbf{u}(N - 1)]^T \mathbf{Q}(N - 1)[\mathbf{A}(N - 1)\mathbf{x}(N - 1)$$
$$+ \mathbf{B}(N - 1)\mathbf{u}(N - 1)]_{\mathbf{u_o}(N-1)} \tag{8.98}$$

where the subscript $\mathbf{u_o}(N - 1)$ indicates that the expression on the right-hand side of Eq. (8.98) is evaluated for the optimal control input at the $(N - 1)$th sampling instant. Then $J(N-1, N)$ is determined by substituting Eq. (8.98) into $J(N-1, N) = F(N-1) + J_o(N, N)$ to yield

$$J(N - 1, N) = \mathbf{x}^{\mathrm{T}}(N - 1)\mathbf{Q}(N - 1)\mathbf{x}(N - 1) + \mathbf{u}^{\mathrm{T}}(N - 1)\mathbf{R}(N - 1)\mathbf{u}(N - 1)$$
$$+ [\mathbf{A}(N - 1)\mathbf{x}(N - 1) + \mathbf{B}(N - 1)\mathbf{u}(N - 1)]^T \mathbf{Q}(N - 1)$$
$$\times [\mathbf{A}(N - 1)\mathbf{x}(N - 1) + \mathbf{B}(N - 1)\mathbf{u}(N - 1)]|_{\mathbf{u_o}(N-1)} \tag{8.99}$$

The optimal control input at the $(N - 1)$th sampling instant, $\mathbf{u_o}(N - 1)$, can be evaluated by solving

$$\partial J(N - 1, N)/\partial \mathbf{u}(N - 1) = \mathbf{0} \tag{8.100}$$

Since $J(N - 1, N)$ consists of quadratic terms, such as $\mathbf{u}^{\mathrm{T}}\mathbf{R}\mathbf{u}$, and *bilinear* terms, such as $\mathbf{u}^{\mathrm{T}}\mathbf{Q}\mathbf{x}$ and $\mathbf{x}^{\mathrm{T}}\mathbf{Q}\mathbf{u}$, to evaluate the derivative in Eq. (8.100), we must know how to differentiate such *scalar* terms with respect to the *vector*, $\mathbf{u}(N - 1)$. From Appendix B, we can write $\partial(\mathbf{u}^{\mathrm{T}}\mathbf{R}\mathbf{u})/\partial\mathbf{u} = 2\mathbf{R}\mathbf{u}$, $\partial(\mathbf{u}^{\mathrm{T}}\mathbf{Q}\mathbf{x})/\partial\mathbf{u} = \mathbf{Q}\mathbf{x}$, and $\partial(\mathbf{x}^{\mathrm{T}}\mathbf{Q}\mathbf{u})/\partial\mathbf{u} = \mathbf{Q}^{\mathrm{T}}\mathbf{x}$. After carrying out the differentiation in Eq. (8.100), we get

$$2\mathbf{R}(N - 1)\mathbf{u_o}(N - 1) + 2\mathbf{B}^{\mathrm{T}}(N - 1)\mathbf{Q}(N - 1)\mathbf{B}(N - 1)\mathbf{u_o}(N - 1)$$
$$+ [\mathbf{A}^{\mathrm{T}}(N - 1)\mathbf{Q}(N - 1)\mathbf{B}(N - 1)]^T\mathbf{x}(N - 1)$$
$$+ \mathbf{B}^{\mathrm{T}}(N - 1)\mathbf{Q}(N - 1)\mathbf{A}(N - 1)\mathbf{x}(N - 1) = \mathbf{0} \tag{8.101}$$

or

$$2\mathbf{R}(N-1)\mathbf{u_o}(N-1) + 2\mathbf{B}^T(N-1)\mathbf{Q}(N-1)\mathbf{B}(N-1)\mathbf{u_o}(N-1)$$
$$+ 2\mathbf{B}^T(N-1)\mathbf{Q}(N-1)\mathbf{A}(N-1)\mathbf{x}(N-1) = \mathbf{0} \tag{8.102}$$

which yields

$$\mathbf{u_o}(N-1) = -[\mathbf{R}(N-1) + \mathbf{B}^T(N-1)\mathbf{Q}(N-1)\mathbf{B}(N-1)]^{-1}$$
$$+ \mathbf{B}^T(N-1)\mathbf{Q}(N-1)\mathbf{A}(N-1)\mathbf{x}(N-1) \tag{8.103}$$

Comparing Eq. (8.103) with the control law $\mathbf{u}(k) = -\mathbf{K}(k)\mathbf{x}(k)$, the *optimal regulator gain matrix* at the $(N-1)$th sampling instant, $\mathbf{K_o}(N-1)$, is obtained to be

$$\mathbf{K_o}(N-1) = [\mathbf{R}(N-1) + \mathbf{B}^T(N-1)\mathbf{Q}(N-1)\mathbf{B}(N-1)]^{-1}$$
$$\times \mathbf{B}^T(N-1)\mathbf{Q}(N-1)\mathbf{A}(N-1) \tag{8.104}$$

Substituting Eq. (8.103) into Eq. (8.99), we get the following quadratic form for $J_o(N-1, N)$:

$$J_o(N-1, N) = \mathbf{x}^T(N-1)\mathbf{M}(N-1)\mathbf{x}(N-1) \tag{8.105}$$

where $\mathbf{M}(N-1)$ is a symmetric matrix. Continuing in this manner, we can write

$$J_o(N-2, N) = \mathbf{x}^T(N-2)\mathbf{M}(N-2)\mathbf{x}(N-2) \tag{8.106}$$

$$J_o(m-1, N) = \mathbf{x}^T(m-1)\mathbf{M}(m-1)\mathbf{x}(m-1) \tag{8.107}$$

and

$$J_o(m, N) = \mathbf{x}^T(m)\mathbf{M}(m)\mathbf{x}(m) \tag{8.108}$$

Substituting Eq. (8.95) into Eq. (8.108) for $k = m-1$, we can write

$$J_o(m, N) = [\mathbf{A}(m-1)\mathbf{x}(m-1) + \mathbf{B}(m-1)\mathbf{u}(m-1)]^T \mathbf{M}(m)$$
$$\times [\mathbf{A}(m-1)\mathbf{x}(m-1) + \mathbf{B}(m-1)\mathbf{u}(m-1)] \tag{8.109}$$

To obtain the optimal control input at the $(m-1)$th sampling instant, $\mathbf{u_o}(m-1)$, we must minimize $J(m-1, N)$, which is written as

$$J(m-1, N) = J_o(m, N) + F(m-1)$$
$$= [\mathbf{A}(m-1)\mathbf{x}(m-1) + \mathbf{B}(m-1)\mathbf{u}(m-1)]^T \mathbf{M}(m)[\mathbf{A}(m-1)\mathbf{x}(m-1)$$
$$+ \mathbf{B}(m-1)\mathbf{u}(m-1)] + \mathbf{x}^T(m-1)\mathbf{Q}(m-1)\mathbf{x}(m-1)$$
$$+ \mathbf{u}^T(m-1)\mathbf{R}(m-1)\mathbf{u}(m-1) \tag{8.110}$$

Then the minimization of $J(m-1, N)$ with respect to $\mathbf{u}(m-1)$ yields

$$\partial J(m-1, N)/\partial \mathbf{u}(m-1) = 2\mathbf{B}^T(m-1)\mathbf{M}(m)[\mathbf{A}(m-1)\mathbf{x}(m-1) + \mathbf{B}(m-1)\mathbf{u}(m-1)]$$
$$+ 2\mathbf{R}(m-1)\mathbf{u}(m-1) = \mathbf{0} \tag{8.111}$$

or

$$\mathbf{u}_0(m-1) = -[\mathbf{B}^T(m-1)\mathbf{M}(m)\mathbf{B}(m-1)$$
$$+ \mathbf{R}(m-1)]^{-1}\mathbf{B}^T(m-1)\mathbf{M}(m)\mathbf{A}(m-1)\mathbf{x}(m-1) \qquad (8.112)$$

Thus, the optimal regulator matrix at the $(m-1)$th sampling instant is

$$\mathbf{K}_0(m-1) = [\mathbf{B}^T(m-1)\mathbf{M}(m)\mathbf{B}(m-1) + \mathbf{R}(m-1)]^{-1}\mathbf{B}^T(m-1)\mathbf{M}(m)\mathbf{A}(m-1)$$
$$(8.113)$$

Finally, substituting Eq. (8.112) into Eq. (8.110), we get the following expression for $J_o(m-1, N)$:

$$J_o(m-1, N) = \mathbf{x}^T(m-1)[\mathbf{A}_c^T(m-1)\mathbf{M}(m)\mathbf{A}_c(m-1) + \mathbf{Q}(m-1)$$
$$+ \mathbf{K}^T(m-1)\mathbf{R}(m-1)\mathbf{K}(m-1)]\mathbf{x}(m-1) \qquad (8.114)$$

where $\mathbf{A}_c(m-1) = \mathbf{A}(m-1) - \mathbf{B}(m-1)\mathbf{K}(m-1)$. Comparing Eqs. (8.107) and (8.114), it is clear that

$$\mathbf{M}(m-1) = \mathbf{A}_c^T(m-1)\mathbf{M}(m)\mathbf{A}_c(m-1) + \mathbf{Q}(m-1) + \mathbf{K}^T(m-1)\mathbf{R}(m-1)\mathbf{K}(m-1)$$
$$(8.115)$$

Equation (8.115) is a *nonlinear* matrix difference equation, and can be recognized as the digital equivalent of the *matrix Riccati equation*, which must be integrated *backwards* in time, beginning from the *terminal condition* obtained from Eq. (8.97) as $\mathbf{M}(N) = \mathbf{Q}(N)$.

In summary, the digital, linear quadratic optimal regulator problem consists of the *recursive* solution of the following equations:

$$\mathbf{x}(k+1) = [\mathbf{A}(k) - \mathbf{B}(k)\mathbf{K}_0(k)]\mathbf{x}(k) \qquad (8.116)$$

$$\mathbf{K}_0(k) = [\mathbf{B}^T(k)\mathbf{M}(k+1)\mathbf{B}(k) + \mathbf{R}(k)]^{-1}\mathbf{B}^T(k)\mathbf{M}(k+1)\mathbf{A}(k) \qquad (8.117)$$

$$\mathbf{M}(k) = [\mathbf{A}(k) - \mathbf{B}(k)\mathbf{K}_0(k)]^T\mathbf{M}(k+1)[\mathbf{A}(k) - \mathbf{B}(k)\mathbf{K}_0(k)] + \mathbf{Q}(k)$$
$$+ \mathbf{K}_0^T(k)\mathbf{R}(k)\mathbf{K}_0(k) \qquad (8.118)$$

with the *terminal conditions*, $\mathbf{M}(N) = \mathbf{Q}(N)$ and $\mathbf{K}_0(N) = \mathbf{0}$, and given the *initial condition*, $\mathbf{x}(0)$. Since both initial and terminal conditions are specified for solving Eqs. (8.116)–(8.118), these difference equations pose a *two-point boundary value problem*. Note that the minimum value of the objective function, $J(0, N)$, can be obtained from Eq. (8.108) with $m = 0$ as

$$J_o(0, N) = \mathbf{x}^T(0)\mathbf{M}(0)\mathbf{x}(0) \qquad (8.119)$$

While it is possible to solve the two-point boundary value problem posed by Eqs. (8.116)–(8.118) using a nonlinear time marching numerical method (such as the *Runge–Kutta* method discussed in Chapter 4), usually we are interested in a steady state solution, where the terminal time is infinite, or $N \to \infty$. In the limit $N \to \infty$, both the objective function, $J(0, N)$, and the optimal gain matrix, $\mathbf{K}_0(N)$, become constants. It must be pointed out that we are not going to solve Eqs. (8.116)–(8.118) for an *infinite number* of sampling instants, but assume that $J(0, N)$ and $\mathbf{K_o}(N)$ approximately become constants as N becomes large. Of course, the closed-loop regulated system given by Eq. (8.116) must be *asymptotically stable* for the steady state approximation to be used. The steady state approximation of the linear optimal control problem is especially valid for *time-invariant* systems for which the optimal regulator gain matrix approaches a *constant* value after a few sampling instants. Let us therefore consider a time-invariant system with $\mathbf{A}(k)$ and $\mathbf{B}(k)$ replaced by the constant matrices $\mathbf{A}$ and $\mathbf{B}$, respectively. Also, for the time-invariant system, the state and control weighting matrices, $\mathbf{Q}$ and $\mathbf{R}$, in the objective function are also constant. In the steady state, the matrix $\mathbf{M}(k)$ becomes constant, and we can write $\mathbf{M}(k) = \mathbf{M}(k + 1) = \mathbf{M_o}$, where $\mathbf{M_o}$ is a constant matrix. Thus, in the steady state, Eq. (8.118) can be written as

$$\mathbf{M_o} = (\mathbf{A} - \mathbf{BK_o})^T \mathbf{M_o}(\mathbf{A} - \mathbf{BK_o}) + \mathbf{Q} + \mathbf{K_o^T R K_o} \tag{8.120}$$

and Eq. (8.117) becomes

$$\mathbf{K_o} = (\mathbf{B^T M_o B} + \mathbf{R})^{-1} \mathbf{B^T M_o A} \tag{8.121}$$

With the substitution of Eq. (8.121), Eq. (8.120) can be re-written in the following form, which does not contain $\mathbf{K_o}$:

$$0 = \mathbf{M_o} - \mathbf{A^T M_o A} + \mathbf{A^T M_o B}(\mathbf{R} + \mathbf{B^T M_o B})^{-1}\mathbf{B^T MA} - \mathbf{Q} \tag{8.122}$$

Equation (8.122) is the *digital algebraic Riccati equation* – a digital equivalent of Eq. (6.33). The optimal control gain matrix, $\mathbf{K_o}$, is thus obtained from Eq. (8.121) using the solution, $\mathbf{M_o}$, to the algebraic Riccati equation, Eq. (8.122). A set of *sufficient conditions* for the existence of a *unique, positive definite* solution to Eq. (8.122) are as follows:

(a) The state weighting matrix, $\mathbf{Q}$, must be symmetric and *positive semi-definite*.

(b) The control weighting matrix, $\mathbf{R}$, must be symmetric and *positive definite*.

(c) The digital system represented by $\mathbf{A}$ and $\mathbf{B}$ must be *controllable* (or at least, *stabilizable*).

A commonly employed numerical approach to finding the solution to Eq. (8.122) is by defining a *Hamiltonian matrix*, $\mathcal{H}$, as follows:

$$\begin{bmatrix} \mathbf{x}(k + 1) \\ \mathbf{p}(k + 1) \end{bmatrix} = \mathcal{H} \begin{bmatrix} \mathbf{x}(k) \\ \mathbf{p}(k) \end{bmatrix} \tag{8.123}$$

where $\mathbf{p}(k)$ is called the *costate vector* and $\mathcal{H}$ is the following:

$$\mathcal{H} = \begin{bmatrix} \mathbf{A} + \mathbf{BR^{-1}B^T(A^{-1})^T Q} & -\mathbf{BR^{-1}B^T(A^{-1})^T} \\ -(\mathbf{A^{-1}})^T \mathbf{Q} & (\mathbf{A^{-1}})^T \end{bmatrix} \tag{8.124}$$

Of course, the definition of the Hamiltonian matrix by Eq. (8.124) requires that $\mathbf{A}$ must be *non-singular*. The Hamiltonian matrix and the costate vector are creatures residing in an *alternative formulation* of the linear optimal control problem, called the *minimum principle* [3], which requires minimizing the Hamiltonian, $\mathcal{H} = 1/2[\mathbf{x}^\mathrm{T}(k)\mathbf{Q}\mathbf{x}(k) + \mathbf{u}^\mathrm{T}(k)\mathbf{R}\mathbf{u}(k)] + \mathbf{p}^\mathrm{T}(k+1)[\mathbf{A}\mathbf{x}(k) + \mathbf{B}\mathbf{u}(k)]$ with respect to the control input, $\mathbf{u}(k)$. Note that $\mathcal{H}$ includes the term $\mathbf{p}^\mathrm{T}(k+1)[\mathbf{A}\mathbf{x}(k) + \mathbf{B}\mathbf{u}(k)]$ as a *penalty* for deviating from the system's state-equation, $\mathbf{x}(k+1) = \mathbf{A}\mathbf{x}(k) + \mathbf{B}\mathbf{u}(k)$, which was a constraint in the minimization of the objective function, $J(0, N)$. The optimal control input resulting from the minimization of $\mathcal{H}$ is exactly the same as that obtained earlier in this section. In terms of the costate vector, $\mathbf{p}(k)$, the optimal control input can be expressed as follows [3]:

$$\mathbf{u}_\mathbf{o}(k) = -\mathbf{R}^{-1}\mathbf{B}^\mathrm{T}\mathbf{p}(k+1) \tag{8.125}$$

Now, since Eq. (8.125) requires marching *backwards* in time, it would be mores useful to express Eq. (8.123) as follows:

$$\begin{bmatrix} \mathbf{x}(k) \\ \mathbf{p}(k) \end{bmatrix} = \mathcal{H}^{-1} \begin{bmatrix} \mathbf{x}(k+1) \\ \mathbf{p}(k+1) \end{bmatrix} \tag{8.126}$$

It can be shown easily that the inverse of the Hamiltonian matrix is given by

$$\mathcal{H}^{-1} = \begin{bmatrix} \mathbf{A}^{-1} & \mathbf{A}^{-1}\mathbf{B}\mathbf{R}^{-1}\mathbf{B}^\mathrm{T} \\ \mathbf{Q}\mathbf{A}^{-1} & \mathbf{A}^\mathrm{T} + \mathbf{Q}\mathbf{A}^{-1}\mathbf{B}\mathbf{R}^{-1}\mathbf{B}^\mathrm{T} \end{bmatrix} \tag{8.127}$$

The Hamiltonian matrix has an interesting property that if λ is an eigenvalue of $\mathcal{H}$, then $1/\lambda$ is an eigenvalue of $\mathcal{H}^{-1}$. Also, the eigenvalues of $\mathcal{H}$ are the eigenvalues of $\mathcal{H}^{-1}$. Hence, it follows that the eigenvalues of $\mathcal{H}$ (or $\mathcal{H}^{-1}$) must occur in *reciprocal pairs*, i.e. λ_1, $1/\lambda_1$, λ_2, $1/\lambda_2$, etc. If the eigenvalues of $\mathcal{H}$ (or $\mathcal{H}^{-1}$) are *distinct*, then we can *diagonalize* the state-equations, (8.126) (see Chapter 3). The state-equations given by Eq. (8.126) can be diagonalized by a transformation matrix, $\mathbf{T} = \mathbf{V}^{-1}$, where $\mathbf{V}$ is a *modal matrix* whose columns are the eigenvectors of $\mathcal{H}^{-1}$. If we partition $\mathbf{V}$ into *four* $(n \times n)$ sized blocks (where n is the order of the system) as follows:

$$\mathbf{V} = \begin{bmatrix} \mathbf{V}_{11} & \mathbf{V}_{12} \\ \mathbf{V}_{21} & \mathbf{V}_{22} \end{bmatrix} \tag{8.128}$$

such that

$$\mathbf{D} = \mathbf{V}^{-1}\mathcal{H}^{-1}\mathbf{V} = \begin{bmatrix} \Lambda & \mathbf{0} \\ \mathbf{0} & \Lambda^{-1} \end{bmatrix} \tag{8.129}$$

where Λ is the *diagonal* matrix consisting of the eigenvalues of $\mathcal{H}$ (or $\mathcal{H}^{-1}$) that lie *inside* the unit circle in the z-plane. Clearly, Λ^{-1} is the *diagonal* matrix consisting of the eigenvalues of $\mathcal{H}$ (or $\mathcal{H}^{-1}$) that lie *outside* the unit circle in the z-plane. Equation (8.129) suggests that the modal matrix, $\mathbf{V}$, is partitioned into *stable* and *unstable* eigenvalues of $\mathcal{H}$ (or $\mathcal{H}^{-1}$). Comparing Eqs. (8.125) and (8.112) in the limit $k \to \infty$ it can be shown [2] that

$$\mathbf{M}_\mathbf{o} = \mathbf{V}_{21}\mathbf{V}_{11}^{-1} \tag{8.130}$$

Equation (8.130) gives us a direct method of calculating the solution to the algebraic Riccati equation, $\mathbf{M}_\mathbf{o}$. However, use of Eq. (8.130) imposes an *additional* condition to the sufficient conditions for the existence of a unique positive definite solution of the algebraic Riccati equation, namely that $\mathbf{A}$ must be *non-singular*. We can use Eq. (8.130) by constructing the Hamiltonian matrix, finding the eigenvectors of its inverse, and partitioning the modal matrix, $\mathbf{V}$, according to Eq. (8.128), or simply use the specialized MATLAB (CST) command *dlqr* which does the same thing. The command *dlqr* is the digital equivalent of the command *lqr* for solving the analog algebraic Riccati equation, and is used as follows:

```
>>[Ko,Mo,E] = dlqr(A,B,Q,R) <enter>
```

where the matrices $\mathbf{A}$, $\mathbf{B}$, $\mathbf{Q}$, $\mathbf{R}$, $\mathbf{K_o}$, and $\mathbf{M_o}$ are the same as in the foregoing discussion, while $\mathbf{E}$ is the returned matrix of the eigenvalues of $\mathbf{A} - \mathbf{BK_o}$.

Example 8.20

Let us design an optimal regulator for a digital turbo-generator with the following *digital*, linear, time-invariant state-space representation:

$$\mathbf{A} = \begin{bmatrix} 0.1346 & 0.1236 & -0.0361 & 0.0037 & 0.0004 & -0.0003 \\ -0.1091 & 0.5412 & 0.3851 & -0.0631 & -0.0520 & 0.0152 \\ 0.0426 & 0.1052 & 0.7915 & 0.0700 & 0.0504 & -0.0172 \\ -0.0045 & 0.0205 & -0.0542 & 0.7932 & -0.5687 & 0.0025 \\ 0.0022 & -0.0150 & 0.0384 & 0.5681 & 0.7526 & 0.0357 \\ 0.0002 & -0.0162 & 0.0142 & 0.0004 & -0.0299 & 0.9784 \end{bmatrix}$$

$$\mathbf{B} = \begin{bmatrix} -0.0121 & 0.0117 \\ -0.0046 & -0.4960 \\ -0.0150 & 0.5517 \\ 0.0095 & -0.1763 \\ -0.0055 & -1.0216 \\ 0.0025 & 1.3944 \end{bmatrix}$$

$$\mathbf{C} = \begin{bmatrix} 0.5971 & -0.7697 & 4.8850 & 4.8608 & -9.8177 & -8.8610 \\ 3.1013 & 9.3422 & -5.6000 & -0.7490 & 2.9974 & 10.5719 \end{bmatrix} \quad (8.131)$$

$$\mathbf{D} = \begin{bmatrix} 0 & 0 \\ 0 & 0 \end{bmatrix}$$

The two inputs are the *throttle-valve position*, $u_1(k)$, and the *loading torque*, $u_2(k)$, while the two outputs are the *deviation from the desired generated voltage* (or *voltage error*), $y_1(k)$ (volt), and the *deviation of the generator load's angular position*, (or *load position error*), $y_2(k)$ (radians). For this sixth-order system, it is desired that the closed-loop initial response to the initial condition, $\mathbf{x}(0) = [0.1; 0; 0; 0; 0; 0]^T$ should decay to zero in about 20 sampling instants, with a maximum overshoot in $y_1(k)$ of 0.1 V and in $y_2(k)$ of 0.35 rad. For the plant, the first output takes about 50 samples to settle to zero. We select the weighting matrices for the optimal control problem as $\mathbf{Q} = \mathbf{I}$ and $\mathbf{R} = \mathbf{I}$. The inverse of the Hamiltonian matrix, $\mathcal{H}^{-1}$, and its eigenvalues are calculated as follows:

```
>>Q=eye(6); R=eye(2); <enter>

>>Ainv=inv(A); Rp=B*inv(R)*B'; Hinv = [Ainv Ainv*Rp; Q*Ainv A'
  +Q*Ainv*Rp] <enter>

Hinv =

Columns 1 through 7
  5.8602    -1.5384     1.0209    -0.0739    -0.2316     0.0523     0.0211
  1.5564     1.6366    -0.7308     0.0475     0.1953    -0.0451    -0.0170
 -0.5237    -0.1360     1.3087    -0.0359    -0.1227     0.0295     0.0114
 -0.0100    -0.0193     0.0356     0.8172     0.6128    -0.0236    -0.0091
  0.0461     0.0572    -0.1095    -0.6130     0.8754    -0.0332    -0.0107
  0.0338     0.0311    -0.0347    -0.0178     0.0316     1.0199     0.0159
  5.8602    -1.5384     1.0209    -0.0739    -0.2316     0.0523     0.1557
  1.5564     1.6366    -0.7308     0.0475     0.1953    -0.0451     0.1065
 -0.5237    -0.1360     1.3087    -0.0359    -0.1227     0.0295    -0.0247
 -0.0100    -0.0193     0.0356     0.8172     0.6128    -0.0236    -0.0054
  0.0461     0.0572    -0.1095    -0.6130     0.8754    -0.0332    -0.0103
  0.0338     0.0311    -0.0347    -0.0178     0.0316     1.0199     0.0156

Columns 8 through 12
 -0.8515     0.9488    -0.3035    -1.7544     2.3950
  0.7278    -0.8093     0.2585     1.4991    -2.0461
 -0.4742     0.5277    -0.1687    -0.9768     1.3333
  0.3838    -0.4270     0.1365     0.7906    -1.0792
  0.4567    -0.5079     0.1622     0.9408    -1.2840
 -0.6740     0.7497    -0.2395    -1.3883     1.8950
 -0.9606     0.9914    -0.3080    -1.7521     2.3951
  1.2691    -0.7041     0.2790     1.4842    -2.0623
 -0.0890     1.3193    -0.2229    -0.9384     1.3476
  0.3207    -0.3571     0.9297     1.3588    -1.0787
  0.4047    -0.4574    -0.4064     1.6934    -1.3139
 -0.6588     0.7325    -0.2370    -1.3526     2.8733

>>eig(Hinv) <enter>
ans =
  5.7163 + 1.8038i
  5.7163 - 1.8038i
  2.0510
  1.0596 + 0.6776i
  1.0596 - 0.6776i
  0.1591 + 0.0502i
  0.1591 - 0.0502i
  0.6698 + 0.4283i
  0.6698 - 0.4283i
  0.4876
  1.1062
  0.9040
```

Note that the eigenvalues of $\mathcal{H}^{-1}$ occur in reciprocal stable and unstable pairs, as expected. Then, the steady state solution to the algebraic Riccati equation, $\mathbf{M_o}$, the optimal regulator gain matrix, $\mathbf{K_o}$, and the set of closed-loop eigenvalues, $\mathbf{E}$, can be obtained using *dlqr* as follows:

```
>>[Ko, Mo, E] = dlqr(A, B, Q, R) <enter>

Ko =
 -0.0027    -0.0267    -0.0886     0.0115    -0.0375     0.0216
  0.0024    -0.0025     0.0223     0.0514    -0.4316     0.3549

Mo =
  1.0375    -0.0491     0.0001     0.0495     0.0678    -0.0246
 -0.0491     1.8085     1.6151     0.1214    -0.2535    -0.3059
  0.0001     1.6151     5.8619     0.7194     0.3663    -1.0029
  0.0495     0.1214     0.7194     5.2730    -0.5666     1.6496
  0.0678    -0.2535     0.3663    -0.5666     4.1756     0.8942
 -0.0246    -0.3059    -1.0029     1.6496     0.8942     3.2129

E =
 0.1591 + 0.0502i
 0.1591 - 0.0502i
 0.4876
 0.6698 + 0.4283i
 0.6698 - 0.4283i
 0.9040
```

The closed-loop initial response is obtained using *dinitial* as follows:

```
>>dinitial(A-B*Ko,B,C,D,[0.1 zeros(1,5)]') <enter>
```

The resulting outputs, $y_1(k)$ and $y_2(k)$, are plotted in Figure 8.13. Note that the requirement of both outputs settling to zero in about 20 sampling instants, with specified maximum overshoots, has been met.

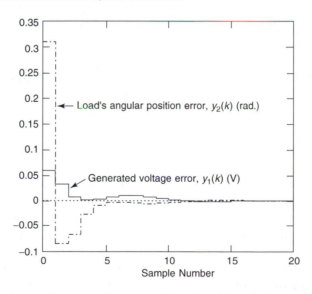

Figure 8.13   Initial response of the optimally regulated digital turbo-generator system

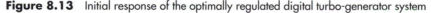

As in Chapter 6, we can formulate the digital equivalent of the *output weighted* linear, optimal control, where the weightage is placed on the output rather than on the state-vector. The objective function for such a problem for the time-invariant case can be written as

$$J(0, N) = \sum_{k=0}^{N} \mathbf{y}^{\mathrm{T}}(k)\mathbf{Q}\mathbf{y}(k) + \mathbf{u}^{\mathrm{T}}(k)\mathbf{R}\mathbf{u}(k) \tag{8.132}$$

which, by the substitution of the output equation, $\mathbf{y}(k) = \mathbf{C}\mathbf{x}(k) + \mathbf{D}\mathbf{u}(k)$, becomes of the following form:

$$J(0, N) = \sum_{k=0}^{N} \mathbf{x}^{\mathrm{T}}(k)\mathbf{Q}^*\mathbf{x}(k) + 2\mathbf{u}^{\mathrm{T}}(k)\mathbf{N}\mathbf{x}(k) + \mathbf{u}^{\mathrm{T}}(k)\mathbf{R}\mathbf{u}(k) \tag{8.133}$$

Equation (8.133) is the general quadratic form of the quadratic objective function, and can be minimized using *dlqr* with a fourth additional input argument, the cross-weighting matrix, $\mathbf{N}$. You can derive the algebraic Riccati equation for the steady state solution of the output weighted problem along the lines of the foregoing discussion. The MATLAB (CST) directly provides the solution for the output weighted, steady state, linear, time-invariant, optimal control problem through its command *dlqry* which is used as follows:

```
>>[Ko,Mo,E] = dlqry(A,B,C,D,Q,R) <enter>
```

where the symbols have their usual meanings.

The solution to the digital terminal-time weighted linear optimal control problem for a time-invariant plant can be obtained along the lines of Section 6.5, with suitable modifications. You are encouraged to write a computer program – a digital equivalent of *tpbvlti.m* – for solving the two-point boundary value problem for the terminal-time weighted digital optimal control.

8.12 Stochastic Digital Systems, Digital Kalman Filters, and Optimal Digital Compensators

There is no fundamental difference in the way random (or stochastic) variables are handled in modeling and control of digital and analog systems. Hence, the methods of Chapter 7 can be applied on a one-to-one basis to stochastic systems, of course with the understanding that the s-plane is transformed to the z-plane, and an analog state-space representation is transformed into a digital representation. However, there is a fundamental difference in the manner in which digital and analog systems respond to noise. As we have seen earlier in this chapter, a digital control system's frequency response is limited to signals whose frequencies are *less than* the Nyquist frequency, π/T. Hence, a digital control system is inherently *robust* with respect to high-frequency noise. This is precisely why digital control systems have become so popular in applications where high-frequency noise is present. For example, a digital video disc (DVD) player shows much clearer pictures than the analog video cassette player (VCP), primarily due to this inherent robustness.

Since any noise with frequency higher than the Nyquist frequency gets automatically cut-off by a digital control system, we have to only worry about noise with frequency contents smaller than the Nyquist frequency. In a manner similar to analog filters, we can define a *digital filter* as any digital system through which a noise is input. For single-input, single-output applications, a digital filter has a pulse transfer function, $G(z)$. Depending on the application, we can design *low-pass*, *band-pass*, or *high-pass* digital filters with appropriate pulse transfer functions using an extension of the techniques of Chapter 7.

Example 8.21

Let us construct a first order low-pass digital filter to cut-off a uniformly distributed random noise superimposed on a signal, $\sin(10t)$. The pulse transfer function of the filter is given by $G(z) = a/(z + b)$, where a and b are constants. Let us also initially select a sampling interval, $T = 0.1$ s. The Nyquist frequency is $\pi/T = 31.4$ rad/s. Hence, the noise with frequencies higher than 31.4 rad/s gets automatically cut-off. Since the input is a continuous time signal, the filter is selected to be a sampled-data analog filter with transfer function, $G(s) = 15/(s + 15)$ and a zero-order hold. Then $G(z)$ is obtained using *c2dm* as follows:

```
>>num = 15; den = [1 15]; [numd, dend] = c2dm(num,den,0.1,'zoh')
<enter>

numd2 =
     0      0.7769

dend2 =
   1.0000   -0.2231
```

Hence, $G(z) = 0.7769/(z - 0.2231)$. The response of this filter to the noisy signal can be simulated with the following command:

```
>>t=0:0.05:1; u = sin(10*t)+randn(size(t));
  y = dlsim(numd,dend,u); <enter>
```

The most important extension of the analog stochastic concepts of Chapter 7 for digital stochastic systems is the *digital Kalman filter*. The digital Kalman filter is an *optimal digital observer* for a noisy plant. Consider a linear, time varying plant with the following digital state-space representation:

$$\mathbf{x}(k + 1) = \mathbf{A}(k)\mathbf{x}(k) + \mathbf{B}(k)\mathbf{u}(k) + \mathbf{F}(k)\mathbf{v}(k) \tag{8.134}$$

$$\mathbf{y}(k) = \mathbf{C}(k)\mathbf{x}(k) + \mathbf{D}(k)\mathbf{u}(k) + \mathbf{z}(k) \tag{8.135}$$

where $\mathbf{v}(k)$ is the *process noise vector*, which may arise due to modeling errors such as neglecting nonlinear or higher-frequency dynamics, and $\mathbf{z}(k)$ is the *measurement noise vector*. By assuming $\mathbf{v}(k)$ and $\mathbf{z}(k)$ to be discrete, *uncorrelated white noises*, we will only be extending the methodology of Chapter 7 for a description of the digital stochastic plant. The *correlation matrices* of *non-stationary* discrete white noises, $\mathbf{v}(k)$ and $\mathbf{z}(k)$, can be expressed as follows:

$$\mathbf{R_v}(j, k) = \mathbf{V}(k)\delta_{jk} \tag{8.136}$$

$$\mathbf{R_z}(j, k) = \mathbf{Z}(k)\delta_{jk} \tag{8.137}$$

where $\mathbf{V}(k)$ and $\mathbf{Z}(k)$ are the power spectral density matrices (also the covariance matrices) of $\mathbf{v}(k)$ and $\mathbf{z}(k)$, respectively, and δ_{jk} is the *Kronecker delta function* ($\delta_{jk} = 0$ when $j \neq k$, and $\delta_{jk} = 1$ when $j = k$). The digital, time-varying Kalman filter's state-equation is written as that of the *current observer* (Eq. (8.86)) in the following manner:

$$\mathbf{x_o}(k + 1) = [\mathbf{A}(k) - \mathbf{L}^o(k)\mathbf{C}(k)\mathbf{A}(k)]\mathbf{x_o}(k) + [\mathbf{B}(k) - \mathbf{L}^o(k)\mathbf{C}(k)\mathbf{B}(k)]\mathbf{u}(k)$$
$$- \mathbf{L}^o(k)\mathbf{D}(k)\mathbf{u}(k + 1) + \mathbf{L}^o(k)\mathbf{y}(k + 1) \tag{8.138}$$

where the optimal gain matrix, $\mathbf{L}^o(k)$, is to be determined by *minimizing* the *covariance* of the estimation error vector, $\mathbf{e_o}(k)$. Since the estimation error vector, $\mathbf{e_o}(k)$, is a discrete random variable, we can use the methods of Section 7.4 for obtaining a corresponding difference equation to be satisfied by the optimal covariance matrix. However, recall that in Section 7.4, we had only minimized a *conditional covariance matrix* of estimation error (i.e. the covariance matrix based on a *finite record* of the output). The digital equivalent of the conditional covariance matrix is the covariance matrix of the *prediction error vector*, $\mathbf{e_1}(k + 1) = [\mathbf{x}(k + 1) - \mathbf{x_1}(k + 1)]$, which is based on the measurement of the output at the *previous* sampling instant, $\mathbf{y}(k)$ (see Eq. (8.87)). Recall that the current observer state-equation can be alternatively expressed as two *predictor-corrector* equations, Eqs. (8.87) and (8.88). For the optimal current observer given by Eq. (8.138), the predictor-corrector difference equations are the following:

$$\mathbf{x_1}(k + 1) = \mathbf{A}(k)\mathbf{x_o}(k) + \mathbf{B}(k)\mathbf{u}(k) \tag{8.139}$$

$$\mathbf{x_o}(k + 1) = \mathbf{x_1}(k + 1) + \mathbf{L}^o(k + 1)[\mathbf{y}(k + 1) - \mathbf{D}(k + 1)\mathbf{u}(k + 1)$$
$$- \mathbf{C}(k + 1)\mathbf{x_1}(k + 1)] \tag{8.140}$$

where $\mathbf{x_1}(k + 1)$ is the *predicted state-vector* at the $(k + 1)$th sampling instant based upon the quantities at the *previous* sampling instant, and $\mathbf{x_o}(k + 1)$ is the *corrected state-vector* estimate due to the measurement of the output at the current sampling instant, $\mathbf{y}(k + 1)$. Defining $\mathbf{R_e^o}(k)$ as the *optimal covariance matrix* of the *prediction error*, we can write

$$\mathbf{R_e^o}(k) = \mathbf{E}[\mathbf{e_1}(k)\mathbf{e_1^T}(k)] = \mathbf{E}[\{\mathbf{x}(k) - \mathbf{x_1}(k)\}\{\mathbf{x}(k) - \mathbf{x_1}(k)\}^T] \tag{8.141}$$

whereas the optimal covariance matrix of the *corrected* (or *true*) estimation error is written as

$$\mathbf{P}^o(k) = \mathbf{E}[\mathbf{e_o}(k)\mathbf{e_o^T}(k)] = \mathbf{E}[\{\mathbf{x}(k) - \mathbf{x_o}(k)\}\{\mathbf{x}(k) - \mathbf{x_o}(k)\}^T] \tag{8.142}$$

You can show, using these definitions and the method of Section 7.4, that for the linear, time-varying, digital Kalman filter, the following difference equations must be satisfied by the optimal covariance matrix, $\mathbf{R}_e^o(k)$:

$$\mathbf{R}_e^o(k+1) = \mathbf{A}(k)[\mathbf{I} - \mathbf{L}^o(k)\mathbf{C}(k)]\mathbf{R}_e^o(k)\mathbf{A}^T(k) + \mathbf{F}(k)\mathbf{V}(k)\mathbf{F}^T(k) \tag{8.143}$$

where the Kalman filter gain, $\mathbf{L}^o(k)$, is given by

$$\mathbf{L}^o(k) = \mathbf{R}_e^o(k)\mathbf{C}^T(k)[\mathbf{C}(k)\mathbf{R}_e^o(k)\mathbf{C}^T(k) + \mathbf{Z}(k)]^{-1} \tag{8.144}$$

For a linear time invariant plant in the steady state, the state coefficient matrices, the power spectral densities of the noise processes, the Kalman filter gain, and the optimal covariance matrix, all become constants. In such a case, Eq. (8.143) becomes

$$\mathbf{R}_e^o = \mathbf{A}(\mathbf{I} - \mathbf{L}^o\mathbf{C})\mathbf{R}_e^o\mathbf{A}^T + \mathbf{FVF}^T \tag{8.145}$$

where

$$\mathbf{L}^o = \mathbf{R}_e^o\mathbf{C}^T(\mathbf{C}\mathbf{R}_e^o\mathbf{C}^T + \mathbf{Z})^{-1} \tag{8.146}$$

Substituting Eq. (8.146) into Eq. (8.145) we get the following explicit equation in terms of the unknown matrix, $\mathbf{R}_e^o$:

$$\mathbf{R}_e^o = \mathbf{A}\mathbf{R}_e^o\mathbf{A}^T - \mathbf{A}\mathbf{R}_e^o\mathbf{C}^T(\mathbf{C}\mathbf{R}_e^o\mathbf{C}^T + \mathbf{Z})^{-1}\mathbf{C}\mathbf{R}_e^o\mathbf{A}^T + \mathbf{FVF}^T \tag{8.147}$$

Equations (8.146) and (8.147) should be compared with Eqs. (8.121) and (8.122), respectively, for the digital optimal regulator problem. It can be seen that the two sets of equations are *identical* in form, with the corresponding matrices compared in Table 8.2. Due to this similarity, the Kalman filter and optimal regulator are called *dual* problems. An alternative form of Eqs. (8.146) and (8.147) in terms of the optimal covariance matrix of the estimation error, $\mathbf{P}^o$, can be written as follows:

$$\mathbf{L}^o = \mathbf{P}^o\mathbf{C}^T\mathbf{Z}^{-1} \tag{8.148}$$

$$\mathbf{R}_e^o = \mathbf{A}\mathbf{P}^o\mathbf{A}^T + \mathbf{FVF}^T \tag{8.149}$$

where

$$\mathbf{P}^o = \mathbf{R}_e^o - \mathbf{R}_e^o\mathbf{C}^T(\mathbf{C}\mathbf{R}_e^o\mathbf{C}^T + \mathbf{Z})^{-1}\mathbf{C}\mathbf{R}_e^o \tag{8.150}$$

From Table 8.2 it is clear that Eq. (8.147) is none other than the *algebraic Riccati equation*, which can be solved by the techniques of the previous section. After the solution,

Table 8.2 Comparison of analogous matrices for the digital Kalman filter and the digital optimal regulator

Digital Optimal Regulator	Digital Kalman Filter
$\mathbf{A}$	$\mathbf{A}^T$
$\mathbf{B}$	$\mathbf{C}^T$
$\mathbf{M}_o$	$\mathbf{R}_e^o$
$\mathbf{Q}$	$\mathbf{FVF}^T$
$\mathbf{R}$	$\mathbf{Z}$
$\mathbf{K}$	$\mathbf{P}^o = \mathbf{R}_e^o - \mathbf{R}_e^o\mathbf{C}^T(\mathbf{C}\mathbf{R}_e^o\mathbf{C}^T + \mathbf{Z})^{-1}\mathbf{C}\mathbf{R}_e^o$

$\mathbf{R}_e^o$, of the algebraic Riccati equation is obtained, it can be substituted into Eq. (8.146) to get the Kalman filter gain matrix, $\mathbf{L}^o$.

The inverse of the Hamiltonian matrix for the steady state Kalman filter is given by

$$\mathcal{H}^{-1} = \begin{bmatrix} (\mathbf{A}^T)^{-1} & (\mathbf{A}^T)^{-1}\mathbf{C}^T\mathbf{Z}^{-1}\mathbf{C} \\ \mathbf{F}\mathbf{V}\mathbf{F}^T(\mathbf{A}^T)^{-1} & \mathbf{A} + \mathbf{F}\mathbf{V}\mathbf{F}^T(\mathbf{A}^T)^{-1}\mathbf{C}^T\mathbf{Z}^{-1}\mathbf{C} \end{bmatrix} \qquad (8.151)$$

Then the steps given by Eqs. (8.127)–(8.130) can be extended for obtaining the solution to the algebraic Riccati equation. We can either use the MATLAB (CST) command *dlqr* with the appropriate substitution of the regulator matrices given by Table 8.2 for obtaining the Kalman filter gain, or the specialized CST command *dlqe*, which directly uses the Kalman filter matrices. The command *dlqe* is used as follows:

```
>>[Lo,Ro,Po,E] = dlqe(A,F,C,V,Z) <enter>
```

where $\mathbf{A}$, $\mathbf{F}$, and $\mathbf{C}$ are the state coefficient matrices of the plant, $\mathbf{V}$ and $\mathbf{Z}$ are the covariance matrices of the process and measurement noise, and $\mathbf{Lo} = \mathbf{L}^o$, the returned Kalman filter gain, $\mathbf{Ro} = \mathbf{R}_e^o$, the returned optimal covariance matrix of the predicted estimation error, $\mathbf{Po} = \mathbf{P}^o$, the returned optimal covariance matrix of the estimation error, and $\mathbf{E}$ is a vector containing the eigenvalues of the digital Kalman filter (i.e. the eigenvalues of $\mathbf{A} - \mathbf{L}^o\mathbf{C}\mathbf{A}$). When used with a *fifth* input argument, $\mathbf{N}$, *dlqe* solves the Kalman filter gain when the process and measurement noises are *correlated*, with the cross-covariance matrix, $\mathbf{N} = \mathbf{E}[\mathbf{v}(k)\mathbf{z}^T(k)]$. MATLAB also provides the command *dlqew* to solve for the Kalman filter gain when the process noise directly affects the output, and the process and measurement noises are uncorrelated.

An optimal digital compensator can be formed with the optimal regulator and the Kalman filter using the methodology of Section 8.10. Such a compensator is called the *digital, linear quadratic gaussian* (DLQG) compensator, and is the digital counterpart of the LQG compensator of Chapter 7.

Example 8.22

Let us derive a Kalman filter for the turbo-generator of Example 8.20, and combine the optimal regulator and the Kalman filter into an optimal digital compensator. Assuming that the process noise, $\mathbf{v}(k)$, consists of two variables, with the coefficient matrix, $\mathbf{F} = \mathbf{B}$, and the covariance matrix, $\mathbf{V} = \mathbf{I}$. The measurement noise is assumed to have a covariance matrix, $\mathbf{Z} = \mathbf{I}$. The Kalman filter gain and the optimal covariance matrices of the prediction and estimation errors are obtained using *dlqe* as follows:

```
>>[Lo, Ro, Po, E]=dlqe(A, B, C, eye(2), eye(2)) <enter>

Lo =
  -0.0011   -0.0037
  -0.0041   -0.0666
   0.0112    0.0881
  -0.0054   -0.0118
  -0.0503   -0.1789
   0.0219    0.2481
```

```
Ro =
     0.0025   -0.0044    0.0000   -0.0058    0.0066   -0.0102
    -0.0044    0.2479   -0.2770    0.0864    0.5165   -0.7057
     0.0000   -0.2770    0.3263   -0.0859   -0.6216    0.8529
    -0.0058    0.0864   -0.0859    0.0448    0.1470   -0.1939
     0.0066    0.5165   -0.6216    0.1470    1.2086   -1.6557
    -0.0102   -0.7057    0.8529   -0.1939   -1.6557    2.2813

Po =
     0.0020   -0.0122    0.0102   -0.0071   -0.0142    0.0187
    -0.0122    0.1067   -0.0905    0.0617    0.1393   -0.1801
     0.0102   -0.0905    0.0800   -0.0532   -0.1233    0.1589
    -0.0071    0.0617   -0.0532    0.0404    0.0806   -0.1017
    -0.0142    0.1393   -0.1233    0.0806    0.1994   -0.2521
     0.0187   -0.1801    0.1589   -0.1017   -0.2521    0.3255

E =
  0.0577
 -0.0957
 -0.3994
  0.6125 + 0.5153i
  0.6125 - 0.5153i
  0.8824
```

All the Kalman filter eigenvalues are seen to be inside the unit circle. An optimal regulator for this plant was designed in Example 8.20 with the following regulator gain matrix:

$$\mathbf{K_o} = \begin{bmatrix} -0.0027 & -0.0267 & -0.0886 & 0.0115 & -0.0375 & 0.0216 \\ 0.0024 & -0.0025 & 0.0223 & 0.0514 & -0.4316 & 0.3549 \end{bmatrix}$$

We can use this regulator matrix and the Kalman filter gain calculated above to form a digital optimal compensator as follows:

```
>>[ac,bc,cc,dc] = dreg(A,B,C,D,Ko,Lo)%
  digital compensator's state-space model <enter>

ac =
   0.1788    0.2529   -0.1088   -0.0003    0.0329    0.1310
  -0.4141   -0.3308    0.8289   -0.0642   -0.3533   -0.6335
   0.1827    0.5037    0.5889    0.0634    0.3088    0.1481
  -0.3818   -1.0436    0.4609    0.7491   -0.7100   -0.9088
  -0.1208   -0.4202    0.3444    0.6989    0.0907   -0.1336
  -0.0477   -0.1929    0.1706    0.0292    0.3426    0.1362

bc =
  -0.0014   -0.0140
   0.0201    0.0949
  -0.0111   -0.0435
   0.0288    0.1163
  -0.0096    0.0423
  -0.0179    0.0178
```

```
cc =
 -0.0219   -0.0809   -0.0624    0.0091   -0.0413   -0.0284
 -0.5325   -1.5382    0.8126    0.0332   -0.6425   -1.1477

dc =
  0.0014    0.0059
  0.0294    0.1668

>>sysc = ss(ac,bc,cc,dc); sysp = ss(A,B,C,D); <enter>

>>syss = series(sysc,sysp);
  sysCL = feedback(syss,eye(size(syss))); <enter>

>>[aCL, bCL, cCL, dCL] = ssdata(sysCL); %
  closed-loop system's state-space model <enter>
```

The closed-loop initial response (with the same initial condition as in Example 8.20) is now calculated and plotted in Figure 8.14 using the following command:

```
>>dinitial(aCL,bCL,cCL,dCL,[0.1 zeros(1,11)]') <enter>
```

Comparing Figures 8.13 and 8.14, we find that the DLQG compensated system has a slightly deteriorated performance when compared to the full-state feedback regulated system of Example 8.20 (the overshoots and steady-state error are larger).

Finally, we test the robustness of the DLQG compensated system to white noise disturbances occurring at the compensator's input. For this purpose, we construct a SIMULINK block diagram shown in Figure 8.15, consisting of the plant connected in closed-loop with the DLQG compensator. A two channel simultaneous white

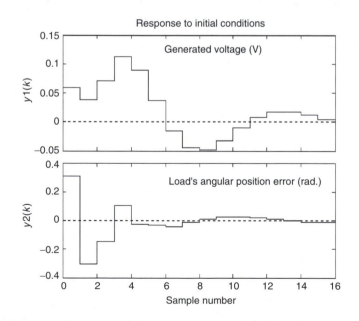

Figure 8.14 Initial response of the DLQG compensated digital turbo-generator system

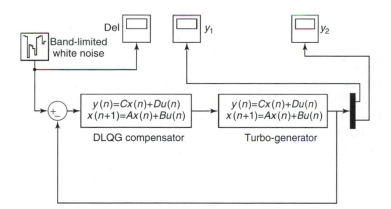

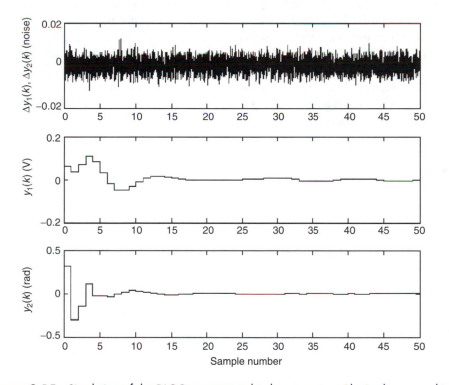

Figure 8.15 Simulation of the DLQG compensated turbo-generator with simultaneous white noise disturbances in each output channel

noise input to the control system is provided through the *band-limited white noise* block, with the noise power set at 10^{-6} at a sample interval of 0.01 s. The simulated initial response of $y_1(k)$ and $y_2(k)$ is seen in Figure 8.15 to be quite unaffected by the applied disturbances, which are of very high frequency. This simulation illustrates the built-in robustness of digital systems to high-frequency noise.

Robustness of digital, optimally compensated (DLQG) systems can be ensured in a manner similar to the analog multivariable systems presented in Section 7.6 using loop-transfer recovery (LTR), with the difference that the singular values are obtained for digital systems rather than analog systems. MATLAB's *Robust Control Toolbox* provides the command *dsigma* for computing singular values of a digital system. The command *dsigma* is used in precisely the same manner as the command *sigma* for analog systems, with the input arguments being the digital state-space coefficient matrices (or the numerator and denominator polynomials of single-input, single-output pulse transfer function). In a manner similar to the digital magnitude plot obtained by *dbode*, the command *dsigma* computes the singular values only upto the Nyquist frequency, π/T. As noted earlier, a digital system is unable to respond to frequencies above the Nyquist frequency, and thus has an inherent robustness with respect to high-frequency noise. As an exercise, you should carry out loop-transfer recovery for the plant of Example 2.22 using the techniques of Section 7.6 and the command *dsigma*.

Whereas digital control systems are quite robust with respect to high-frequency noise, their implementation using *integer*, or *binary* (rather than *real*) arithmetic results in some robustness problems that are not encountered in analog systems. The effects of *round-off, quantization*, and *overflow* when handling integer arithmetic on a digital computer appear as *additional noise* in digital systems, and robustness of a digital control system to such noise–referred to as *finite-word length effects* becomes important. A discussion of these topics is beyond the scope of this book, and you may refer to a textbook on digital control [2, 4] for this purpose.

Exercises

8.1. Find the z-transform, $F(z)$, of each of the following digital functions:

(a) $f(kT) = e^{-akT} u_s(kT)$

(b) $f(kT) = (kT)^2 e^{-akT} u_s(kT)$

(c) $f(kT) = u_s(kT) \cos(akT)$

8.2. Find the inverse z-transform, $f(kT)$, of each of the following functions:

(a) $F(z) = z/(z^2 + 0.99)$

(b) $F(z) = z(z+1)/(z^3 - 2z^2 - 5z - 1)$

(c) $F(z) = z(z+1.1)(z-0.9)/(z^4 + 5z^3 - 2.2z^2 - 3)$

(d) $F(z) = (z+1)/[(z^2 - 1.14z + 0.65)(z - 0.9)]$

8.3. Find the pulse transfer function, $G(z)$, for each of the following sampled-data analog systems with transfer function, $G(s)$, and sampling interval, $T = 0.2$ second:

(a) $G(s) = (s+2)/[s(s+1)]$

(b) $G(s) = s/(s^2 - 2s + 6)$

(c) $G(s) = 100/(s^3 - 4s + 25)$

(d) $G(s) = (s + 1)(s - 2)/[(s + 10)(s^3 + 5s^2 + 12s + 55)]$

8.4. Find the pulse transfer function of the closed-loop system consisting of an A/D converter with z.o.h and an analog plant of transfer function, $G(s)$, in the configuration shown in Figure 8.4, for each $G(s)$ given in Exercise 8.3.

8.5. Plot the digital Bode diagrams for each of the sampled-data analog systems in Exercise 8.3, and analyze their stability. What is the range of the sampling interval, T, for which each of the systems are stable?

8.6. Plot the digital Bode diagrams for each of the closed-loop digital systems in Exercise 8.4, and analyze their stability. What is the range of the sampling interval, T, for which each of the systems are stable?

8.7. Compute and plot the digital step responses of the sampled-data analog systems in Exercise 8.3.

8.8. Compute and plot the digital step responses of the closed-loop digital systems in Exercise 8.4.

8.9. Find the steady-state value of the output of each of the digital systems in Exercises 8.3 and 8.4 if the input is a *unit ramp function, $tu_s(t)$*.

8.10. Find the steady-state value of the output of each of the digital systems in Exercises 8.3 and 8.4 if the input is a *unit parabolic function, $t^2u_s(t)$*.

8.11. A robotic welding machine has the transfer function, $G(s) = 10/(s^3 + 9s^2 + 11\,000)$. A computer based closed-loop digital controller is to be designed for controlling the robot with the configuration shown in Figure 8.8. Let the computer act as an ideal sampler and z.o.h with a sampling interval of $T = 0.1$ second, and the digital controller be described by a *constant* pulse transfer function, $H(z) = K$. Find the range of K for which the closed-loop system is stable.

8.12. The pitch dynamics of a satellite launch vehicle is described by the analog transfer function, $G(s) = 1/(s^2 - 0.03)$. A digital control system is to be designed for controlling this vehicle with a z.o.h, a sampling interval $T = 0.02$ second, and a constant controller pulse transfer function, $H(z) = K$. Find the range of K for which the closed-loop system is stable. What is the value of K for which the closed-loop step response settles in less than five seconds with a maximum overshoot less than 10 percent?

8.13. For the hard-disk read/write head of Example 2.23 with the analog transfer function given by Eq. (2.165), it is desired to design a computer based closed-loop digital system shown in Figure 8.8. The computer is modeled as an A/D converter with z.o.h and a sampling interval of $T = 0.05$ second, and the digital controller is a *lag-compensator* given by the pulse transfer function, $H(z) = K(z - z_o)/(z - z_p)$, where K is a constant, $z_o = (2/T - \alpha\omega_o)/(2/T + \alpha\omega_o)$, and $z_p = (2/T - \omega_o)/(2/T + \omega_o)$, with $\alpha > 1$. Find the values of K, α, and ω_o such that the closed-loop system has a maximum overshoot less than 20 percent and a steady-state error less than 1 percent if the desired output is a unit step function.

8.14. For the roll dynamics of a fighter aircraft described in Example 2.24 with the analog transfer function given by Eq. (2.181), it is desired to design a computer based closed-loop digital system shown in Figure 8.8. The computer is modeled as an A/D converter with z.o.h and a sampling interval of $T = 0.2$ second, and the digital controller is a *lead-compensator* given by the pulse transfer function, $H(z) = K(z - z_o)/(z - z_p)$, where K is a constant, $z_o = (2/T - \alpha\omega_o)/(2/T + \alpha\omega_o)$, and $z_p = (2/T - \omega_o)/(2/T + \omega_o)$, with $\alpha < 1$. Find the values of K, α, and ω_o such that the closed-loop system has a maximum overshoot less than 5 percent with a settling time less than two seconds and a zero steady-state error, if the desired output is a unit step function.

8.15. Derive a digital state-space representation for each of the sampled-data analog systems in Exercise 8.3.

8.16. Derive a digital state-space representation for each of the closed-loop digital systems in Exercise 8.4.

8.17. Find a state-space representation for each of the digital systems described by the following difference equations, with $y(k)$ as the output and $u(k)$ as the input at the kth sampling instant:

(a) $y(k + 3) = 2y(k + 2) - 1.3y(k + 1) - 0.8y(k) + 0.2u(k)$

(b) $y(k + 2) = 4y(k + 1) - 3y(k) + 2u(k + 1) - u(k)$

(c) $y(k + 4) = 21y(k + 3) - 15y(k + 2) - y(k) + 3u(k + 3) - u(k + 1) + 2.2u(k)$

8.18. Find the controller companion form, the observer companion form, and the Jordan canonical form state-space representations of the digital systems in Exercise 8.17.

8.19. A multivariable digital system is described by the following difference equations:

$$y_1(k + 2) + y_1(k + 1) - y_2(k) = u_1(k) + u_2(k)$$

$$y_2(k + 1) + y_2(k) - y_1(k) = u_1(k)$$

where $u_1(k)$ and $u_2(k)$ are the inputs and $y_1(k)$ and $y_2(k)$ are the outputs at the kth sampling instant. Derive a state-space representation of the system.

8.20. For the distillation column whose analog state-space representation is given in Exercises 5.4 and 5.15, derive a digital state-space representation with z.o.h and a sampling interval of $T = 0.2$ second.

8.21. For the aircraft lateral dynamics described by Eq. (4.97) in Exercise 4.3, derive a digital state-space representation with z.o.h and a sampling interval of $T = 0.2$ second.

8.22. Repeat Exercises 8.20 and 8.21 with a first-order hold. Compare the respective z-plane pole locations with those in Exercises 8.20 and 8.21.

8.23. Repeat Exercises 8.20 and 8.21 with a bilinear transformation. Compare the respective z-plane pole locations with those in Exercises 8.20 and 8.21.

8.24. For the distillation column with digital state-space representation derived in Exercise 8.20, design a full-state feedback regulator to place closed-loop poles at $z_{1,2} = 0.9 \pm 0.1i$, $z_3 = 0.37$, and $z_4 = 0.015$. Find the initial response of the regulated system to the initial condition $\mathbf{x}(0) = [1; 0; 0; 0]^T$.

8.25. For the aircraft lateral dynamics with digital state-space representation derived in Exercise 8.21, design a full-state feedback regulator to place closed-loop poles at $z_{1,2} = 0.9 \pm 0.1i$, $z_3 = 0.92$, and $z_4 = 0.22$. Find the initial response of the regulated system to the initial condition $\mathbf{x}(0) = [0.5; 0; 0; 0]^T$.

8.26. Using the MATLAB (CST) command *place*, design a full-order, two-output, current observer for the digitized distillation column of Exercise 8.20 such that the observer poles are placed at $z_{1,2} = 0.8 \pm 0.2i$, $z_3 = 0.3$, and $z_4 = 0.015$. Combine the observer with the regulator designed in Exercise 8.24 to obtain a compensator. Find the initial response of the compensator to the initial condition $\mathbf{x}(0) = [1; 0; 0; 0]^T$. Compare the required control inputs of the compensated system for the initial response to those required by the regulator in Exercise 8.24.

8.27. Using the MATLAB (CST) command *place*, design a full-order, two-output, current observer for the digitized aircraft lateral dynamics of Exercise 8.21 such that the observer poles are placed at $z_{1,2} = 0.8 \pm 0.2i$, $z_3 = 0.8$, and $z_4 = 0.1$. Combine the observer with the regulator designed in Exercise 8.25 to obtain a compensator. Find the initial response of the compensator to the initial condition $\mathbf{x}(0) = [0.5; 0; 0; 0]^T$. Compare the required inputs of the compensated system for the initial response to those required by the regulator in Exercise 8.25.

8.28. Repeat the compensator designs of Exercises 8.26 and 8.27 with *all* closed-loop poles at $z = 0$. Plot the deadbeat initial response of the designed closed-loop systems.

8.29. For the compensators designed in Exercise 8.28, simulate the response to measurement noise appearing simultaneously in all the output channels, using SIMULINK. (Use a *band-limited white noise* of power 10^{-6}.)

8.30. For the longitudinal dynamics of a flexible bomber described in Example 4.7, design a full-state feedback, digital optimal regulator with z.o.h and a sampling interval of $T = 0.1$ second, which would produce a maximum overshoot of less than ± 2 m/s^2 in the normal-acceleration, $y_1(t)$, and of less than ± 0.03 rad/s in pitch-rate, $y_2(t)$, and a settling time less than 5 s, while requiring elevator and canard deflections *not exceeding* ± 0.1 rad. (5.73°), if the initial condition is 0.1 rad/s perturbation in the pitch-rate, i.e. $\mathbf{x}(0) = [0; 0.1; 0; 0; 0; 0]^T$.

8.31. For the flexible bomber airplane of Exercise 8.30, design a digital Kalman filter based on the measurement of only the normal acceleration, $y_1(t)$, and combine it with the digital optimal regulator designed in Exercise 8.30 to form a digital optimal compensator such that the closed-loop performance requirements of Exercise 8.30 are met. Simulate the response of the digital closed-loop system to a measurement noise in both the output channels, using SIMULINK with *band-limited white noise* of power 10^{-8}.

8.32. For the digital optimal compensator designed in Exercise 8.31, compare the digital singular values of the return ratio matrix at the plant's input with those of the full-state feedback system of Exercise 8.30. How robust is the digital compensator?

8.33. Re-design the digital Kalman filter in Exercise 8.31 such that the loop-transfer recovery at the plant's input occurs in the frequency range 1–100 rad/s.

8.34. Design a digital optimal tracking system for the tank-gun turret described in Exercise 6.11 with z.o.h, a sampling interval of $T = 0.2$ second, and full-state feedback for achieving a constant desired state $\mathbf{x_d^c} = [1.57; 0; 0; 0; 0.5; 0; 0; 0]^T$, in less than seven seconds, if the initial condition is zero, i.e. $\mathbf{x}(0) = \mathbf{0}$, with the control inputs not exceeding five units.

8.35. Write a MATLAB computer program – a digital equivalent of *tpbvlti.m* – for solving the two-point boundary value problem for the terminal-time weighted digital optimal control.

References

1. *Control System Toolbox-5.0®–User's Guide*. The Math Works Inc., Natick, MA, 2000.
2. Franklin, G.F., Powell, J.D. and Workman, M.L. *Digital Control of Dynamic Systems*. Addison Wesley, Reading, MA, 1990.
3. Bellman, R. and Dreyfuss, S.E. *Applied Dynamic Programming*. Princeton University Press, 1962.
4. Phillips, C.L. and Nagle, H.T. *Digital Control System Analysis and Design*. Prentice-Hall, 1990.

9

Advanced Topics in Modern Control

9.1 Introduction

The previous chapters saw coverage of topics ranging from classical control to modern digital control. In many modern control applications, certain other techniques are employed which are grouped here as *advanced topics*. A detailed coverage of the advanced topics – such as H_∞ *control*, *input shaping*, and *nonlinear control* – is beyond the scope of the present text. However, you can get a flavor of each topic in this chapter, and for details you are encouraged to look at the many references given at the end of the chapter.

9.2 H_∞ Robust, Optimal Control

In Chapter 6, we derived linear optimal controllers with full-state feedback that minimized a quadratic objective function, and in Chapter 7 we studied Kalman filters as optimal observers in the presence of white noise. The resulting compensator with an optimal regulator (or tracking system) and a Kalman filter was referred to as an LQG (Linear, Quadratic, Gaussian) controller. While LQG controllers exhibit good performance, their robustness to process and measurement noise can only be indirectly ensured by iterative techniques, such as loop-transfer recovery (LTR) covered in Chapter 7. The H_∞ (pronounced *H-infinity*) *optimal control* design technique, however, directly address the problem of robustness by deriving controllers which maintain system response and error signals to within prescribed tolerances, despite the presence of noise in the system. Figure 9.1 shows a plant with transfer matrix, $\mathbf{G}(s)$, input vector, $\mathbf{U}(s)$, and output vector, $\mathbf{Y}(s)$, being controlled by a feedback compensator with transfer matrix, $\mathbf{H}(s)$. The vector $\mathbf{w}(s)$ contains all inputs *external* to the closed-loop system, i.e. process and measurement noise vectors, as well as the desired output vector. The vector $\mathbf{e}(s)$ contains all the *errors* that determine the behavior of the closed-loop system, i.e. the estimation error and the tracking error vectors.

The plant's transfer matrix can be partitioned as follows:

$$\mathbf{G}(s) = \begin{bmatrix} \mathbf{G}_{11}(s) & \mathbf{G}_{12}(s) \\ \mathbf{G}_{21}(s) & \mathbf{G}_{22}(s) \end{bmatrix} \qquad (9.1)$$

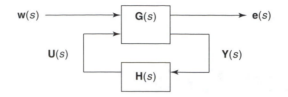

Figure 9.1 Multivariable closed-loop control system with plant, $\mathbf{G}(s)$, controller, $\mathbf{H}(s)$, external input vector, $\mathbf{w}(s)$, error vector, $\mathbf{e}(s)$, plant input vector, $\mathbf{U}(s)$, and plant output, $\mathbf{Y}(s)$

such that

$$\mathbf{e}(s) = \mathbf{G}_{11}(s)\mathbf{w}(s) + \mathbf{G}_{12}(s)\mathbf{U}(s) \tag{9.2}$$

and

$$\mathbf{Y}(s) = \mathbf{G}_{21}(s)\mathbf{w}(s) + \mathbf{G}_{22}(s)\mathbf{U}(s) \tag{9.3}$$

Substituting the control-law, $\mathbf{U}(s) = \mathbf{H}(s)\mathbf{Y}(s)$, into Eqs. (9.2) and (9.3), we can write the following relationship between the error vector and the external inputs vector:

$$\mathbf{e}(s) = [\mathbf{G}_{11}(s) + \mathbf{G}_{12}(s)\mathbf{H}(s)\{\mathbf{I} - \mathbf{G}_{22}(s)\mathbf{H}(s)\}^{-1}\mathbf{G}_{21}(s)]\mathbf{w}(s) \tag{9.4}$$

The transfer matrix multiplying $\mathbf{w}(s)$ on the right-hand side of Eq. (9.4) can be denoted as $\mathbf{F}(s) = [\mathbf{G}_{11}(s) + \mathbf{G}_{12}(s)\mathbf{H}(s)\{\mathbf{I} - \mathbf{G}_{22}(s)\mathbf{H}(s)\}^{-1}\mathbf{G}_{21}(s)]$ for simplicity of notation, and we can re-write Eq. (9.4) as

$$\mathbf{e}(s) = \mathbf{F}(s)\mathbf{w}(s) \tag{9.5}$$

The H_∞ *optimal control* synthesis procedure consists of finding a *stabilizing* controller, $\mathbf{H}(s)$, such that the H_∞-*norm* of the closed-loop transfer matrix, $\mathbf{F}(s)$, is *minimized*. The H_∞-norm is a *scalar* assigned to a *matrix* (an animal of the same species as the *singular value* discussed in Chapters 2 and 7) and is defined as follows:

$$\|\mathbf{F}(i\omega)\|_\infty = \sup_\omega [\sigma_{\max}(\mathbf{F}(i\omega))] \tag{9.6}$$

where $\sigma_{\max}(\mathbf{F}(i\omega))$ denotes the *largest singular value* of $\mathbf{F}(i\omega)$ (see Chapter 7 for the definition and calculation of singular values), while $\sup_\omega[\cdot]$ is called the *supremum function*, and denotes the *largest value* of the function within the square brackets encountered as the frequency, ω, is varied. Clearly, obtaining the H_∞-norm of a transfer matrix requires calculating the singular values of the transfer matrix with $s = i\omega$ at a range of frequencies, and then obtaining the *maximum* of the *largest singular value* over the given frequency range. Using the MATLAB's *Robust Control Toolbox* command *sigma* for computing the singular values, you can easily write an M-file for calculating the H_∞-norm over a specified frequency range (or *bandwidth*). The LQG problem can be expressed in a similar manner as the *minimization* of another matrix norm, called the H_2-*norm* [1].

To better understand the H_∞ optimal control, let us consider a regulator problem (i.e. a zero desired output) with a *process noise*, $\mathbf{p}(s)$. Then, $\mathbf{w}(s)$ contains only the process noise, and comparing Figures 9.1 and 7.10, we can write $\mathbf{w}(s) = \mathbf{p}(s)$. In Section 7.6, we saw that the *sensitivity* of the *output*, $\mathbf{Y}(s)$, with respect to process noise depends upon the matrix $[\mathbf{I} + \mathbf{G}(s)\mathbf{H}(s)]^{-1}$, which we call the *sensitivity matrix of the output*, $\mathbf{S}(s) = [\mathbf{I} + \mathbf{G}(s)\mathbf{H}(s)]^{-1}$. If a scalar norm of this matrix is minimized, we can be assured of the robustness of the closed-loop system with respect to process noise. Such a scalar norm is the H_∞-norm. Thus, our robust optimal control problem consists of finding a stabilizing controller, $\mathbf{H}(s)$, such that the H_∞-norm of the sensitivity matrix at the output, $\|\mathbf{S}(i\omega)\|_\infty$, is minimized. Note that, in this case, $\mathbf{F}(s) = \mathbf{S}(s)$. However, instead of minimizing $\|\mathbf{S}(i\omega)\|_\infty$ over all frequencies, which will increase sensitivity to high-frequency *measurement noise* and poor stability margins, we should minimize $\|\mathbf{S}(i\omega)\|_\infty$ over only those frequencies where the largest magnitudes of the process noise occur. This is practically achieved by defining a *frequency weighting matrix*, $\mathbf{W}(i\omega)$, such that the *largest singular value* of $\mathbf{W}(i\omega)$ is close to unity in a specified frequency range, $0 \leq \omega \leq \omega_o$, and rapidly decay to zero for higher frequencies, $\omega > \omega_o$. The frequency, ω_o, is thus regarded as a *cut-off frequency* below which the sensitivity to process noise is to be minimized. The robust, optimal control problem is then solved by finding a stabilizing controller which minimizes $\|\mathbf{W}(i\omega)\mathbf{S}(i\omega)\|_\infty$. The H_∞-optimal control problem for a regulator is thus the *weighted sensitivity minimization* problem. Equating $\mathbf{F}(s) = \mathbf{W}(s)\mathbf{S}(s)$ for the weighted sensitivity minimization, we get the following partition matrices for the *augmented plant*:

$$\mathbf{G}_{11}(s) = \mathbf{W}(s); \quad \mathbf{G}_{12}(s) = -\mathbf{W}(s)\mathbf{G}(s); \quad \mathbf{G}_{21}(s) = \mathbf{I}; \quad \mathbf{G}_{22}(s) = -\mathbf{G} \qquad (9.7)$$

which, substituted into Eqs. (9.2) and (9.3), make $\mathbf{e}(s) = \mathbf{W}(s)\mathbf{Y}(s)$.

We learned in Section 7.6 that the requirements of robustness conflict with the requirements of optimal control (i.e. minimal control input magnitudes) and tracking performance (i.e. increased sensitivity to a changing desired output). While robustness to process noise is obtained by minimizing the largest singular value of a frequency weighted sensitivity matrix, $\mathbf{S}(s)$ (as seen above), optimal control requires *minimizing* the *largest* singular value of $\mathbf{H}(s)\mathbf{S}(s)$, and good tracking performance to a changing desired output requires *maximizing* the *smallest* singular value of the *complementary sensitivity matrix*, $\mathbf{T}(s) = \mathbf{I} - \mathbf{S}(s)$. However, high-frequency measurement noise rejection requires *minimizing* the *largest* singular value of $\mathbf{T}(s)$ at high frequencies. Such conflicts can be resolved (as in Section 7.6) by choosing different frequency ranges for maximizing and minimizing the different singular values (or H_∞-norms). The different frequency ranges for the various optimizations can be specified as *three different* frequency weighting matrices, $\mathbf{W}_1(i\omega)$, $\mathbf{W}_2(i\omega)$, and $\mathbf{W}_3(i\omega)$ such that the H_∞-norm of the *mixed-sensitivity matrix*, $\|\mathbf{M}(i\omega)\|_\infty$, is *minimized* where

$$\mathbf{M}(i\omega) = \begin{bmatrix} \mathbf{W}_1(i\omega)\mathbf{S}(i\omega) \\ \mathbf{W}_2(i\omega)\mathbf{H}(i\omega)\mathbf{S}(i\omega) \\ \mathbf{W}_3(i\omega)\mathbf{T}(i\omega) \end{bmatrix} \qquad (9.8)$$

Formulating the H_∞-optimal control problem in this fashion ensures the specification of both performance and robustness of the desired closed loop system by the three frequency weighting matrices, such that

$$\sigma_{\max}(\mathbf{S}(i\omega)) \leq \sigma_{\max}(\mathbf{W}_1^{-1}(i\omega)) \qquad (9.9)$$

$$\sigma_{\max}(\mathbf{H}(i\omega)\mathbf{S}(i\omega)) \leq \sigma_{\max}(\mathbf{W}_2^{-1}(i\omega)) \qquad (9.10)$$

and

$$\sigma_{\max}(\mathbf{T}(i\omega)) \leq \sigma_{\max}(\mathbf{W_3^{-1}}(i\omega)) \tag{9.11}$$

The mixed-sensitivity H_∞-optimal control problem posed above is difficult to solve for a general system, because the existence of a *stabilizing* solution, $\mathbf{H}(s)$, requires extra conditions [1], apart from those obeyed by the LQG controllers of Chapter 7. However, the advantages of an H_∞ controller lie in its automatic loop shaping as function of the weighting matrices, and hence it is a direct, one-step procedure for addressing both performance and robustness. Here, the weighting matrices are the design parameters to play with. Glover and Doyle [1] provide an efficient algorithm for solving the mixed-sensitivity H_∞-optimal control problem, which results in *two* algebraic Riccati equations. The algorithm imposes the restriction

$$\|\gamma \mathbf{M}(i\omega)\|_\infty \leq 1 \tag{9.12}$$

where γ is a scaling factor, to be determined by the optimization process. The *Robust Control Toolbox* [2] of MATLAB contains the algorithm of Glover and Doyle [1] in an M-file called *hinfopt.m*, which iterates for γ until a stabilizing solution satisfying Eq. (9.12) is obtained.

Example 9.1

Consider the design of an *active flutter-suppression* system for a flexible aircraft wing [3] with the aeroelastic plant's linear, time-invariant model given by the following transfer function:

$Y(s)/U(s)$

$$= \frac{[-79.67s^{12} - 2242s^{11} - 4.160s^{10} - 630\,500s^9 - 5.779 \times 10^6 s^8 - 4.007 \times 10^7 s^7 - 1.933}{[s^{12} + 39.27s^{11} + 21\,210s^{10} + 4.677 \times 10^5 s^9 + 5.682 \times 10^6 s^8 + 5.233 \times 10^7 s^7 + 3.485 \times 10^8 s^6}$$
$$\frac{\times 10^8 s^6 - 5.562 \times 10^8 s^5 - 8.377 \times 10^8 s^4 - 6.098 \times 10^8 s^3 - 1.698 \times 10^8 s^2 + 0.0001853s]}{+ 1.528 \times 10^9 s^5 + 4.289 \times 10^9 s^4 + 7.636 \times 10^9 s^3 + 8.155 \times 10^9 s^2 + 4.683 \times 10^9 s + 1.101x10^9]}$$

The single-input, single-output, 12th order plant has a *trailing-edge control-surface deflection* as the input, $U(s)$, and *normal acceleration* at a sensor location as the output, $Y(s)$. The design objective is to minimize the sensitivity matrix at frequencies below 1 rad/sec, while achieving approximately 20 dB/decade roll-off above 8 rad/sec, for suppression of *flutter*. (*Flutter* is a destructive structural instability of aircraft wings.) In the present case, the weighting matrices are chosen as follows

$$\mathbf{W_1}(s) = (s^2 + 2s + 1)/(s^2 + 60s + 900) \tag{9.13}$$

$$\mathbf{W_2}(s) = (0.01s + 0.1)/(s + 0.1) \tag{9.14}$$

$$\mathbf{W_3}(s) = (0.010\,533s + 3.16)/(0.1s + 1) \tag{9.15}$$

Note that the weighting matrices are scalar functions. The stable solution to the mixed-sensitivity H_∞-optimal control problem is solved by the MATLAB's Robust Control Toolbox [2] function *hinfopt*, which iteratively searches for the optimum value of γ, and is carried out by the following MATLAB statements:

```
>>sysp=ss(sysp); [a,b,c,d]=ssdata(sys); % plant's state-space model
  <enter>

>> w1=[1 2 1;1 60 900]; w2=[0.01 0.1;1 0.1]; w3=[0.010533 3.16;0.1 1];
  % frequency weights <enter>

>>[A,B1,B2,C1,C2,D11,D12,D21,D22]=augtf(a,b,c,d,w1,w2,w3);
  % 2-port augmented system <enter>

>>[gamopt,acp,bcp,ccp,dcp,acl,bcl,ccl,dcl]=hinfopt
  (A,B1,B2,C1,C2,D11,D12,D21,D22); <enter>

<<H-Infinity Optimal Control Synthesis >>
No   Gamma      D11<=1  P-Exist  P>=0  S-Exist  S>=0  lam(PS)<1  C.L.
-----------------------------------------------------------------------
1    1.0000e+000  OK  OK       OK    OK       FAIL  OK         UNST
2    5.0000e-001  OK  OK       OK    OK       FAIL  OK         UNST
3    2.5000e-001  OK  OK       OK    OK       OK    OK         STAB
4    3.7500e-001  OK  OK       OK    OK       OK    OK         STAB
5    4.3750e-001  OK  OK       OK    OK       FAIL  OK         UNST
6    4.0625e-001  OK  OK       OK    OK       OK    OK         STAB
7    4.2188e-001  OK  OK       OK    OK       FAIL  OK         UNST
8    4.1406e-001  OK  OK       OK    OK       FAIL  OK         UNST
9    4.1016e-001  OK  OK       OK    OK       FAIL  OK         UNST

Iteration no. 6 is your best answer under the tolerance: 0.0100.
```

Hence, the optimum value of $\gamma = 0.40\,625$ for a *design tolerance* of 0.01, is obtained in nine iterations. A stable closed-loop initial response (Figure 9.2), and a 2.2 percent increase in the flight-velocity at which flutter occurs at standard sea-level are the results of this H_∞ design [3].

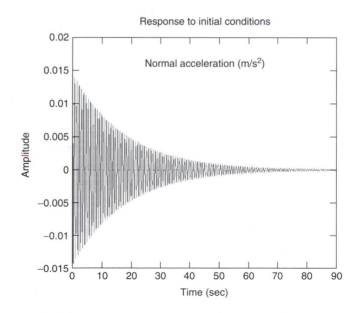

Figure 9.2 Closed-loop initial response of the H_∞-based active flutter suppression system

9.3 Structured Singular Value Synthesis for Robust Control

In Chapter 7, as well as in the previous section, we treated variations in a plant's model – due to factors such as unmodeled dynamics, linearization errors, and parametric uncertainties – as the random *process noise*. However, in certain cases (just as there is a method in madness), it is possible to identify a *structure* in the *uncertainty* in the plant's transfer matrix, $\mathbf{G}(s)$. The plant's *actual transfer matrix*, $\mathbf{G}(s)$, can be then written as

$$\mathbf{G}(s) = \mathbf{G_o}(s) + \Delta_{\mathbf{A}}(s) \tag{9.16}$$

or

$$\mathbf{G}(s) = \mathbf{G_o}(s)[\mathbf{I} + \Delta_{\mathbf{MI}}(s)] \tag{9.17}$$

or

$$\mathbf{G}(s) = [\mathbf{I} + \Delta_{\mathbf{Mo}}(s)]\mathbf{G_o}(s) \tag{9.18}$$

where $\mathbf{G_o}(s)$ is the *nominal plant transfer matrix* (arrived at by certain physical modeling), $\Delta_{\mathbf{A}}(s)$ is the *additive structured uncertainty*, $\Delta_{\mathbf{MI}}(s)$ is the *multiplicative structured uncertainty occurring at the plant's input*, and $\Delta_{\mathbf{Mo}}(s)$ is the *multiplicative structured uncertainty occurring at the plant's output*. Equations (9.16)–(9.18) provide *three* alternative ways in which a plant with structured uncertainty can be expressed. In the LQG/LTR method of Chapter 7 and H_∞-optimal control of Section 9.2, we had assumed that the uncertainty in the plant's model (denoted by the process noise) was random, or *unstructured*. A compensator design made robust with respect to an unstructured uncertainty is unduly conservative, because the uncertainty in a physical plant model is usually structured. It is a task for the designer to identify the structure in the uncertainty ($\Delta_{\mathbf{A}}(s)$, $\Delta_{\mathbf{MI}}(s)$, or $\mathbf{\Delta_{Mo}}(s)$), and make the compensator robust with respect to the structured uncertainty. The most common structured uncertainty model is a *block-diagonal* structure for the additive uncertainty, given by

$$\Delta_{\mathbf{A}}(s) = \mathbf{diag}\{\Delta_{\mathbf{1}}(s), \ldots, \Delta_{\mathbf{m}}(s)\} \tag{9.19}$$

which indicates a matrix with m *blocks*, $\Delta_{\mathbf{1}}(s), \ldots, \Delta_{\mathbf{m}}(s)$, occurring on the *diagonal*. Each block, $\Delta_j(s)$, represents a *different* structure associated with a particular uncertainty, and could be either a scalar or a matrix. By employing a block-diagonal structure, it is assumed that the uncertainty of *each set of elements* of $\mathbf{G_o}(s)$ is independent of the uncertainty for the others.

Robustness of closed-loop systems with respect to the structured uncertainty in the plant is most directly addressed by the *structured singular value*, introduced by Doyle [4]. For a plant with an additive uncertainty matrix, $\mathbf{\Delta_A}(s)$, which is assumed to have a block-diagonal structure given by Eq. (9.19), with each uncertainty block being *stable* and obeying

$$\|\Delta_{\mathbf{j}}(i\omega)\|_\infty \le \partial \tag{9.20}$$

where ∂ is a selected positive number less than unity.

We can treat the uncertain closed-loop system as the *compensated nominal plant* with transfer matrix, $\mathbf{P}_{11}(s) = \mathbf{G_o}(s)\mathbf{H}(s)[\mathbf{I} + \mathbf{G_o}(s)\mathbf{H}(s)]^{-1}$ (where $\mathbf{H}(s)$ is the compensator's transfer matrix), and an additional *feedback loop* with a *fictitious controller*, $\Delta_\mathbf{A}(s)$. The additional feedback loop consists of the uncertain variables, $\mathbf{U}_1(s)$ and $\mathbf{Y}_1(s)$, that are the input and the output, respectively, of the fictitious controller, $\Delta_\mathbf{A}(s)$. Then we can write the uncertain closed-loop system in the following partitioned form:

$$\begin{bmatrix} \mathbf{Y}(s) \\ \mathbf{U}_1(s) \end{bmatrix} = \begin{bmatrix} \mathbf{P}_{11}(s) & \mathbf{P}_{12}(s) \\ \mathbf{P}_{21}(s) & \mathbf{P}_{22}(s) \end{bmatrix} \begin{bmatrix} \mathbf{Y_d}(s) \\ \mathbf{Y}_1(s) \end{bmatrix} \tag{9.21}$$

The *structured singular value* of the matrix, $\mathbf{P}_{22}(s)$, in the frequency domain is defined as

$$\mu(\mathbf{P}_{22}(i\omega)) = \begin{bmatrix} 0; & \text{if } \det[\mathbf{I} - \mathbf{P}_{22}(i\omega)\Delta_\mathbf{j}(i\omega)] \neq 0 \text{ for any } \Delta_\mathbf{j}(i\omega) \\ 1/\min_j[\sigma_{\max}(\Delta_\mathbf{j}(i\omega)]; & \text{if } \det[\mathbf{I} - \mathbf{P}_{22}(i\omega)\Delta_\mathbf{j}(i\omega)] = 0 \end{bmatrix} \tag{9.22}$$

where $\det[\cdot]$ denotes the determinant. Computation of $\mu(\mathbf{P}_{22}(i\omega))$ is made possible by the following lower and upper bounds [4]:

$$\max_U \rho(\mathbf{U}\mathbf{P}_{22}(i\omega)) \leq \mu(\mathbf{P}_{22}(i\omega)) \leq \inf_D[\sigma_{\max}(\mathbf{D}\mathbf{P}_{22}(i\omega)\mathbf{D}^{-1})] \tag{9.23}$$

where $\mathbf{U}$ is a *unitary matrix* (i.e. a matrix with the property $\mathbf{U}\mathbf{U}^\mathbf{H} = \mathbf{I}$) of the *same* block-diagonal structure as $\Delta_\mathbf{A}(s)$, $\mathbf{D}$ is a real, diagonal positive definite matrix, with the following structure:

$$\mathbf{D} = \mathbf{diag}\{d_1\mathbf{I}, \ldots, d_m\mathbf{I}\} \tag{9.24}$$

(where $d_j > 0$), $\inf_D[\cdot]$ is the *infimum function*, and denotes the *minimum value* of the matrix within the square brackets, *with respect to* the matrix $\mathbf{D}$, and $\rho(\mathbf{P})$ denotes the *spectral radius of a square matrix* $\mathbf{P}$, defined as

$$\rho(\mathbf{P}) = \max_j |\lambda_j(\mathbf{P})| \tag{9.25}$$

with $\lambda_j(\mathbf{P})$ denoting an eigenvalue of $\mathbf{P}$. It can be shown using Eq. (9.25) and the definition of the singular values that the spectral radius obeys the following inequality:

$$|\lambda(\mathbf{P})| \leq \rho(\mathbf{P}) \leq \sigma_{\max}(\mathbf{P}) \tag{9.26}$$

Doyle [4] shows that the lower bound in Eq. (9.23) is actually an *equality*, and the upper bound is an equality *if* there are *no more* than three blocks in $\Delta_\mathbf{A}(s)$ with *no repetitions*. Equation (9.26) thus provides *two different* methods of computing $\mu(\mathbf{P}_{22}(i\omega))$, which is, however, a formidable task requiring *nonlinear optimization*.

The objective of the *structural singular value synthesis* (also called *μ-synthesis*) is to find a stabilizing controller, $\mathbf{H}(s)$, and a diagonal scaling matrix, $\mathbf{D}$, such that

$$\|\mathbf{D}\mathbf{P}_{22}(i\omega)\mathbf{D}^{-1}\|_\infty < 1 \tag{9.27}$$

However, it is more practical to assign a mixed-sensitivity frequency weighted control of the type given in the previous section, and replace $\mathbf{P}_{22}(i\omega)$ by the mixed-sensitivity matrix, $\mathbf{M}(i\omega)$, given by Eq. (9.8), such that

$$\|\mathbf{D}\mathbf{M}(i\omega)\mathbf{D}^{-1}\|_\infty < 1 \tag{9.28}$$

Since $\|\mathbf{D}\mathbf{M}(i\omega)\mathbf{D}^{-1}\|_\infty$ is the upper bound of $\mu(\mathbf{M}(i\omega))$ (Eq. (9.23)), it implies that Eq. (9.28) is sufficient to ensure both stability and performance robustness.

The μ-synthesis problem requires nonlinear optimization, and MATLAB provides the M-file *musyn.m* in the Robust Control Toolbox [2] for solving the combined structured/unstructured uncertainty problem. Also, a dedicated toolbox called the *μ-Analysis and Synthesis Toolbox* [5] is available for use with MATLAB for the computation of structured singular values, as well as for analyzing and designing robust control systems for plants with uncertainty using structured singular values. The iterative procedure used in μ-synthesis is based upon repetitive solution for the stabilizing H_∞ controller, and the associated diagonal scaling matrix, $\mathbf{D}$, which minimizes the structured singular value given by Eq. (9.22).

Example 9.2

Reconsider the active flutter-suppression system of Example 9.1. For μ-synthesis, two alternative choices of the disturbance frequency weight, $\mathbf{W_1}(s)$, are:

$$\mathbf{W_1}(s) = (s^2 + 2s + 1)/(s^2 + 60s + 900) \tag{9.29}$$

and

$$\mathbf{W_1}(s) = (0.01s + 1)/(0.1s + 1) \tag{9.30}$$

while $\mathbf{W_2}(s)$ and $\mathbf{W_3}(s)$ are taken to be zeros. While the two performance specifications are similar at frequencies above the open-loop flutter frequency (8.2 rad/s), their shapes are different at frequencies less than 5 rad/s. Let us begin with the first expression for $\mathbf{W_1}(s)$ and obtain an optimal controller by μ-synthesis, using the Robust Control Toolbox functions *augtf* to construct the two-port augmented state-space plant model, and *musyn* to iterate for μ-synthesis as follows:

```
>>[A,B1,B2,C1,C2,D11,D12,D21,D22]=augtf(a,b,c,d,[1 2 1;1 60 900],[],[]);
  <enter>

>>[acp,bcp,ccp,dcp,mu,logd,ad,bd,cd,dd,gam]=
  musyn(A,B1,B2,C1,C2,D11,D12,D21,D22,logspace(-2,2)); <enter>
```

```
        << H-Infinity Optimal Control Synthesis >>
No  Gamma         D11<=1  P-Exist  P>=0   S-Exist  S>=0  lam(PS)<1  C.L.
-----------------------------------------------------------------------
1   1.0000e+000   OK      OK       OK     OK       OK    OK         UNST
2   5.0000e-001   OK      FAIL     FAIL   OK       OK    OK         STAB
3   2.5000e-001   OK      OK       OK     OK       OK    OK         UNST
4   1.2500e-001   OK      FAIL     FAIL   OK       OK    OK         UNST
5   6.2500e-002   OK      OK       OK     OK       OK    OK         UNST
6   3.1250e-002   OK      FAIL     OK     OK       OK    OK         UNST
7   1.5625e-002   OK      OK       OK     OK       OK    OK         UNST
```

```
8   7.8125e-003  OK    FAIL    OK    OK    OK    OK      UNST
9   3.9063e-003  OK    OK      OK    OK    OK    OK      UNST
10  1.9531e-003  OK    OK      OK    OK    OK    OK      UNST
11  9.7656e-004  OK    OK      OK    OK    OK    OK      UNST
12  4.8828e-004  OK    FAIL    OK    OK    OK    OK      UNST
13  2.4414e-004  OK    FAIL    FAIL  OK    OK    OK      STAB
14  1.2207e-004  OK    OK      OK    OK    OK    OK      STAB
15  1.8311e-004  OK    OK      OK    OK    OK    OK      UNST
16  1.5259e-004  OK    FAIL    FAIL  OK    OK    OK      UNST
17  1.3733e-004  OK    OK      FAIL  OK    OK    OK      UNST
18  1.2970e-004  OK    OK      OK    OK    OK    OK      UNST
19  1.2589e-004  OK    FAIL    FAIL  OK    OK    OK      STAB
20  1.2398e-004  OK    OK      OK    OK    OK    OK      STAB
21  1.2493e-004  OK    OK      OK    OK    OK    OK      UNST
```

```
Iteration no. 20 is your best answer under the tolerance: 0.0100.

Executing SSV.....................Done SSV
```

Hence, the optimum γ value for the first value of $\mathbf{W}_1(s)$ (Eq. (9.29)) is $\gamma = 1.2398 \times 10^{-4}$. Next we design a μ-synthesis controller for the second value of $\mathbf{W}_1(s)$ (Eq. (9.30)) as follows:

```
>>[A,B1,B2,C1,C2,D11,D12,D21,D22]=augtf(a,b,c,d,[0.01 1;0.1 1],[],[]);
    <enter>
```

```
>>[acp,bcp,ccp,dcp,mu,logd,ad,bd,cd,dd,gam]=
    musyn(A,B1,B2,C1,C2,D11,D12,D21,D22,logspace(-2,2)); <enter>
```

```
<<H-Infinity Optimal Control Synthesis >>
No   Gamma         D11<=1  P-Exist  P>=O  S-Exist  S>=0  lam(PS)<1  C.L.
------------------------------------------------------------------------
1    1.0000e+000   OK      OK       OK    OK       OK    OK         UNST
2    5.0000e-001   OK      OK       OK    OK       OK    OK         STAB
3    7.5000e-001   OK      OK       OK    OK       OK    OK         STAB
4    8.7500e-001   OK      OK       FAIL  OK       OK    OK         UNST
5    8.1250e-001   OK      OK       OK    OK       OK    OK         STAB
6    8.4375e-001   OK      FAIL     FAIL  OK       OK    OK         UNST
7    8.2813e-001   OK      OK       FAIL  OK       OK    OK         UNST
8    8.2031e-001   OK      FAIL     FAIL  OK       OK    OK         UNST
```

```
Iteration no. 5 is your best answer under the tolerance: 0.0100.

Executing SSV..........................Done SSV
```

Hence, the second design has $\gamma = 0.8125$. Let us compare the structured singular value plots of the two designs in Figure 9.3, by plotting the returned *mu* vectors in dB on a semilog frequency scale. We see that both the designs achieve the same structured singular value at higher frequencies, while at lower frequencies the second design ($\gamma = 0.8125$) produces a slightly smaller value of μ. However, the range of variation of μ is very small in Figure 9.3, which indicates that the two designs converge to the same result.

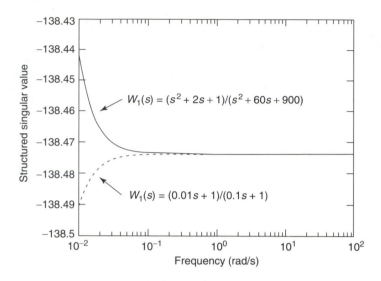

Figure 9.3   Structured singular value (dB) plots of the two optimal μ-synthesis controllers for active flutter-suppression of an aircraft wing

Many illustrative examples of how various control design techniques handle the problem of structured singular value synthesis, especially for robust flight control, can be found in Magni, et al. [6].

9.4 Time-Optimal Control with Pre-shaped Inputs

A topic of current interest is the development of *pre-shaped* inputs for controlling well modeled plants. The term *pre-shaped inputs* (or *input shaping*) refers to an open-loop control system, where the input vector, $\mathbf{u}(t)$, is pre-determined from certain strategy other than feedback. As an alternative to feedback control, input shaping offers a simpler control mechanism for implementation. However, being an open-loop design, input shaping is more sensitive to errors in modeling the plant dynamics when compared to a feedback design. Hence, input shaping requires a better modeling of a physical plant compared to the closed-loop control systems.

A common strategy for generating pre-shaped inputs is the *time-optimal control*. Consider a linear, time-invariant plant described by the following state-equation:

$$\mathbf{x}^{(1)}(t) = \mathbf{A}\mathbf{x}(t) + \mathbf{B}\mathbf{u}(t) \tag{9.31}$$

The time-optimal control refers to moving the system from an *initial state*, $\mathbf{x}(0)$, to a *final state*, $\mathbf{x}(t_f)$, such that the time taken, t_f, is the minimum. In other words, the time-optimal control input vector, $\mathbf{u}(t)$, minimizes the following objective function:

$$J(0, t_f) = \mathbf{x}^{\mathbf{T}}(t_f)\mathbf{V}\mathbf{x}(t_f) + \int_0^{t_f} d\tau \tag{9.32}$$

subject to the constraint that the state-vector, $\mathbf{x}(t)$, must obey Eq. (9.31). Since there is usually a physical limit on the magnitude of control inputs that can be generated by a given actuator, each element, $u_j(t)$, of the input vector, $\mathbf{u}(t)$, must obey an additional constraint given by

$$|u_j(t)| \leq m_j \tag{9.33}$$

where m_j denotes the upper bound on $u_j(t)$. A solution to the time-optimal control problem given by Eqs. (9.31)–(9.33) is obtained by applying the *minimum-principle* [7] (also see Section 8.11), which minimizes $J(0, t_f)$ only at a *finite number* of *discrete* time points [8] (rather than in continuous time). Hence, the solution to the time-optimal control problem posed above results in a *digital* input – even for an analog plant. If the final state is zero, i.e. $\mathbf{x}(t_f) = \mathbf{0}$, then each element of the time-optimal control input, $\mathbf{u}^*(t)$, can be written as follows [8]:

$$u_j^*(t) = -m_j \ \text{sgn}[\mathbf{B}_j^{\mathrm{T}}\mathbf{p}^*(t)] \tag{9.34}$$

where sgn[·] denotes the *vector signum function* (defined as 1 when the corresponding element of the vector within square brackets is *positive*, and −1 when the corresponding element of the vector within square brackets is *negative*), $\mathbf{B}_j^{\mathrm{T}}$ denotes the jth row of $\mathbf{B}^{\mathrm{T}}$, and $\mathbf{p}^*(t)$ is the optimal *co-state vector*, which is defined as the solution of the following *co-state equation*:

$$\mathbf{p}^{*(1)}(t) = -\mathbf{A}^{\mathrm{T}}\mathbf{p}^*(t) \tag{9.35}$$

Equation (9.34) indicates that each input vector element, $u_j^*(t)$, assumes values of $\pm m_j$ depending upon the sign of $\mathbf{B}_j^{\mathrm{T}}\mathbf{p}^*(t)$, which oscillates about 0. Such a time-optimal control thus exerts the *maximum possible* control input magnitude until $\mathbf{x}(t_f) = \mathbf{0}$ is reached, and is given the name *bang-bang input*. An alternative way of expressing the bang-bang input sequence given by Eq. (9.34) is by a *sampling switch* (see Section 8.1), which turns on and off at a fixed sampling interval, T. The bang-bang input sequence is thus regarded as an *ideal sampler* which produces a series of *pulses*, $u_j^*(t)$, weighted by the input value, $f_j(kT)$, according to Eq. (8.2), re-written as follows:

$$u_j^*(t) = \sum_{k=0}^{\infty} f_j(kT)\delta(t - kT) \tag{9.36}$$

where $f_j(kT) = -m_j \ \text{sgn}[\mathbf{B}_j^{\mathrm{T}}\mathbf{p}^*(kT)]$.

Considering the fact that the bang-bang input is constant within each sampling interval, Eq. (9.36) is said to *convolve* a series of impulses, $A(k)\delta(t - kT)$, with a step input, $m_j u_s(t)$, where $A(k)$ is the amplitude of the impulse applied at the kth sampling instant. Such a statement can be used with the MATLAB function *kron* which multiplies *two sequences*, in order to obtain the bang-bang input sequence as follows:

$$u_j^*(t) = m_j u_s(t)^* A(k)\delta(t - kT) \tag{9.37}$$

where * denotes the *convolution* of the step function, $m_j u_s(t)$, with the weighted impulse sequence, $A(k)\delta(t - kT)$, and $A(k)$ is given by

$$A(k) = 1; \quad k = 0, t_f/(kT)$$

$$A(k) = 2(-1)^k; \quad k = 1, \ldots, t_f/(kT) - 1 \tag{9.38}$$

It can be shown [8] that if the eigenvalues of $\mathbf{A}$ are real, then each component of the optimal control vector, $\mathbf{u}^*(t)$, can switch from m_j to $-m_j$, or from $-m_j$ to m_j, *at most* $(n-1)$ times, where n is the order of the plant. For example, a second order plant requires an optimal bang-bang input sequence that switches only once between $t = 0$ to $t = t_f$. Since the sampling interval is a constant for the bang-bang input, it follows that for a second order plant, $T = t_f/2$. Hence, switching occurs at the middle of the control input sequence, and the impulse magnitudes at the various values of k are given as follows for a second order plant:

$$\begin{bmatrix} k \\ A(k) \end{bmatrix} = \begin{bmatrix} 0 & 1 & 2 \\ 1 & -2 & 1 \end{bmatrix} \tag{9.39}$$

Note that the bang-bang input sequence only moves the plant to a final *zero* state. If the final state, $\mathbf{x}(t_f)$, is *non-zero*, the bang-bang input sequence must be modified.

A practical application of the time-optimal pre-shaped inputs is in the control of flexible structures, where it is required to make the *vibration* in the structure (i.e. *velocity* and *acceleration* of flexible modes), zero at the end of the input sequence. There are many possible ways to develop such vibration suppression pre-shaped inputs, which has been the subject of active research for many years. However, since input shaping is an open-loop control, only those input shapes can be successfully implemented that are robust with respect to variations in the plant's model. Singer and Seering [9] proposed the following *robust*, weighted impulse sequence, $A(k)\delta(t - kT)$, based on linear dynamics to reduce the residual vibration of a flexible structure with one rigid-body mode and N *flexible modes*, and of total mass, m:

$$u_j^*(t) = m_j u_s(t)^* A(k_0)\delta(t - k_0 T_0)^* A(k_1)\delta(t - k_1 T_1)^* A(k_2)\delta(t - k_2 T_2)^* \ldots^*$$
$$A(k_N)\delta(t - k_N T_N) \tag{9.40}$$

where k_0 and $A(k_0)$ are the *bang-bang* impulse sequence given by Eq. (9.39), and k_i and $A(k_i)$ denote the sampling instants and impulse amplitudes, respectively, for the ith *flexible mode* of the structure given by

$$\begin{bmatrix} k_i \\ A(k_i) \end{bmatrix} = \begin{bmatrix} 0 & 1 & 2 \\ 1/4 & 1/2 & 1/4 \end{bmatrix} \tag{9.41}$$

$T_0 = t_f/2$, and $T_i = 2\pi/\omega_i$, where ω_i is the natural frequency of the ith flexible mode. Clearly, t_f is a summation of all the sampling instants of the input sequence given by Eq. (9.40). The convolutions required in Eq. (9.40) can be carried out using MATLAB with the function *kron*. An M-file which generates the input sequence of Eq. (9.40) for a *single-input* plant, assuming a maximum input magnitude of unity, is *inshape.m*, listed in Table 9.1. The M-file *inshape.m* requires the natural frequencies of the structure as a *column vector*, $\mathbf{w}$, and the final time, tf, and returns the input sequence as an amplitude vector, $\mathbf{u}$, and the discrete time vector, $\mathbf{t}$:

```
>>[u,t]=inshape(w,tf) <enter>
```

The M-file *inshape.m* assumes a maximum input magnitude of unity. If the maximum input magnitude is not unity, you should multiply the input sequence, $\mathbf{u}$, by the maximum

Table 9.1 Listing of the M-file *inshape.m* for generating pre-shaped inputs for vibration suppression of a linear flexible structure with a rigid body displacement

---

inshape.m

```
function [a,t]=inshape(w,tf)
%Input shaping for robust, time-optimal control of nth order system
% w= vector of natural frequencies of the structure (rad/s)
% (first element of w must be zero, denoting the rigid body mode)
% tf=final time
% a = returned vector of input magnitudes
% t = returned vector of discrete times for the input sequence
% Copyright (c) 2000 by Ashish Tewari
%size of w vector is n
n=size(w,1);
p=n-1; % p is the no. of flexible modes
% Calculate the time-periods of the flexible modes:-
for i=2:n;
T(i-1)=2*pi/w(i);
end
% Assign the flexible mode impulse sequences:-
for i=1:p;
F1(i,:)=[0 0.5*T(i) T(i)];
ef1(i,:)=exp(F1(i,:));
F2=[0.25 0.5 0.25];
end
% Convolve the impulse sequences for flexible modes:-
a1=zeros(1,3^p);
et=zeros(1,3^p);
size(a1)
if p>1
a1=[kron(F2,F2) zeros(1,3^p-9)];
et=[kron(ef1(1,:),ef1(2,:)) zeros(1,3^p-9)];
else
a1=F2;
et=ef1(1,:);
end
for i=2:p-1;
a1=[kron((nonzeros(a1))',F2) zeros(1,3^p-3^(i+1))];
et=[kron((nonzeros(et))',ef1(i+1,:)) zeros(1,3^p-3^(i+1))];
end
% Assign impulse sequence for the rigid-body mode:-
Ftf=[0 0.5*tf tf];
Fa=[1 -2 1];
% Convolve rigid-body impulse sequence with that of the flexible
% modes:- a0=kron(a1,Fa)';
eft=exp(Ftf);
et1=kron(et,eft);
% Generate input sequence:-
t0=log(et1)';
[t0,i]=sort(t0);
```

(continued overleaf)

Table 9.1   (*continued*)

inshape.m

```
a0=a0(i);
m=size(t0,1);
a2(1)=0;
for j=2:m+1;
a2(j)=a2(j-1)+a0(j-1);
end
eps=1e-10;
t(1)=0;
t(2)=eps;
for j=2:m-1;
t(2*j-1)=t0(j)-eps;
t(2*j)=t0(j);
end
t(2*m-1)=t0(m);
t(2*m)=t0(m)+eps;
a(1)=a2(1);
a(2)=a2(2);
for j=2:m;
a(2*j-1)=a2(j);
a(2*j)=a2(j+1);
end
```

magnitude. It is not required that the frequency vector, **w**, for *inshape.m* should contain *all* the natural frequencies of the plant. Instead, it is a common practice to select a few dominant modes (usually the ones with the *smallest* natural frequencies) for input shaping. However, the higher frequency modes will not be suppressed by an input shaper that is based only on the first few modes. Let us examine the multi-mode input shaping for a simple system with a smooth frequency spectrum – a *train* of *n* equal masses.

Example 9.3

Consider a simple system of a train of n identical masses, m, connected by identical springs of stiffness, k. A force, $f(t)$, applied on the *first* mass is the input to the system, while the output is the displacement of the center of mass, $x = 1/n \sum x_i$, where x_i is the displacement of the ith mass. A state-space representation of the plant can be obtained with the state-vector $\mathbf{x}(t) = [x_1(t); x_2(t); \dots x_n(t); x_1^{(1)}(t); x_2^{(1)}(t); \dots; x_n^{(1)}(t)]^T$. An M-file called *nmass.m*, which generates the state-space representation for a train of n-masses, is given in Table 9.2. Let us select $n = 13$, $m = 1$ kg and $k = 10^5$ N/m, and generate the state-space model and the natural frequencies of the plant in increasing magnitudes as follows:

```
>>[A,B,C,D]=nmass(13,1e5); freq=sort(damp(A))' <enter>

freq =
  Columns 1 through 7
  0.0000 0.0000 76.2341 76.2341 151.3565 151.3565 224.2718
```

Table 9.2 Listing of the M-file *nmass.m* for the state-space model of a train of n identical masses, connected by (n − 1) identical springs

nmass.m

```
function [a,b,c,d]=nmass(n,k)
% system of n unit masses connected with n-1 springs of
% stiffness k, with input force applied to the first mass
% and displacement of center of mass as the output
K=zeros(n);
K(1,:)=k*[1 -1 zeros(1,n-2)];
K(n,:)=k*[zeros(1,n-2) -1 1];
for i=2:n-1;
K(i,:)=k*[zeros(1,i-2) -1 2 -1 zeros(1,n-i-1)];
end
a=[zeros(n) eye(n);-K zeros(n)];
b=[zeros(1,n) 1 zeros(1,n-1)]';
c=[ones(1,n) zeros(1,n)]/n;
d=0;
```

```
Columns 8 through 14
224.2718 293.9167 293.9167 359.2757 359.2757 419.3956 419.3956

Columns 15 through 21
473.3998 473.3998 520.5007 520.5007 560.0116 560.0116 591.3562

Columns 22 through 26
591.3562 614.0775 614.0775 627.8442 627.8442
```

The 26th order plant has one rigid and 12 flexible second-order modes with frequencies 76.2, 151.4, 224.3, 293.9, 359.3, 419.4, 473.4, 520.5, 560.0, 591.4, 614.1, and 627.8 rad/s. Since there is a double eigenvalue at $s = 0$ (denoting the rigid-body mode), the plant is unstable. The damping ratios associated with the flexible modes are all zeros, indicating an undamped plant. Although the plant is undamped, a Bode gain plot of the system would be *smooth* with no resonant peaks. This is because the plant is *minimum phase*, and has all imaginary-axis poles cancelled by the transmission zeros. The control objective is to cancel the residual velocity and acceleration at the end of the input sequence, while keeping the motion of the center of mass, $y(t)$, negligible.

To see the vibration suppression achieved by an input shaper obtained by taking only the first few modes in Eq. (9.40) instead of all 12 flexible modes, let us calculate the input sequence with only the first two flexible modes for $t_f = 0.1$ s, assuming the maximum input magnitude of unity:

```
>>w=freq(1:2:6)'<enter>

w =
     0.0000
    76.2341
   151.3565
```

```
>>[u,t]=inshape(w, 0.1); % input profile with only first
    two flexible modes <enter>
```

The displacement of the center of mass to the pre-shaped input sequence is obtained using *lsim* as follows:

```
>>sys=ss(A,B,C,D); [y,t,X]=lsim(sys,u,t); <enter>
```

whereas the velocity of the first mass is calculated as follows:

```
>>sys1=ss(A,B,[zeros(1,13) 1 zeros(1,12)], 0);
  [x1,t,X]=lsim(sys1,u,t); <enter>
```

The pre-shaped input sequence, $u(t)$, the resulting center of mass displacement, $y(t)$, and the velocity of the first mass, $dx_1(t)/dt$, are plotted in Figure 9.4. Note that the vibration suppression is evident with zero velocity of the first mass at the end of the input sequence. In a practical implementation, input shaping can be combined with a *feedback* tracking system which moves the train by a given displacement, or achieves a given velocity of the center of mass, while suppressing the residual vibration using pre-shaped inputs as a *feedforward controller*.

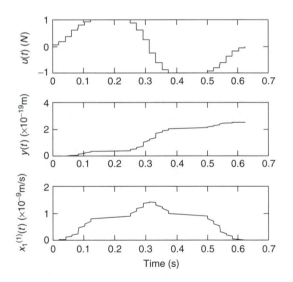

Figure 9.4 Pre-shaped input, $u(t)$, center of mass displacement, $y(t)$, and velocity of the first mass, $x_1^{(1)}(t)$, for a train of 13 identical masses connected by springs

Many investigators [10,11] have applied the input shaping approach to practical problems. Since robustness of input sequences with variations in the plant model is crucial, Singhose, Derezinski and Singer [12] derived an *extra-insensitive* input shaper using on-off reaction jets for time-optimal maneuvers of spacecraft. The robustness of the shaper was addressed [12] by minimizing the *percentage residual vibration* at the end of the input sequence with respect to the frequency of each flexible mode considered.

Input shaping is not limited to linear systems. The time-optimal control principles can be extended to a nonlinear plant for arriving at the pre-shaped inputs [13,14]. Banerjee and Singhose [14] presented an application of time-optimal input shaping to reduce nonlinear vibration in a *two-link flexible robotic manipulator*, while using either a linear or a nonlinear feedback control for tracking. The input shaping was based on two flexible modes. Combining a feedback controller with a feedforward input shaping is a practical method of compensating for the loss of information about nonlinear system dynamics when only a few flexible modes are included in input shaping. However, the implementation of such compensators is greatly dependent on the robustness of the feedback controller. Also, while optimizing for the hybrid feedback/input-shaping controllers, time optimality cannot be guaranteed.

In the several approaches presented in the literature, the adequacy of input shaping for time-optimal control with one or two flexible modes has been demonstrated. Expectedly, by including a larger number of flexible modes, N, a smoother input profile and a reduced transient vibration can be obtained. However, if input profiles must be stored, memory could become a factor when the number of flexible modes is large. The size of the impulse sequence of Eq. (9.39) is 3^N, where N is the number of flexible modes retained in input shaping. In Example 9.3, an input sequence of nine impulse sequence was obtained using two flexible modes.

9.5 Output-Rate Weighted Linear Optimal Control

The optimal linear quadratic regulator (LQR) problem (Chapter 6) is the at the heart of many modern optimal, robust control design methods, such as the linear quadratic Gaussian procedure with loop-transfer recovery (LQG/LTR) (see Chapter 7), and H_∞ control (Section 9.2). The output-weighted LQR problem (referred to as LQRY) (see Section 6.4) minimizes an objective function containing the quadratic form of the measured output as an integrand. However, in several applications it may be more desirable to minimize the *time-rate of change* of the output, rather than the measured output itself. Examples of such active control applications are *vibration reduction* of flexible structures, *flutter-suppression* (Examples 9.1, 9.2), *gust and maneuver load alleviation* of aircraft and *ride-quality augmentation* of any vehicle. In these cases, the measured output is usually the *normal acceleration*, while it is necessary from considerations of passenger/crew comfort (as well as issues such as weapons aiming and delivery) to have an optimal controller that minimizes the *roughness* of the motion, which can be defined as the time-rate of change of normal acceleration. This mechanical analogy can be extended to other physical systems, where sensor limitations prevent the measurement of time-rate of change of an available signal, and even to economic models. Also, it is well known that certain *jerky, nonlinear* motions, if uncorrected, can lead to chaos [15]. Optimal control of such motions may require *output-rate weighted* (ORW) objective functions.

Consider a linear, time-invariant system described by the following state and output equations:

$$\mathbf{x}^{(1)}(t) = \mathbf{A}\mathbf{x}(t) + \mathbf{B}\mathbf{u}(t) \tag{9.42}$$

$$\mathbf{y}(t) = \mathbf{C}\mathbf{x}(t) + \mathbf{D}\mathbf{u}(t) \tag{9.43}$$

The regulator design problem is finding an optimal feedback gain matrix, $\mathbf{K}$, which obeys the control-law:

$$\mathbf{u}(t) = -\mathbf{K}\mathbf{x}(t) \tag{9.44}$$

such that the following infinite-time, *output-rate weighted* objective function is minimized:

$$J = \int_0^\infty [\mathbf{y}^{(1)}(t)]^T \mathbf{Q} \mathbf{y}^{(1)}(t) + \mathbf{u}^T(t)\mathbf{R}\mathbf{u}(t)] \, dt \tag{9.45}$$

Equation (9.45) can be re-written as

$$J = \int_0^\infty [\mathbf{x}^T(t)\mathbf{Q_c}\mathbf{x}(t) + \mathbf{x}^T(t)\mathbf{S}^T\mathbf{U}(t) + \mathbf{U}^T(t)\mathbf{S}\mathbf{x}(t) + \mathbf{U}^T(t)\mathbf{R_c}\mathbf{U}(t)] \, dt \tag{9.46}$$

where

$$\mathbf{Q_c} = \mathbf{A}^T\mathbf{C}^T\mathbf{Q}\mathbf{C}\mathbf{A}; \quad \mathbf{S} = \mathbf{D_c^T}\mathbf{Q}\mathbf{C}\mathbf{A}; \quad \mathbf{R_c} = \mathbf{R} + \mathbf{D_c^T}\mathbf{Q}\mathbf{D_c}; \quad \mathbf{D_c} = [\mathbf{CB}; \mathbf{D}];$$

$$\mathbf{U}(t) = [\mathbf{u}(t); \quad \mathbf{u}^{(1)}(t)]^T \tag{9.47}$$

It is to be noted that the output weighted LQRY control (Section 6.4) is a sub-case of the general *output-rate weighted* (ORW) control described by Eqs. (9.42)–(9.47). For *strictly proper* plants ($\mathbf{D} = \mathbf{0}$), the performance integral in Eq. (9.46) reduces to a form similar to of the general LQRY problem. Using steps similar to those of Section 6.4, the optimal regulator gain matrix, $\mathbf{K}$, which minimizes the objective function of Eq. (9.46) subject to Eq. (9.40), is given by

$$\mathbf{K} = \mathbf{R}^{-1}(\mathbf{B}^T\mathbf{M} + \mathbf{S}) \tag{9.48}$$

where $\mathbf{M}$ is the solution of the following *algebraic Riccati equation*:

$$\mathbf{MA_c} + \mathbf{A_c}^T\mathbf{M} - \mathbf{MBR}^{-1}\mathbf{B}^T\mathbf{M} + \mathbf{Q_c} - \mathbf{S}^T\mathbf{RS} = \mathbf{0} \tag{9.49}$$

in which

$$\mathbf{A_c} = \mathbf{A} - \mathbf{BR}^{-1}\mathbf{S} \tag{9.50}$$

is the regulator state-dynamics matrix. After a regulator is designed by the above procedure, an observer (or Kalman filter) can be designed with the usual procedure of Chapters 5 or 7, and combined to form a compensator using the separation principle.

Example 9.5

The longitudinal dynamics of the flexible bomber airplane (Example 4.7) is selected as an example to demonstrate output-rate weighted (ORW) optimal control. For the bomber airplane, it is necessary that all transient motion defined by the normal acceleration, $y_1(t)$, and the pitch-rate, $y_2(t)$, should quickly decay to zero when a vertical gust is encountered, since the stability of the bombing platform is crucial for weapons aiming and delivery. The ORW controller minimizes the *time-rate of change of normal acceleration*, $y_1^{(1)}(t)$, and the *pitch-acceleration*, $y_2^{(1)}(t)$. For this strictly proper plant, the best case ORW regulator is obtained by taking $\mathbf{Q} = 10^{-8}\mathbf{I}$ and $\mathbf{R} = \mathbf{I}$, and the optimal ORW regulator gain matrix, $\mathbf{K}$, the solution to the algebraic Riccati equation, $\mathbf{M}$, and the closed-loop eigenvalues, $\mathbf{E}$, are calculated using the MATLAB command *lqry* as follows:

```
>>Dc=C*B; Cc=C*A; [K,M,E]=lqry(A,B,Cc,Dc, 1e-8*eye(2),eye(2))
  <enter>

K=
    0.0297    0.0070    0.0010    0.0001   -0.7041   -0.4182
    0.0557   -0.0074    0.0015   -0.0004   -0.2539   -0.1670

M=
    0.1531    0.0023    0.0044   -0.0002    0.1084    0.0480
    0.0023    0.0036    0.0001    0.0001   -0.0137   -0.0061
    0.0044    0.0001    0.0001    0.0000    0.0029    0.0013
   -0.0002    0.0001    0.0000    0.0000   -0.0006   -0.0003
    0.1084   -0.0137    0.0029   -0.0006    0.1945    0.0871
    0.0480   -0.0061    0.0013   -0.0003    0.0871    0.0390

E=
   -94.2714
    -9.6068
    -1.3166+5.0317i
    -1.3166-5.0317i
    -0.4260+1.8738i
    -0.4260-1.8738i
```

You can compare the closed-loop initial response of the ORW regulator designed above, with that of the traditional LQRY regulator designed with the same weighting matrices, **Q** and **R**. Generally, when applied to flexible structures, the ORW regulator produces a much *smoother response* – with smaller overshoots – which decays faster than that of the corresponding LQRY regulator [16, 17]. The application of ORW optimal control is not limited to flexible structures, and can be extended to any plant where smoothening of the transient response is critical. However, in some applications the sensitivity to noise may increase with ORW based compensators, and must be carefully studied before implementing such controllers.

9.6 Nonlinear Optimal Control

Up to this point, we have confined our attention to the design of linear control systems – such as the classical approach of Chapter 2, the pole-placement state-space design of Chapter 5, and the optimal, linear, state-space design of Chapter 6. In Chapter 2, we had seen how nonlinear systems can be linearized by assuming *small amplitude* motion about an *equilibrium point*. Hence, linear control design techniques can be used for controlling nonlinear plants *linearized* about an equilibrium point. However, when we are interested in *large amplitude* motion of a nonlinear plant – either about an equilibrium point, or *between* two equilibrium points (such as the motion of a pendulum going from the equilibrium point at $\theta = 0°$ to that at $\theta = 180°$) – the linearization of the plant is invalid, and one has to grapple with the nonlinear model of the plant. The control design strategy for a nonlinear plant can be based upon either *linear* or *nonlinear* feedback control laws.

Since we cannot talk about the *poles* of a nonlinear plant, there can be no *nonlinear* pole-placement design approach analogous to the methods of Chapter 5. A possible nonlinear control strategy could be to come-up with a nonlinear feedback control law by *trial and error*, that meets the closed-loop design requirements verified by carrying out a nonlinear simulation (using the techniques of Section 4.6). Surely, such an approach cannot be called a *design strategy*, due to its *ad hoc* nature. An alternative design approach is to *transform* the nonlinear plant into a linear system using an appropriate *feedback* control law, and then treat the linearized system by the *linear control design* strategies covered up to this point. Such a procedure is referred to as *feedback linearization*. Feedback linearization requires a very complicated (generally nonlinear) feedback control law, which is quite sensitive to parametric uncertainties [18]. For a time-varying nonlinear plant, the use of feedback linearization requires time-dependent scheduling of the feedback linearization control law – called *adaptive feedback linearization* [19]. It is clear that the success of feedback linearization is limited to those nonlinear plants which are feedback linearizable. Fortunately, there is another nonlinear control design strategy, called *nonlinear optimal control*, which can be applied generally to control a nonlinear plant.

Nonlinear optimal control is carried out in a manner similar to the linear optimal control of Chapter 6 by minimizing an objective function formulated in terms of the energy of motion and the control input energy. However, nonlinear optimal control – owing to the nonlinear nature of the governing differential equations and control laws – is mathematically more complex than the linear optimal control of Chapter 6. It is beyond the scope of this book to give details of the nonlinear optimal control theory, and you are referred to Kirk [20] and Bellman [21] for the general derivation of the nonlinear optimal control problem.

Let us consider a nonlinear plant with the following state-equation:

$$\mathbf{x}^{(1)}(t) = \mathbf{f}\{\mathbf{x}(t), \mathbf{u}(t), t\} \tag{9.51}$$

where $\mathbf{x}(t)$ is the state-vector, $\mathbf{u}(t)$ is the input vector, and $\mathbf{f}\{\mathbf{x}(t), \mathbf{u}(t), t\}$ denotes a nonlinear vector function involving the state variables, the inputs, and time, t. The solution of Eq. (9.51) with a known input vector, $\mathbf{u}(t)$, was discussed in Section 4.6, with some special conditions to be satisfied by the nonlinear function, $\mathbf{f}\{\mathbf{x}(t), \mathbf{u}(t), t\}$ for the existence of the solution, $\mathbf{x}(t)$, such as the *continuity in time* and the *Lipschitz condition* given by Eq. (4.85). Suppose such conditions are satisfied, and we can solve Eq. (9.51) for $\mathbf{x}(t)$ if we are specified $\mathbf{u}(t)$ and the initial-conditions, $\mathbf{x}(0)$. For simplicity, let us assume that the nonlinear plant is *time-invariant*, i.e. $\mathbf{f}\{\mathbf{x}(t), \mathbf{u}(t), t\} = \mathbf{f}\{\mathbf{x}(t), \mathbf{u}(t)\}$, and the state-equation does not explicitly depend upon the time, t, in Eq. (9.51). Furthermore, let us assume for simplicity that the nonlinear function, $\mathbf{f}\{\mathbf{x}(t), \mathbf{u}(t)\}$, can be expressed in the following form:

$$\mathbf{f}\{\mathbf{x}(t), \mathbf{u}(t)\} = \mathbf{A}\mathbf{x}(t) + \mathbf{B}\mathbf{u}(t) + \mathbf{F}\{\mathbf{x}(t)\} \tag{9.52}$$

where $\mathbf{F}\{\mathbf{x}(t)\}$ is a nonlinear vector function that depends *only* upon the state-vector, $\mathbf{x}(t)$. Equation (9.52) implies that there are no nonlinear terms involving the control input, $\mathbf{u}(t)$, in the state-equation. Such nonlinear plants are fairly common in applications such as robotics, spacecraft attitude control, bio-chemical dynamics, and economics [18]. Then a nonlinear *regulator* problem for *infinite-time* can be posed by finding an optimal control input, $\mathbf{u}(t)$, such that the following objective function is minimized:

$$J = \int_0^\infty [q(\mathbf{x}(t)) + \mathbf{u}^T(t)\mathbf{R}\mathbf{u}(t)]\, dt \qquad (9.53)$$

where $q\{\mathbf{x}(t)\}$ is a *positive semi-definite* function denoting the cost associated with the transient response, $\mathbf{x}(t)$, and $\mathbf{u}^T(t)\mathbf{R}\mathbf{u}(t)$ is the *quadratic* cost associated with the control input, $\mathbf{u}(t)$, with the matrix, $\mathbf{R}$, being *symmetric* and *positive definite*. It can be shown by the *minimum principle* [7] that if *all* the *derivatives* of $\mathbf{F}\{\mathbf{x}(t)\}$ *with respect to* $\mathbf{x}(t)$ are *continuous* in the space formed by the elements of $\mathbf{x}(t)$, then the minimization of the objective function, J, given by Eq. (9.53) with respect to the control input vector, $\mathbf{u}(t)$, *subject to the constraint* that the state-vector, $\mathbf{x}(t)$, satisfies Eq. (9.51), is *equivalent* to the minimization of the following scalar function, called the *Hamiltonian*, with respect to the control input, $\mathbf{u}(t)$:

$$\mathcal{H} = q\{\mathbf{x}(t)\} + \mathbf{u}^T(t)\mathbf{R}\mathbf{u}(t) + [dV\{\mathbf{x}(t)\}/d\mathbf{x}(t)][\mathbf{A}\mathbf{x}(t) + \mathbf{B}\mathbf{u}(t) + \mathbf{F}\{\mathbf{x}(t)\}] \qquad (9.54)$$

where $V\{\mathbf{x}(t)\}$ is a *positive semi-definite* function with the property $V\{\mathbf{0}\} = 0$, called the *Lyapunov function*. Note that J, H, $q\{\mathbf{x}(t)\}$, and $V\{\mathbf{x}(t)\}$ are all *scalars*. (We have seen scalar functions of a vector in Chapter 6, such as $\mathbf{x}^T(t)\mathbf{Q}\mathbf{x}(t)$ and $\mathbf{u}^T(t)\mathbf{R}\mathbf{u}(t)$.) However, $dV\mathbf{x}(t)/d\mathbf{x}(t)$ is a *row-vector* (sometimes expressed as $dV\mathbf{x}(t)/d\mathbf{x}(t) = \mathbf{p}^T(t)$, where $\mathbf{p}(t)$ is called the *co-state vector* of the nonlinear system); the derivative of a scalar, $V\{\mathbf{x}(t)\}$, with respect to a vector, $\mathbf{x}(t)$, means the differentiation of the scalar, $V\{\mathbf{x}(t)\}$, by each element of the vector, $\mathbf{x}(t)$, and storing the result as the corresponding element of a vector of the same size as $\mathbf{x}(t)$ (see Appendix B). Note that the Hamiltonian, $\mathcal{H}$, includes the term $[\partial V\{\mathbf{x}(t)\}/\partial\mathbf{x}(t)][\mathbf{A}\mathbf{x}(t) + \mathbf{B}\mathbf{u}(t) + \mathbf{F}\{\mathbf{x}(t)\}]$, which can be seen as imposing a *penalty* on deviating from the state-equation, Eq. (9.51). Thus, the constraint of Eq. (9.51) is implicitly satisfied by minimizing the Hamiltonian, $\mathcal{H}$. The *necessary conditions* of optimal control can be expressed as follows [20]:

$$\partial\mathcal{H}/\partial\mathbf{u}(t) = \mathbf{0} \qquad (9.55)$$

$$\mathcal{H}_{\text{min}} = 0 \qquad (9.56)$$

$$\mathbf{x}^{(1)}(t) = \partial\mathcal{H}/\partial\mathbf{p}(t)|_{\mathbf{u}(t)=\mathbf{u}^*(t)} \qquad (9.57)$$

$$\mathbf{p}^{(1)}(t) = -\partial\mathcal{H}/\partial\mathbf{x}(t)|_{\mathbf{u}(t)=\mathbf{u}^*(t)} \qquad (9.58)$$

where $\mathbf{p}(t) = [dV\{\mathbf{x}(t)\}/d\mathbf{x}(t)]^T$ (the co-state vector), and $\mathbf{u}^*(t)$ denotes the optimal control input (which minimizes $\mathcal{H}$). The result of Eq. (9.55) is the following expression for the *optimal* control input:

$$\mathbf{u}^*(t) = -\mathbf{R}^{-1}\mathbf{B}^T[\partial V\{\mathbf{x}(t)\}/\partial\mathbf{x}(t)]^T \qquad (9.59)$$

It is clear from Eq. (9.59) that we must know the *Lyapunov function*, $V\{\mathbf{x}(t)\}$, (or the co-state vector $\mathbf{p}(t) = [dV\{\mathbf{x}(t)\}/d\mathbf{x}(t)]^T$) if we have any chance of finding the optimal control input, $\mathbf{u}^*(t)$. The Lyapunov function (or the co-state vector) depends upon the characteristics of the nonlinear plant. Mathematically, $\mathbf{p}(t)$, can be obtained from the coupled solution to the two-point boundary-value problem posed by Eqs. (9.57) and (9.58), which is not always possible to obtain analytically. However, since $\mathbf{p}(t)$ must also satisfy Eq. (9.56), which is an algebraic equation analogous to the *algebraic Riccati* equation for the linear problem, Eq. (9.56) gives a practical method of finding $\mathbf{p}(t) = [dV\{\mathbf{x}(t)\}/d\mathbf{x}(t)]^T$. Equation (9.56) is generally known as the *Hamilton–Bellman–Jacobi* equation (or, in short, the *Bellman* equation [21]) for the infinite-time problem. (For *finite-time* control, the Bellman equation becomes a *partial differential* equation analogous to the *matrix Riccati* equation for the linear problem.)

Once we have posed the optimal control problem by Eqs. (9.54)–(9.59), we can look at a workable solution procedure [22] which uses Eq. (9.56) to derive $dV\{\mathbf{x}(t)\}/d\mathbf{x}(t)$. However, in using such a procedure, we must specify a form for $V\{\mathbf{x}(t)\}$. Let us begin by expressing the transient response energy, $q\{\mathbf{x}(t)\}$, in the following form:

$$q\{\mathbf{x}(t)\} = (1/2)[\mathbf{x}^T(t) \ \{\mathbf{x}^2(t)\}^T \dots \{\mathbf{x}^n(t)\}^T] \begin{bmatrix} \mathbf{Q}_{11} & \mathbf{Q}_{12} & \dots & \mathbf{Q}_{1n} \\ \mathbf{Q}_{21} & \mathbf{Q}_{22} & \dots & \mathbf{Q}_{2n} \\ \cdot & \cdot & \dots & \cdot \\ \mathbf{Q}_{n1} & \mathbf{Q}_{n2} & \dots & \mathbf{Q}_{nn} \end{bmatrix} \begin{bmatrix} \mathbf{x}(t) \\ \mathbf{x}^2(t) \\ \cdot \\ \mathbf{x}^n(t) \end{bmatrix} \quad (9.60)$$

where $\mathbf{Q}_{ij} = \mathbf{Q}_{ji}$, and $\mathbf{x}^k(t)$ denotes a vector formed by raising each element of $\mathbf{x}(t)$ to the power k. This definition of $q\{\mathbf{x}(t)\}$ makes it positive semi-definite, as required, with a proper selection of $\mathbf{Q}_{ij}$. Note that $(2n - 1)$ is the highest power of $\mathbf{x}(t)$ which appears in the nonlinear function, $\mathbf{F}\{\mathbf{x}(t)\}$, which can be written in the following form:

$$\mathbf{F}\{\mathbf{x}(t)\} = \sum_{k=2}^{(2n-1)} \mathbf{F}_{\mathbf{k}}\{\mathbf{x}(t)\} \quad (9.61)$$

where $\mathbf{F}_k\{\mathbf{x}(t)\}$ denotes a *nonlinearity* of power k. Similarly, Eq. (9.60) can be re-written as

$$q\{\mathbf{x}(t)\} = \sum_{k=2}^{2n} q_k\{\mathbf{x}(t)\} \quad (9.62)$$

where

$$q_2\{\mathbf{x}(t)\} = \mathbf{x}^T(t)\mathbf{Q}_{11}\mathbf{x}(t); \quad q_3\{\mathbf{x}(t)\} = \mathbf{x}^T(t)\mathbf{Q}_{12}\mathbf{x}^2(t) + \{\mathbf{x}^2(t)\}^T\mathbf{Q}_{21}\mathbf{x}(t);$$

$$q_4\mathbf{x}(t) = \mathbf{x}^T(t)\mathbf{Q}_{13}\mathbf{x}^3(t) + \{\mathbf{x}^2(t)\}^T\mathbf{Q}_{22}\mathbf{x}^2(t) + \{\mathbf{x}^3(t)\}^T\mathbf{Q}_{31}\mathbf{x}(t);$$

$$\dots \quad (9.63)$$

$$q_{2n}\{\mathbf{x}(t)\} = \{\mathbf{x}^n(t)\}^T\mathbf{Q}_{nn}\mathbf{x}^n(t)$$

To determine a structure for the Lyapunov function, $V\{\mathbf{x}(t)\}$, that ensures its positive semi-definiteness, and satisfies the property $V\{\mathbf{0}\} = 0$, it is assumed that $V\{\mathbf{x}(t)\}$ has the *same form* as $q\{\mathbf{x}(t)\}$, i.e.

$$V\{\mathbf{x}(t)\} = (1/2)[\mathbf{x}^T(t) \ \{\mathbf{x}^2(t)\}^T \dots \{\mathbf{x}^n(t)\}^T] \begin{bmatrix} \mathbf{P}_{11} & \mathbf{P}_{12} & \dots & \mathbf{P}_{1n} \\ \mathbf{P}_{21} & \mathbf{P}_{22} & \dots & \mathbf{P}_{2n} \\ \cdot & \cdot & \dots & \cdot \\ \mathbf{P}_{n1} & \mathbf{P}_{n2} & \dots & \mathbf{P}_{nn} \end{bmatrix} \begin{bmatrix} \mathbf{x}(t) \\ \mathbf{x}^n(t) \\ \cdot \\ \mathbf{x}^2(t) \end{bmatrix}$$

$$(9.64)$$

where $\mathbf{P}_{ij} = \mathbf{P}_{ji}$. $V\{\mathbf{x}(t)\}$ can also be expressed as

$$V\{\mathbf{x}(t)\} = \sum_{k=2}^{2n} V_k\{\mathbf{x}(t)\} \tag{9.65}$$

where

$$V_2\{\mathbf{x}(t)\} = \mathbf{x}^T(t)\mathbf{P}_{11}\mathbf{x}(t); \quad V_3\{\mathbf{x}(t)\} = \mathbf{x}^T(t)\mathbf{P}_{12}\mathbf{x}^2(t) + \{\mathbf{x}^2(t)\}^T\mathbf{P}_{21}\mathbf{x}(t);$$

$$V_4\{\mathbf{x}(t)\} = \mathbf{x}^T(t)\mathbf{P}_{13}\mathbf{x}^3(t) + \{\mathbf{x}^2(t)\}^T\mathbf{P}_{22}\mathbf{x}^2(t) + \{\mathbf{x}^3(t)\}^T\mathbf{P}_{31}\mathbf{x}(t);$$

$$\cdots \tag{9.66}$$

$$V_{2n}\{\mathbf{x}(t)\} = \{\mathbf{x}^n(t)\}^T\mathbf{P}_{nn}\mathbf{x}^n(t)$$

Based on the expressions of $\mathbf{F}\{\mathbf{x}(t)\}$, $q\{\mathbf{x}(t)\}$, and $V\{\mathbf{x}(t)\}$ given by Eqs. (9.61), (9.62) and (9.65), the minimum value of the Hamiltonian can be written using Eq. (9.56) as

$$\mathcal{H}_{\min} = \sum_{k=2}^{2n} H_k = 0 \tag{9.67}$$

where

$$H_2 = q_2\{\mathbf{x}(t)\} - [dV_2\{\mathbf{x}(t)\}/d\mathbf{x}(t)]\mathbf{B}\mathbf{R}^{-1}\mathbf{B}^T[dV_2\{\mathbf{x}(t)\}/d\mathbf{x}(t)]^T$$

$$+ [dV_2\{\mathbf{x}(t)\}/d\mathbf{x}(t)]\mathbf{A}\mathbf{x}(t) = 0 \tag{9.68}$$

$$H_3 = q_3\{\mathbf{x}(t)\} - [dV_2\{\mathbf{x}(t)\}/d\mathbf{x}(t)]\mathbf{B}\mathbf{R}^{-1}\mathbf{B}^T[dV_3\{\mathbf{x}(t)\}/d\mathbf{x}(t)]^T$$

$$- [dV_3\{\mathbf{x}(t)\}/d\mathbf{x}(t)]\mathbf{B}\mathbf{R}^{-1}\mathbf{B}^T[dV_2\{\mathbf{x}(t)\}/d\mathbf{x}(t)]^T + [dV_3\{\mathbf{x}(t)\}/d\mathbf{x}(t)]\mathbf{A}\mathbf{x}(t)$$

$$+ [dV_2\{\mathbf{x}(t)\}/d\mathbf{x}(t)]\mathbf{F}_2\{\mathbf{x}(t)\} = 0 \tag{9.69}$$

$$H_4 = q_4\{\mathbf{x}(t)\} - [dV_2\{\mathbf{x}(t)\}/d\mathbf{x}(t)]\mathbf{B}\mathbf{R}^{-1}\mathbf{B}^T[dV_4\{\mathbf{x}(t)\}/d\mathbf{x}(t)]^T$$

$$- [dV_3\{\mathbf{x}(t)\}/d\mathbf{x}(t)]\mathbf{B}\mathbf{R}^{-1}\mathbf{B}^T[dV_3\{\mathbf{x}(t)\}/d\mathbf{x}(t)]^T$$

$$- [dV_4\{\mathbf{x}(t)\}/d\mathbf{x}(t)]\mathbf{B}\mathbf{R}^{-1}\mathbf{B}^T[dV_2\{\mathbf{x}(t)\}/d\mathbf{x}(t)]^T + [dV_4\mathbf{x}(t)/d\mathbf{x}(t)]\mathbf{A}\mathbf{x}(t)$$

$$+ [dV_3\mathbf{x}(t)/d\mathbf{x}(t)]\mathbf{F}_2\mathbf{x}(t) + [dV_2\mathbf{x}(t)/d\mathbf{x}(t)]\mathbf{F}_3\mathbf{x}(t) = 0 \tag{9.70}$$

$$\cdots$$

$$H_{2n} = q_{2n}\{\mathbf{x}(t)\} - [dV_2\{\mathbf{x}(t)\}/d\mathbf{x}(t)]\mathbf{B}\mathbf{R}^{-1}\mathbf{B}^T[dV_{2n}\{\mathbf{x}(t)\}/d\mathbf{x}(t)]^T - \cdots$$

$$- [dV_{2n-1}\{\mathbf{x}(t)\}/d\mathbf{x}(t)]\mathbf{B}\mathbf{R}^{-1}\mathbf{B}^T[dV_3\{\mathbf{x}(t)\}/d\mathbf{x}(t)]^T$$

$$- [dV_{2n}\{\mathbf{x}(t)\}/d\mathbf{x}(t)]\mathbf{B}\mathbf{R}^{-1}\mathbf{B}^T[dV_2\{\mathbf{x}(t)\}/d\mathbf{x}(t)]^T + [dV_{2n}\{\mathbf{x}(t)\}/d\mathbf{x}(t)]\mathbf{A}\mathbf{x}(t)$$

$$+ [dV_{2n-1}\{\mathbf{x}(t)\}/d\mathbf{x}(t)]\mathbf{F}_2\{\mathbf{x}(t)\} + \cdots + [dV_2\{\mathbf{x}(t)\}/d\mathbf{x}(t)]\mathbf{F}_{2n-1}\{\mathbf{x}(t)\} = 0 \tag{9.71}$$

The Lyapunov parameters $\mathbf{P_{ij}}$ are determined by the following procedure [22]:

(a) Equation (9.68) – which has the *linear part* of the Hamiltonian, H_2 – can be expressed as the following *algebraic Riccati* equation:

$$\mathbf{A^T P_{11} + P_{11} A + Q_{11} - P_{11} B R^{-1} B^T P_{11} = 0} \qquad (9.72)$$

Equation (9.72) is solved by standard procedures of Chapter 6 to get $\mathbf{P_{11}}$. Note that the *linear feedback* control input for the plant is given by $\mathbf{u}(t) = -\mathbf{R^{-1} B^T P_{11} x}(t)$.

(b) $\mathbf{P_{11}}$ is substituted into Eq. (9.69) to determine $\mathbf{P_{12}}(= \mathbf{P_{21}})$.

(c) $\mathbf{P_{11}}$ and $\mathbf{P_{12}}$ are substituted into Eq. (9.70) to determine $\mathbf{P_{13}}(= \mathbf{P_{31}})$ and $\mathbf{P_{22}}$.

(d) Continue successive substitution of known $\mathbf{P_{ij}}$ into Eqs. (9.70)–(9.71) until *all parameters*, including $\mathbf{P_{nn}}$, are found.

When all Lyapunov parameters are calculated, they are substituted into Eq. (9.64) to determine $V\{\mathbf{x}(t)\}$, which is then differentiated with respect to $\mathbf{x}(t)$ and substituted into Eq. (9.59) to determine the nonlinear optimal feedback control input, $\mathbf{u^*}(t)$.

Example 9.6

Consider the wing-rock dynamics of a fighter airplane described in Example 4.13 by nonlinear state-equations, Eq. (4.94), and programmed in the M-file *wrock.m* which is listed in Table 4.8. The state-space matrices (Eq. (9.52)) of the time-invariant, nonlinear plant are as follows:

$$\mathbf{A} = \begin{bmatrix} 0 & 1 & 0 & 0 & 0 \\ -0.02013 & 0.0105 & 1 & -0.02822 & -0.1517 \\ 0 & 0 & -20.202 & 0 & 0 \\ 0 & 0 & 0 & 0 & 1 \\ 0 & 0.0629 & 0 & -1.3214 & -0.2491 \end{bmatrix};$$

$$\mathbf{B} = \begin{bmatrix} 0 \\ 0 \\ 20.202 \\ 0 \\ 0 \end{bmatrix} \qquad (9.73)$$

$$\mathbf{F}\{\mathbf{x}(t)\} = \begin{bmatrix} 0; & \{0.026 x_2^3(t) - 0.1273 x_1^2(t) x_2(t) + 0.5197 x_1(t) x_2^2(t)\}; & 0; & 0; & 0 \end{bmatrix}^T$$

The *limit cycle* wing-rock motion when the airplane is excited by the initial condition $\mathbf{x}(0) = [0.2; 0; 0; 0; 0]^T$, was illustrated in Figures 4.17–4.19. Since the highest power of the elements of $\mathbf{x}(t)$ in the expression for $\mathbf{F}\{\mathbf{x}(t)\}$ in the present wing-rock model is 3, it follows that $n = 2$, and only Eqs. (9.68)–(9.70) are required for determining the unknown Lyapunov parameters $\mathbf{P_{11}}, \mathbf{P_{12}}(= \mathbf{P_{21}})$, and $\mathbf{P_{22}}$. Here $V_4\{\mathbf{x}(t)\} = (1/2)\{\mathbf{x}^2(t)\}^T \mathbf{P_{22} x}^2(t)$, which implies $\mathbf{P_{13}} = \mathbf{P_{31}} = \mathbf{0}$. Following the nonlinear control derivation procedure of the previous section, the non-zero elements of the nonlinear control Lyapunov parameters are determined to be the following [23]:

$$\mathbf{P}_{12}(3, 1) = [0.02013\mathbf{Q}_{12}(1, 3) + \mathbf{Q}_{12}(1, 1)]/[416.34\mathbf{P}_{11}(3, 1)$$

$$+ 8.2162\mathbf{P}_{11}(3, 3) + 0.4067] \tag{9.74}$$

$$\mathbf{P}_{12}(3, 4) = -\mathbf{P}_{11}(3, 4)\mathbf{P}_{12}(3, 1)/\mathbf{P}_{11}(3, 1) \tag{9.75}$$

$$\mathbf{P}_{12}(3, 5) = \mathbf{P}_{11}(3, 5)\mathbf{P}_{12}(3, 4)/\mathbf{P}_{11}(3, 4) \tag{9.76}$$

$$\mathbf{P}_{12}(2, 1) = 49.6771[\mathbf{Q}_{12}(1, 1) - 408.12\mathbf{P}_{11}(3, 1)\mathbf{P}_{12}(3, 1)] \tag{9.77}$$

$$\mathbf{P}_{12}(3, 2) = [\mathbf{P}_{12}(3, 1)(1 - 408.12\mathbf{P}_{11}(3, 2)$$

$$+ \mathbf{P}_{12}(2, 1)]/(408.12\mathbf{P}_{11}(3, 1) + 0.02013) \tag{9.78}$$

$$\mathbf{P}_{12}(3, 3) = [\mathbf{Q}_{12}(3, 1) - 40.404\mathbf{P}_{12}(3, 1)(20.202\mathbf{P}_{11}(3, 3) + 1)$$

$$- 0.02013\mathbf{P}_{12}(3, 2)]/[1224.4\mathbf{P}_{11}(3, 1) \tag{9.79}$$

$$\mathbf{P}_{22}(3, 1) = -\mathbf{P}_{12}(3, 1)^2/\mathbf{P}_{11}(3, 1) \tag{9.80}$$

$$\mathbf{P}_{22}(3, 2) = [0.026\mathbf{P}_{11}(3, 3) - 816.24\mathbf{P}_{12}(3, 2)\mathbf{P}_{12}(3, 1)]/(816.24\mathbf{P}_{11}(3, 1)) \tag{9.81}$$

$$\mathbf{P}_{22}(3, 3) = -3\mathbf{P}_{12}(3, 1)\mathbf{P}_{12}(3, 3)/\mathbf{P}_{11}(3, 1) \tag{9.82}$$

$$\mathbf{P}_{22}(3, 4) = -\mathbf{P}_{12}(3, 1)\mathbf{P}_{12}(3, 4)/\mathbf{P}_{11}(3, 1) \tag{9.83}$$

$$\mathbf{P}_{22}(3, 5) = -\mathbf{P}_{12}(3, 1)\mathbf{P}_{12}(3, 5)/\mathbf{P}_{11}(3, 1) \tag{9.84}$$

The nonlinear optimal control input is then obtained to be the following:

$$\mathbf{u}^*(t) = -20.202[\mathbf{P}_{11}(3, 1)x_1(t) + \mathbf{P}_{11}(3, 2)x_2(t) + \mathbf{P}_{11}(3, 3)x_3(t) + \mathbf{P}_{11}(3, 4)x_4(t)$$

$$+ \mathbf{P}_{11}(3, 5)x_5(t) + \mathbf{P}_{12}(3, 1)x_1^2(t) + \mathbf{P}_{12}(3, 2)x_2^2(t) + 2\mathbf{P}_{12}(3, 1)x_1(t)x_3(t)$$

$$+ 2\mathbf{P}_{12}(3, 2)x_2(t)x_3(t) + 3\mathbf{P}_{12}(3, 3)x_3^2(t) + 2\mathbf{P}_{12}(3, 4)x_3(t)x_4(t)$$

$$+ 2\mathbf{P}_{12}(3, 5)x_3(t)x_5(t) + \mathbf{P}_{12}(3, 4)x_4^2(t) + \mathbf{P}_{12}(3, 5)x_5^2(t)$$

$$+ 2\mathbf{P}_{22}(3, 1)x_1^2(t)x_3(t) + 2\mathbf{P}_{22}(3, 2)x_2^2(t)x_3(t) + 2\mathbf{P}_{22}(3, 3)x_3^3(t)$$

$$+ 2\mathbf{P}_{22}(3, 4)x_3(t)x_4^2(t) + 2\mathbf{P}_{22}(3, 5)x_3(t)x_5^2(t)] \tag{9.85}$$

The cost parameters $\mathbf{Q}_{11}$ and $\mathbf{Q}_{12}$ are selected such that for a *large initial condition*, such as $\mathbf{x}(0) = [1; 1; 0; 0; 0]^T$, the resulting closed-loop aileron response is limited to $\pm 35°$ and all the transients subside within 50 s. Furthermore, stability and performance robustness with respect to a 10 per cent variation in nonlinear aerodynamic and actuator parameters [23] must be ensured. The $\mathbf{Q}_{11}$, $\mathbf{Q}_{12}$ and $\mathbf{Q}_{22}$ matrices to achieve these specifications are the following:

$$\mathbf{Q}_{11} = \begin{bmatrix} 0.01 & 0 & 0 & 0 & 0 \\ 0 & 0.01 & 0 & 0 & 0 \\ 0 & 0 & 1 & 0 & 0 \\ 0 & 0 & 0 & 0.001 & 0 \\ 0 & 0 & 0 & 0 & 0.001 \end{bmatrix}; \quad \mathbf{Q}_{12} = 0.005\mathbf{I}_{3\times3};$$

$$\mathbf{Q}_{22} = 0.1\mathbf{I}_{2\times2} \tag{9.86}$$

where $\mathbf{I}_k$ denotes a $(k \times k)$ identity matrix. The resulting solution, $\mathbf{P}_{11}$, of the algebraic Riccati equation (Eq. (9.73)) is the following [23]:

$$\mathbf{P}_{11} = \begin{bmatrix} 0.0499 & 0.1081 & 0.0037 & 0.0124 & -0.0076 \\ 0.1081 & 0.6927 & 0.0241 & 0.0952 & -0.0295 \\ 0.0037 & 0.0241 & 0.0213 & 0.0033 & -0.0010 \\ 0.0124 & 0.0952 & 0.0033 & 0.0189 & -0.0034 \\ -0.0076 & -0.0295 & -0.0010 & -0.0034 & 0.0057 \end{bmatrix} \qquad (9.87)$$

The nonlinear optimal feedback closed-loop response for the initial condition $\mathbf{x}(0) = [1; 1; 0; 0; 0]^T$ is shown in Figure 9.5. For this initial condition, the linear feedback control input given by $\mathbf{u}(t) = -\mathbf{R}^{-1}\mathbf{B}^T\mathbf{P}_{11}\mathbf{x}(t)$ fails to stabilize the plant [23]. In these figures, comparison is made for 10 per cent variation in the nonlinear aerodynamic parameters of $\mathbf{F}\{\mathbf{x}(t)\}$ from their nominal values given in Eq. (9.73). It is observed that the system's response is stable and within the specified performance limits for 10 per cent uncertainty in the aerodynamic parameters. Tewari [23] also shows robustness with respect to 10 per cent variation in the actuator model.

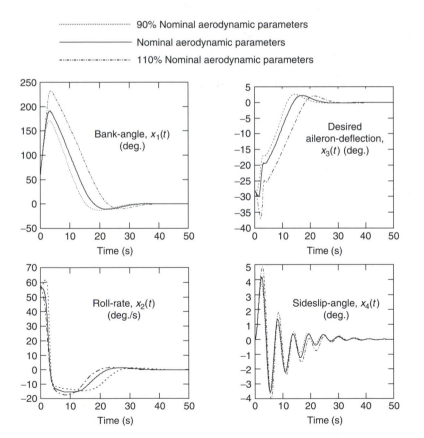

Figure 9.5 Closed-loop response of the nonlinear optimal feedback control system for the wing-rock suppression of a fighter airplane for a large initial condition

Shue, Swan and Rokhsaz [22] introduced the above approach of selecting a positive definite Lyapunov function, and applied it to the wing-rock suppression in a second order system consisting of pure rolling, while Tewari [23] extended the approach to the fifth order system considered in Example 9.6, with additional dynamics of an actuator and the yawing motion – a more realistic model than that of Shue, Swan and Rokhsaz [22]. Assuming a structure for the Lyapunov function that is same as that of the cost function in the performance index [22,23] makes the task of selecting the Lyapunov function easier. The need for nonlinear optimal feedback controller is highlighted in both references [22,23], when it is found that wing-rock suppression by linear feedback control is restricted to only small initial conditions.

Robustness properties of the nonlinear controller with respect to uncertain parameters is an important issue when it is recognized that the nonlinear wing rock aerodynamic model may have significant errors. Such errors generally tend to get amplified by a feedback controller, resulting in a lack of performance and/or stability when implemented in actual conditions. While Tewari [23] ensured robustness with respect to parametric uncertainty by iteratively selecting the controller cost parameters, such that a small variation in the values of aerodynamic and actuator parameters does not lead to a large deviation from nominal performance, more formal methods of guaranteeing robustness in nonlinear optimal control are also available, such as the nonlinear H_∞-optimal control derivation of Wise and Sedwick [24] and van der Schaft [25].

Exercises

9.1. Write a MATLAB M-file for calculating the H_∞-norm of a transfer matrix, $\mathbf{G}(s)$, using the MATLAB command *sigma*. Use the M-file to compute the H_∞-norms of the following transfer matrices:

(a) $\mathbf{G}(s) = [(s+1)/(s^2 + 2s + 3); \quad 1/(s^3 + 7)]$.

(b) $\mathbf{G}(s) = \begin{bmatrix} 1/(s+2) & 0 \\ -1/(s^2 + 3s - 1) & (s+4)/(s+7) \end{bmatrix}$

(c) $\mathbf{G}(s) = \begin{bmatrix} 10(s+1)/(s+10) & 0 & 0 \\ 0 & (10s+1)/[s(s+1)] & 0 \\ 0 & 0 & s/(s+1) \end{bmatrix}$

9.2. Write a MATLAB M-file for estimating the structured singular value of a transfer matrix, $\mathbf{G}(s)$, using the upper bound of Eq. (9.23), when the uncertainty matrix, $\mathbf{\Delta}_A(s)$, has a block-diagonal structure with distinct blocks [4]. Use the M-file to calculate $\mu(\mathbf{G}(i\omega))$ for the following *constant* transfer matrix:

$$\mathbf{G}(i\omega) = \begin{bmatrix} 4 - 2i & -(1+i)/2 & -10 \\ -24 + 6i & 3i & 60 - 80i \\ -6/5 & -(1+i)/5 & 2(1+i) \end{bmatrix}$$

for each of the following block-diagonal structures for a *constant* uncertainty matrix, $\mathbf{\Delta}_A(s)$:

(a) $\Delta_A(s) = \begin{bmatrix} \Delta_1 & 0 & 0 \\ 0 & \Delta_2 & 0 \\ 0 & 0 & \Delta_3 \end{bmatrix}$

(b) $\Delta_A(s) = \begin{bmatrix} \Delta_1 & 0 & 0 \\ 0 & 0 & \Delta_2 \\ 0 & 0 & 0 \end{bmatrix}$

(c) $\Delta_A(s) = \begin{bmatrix} \Delta_1 & 0 & 0 \\ \Delta_2 & 0 & 0 \\ 0 & 0 & \Delta_3 \end{bmatrix}$

(d) $\Delta_A(s) = \begin{bmatrix} \Delta_1 & 0 & 0 \\ \Delta_2 & \Delta_3 & 0 \\ 0 & 0 & \Delta_4 \end{bmatrix}$

where Δ_1, Δ_2, Δ_3, Δ_4 are scalar constants. Compare your results with those given in Doyle [4].

9.3. Using the M-file developed in Exercise 9.2, calculate and plot the structured singular value, $\mu(\mathbf{P_{22}}(i\omega))$, as a function of frequency, ω, for a closed-loop system with the following values of the nominal plant transfer matrix, $\mathbf{G_o}(s)$, compensator transfer matrix, $\mathbf{H}(s)$, and the additive uncertainty matrix, $\Delta_A(s)$:

$$\mathbf{G_o}(s) = \begin{bmatrix} 9/(s+1) & -10/(s+1) \\ -8/(s+2) & 9/(s+2) \end{bmatrix}; \quad \mathbf{H}(s) = 1/(0.0159s)\begin{bmatrix} 9(s+1) & 10(s+2) \\ 8(s+1) & 9(s+2) \end{bmatrix}$$

$$\Delta_A(s) = \begin{bmatrix} \Delta_1 & 0 \\ 0 & \Delta_2 \end{bmatrix}$$

where Δ_1, Δ_2 are scalar constants. What is the *maximum value* of $\mu(\mathbf{P_{22}}(i\omega))$ and what is the value of frequency at which it occurs?

9.4. For the rotating spacecraft of Example 6.2, compute a pre-shaped input sequence assuming the only input to be the hub torque, $u_1(t)$, (i.e. $u_2(t) = u_3(t) = 0$), and retaining only the first two flexible modes in Eq. (9.40), with $t_f = 0.1$ second. Plot the hub-rotation angular displacement and velocity due to the pre-shaped input.

9.5. Repeat Exercise 9.4 with $t_f = 1$ second.

9.6. Repeat Exercises 9.4 and 9.5 with the first six flexible modes retained in the input sequence.

9.7. Devise a nonlinear optimal control-law for stabilizing the inverted pendulum on a moving cart with the nonlinear plant's state-equations given by Eqs. (3.17) and (3.18), when the angular motion of the pendulum is large. Calculate and plot the closed-loop initial response if the initial condition is $x(0) = 0$, $x^{(1)}(0) = 0$, $\theta(0) = 1.0$ rad, $\theta^{(1)}(0) = 0.1$ rad/s.

References

1. Glover, K. and Doyle, J.C. State space formulae for all stabilizing controllers that satisfy an H_∞ norm bound and relations to risk sensitivity. *Systems and Control Letters*, Vol. 11, 1988, pp. 167–172.
2. Chiang, R.Y. and Safonov, M.G. *Robust Control Toolbox*. The Math Works Inc., Natick, MA, 2000.
3. Tewari, A. Robust optimal controllers for active flutter suppression. AIAA Paper 98-4142, *AIAA Guidance, Navigation, and Controls Conference*, Boston, MA, August 10–12 1998.
4. Doyle, J.C. Structured uncertainty in control system design. *Proc. 24th IEEE Conf. on Decision and Control*, Ft. Lauderdale, FL, December 1985, pp. 260–265.
5. Balas, G.J., Doyle, J.C., Glover, K., Packard, A. and Smith, R. *μ-Analysis and Synthesis Toolbox*. The Math Works Inc., Natick, MA, 2000.
6. Magni, J.-F., Bennani, S. and Terlow, J. (eds.) *Robust Flight Control – A Design Challenge*. Lecture Notes in Control and Information Sciences, 224, Springer-Verlag, 1997.
7. Pontryagin, L.S., Boltyanskii, V., Gamkrelidze, R. and Mishchenko, E. *The Mathematical Theory of Optimal Processes*. Interscience, New York, 1962.
8. Meirovitch, L. *Dynamics and Control of Structures*. Wiley, Singapore, 1992.
9. Singer, N.C. and Seering, W.P. Using acausal shaping techniques to reduce robot vibration. *Proc. IEEE Int. Conference on Robotics and Automation*, Philadelphia, PA, April 1988.
10. Banerjee, A.K. Dynamics and control of the WISP shuttle-antennae system. *J. Astronautical Sciences*, Vol. 41, No. 1, 1993, pp. 73–90.
11. Magee, D.P. and Book, W.J. Filtering Schilling manipulator commands to prevent flexible structure vibration. *American Control Conference*, Baltimore, MD, 1994, pp. 2538–2542.
12. Singhose,W., Derezinski, S. and Singer, N.C. Extra-insensitive input shapers for controlling flexible spacecraft. *J. Guidance, Control, and Dynamics*, Vol. 19, No. 2, 1996, pp. 385–391.
13. Gorinevsky, D. and Vukovich, G. Nonlinear input shaping control of flexible spacecraft reorientation maneuver. *J. Guidance, Control, and Dynamics*, Vol. 21, No. 2, 1998, pp. 264–270.
14. Banerjee, A.K. and Singhose, W.E. Command shaping for nonlinear tracking of a two-link flexible manipulator. *Proc. of AAS/AIAA Astrodynamics Conference*, Sun Valley, ID, August 4–7 1997.
15. Linz, S.J. Nonlinear dynamical models and jerky motion. *American J. Physics*, Vol. 65, No. 6, June 1997, pp. 523–526.
16. Tewari, A. Output rate weighted optimal control of aeroelastic systems. *J. Guidance, Control, and Dynamics*, Vol. 24, No. 2, March-April 2001, pp. 409–411.
17. Tewari, A. Output rate weighted active flutter suppression. AIAA Paper 99-4312, *Proc. of AIAA Guidance, Navigation, and Controls Conference*, August 9–11 1999.
18. Nijmeijer, H. and van der Schaft, A.J. *Nonlinear dynamical control systems*. Springer-Verlag, New York, 1990.
19. Singh, S.N. and Steinberg, M. Adaptive control of feedback linearizable nonlinear systems with application to flight control. *J. Guidance, Control, and Dynamics*, Vol. 19, No. 4, 1996, pp. 871–877.
20. Kirk, D.E. *Optimal Control Theory*. Prentice-Hall, Englewood Cliffs, NJ, 1970.
21. Bellman, R. *Dynamic Programming*. Princeton University Press, Princeton, NJ, 1957.
22. Shue, S.P., Swan, M.E. and Rokhsaz, K. Optimal feedback control of nonlinear system: wing rock example. *J. Guidance, Control, and Dynamics*, Vol. 19, No. 1, 1996, pp. 166–171.

23. Tewari, A. Nonlinear optimal control of wing rock including yawing motion. Paper No. AIAA-2000-425, *Proc. of AIAA Guidance, Navigation, and Controls Conference*, Denver, CO, August 14–17 2000.
24. Wise, K.A. and Sedwick, J.L. Nonlinear H_∞ optimal control for agile missiles. *J. Guidance, Control, and Dynamics*, Vol. 19, No. 1, 1996, pp. 157–165.
25. van der Schaft, A.J. L_2-gain analysis of nonlinear systems and nonlinear state feedback H_∞ control, *IEEE Trans. on Auto. Control*, Vol. 37, No. 6, 1992, pp. 770–784.

Appendix A

Introduction to MATLAB, SIMULINK and the Control Systems Toolbox

MATLAB, a registered trademark of MathWorks, Inc. [1], is a high-level programming language which uses *matrices* as the basic numerical entities (rather than *scalars*, as in the low-level programming languages such as BASIC, FORTRAN, PASCAL, and C). In other words, MATLAB allows us to *directly* manipulate matrices – such as adding, multiplying, inverting matrices, and solving for eigenvalues and eigenvectors of matrices (see Appendix B). If similar tasks were to be performed by a low-level programming language, many programming statements constituting scalar operations would be required for even the simplest matrix operations. Hence, MATLAB is ideally suited for linear algebraic computations involving matrices, such as multivariable control design and analysis. Furthermore, MATLAB contains a library of many useful functions – both basic functions (such as trigonometric, hyperbolic, and exponential functions), and specialized mathematical functions – along with an advanced facility for plotting and displaying the results of computations in various graphical forms. In addition, MATLAB is supplemented by various special application *toolboxes*, which contain additional functions and programs. One such toolbox is the *Control System Toolbox*, which has been used throughout this book for solving numerical examples for the design and analysis of modern control systems. The *Control System Toolbox* is available at a small extra cost when you purchase MATLAB, and is likely to be installed at all the computer centres that have MATLAB. If your university (or organization) has a computer center, you can check with them to find out whether the *Control System Toolbox* has been installed with the MATLAB. If you are an engineering/science student, or a practicing engineer, with interest in solving control problems, it is worth having access to both MATLAB and its *Control System Toolbox*, which are available in *student editions* for most platforms supporting WINDOWS or UNIX. In this book, it is assumed that you have the *Control System Toolbox* installed in your MATLAB directory, and we draw upon the special functions and programs contained in the *Control System Toolbox* for solving the numerical examples and exercises. You can devise your own computer programs in MATLAB (called *M-files*) for solving special problems. Some new M-files have been provided in the book for solving a range of control problems.

A.1 Beginning with MATLAB

Here we will discuss how to familiarize ourselves with MATLAB. For further details, you are referred to the MATLAB *User's Guide* [1]. The User's Guide contains necessary information about system requirements, installing and optimizing MATLAB. Once you have MATLAB installed and running on your computer, a *command line* appears on the screen with the prompt (>>) after which you can issue MATLAB commands. After a command is issued at the prompt, you have to press the <enter> key for the command to be executed. All the commands executed at the command line, and the variables computed in those commands, are stored automatically and can be recalled, unless you end the MATLAB session. A MATLAB command consists of one or several MATLAB *statements*. Each MATLAB statement could be of one of the following forms:

```
>> variable = expression
```

or

```
>> expression
```

A *variable* is usually a matrix to be computed, while the *expression* is the mathematical operation by which the variable is to be computed. A *variable* can have a *name* beginning with a letter, followed by up to 18 letters, digits, or underscores. The names in MATLAB are *case sensitive* (i.e. upper and lower case are distinguished). If we omit the '*variable* =' from a statement, MATLAB automatically creates a variable named *ans*, which is abbreviation for *answer*. For example, consider the following matrix, **A**, of size (4×3) and another matrix, **B**, of size (3×2) which have to be multiplied:

$$\mathbf{A} = \begin{bmatrix} -1 & -4 & 0 \\ 18 & 26 & 7 \\ 9 & 6 & -3 \\ 11 & 0 & 4 \end{bmatrix}; \quad \mathbf{B} = \begin{bmatrix} 20 & 8 \\ 2 & 0 \\ 5 & 13 \end{bmatrix} \tag{A.1}$$

We must assign values to the two matrices at the command line by two separate *statements*. Each statement uses *square brackets* to denote the beginning and end of the matrix, and separates two consecutive elements in each row by a *space*. Two consecutive rows are separated by a *semi-colon*. The entire command assigning values to the matrices **A** and **B** is thus issued as follows:

```
>> A=[-1 -4 0; 18 26 7; 9 6 -3; 11 0 4], B=[20 8; 2 0; 5 13] <enter>
```

Note that the two statements in the above command line have been separated by a *comma*. This command produces the following result on the screen:

```
A =
   -1   -4   0
   18   26   7
    9    6  -3
   11    0   4
```

```
B =
   20    8
    2    0
    5   13
```

which confirms that the matrices have been correctly entered. If you do not wish to see the results of your command on the screen, you must end *each statement* in the command line by a semi-colon. For example, the following command will store the matrices **A** and **B** in the memory, but would *not* result in their screen print-out:

```
>> A=[-1 -4 0; 18 26 7; 9 6 -3; 11 0 4]; B=[20 8; 2 0; 5 13]; <enter>
```

The multiplication of the matrices **A** and **B** (already entered into the memory of the MATLAB work-space) can be carried out using the symbol *, and the product, **AB**, can be stored as a *third* matrix, **C**, as follows:

```
>> C = A*B <enter>

C =
  -28    -8
  447   235
  177    33
  240   140
```

In this manner, you can carry out all other basic matrix operations, such as addition, subtraction, transposition, inversion, left-division, right-division, raising a matrix to a power, transcendental and elementary matrix functions, which are briefly described in the following section. Each of the *elements* of a matrix could be a MATLAB *expression*, such as

```
>>x = [1/4 25+sqrt(8.7); 0.5*sin(1.3) 7*log(0.68)] <enter>

x =
   0.2500   27.9496
   0.4818   -2.6996
```

where / denotes division, + denotes addition, sqrt(.) denotes the positive square-root, sin(.) denotes the sine function, and log(.) denotes the natural logarithm. An element of a matrix can be referenced with indices inside parentheses, (i, j), indicating ith row and jth column position, such as

```
>>x(2,2) <enter>

ans =
   -2.6996
```

is the (2, 2) element of the matrix, **x**. If we assign a value to an element of a matrix with indices *larger* than the size of the previously stored value of the same matrix, the size of the matrix is automatically *increased* to the new dimension, and all undefined elements are set to zero. For example,

```
>>x(3,3)=10 <enter>
```

results in the following value of **x**:

```
x =
   0.2500   27.9496    0
   0.4818   -2.6996    0
        0        0    10.0000
```

We can extract smaller matrices from the rows and columns of a larger matrix by using the *colon*. For example, a matrix **y** defined as the matrix formed by taking elements contained in the *first two rows* and the *second and third columns* of **x** is formed as follows:

```
>>y = x(1:2, 2:3) <enter>

y =
   27.9496    0
   -2.6996    0
```

The *colon* can also be used to generate elements with equal spacing, such as

```
>> x = 0:5/4:5 <enter>
```

which results in the following row-vector with elements from 0 to 5 with increments of 1.25:

```
x =
      0 1.2500 2.5000 3.7500 5.0000
```

MATLAB accepts numbers in various *formats*, such as the conventional decimal notation, a power-of-ten scale factor, or a complex unit as a suffix. For example, the following assignment of a matrix is acceptable in MATLAB:

```
>>A = [1 -100i 0.0003; 9.87e5 1.5-4.69j 7.213e-21; 3+5e-4i -7.019e-3 2j] <enter>
```

where i (or j) denotes the imaginary part of a complex number (i.e. square root of -1), and e followed by up to three digits denotes the power of 10 to which a number is raised. The accuracy of floating-point arithmetic in MATLAB is about 16 significant digits, with a range between 10^{-308} and 10^{308}. Any number falling outside this range of floating-point arithmetic is called *NaN*, which stands for *not a number*. You can select from various formats available in MATLAB for printing your results, such as *short* (fixed-point format with 5 digits), *long* (fixed-point format with 15 digits), *short e* (floating-point format with 5 digits), *long e* (floating-point format with 15 digits), *hex* (hexadecimal), and *rat* (numbers approximated by ratios of small integers). The variables *ans* (answer) and *eps*

(2^{-52}) are treated as *permanent variables* in MATLAB, and cannot be cleared or re-assigned in the memory. Some built-in functions return commonly used variables, such as *pi*, and *inf* which stand for π, and ∞, and should not be re-assigned other values in a computation. For ease of programming, MATLAB also provides the function matrices called *ones*, *zeros*, and *eye*, which stand for a matrix of *all* elements equal to 1, a matrix with *all zero* elements, and an *identity* matrix, respectively. The sizes of these matrices can be specified by the user, such as *ones*(3,4), *zeros*(2,6), or *eye*(5).

MATLAB supports help on all its commands, and you can receive help by typing

```
>> help  command <enter>
```

or merely type

```
>>help <enter>
```

to receive information on all the topics on which help is available.

You can save all the variables that you have computed in a work-session by typing

```
>> save <enter>
```

before you end a work-session. This command will save all the computed variables in a file on disk named *matlab.mat*. The next time you begin a session and want to use the previously computed variables, just type

```
>> load <enter>
```

and the work-session saved in *matlab.mat* will be loaded in the current memory. You can also use an optional name of the file in which the work-session is to be saved, such as *fname.mat*, and choose selected variables (rather than all the variables to be saved), such as X, Y, Z by typing the following command:

```
>>save fname X Y Z <enter>
```

The *save* command also lets you import and export *ascii* data files.

A.2 Performing Matrix Operations in MATLAB

In the previous section we saw how MATLAB assigns and multiplies two matrices. The transpose of a matrix, **A**, is simply obtained by using the symbol ′ (*prime*) as follows:

```
>> A′ <enter>
```

If **A** is a complex matrix, then **A′** is the *transpose* of the *complex-conjugate* of **A**. Adding and subtracting matrices (of the same size) is performed simply with + and - symbols, respectively:

```
>> A+B <enter>
```

or

```
>> A-B <enter>
```

If we wish to add or subtract a scalar, *a*, from *each element* of a matrix, **A**, then we can simply type

```
>>A+a <enter>
```

or

```
>>A-a <enter>
```

Dividing a matrix by another matrix is supported by *two matrix division* symbols / and \. The command

```
>>X=A\B <enter>
```

solves the linear algebraic equation $\mathbf{AX} = \mathbf{B}$, provided **A** is a non-singular matrix. The command

```
>>X= B/A <enter>
```

solves the linear algebraic equation $\mathbf{XA} = \mathbf{B}$, provided **A** is a non-singular matrix.

Powers of a matrix can be computed using the symbol ^ as follows:

```
>>A^p <enter>
```

where **A** is a square-matrix, and *p* is a scalar.

Transcendental functions of *individual elements* of matrices can be calculated using in-built MATLAB functions, such as *sin, exp, sqrt, cos, tan, asin, acos, atan, sinh, cosh, asinh, acosh, log, log10, conj, abs, real, imag, sign, angle, gcd, lcm*, etc. These commands, used in the following manner

```
>>exp(A) <enter>
```

produce a matrix whose *elements* are the required transcendental function of the *corresponding elements* of the matrix, **A**. Hence, these transcendental functions are called *array operations*, which are performed on *individual elements* of a matrix, rather than on the matrix *as a whole*. Refer to the MATLAB *Reference Guide* [2] for details on all the in-built transcendental functions available in MATLAB, or issue the *help* command. Other *array operations* are *multiplication* and *division* of the elements of one matrix by the *corresponding* elements of another matrix (of the same size), and *element-by-element powers* of a matrix. A *period* (.) preceding an operator (such as * / \ or ^) denotes an array operation. For example, the command

```
>>C = A.*B <enter>
```

denotes that elements of the matrix, **C**, are products of the elements of the matrices, **A** and **B** (both of the same size), and the command

```
>>C = A.^2 <enter>
```

denotes that the elements of the matrix, **C**, are the squares of the elements of the matrix, **A**.

Some special *matrix transcendental* functions, such as *expm*, *logm*, and *sqrtm* are also available only for square-matrices. These matrix commands have special mathematical significance, such as the *matrix exponential*, *expm*, defined in Chapter 4.

Some useful elementary matrix operations are supported by MATLAB, such as *poly* (the characteristic polynomial), *det* (determinant), *rank* (rank of a matrix), *trace* (matrix trace), *kron* (Kronecker tensor product), *inv* (matrix inverse), *eig* (eigenvalues and eigen-vectors), *svd* (singular-value decomposition), *norm* (1-norm, 2-norm, F-norm, ∞-norm), *rcond* (condition number), *conv* (multiplication of two polynomials), *residue* (partial fraction expansion), *roots* (polynomial roots), etc. For a complete list and details of all the in-built matrix operations available in MATLAB, refer to the MATLAB *Reference Guide* [2], or issue the *help* command.

The *relational* operations comparing two matrices are also supported by the following MATLAB operators: < (less than), <= (less than or equal to), > (greater than), >= (greater than or equal to), == (equal to), ~= (not equal to). Comparing two matrices (of the same size) with the relational operators produces a matrix comprising 1 for each pair of elements for which the relationship is *true*, and zero for each pair of elements for which the relationship is *false*. For example, the command

```
>> [1 2; 0 5] ~= [0 2; 3 5] <enter>
```

results in the following answer:

```
ans =
    1  0
    1  0
```

MATLAB provides special functions, such as *find* and *rem*, which are very useful in relational operations. The function *find* finds the indices of the elements of a vector that satisfy a particular relational condition. When used on a matrix, *find* indexes the elements of the matrix by arranging *all the rows* in a long *column vector*, beginning with the first element of the first row, and ending with the last element of the last row. For example, if we wish to find elements of the matrix, **A**, defined in Eq. (A.1) that are greater than or equal to zero, we can simply issue the following command:

```
>> i=find(A>= 0), A(i)' <enter>

ans =
    18   9   11   26   6   0   0   7   4
```

Note that the vector **i** contains the indices of the elements of **A** that are greater than or equal to zero. Also note that we have printed-out the *transpose* of **A(i)** to save space. The function *rem* is another useful relational function. The command *rem*(A, p) produces a matrix formed by the *remainders* of the elements of a matrix, **A**, when divided by the scalar, p. Suppose we wish to find the locations of the elements of matrix, **A**, defined in Eq. (A.1) which are exactly divisible by 3, and mark these locations by a matrix of ones and zeros, with 1 standing for each element of **A** which is divisible by 3, and 0 standing for those elements that are not divisible by 3. This is simply achieved using the following command:

```
>> rem(A,3)==0 <enter>
```

which results in the following answer:

```
ans =
     0   0   1
     1   0   0
     1   1   1
     0   1   0
```

Other useful relational functions are *isnan* (detect NaNs in a matrix), *isinf* (detect infinities in a matrix), and *finite* (detect finite values in a matrix).

The relational operations in MATLAB are based on the *logical operators*, namely & (and), | (or), ~(not). The logical operations denote *true* by 1 and *false* by 0. For example, the logical statement

```
>> ~A <enter>
```

will produce a matrix which has 1 at all locations where the corresponding elements in **A** are zeros, and 0 at all locations where **A** has non-zero elements. The logical functions *any* and *all* come in handy in many logical operations. The function *any*(**A**) produces 1 for each column of the matrix **A** that has a non-zero element, and 0 for the columns which have all zero elements. The function *all*(**A**) produces 1 for each column of the matrix **A** that has all non-zero elements, and 0 for the columns which have at least one zero element. With the matrix **A** of Eq. (A.1) the *any* and *all* commands produce the following results:

```
>>any(A) <enter>

ans =
     1   1   1

>>all(A) <enter>

ans =
     1   0   0
```

A logical function called *exist* can be used to find out whether a *variable* with a particular name exists in the work-space. For greater information on relational and logical operators, refer to the MATLAB *Reference Guide* [2].

A.3 Programming in MATLAB: Control Flow and M-Files

Instead of issuing individual MATLAB commands at the command-line, you can group a set of commands to be executed in a MATLAB *program*, called an *M-file*. The M-files have extension. *m*, and are of two types: *script files*, and *function files*. A *script file* simply executes all the commands listed in the file, and can be invoked by typing the *name* of the file and pressing <enter>. For example, a script file called *use.m* is invoked by typing *use* <enter> at the command-line prompt. After executing a script file, all the variables computed in the file are automatically stored in the work-space. The *function files* differ from *script files* in that all the variables computed inside the file are not communicated to the work-space, and only a *few* specific variables, called *input* and *output* arguments, are communicated between the *function file* and the work-space. Thus, a function file acts like a *subroutine* of a main FORTRAN, PASCAL, or BASIC program. A function file can either be called from the work-space, or from another M-file, and is therefore useful for extending the function *library* of the MATLAB. The function file must contain the word *function* at the beginning of the first line, followed by a list of *output* arguments separated by commas within *square brackets*, followed by the sign =, followed by the *name* of the function file, and finally followed by a list of *input* arguments separated by commas within *parentheses*. For example, the first line of a function file called *fred.m* looks like the following:

```
function [X, Y] = fred(A, B, C, D)
```

where A, B, C, D are *input* arguments to be specified by the *calling program* (either work-space, or another M-file), and X, Y are the *output* arguments to be returned to the calling program. The existence of this function file anywhere in the MATLAB directory defines a new MATLAB function called *fred*. All the existing MATLAB functions are, thus, in the form of function files.

The programming structure in MATLAB need not be limited to flow of information in a *sequence* of line commands. The flow of information within an M-file can be controlled using the *for* and *while* loops, and the logical *if* statements, as in any other programming language (such as DO and FOR loops, and IF statements in FORTRAN). The *for* loop in MATLAB allows a group of statements to be repeated a specified number of times. The group of statements to be repeated must end with an *end* statement. One can have nested *for* loops within *for* loops, each ending with an *end* statement. The general structure of a *for* loop is the following:

```
for i = N1:dN:N2
        statements to be repeated
end
```

where $N1$, dN, and $N2$ denote the initial value, increment, and final value of the indexing integer, i. Note that $N2$ could be *less than* $N1$, in which case dN must be *negative*.

The *while* loop allows a group of statements to be repeated an indefinite number of times, as long as a logical condition is satisfied. The general form of a *while* loop is the following:

```
while   logical expression
            statements to be repeated
end
```

The statements in a *while* loop are executed as long as the logical expression is *true* (i.e. as long as the *all* the elements of the expression matrix are *non-zero*). Usually, the expression is a scalar. An example of a *while* loop is the following:

```
A=[1 0 -1; 0 -1 1; 1 -2 1];
while norm(A, inf)<5
            A=A+0.1;
end
```

where *norm* (A, *inf*) denotes the *infinity* norm of the matrix, **A**.

The *if* and *else* statements allow a group of statements to be executed, if a specified logical expression is *true*, and a *second* group of statements to be executed, if the *same* logical expression is *false*. It is also possible to execute a *third* set of statements, if the specified logical expression in *false*, and *another* logical expression is *true*, using the statement *elseif*. Each *if, else, elseif* block of statements must be followed by the *end* statement. For example, the following program illustrates how a computation can be carried out in three cases, depending upon the value of a scalar, *p*:

```
if p<1.0
            A=B*C;
elseif p==1.0
            A=(B.*B)*C;
else
            A=zeros(size(B*C));
end
```

Using the *if* statement, it is possible to come out of a *for* or *while* loop with the *break* statement. For example, if you *do not* wish to repeat a *while* loop *more* than 100 times, you can use the *if* and *break* combination as follows:

```
n=0;
while   expression
n=n+1
if n>100, break, end
statements to be repeated
end
```

You can create your own online help for the M-files you have programmed by adding *comment statements* immediately after the first line of the file. A comment statement begins with the symbol % and are not executed by MATLAB. Some programs may require *strings* of texts for their execution. A string of text is specified by entering text within single quotes, such as:

```
>>g = 'goodbye' <enter>
```

which results in

```
s =
    goodbye
```

A mathematical expression can be included as a text string, and you can use the function *eval* to evaluate the value of the expression.

There are several advanced ways of providing *input* and *output* data to and from MATLAB, such as using a shell escape to an externally running program, importing and exporting data using *flat files, MEX-files*, and *MAT-files*, or with the MATLAB functions *fopen, fread*, and *fwrite* for disk data files. For further information on data transfer to and from MATLAB, refer to the MATLAB *User's Guide* [1].

Programming in MATLAB is made easy with the availability of the *command-line editor*, which displays error messages if a command is incorrectly used, or with the help of *debugging* commands, such as *dbstop, dbclear, dbcont, dbstack, dbstatus*, etc. A simple way of checking whether your M-file is doing what it is supposed to do, is displaying the results of selected intermediate computations by removing semi-colons at the end of selected statements.

Finally, you can post-process your computations by plotting important variables using MATLAB's extensive graphical capabilities. The most commonly used MATLAB graphical commands are *plot* (X, Y) (for generating a plot of the elements of vector, **Y**, against the vector, **X**). Most of the graphs contained in this book have been generated using the *plot* command. You should carefully study the various options available in executing the plot command [1], and also other graphical commands, such as *semilogx* (a plot with a log scale on the x-axis), *semilogy* (a plot with a log scale on the y-axis), *loglog* (a plot with log scales on both x- and y-axes), *subplot* (for displaying more than one plots at a time), *grid* (for generating a grid for a plot), etc. Other MATLAB 2-D graphical functions include *stairs* (staircase plot), *bar* (bar-chart), *hist* (histogram), *feather* (feather plot for angles and magnitudes of complex numbers), *polar* (plot in polar coordinates), *quiver* (plots of vector magnitudes and directions), *rose* (angle histogram), *fill* (solid polygonal plot), and *fplot* (plot of an evaluated mathematical function). There are also a range of 3D plotting functions available in MATLAB. Refer to the *User's Guide* [1] for details on graphical functions.

A.4 The Control System Toolbox

The *Control System Toolbox* (CST) for use with MATLAB provides *additional* function M-files (apart from the basic MATLAB functions) that are especially useful in the analysis and design of control systems. Most of the function files from CST have been extensively used throughout this book, and you have been provided information on how to invoke the associated commands in the main text. For additional information about the CST functions, such as *bode, dbode, are, c2d, c2dm, damp, ddamp, nyquist, dnyquist, lsim, dlsim, tf, ss, estim, destim, lqr, dlqr, lqe, dlqe, sigma, dsigma, place, acker, ngrid, nichols, reg, dreg, initial, dinitial, step, dstep, impulse, dimpulse, series, parallel, feedback, margin, rlocus,* etc., you may refer to the *User's Guide for Control System Toolbox* [3], or issue the *help* command. Two valuable user-friendly graphical tools are also available with CST: the *LTI Viewer*, which lets you view all necessary information required in analyzing a linear, time-invariant system at the click of the mouse button, and the *SISO design tool*, which leads you step-by-step in the graphical window world of designing single-input, single-output, LTI systems. To access these tools, go to the MATLAB *launch pad*, click on the + sign next to the Control System Toolbox, and select any of the two tools that appear on the selection tree. As you become sufficiently proficient with MATLAB programming, you may find it relatively easier to write your own function files for carrying out many of the control analysis and design tasks detailed in CST, using the basic MATLAB functions and the theoretical background in control systems provided in this book. Some examples of the new function M-files have been listed elsewhere in this book. However, for a beginner in controls, the CST function files are valuable tools for learning the tricks of the trade. Apart from the CST, there are several other toolboxes available for advanced control applications, such as the *Signal Processing Toolbox, System Identification Toolbox, Optimization Toolbox, Robust Control Toolbox, Nonlinear Control Design Toolbox, Neural Network Toolbox*, and *µ-Analysis and Synthesis Toolbox*. As your control applications become advanced, you may wish to add some of these advanced toolboxes to your MATLAB directory. Information on how to order these toolboxes can be obtained from the MathWorks, Inc., 24 Prime Park Way, Natick, MA.

A.5 SIMULINK

SIMULINK is a Graphical User's Interface (GUI) software which works directly with the block-diagram of a control system (rather than differential equations, or transfer functions) to produce a simulation of the system's response to arbitrary inputs and initial conditions. The basic entity in SIMULINK is a *block*, which can be selected from a *library* of commonly used blocks. Alternatively, a user can devise special blocks out of the common blocks, M-files, MEX files, C, or Java-codes through the *S-function* facility. The procedure for carrying out a system's simulation through SIMULINK is the following:

1. Double click on the SIMULINK icon on the MATLAB toolbar, or issue the command *simulink* < enter > at the MATLAB prompt (>>). The SIMULINK *library browser* window will open.

2. Click on the *create a new model* icon on the SIMULINK *toolbar*. A window for the new model will open.

3. Open the *subsystem* library in the general SIMULINK *library browser* by double-clicking on the appropriate icon. The subsystems are: *continuous, discrete, functions & tables, math, nonlinear, signals & systems, sinks*, and *sources*.

4. Select the required blocks from the *subsystems* libraries, and drag them individually to the open *new model* window.

5. Once you have dragged the required blocks to the *new model* window, you can join the *in-ports* and *out-ports* of the adjacent blocks to create a block-diagram as desired. You can double-click on each block in your model to open a *dialog box*, in which the block's *parameters* can be set.

6. Once the block-diagram is complete, you can save it using the *save* button on the new model's toolbar.

7. Now you are ready to begin the simulation of your control system. Just go to the toolbar of the model you have saved and click on the *play* button. If you have created your model correctly, the simulation will start and you can view the results using any of the *sink* blocks in your model. However, one seldom succeeds at first, and SIMULINK prompts you through a *diagnostics dialog box* to tell you what went wrong with the simulation, and also what you should modify in your model for a successful simulation.

8. You can refine your simulation by adjusting the *simulation parameters* that drop down when you click on the *simulation* button on the model's toolbar.

Following the above steps, any practical control system can be simulated accurately using SIMULINK. Let us briefly see the contents of each SIMULINK *subsystem block library*.

Continuous: *transfer function, state-space, integrator, derivative, transport delay, variable transport-delay, memory, zero-pole.* (These block help you construct a continuous-time (analog) system model.)

Discrete: *discrete transfer function, discrete state-space, discrete zero-pole, discrete filter, discrete-time integrator, first-order hold, zero-order hold, unit delay.* (These block help you construct a discrete-time (digital) system model.)

Functions & Tables: this library contains specialized functions and tables blocks useful for creating complicated systems. It includes all MATLAB intrinsic functions, as well as special user created functions through the *S-function* block.

Math: all the mathematical connection blocks (such as *sum junction, gain, matrix gain, product, dot product, abs, floor, trigonometric function*, etc.), and relational and logical operator blocks (such as *and, combinatorial logic*, etc.) are found here. These blocks are indispensable in constructing any control system.

Nonlinear: contains a number of nonlinear system blocks that are very useful in modeling a variety of physical phenomena. Some examples are *backlash, coulomb & viscous friction, dead zone, saturation, rate limiter, relay, switch,* etc.

Signals & Systems: this library contains many specialized blocks used for representing subsystems and operating on signals passing through a system. Some commonly useful blocks are *subsystem* (which allows you to group a number of blocks into a subsystem), *mux, demux, in1,* and *out1*.

Sinks: contains a number of possible ways of output of data from a model, such as *scope, xy graph, to workspace, simout, display*, and *stop simulation*. For example, a *scope* can be used to directly view a simulation variable in a window of the model, and to also store the data in a file.

Sources: provides a variety of input sources for the model, such as *step, ramp, sine wave, pulse generator, random number, repeating sequence, band-limited white noise, chirp signal, signal generator*, etc.

The SIMULINK provides several *simulation parameters* that can be adjusted to achieve a desired accuracy in a simulation. A user can select from a number of time-integration schemes, such as *Runge–Kutta, Adams, Euler, predictor-corrector,* as well as refine the tolerances and time step sizes used for performing the simulation. Useful diagnostics are generated to let a user improve her simulation.

The most useful feature of SIMULINK is that you can use variables specified in the MATLAB work-space as block parameters, and in this manner work seamlessly with all the intrinsic and toolbox functions of MATLAB. For more information on SIMULINK, refer to its user's guide [4], or work interactively with the SIMULINK blocks and models until you get a hang of it. Once understood, SIMULINK modeling can become a powerful tool in the hands of a control systems designer.

References

1. *MATLAB 6.0 User's Guide*. The Math Works Inc., Natick, MA, USA, 2000.
2. *MATLAB 6.0 Reference Guide*. The Math Works Inc., Natick, MA, USA, 2000.
3. *Control System Toolbox 5.0 for Use with MATLAB-User's Guide*. The Math Works Inc., Natick, MA, USA, 2000.
4. *SIMULINK 4.0 User's Guide*. The Math Works Inc., Natick, MA, USA, 2000.

Appendix **B**

Review of Matrices and Linear Algebra

The concept of *matrices*, defined as a set of numbers arranged in various *rows* and *columns*, began with efforts to simultaneously solve a set of linear algebraic equations. Hence, the study of matrices and their properties is referred to as *linear algebra*. For example, consider the following linear equations:

$$x + 2y + 3z = 5$$
$$-x + y + 7z = 0$$
$$3x - 17y + 2z = 21 \tag{B.1}$$

An attempt to solve these equations simultaneously for the unknowns x, y, and z results in their being expressed in the following form:

$$\begin{bmatrix} 1 & 2 & 3 \\ -1 & 1 & 7 \\ 3 & -17 & 2 \end{bmatrix} \begin{bmatrix} x \\ y \\ z \end{bmatrix} = \begin{bmatrix} 5 \\ 0 \\ 21 \end{bmatrix} \tag{B.2}$$

In Eq. (B.2), we denote

$$\mathbf{A} = \begin{bmatrix} 1 & 2 & 3 \\ -1 & 1 & 7 \\ 3 & -17 & 2 \end{bmatrix}; \quad \mathbf{V} = \begin{bmatrix} x \\ y \\ z \end{bmatrix} \quad \mathbf{d} = \begin{bmatrix} 5 \\ 0 \\ 21 \end{bmatrix} \tag{B.3}$$

and re-write Eq. (B.2) as $\mathbf{AV} = \mathbf{d}$, where $\mathbf{A}$ is called a *matrix* of size (3×3) (because it has three rows and three columns), and $\mathbf{V}$ and $\mathbf{d}$ are *matrices* of size (3×1). A matrix with only *one column* (such as $\mathbf{V}$ and $\mathbf{d}$) has a special name – *column vector*, while a matrix with only *one row* is called a *row vector*. A matrix with only *one* row and only *one* column consists of only one number, and is called a *scalar*. The numbers of which a matrix is formed are called the *elements* of the matrix. For example, in Eq. (B.3), the matrix $\mathbf{A}$ has nine elements. In general, a matrix with n rows and m columns would have

nm elements, and is denoted as follows:

$$
\mathbf{A} = \begin{bmatrix}
a_{11} & a_{12} & \cdots & a_{1m} \\
a_{21} & a_{22} & \cdots & a_{2m} \\
\cdots & \cdots & \cdots & \cdots \\
a_{n1} & a_{n2} & \cdots & a_{nm}
\end{bmatrix} \tag{B.4}
$$

The element of the matrix, $\mathbf{A}$, in Eq. (B.4) located in the ith row and jth column is denoted by a_{ij}. Some elementary matrix operations are defined as follows:

Addition: addition of two matrices of the *same size*, $\mathbf{A}$ and $\mathbf{B}$, is defined as addition of *all the corresponding elements*, a_{ij} and b_{ij}, of the matrices $\mathbf{A}$ and $\mathbf{B}$, i.e.

$$
\mathbf{A} + \mathbf{B} = \mathbf{C} \tag{B.5}
$$

where the element c_{ij} of the matrix $\mathbf{C}$ is calculated as $c_{ij} = a_{ij} + b_{ij}$ for all i and j. *Matrix subtraction* is defined in the same manner as the addition, except that the corresponding elements are subtracted rather than added.

Multiplication by a scalar: a matrix, $\mathbf{A}$, is said to be *multiplied* by a scalar, a, if all the elements of $\mathbf{A}$ are multiplied by a.

Multiplication of a row vector with a column vector: a row vector, $\mathbf{r}$, of size $(1 \times n)$ is said to be multiplied with a column vector, $\mathbf{c}$, of size $(n \times 1)$ if the products of the corresponding elements, $r_{1k}c_{k1}$, are summed as follows, resulting in a *scalar*:

$$
\mathbf{rc} = \begin{bmatrix} r_{11} & r_{12} & \cdots & r_{1n} \end{bmatrix} \begin{bmatrix} c_{11} \\ c_{21} \\ \cdots \\ c_{n1} \end{bmatrix} = r_{11}c_{11} + r_{12}c_{21} + \cdots + r_{1n}c_{n1} \tag{B.6}
$$

Multiplication of two matrices: a matrix, $\mathbf{A}$, of size $(n \times p)$ is said to be *multiplied* with a matrix, $\mathbf{B}$, of size $(p \times m)$, resulting in a matrix, $\mathbf{C}$, of size $(n \times m)$, expressed as

$$
\mathbf{AB} = \mathbf{C} \tag{B.7}
$$

such that the element c_{ij} of the matrix, $\mathbf{C}$, is the *multiplication* of the ith *row* of the matrix, $\mathbf{A}$, with the jth *column* of the matrix, $\mathbf{B}$, for all i and j. For example, in Eq. (B.2), a matrix of size (3×3) is multiplied with a column vector of size (3×1) to produce a column vector of size (3×1).

Some special matrices are defined as follows:

Square matrix: a matrix that has *equal* number of rows and columns is called a *square matrix*. The elements, a_{ii}, of a square matrix, $\mathbf{A}$, for all i are said to be the *diagonal elements* of the square matrix.

Symmetric matrix: a square matrix, $\mathbf{A}$, is said to be *symmetric* if the element in the ith row and jth column is *equal* to the element in the jth row and ith column, i.e. $a_{ij} = a_{ji}$, *for all* i and j.

Diagonal matrix: a square matrix that has all the elements, except the *diagonal elements*, equal to zero is called a *diagonal matrix*.

Identity matrix: a diagonal matrix that has *all* the *diagonal elements* equal to *unity* is called an *identity matrix*, and is denoted by $\mathbf{I}$. An identity matrix has the property that if $\mathbf{A}$ is a square matrix of the same size as the identity matrix, $\mathbf{I}$, then

$$\mathbf{AI} = \mathbf{IA} = \mathbf{A} \tag{B.8}$$

Some of the more advanced matrix operations are defined as follows (you may refer to a textbook on linear algebra for details of these operations [1–3]):

Transpose: transpose of a matrix, $\mathbf{A}$, is defined as the matrix, $\mathbf{A}^{\mathbf{T}}$, in which the *rows* and *columns* of the matrix, $\mathbf{A}$, have been *interchanged*. Transpose of a symmetric matrix, $\mathbf{S}$, is equal to $\mathbf{S}$. The transpose of a product of two matrices has the following property:

$$(\mathbf{AB})^{T} = \mathbf{B}^{\mathbf{T}}\mathbf{A}^{\mathbf{T}} \tag{B.9}$$

Trace: *Trace* of a *square* matrix, $\mathbf{A}$, of size $(n \times n)$ is defined as the sum of *all* the *diagonal elements* of $\mathbf{A}$, i.e.

$$trace(\mathbf{A}) = a_{11} + a_{22} + \cdots + a_{nn} \tag{B.10}$$

Determinant and minor: *Determinant* of a square matrix, $\mathbf{A}$, of size $(n \times n)$ is defined as the sum of *all possible* products of n elements, each taken from a *different* column. A more useful (but *recursive*) definition of the determinant of $\mathbf{A}$, denoted by $|\mathbf{A}|$, is the following:

$$|\mathbf{A}| = a_{11}D_{11} - a_{12}D_{12} + a_{13}D_{13} + \cdots + (-1)^{n}a_{1n}D_{1n} \tag{B.11}$$

where D_{ij} is called the *minor* of element, a_{ij}, and defined as the *determinant* of the *square sub-matrix* of size $((n-1) \times (n-1))$ formed out of the matrix $\mathbf{A}$ by deleting the ith row and the jth column. This procedure can be followed recursively (either by hand or a computer program) until we are left with the minors of the *smallest* size, i.e. a *scalar*. Instead of finding the determinant by taking elements from different columns, we can alternatively take the elements from *different rows* of $\mathbf{A}$, and express the determinant as follows:

$$|\mathbf{A}| = a_{11}D_{11} - a_{21}D_{21} + a_{31}D_{31} + \cdots + (-1)^{n}a_{n1}D_{n1} \tag{B.12}$$

If $|\mathbf{A}| = 0$, then the matrix, $\mathbf{A}$, is said to be *singular*. If $|\mathbf{A}| \neq 0$, then the matrix $\mathbf{A}$ is said to be *non-singular*. Determinant has the following property:

$$|\mathbf{AB}| = |\mathbf{A}||\mathbf{B}| \tag{B.13}$$

Cofactor: the cofactor, c_{ij}, of an element, a_{ij}, of a square matrix, $\mathbf{A}$, is defined as

$$c_{ij} = (-1)^{i+j} D_{ij} \tag{B.14}$$

where D_{ij} is the *minor* associated with the element, a_{ij}.

Adjoint: the *transpose* of a matrix whose elements are the *cofactors*, c_{ij}, of a square matrix, $\mathbf{A}$, is called the *adjoint* of $\mathbf{A}$, denoted by $adj(\mathbf{A})$, and expressed as follows

$$adj(\mathbf{A}) = \begin{bmatrix} c_{11} & c_{12} & \cdots & c_{1n} \\ c_{21} & c_{22} & \cdots & c_{2n} \\ \cdots & \cdots & \cdots & \cdots \\ c_{n1} & c_{n2} & \cdots & c_{nn} \end{bmatrix}^T \tag{B.15}$$

Inverse: the inverse of a square matrix, $\mathbf{A}$, is defined as a matrix, $\mathbf{A}^{-1}$, which when multiplied with $\mathbf{A}$, produces an identity matrix, $\mathbf{I}$,

$$\mathbf{A}^{-1}\mathbf{A} = \mathbf{A}\mathbf{A}^{-1} = \mathbf{I} \tag{B.16}$$

The matrix inverse can be calculated by the *Cramer's rule*, expressed as follows:

$$\mathbf{A}^{-1} = adj(\mathbf{A})/|\mathbf{A}| \tag{B.17}$$

From Eq. (B.17), it is clear that $\mathbf{A}^{-1}$ exists only if $\mathbf{A}$ is *non-singular*. Some important properties of matrix inverse are the following:

$$(\mathbf{AB})^{-1} = \mathbf{B}^{-1}\mathbf{A}^{-1} \tag{B.18}$$

$$(\mathbf{A}^{-1})^T = (\mathbf{A}^T)^{-1} \tag{B.19}$$

$$|\mathbf{A}^{-1}| = 1/|\mathbf{A}| \tag{B.20}$$

$$(\mathbf{A} + \mathbf{BCD})^{-1} = \mathbf{A}^{-1} - \mathbf{A}^{-1}\mathbf{B}(\mathbf{C}^{-1} + \mathbf{DA}^{-1}\mathbf{B})^{-1}\mathbf{DA}^{-1} \tag{B.21}$$

Eigenvalues: the *eigenvalues* of a square matrix, $\mathbf{A}$, are defined as the roots, λ_i, of the following *characteristic polynomial equation*:

$$|\lambda\mathbf{I} - \mathbf{A}| = 0 \tag{B.22}$$

where $\mathbf{I}$ is an identity matrix of the same size as $\mathbf{A}$. A matrix of size $(n \times n)$ has n eigenvalues. The *product* of all the eigenvalues of a square matrix is equal to the determinant of the matrix. The *sum* of all the eigenvalues of a square matrix is equal to the trace of the matrix.

Eigenvectors: The *eigenvector*, $\mathbf{v_i}$, of a square matrix, $\mathbf{A}$, associated with the eigenvalue, λ_i, of $\mathbf{A}$ is defined as the vector that satisfies the following equation:

$$\mathbf{A}\mathbf{v_i} = \lambda_i \mathbf{v_i} \tag{B.23}$$

Cayley–Hamilton Theorem: a fundamental relation of linear algebra is the theorem that states that if the characteristic polynomial equation (Eq. (B.22)) of a square matrix, $\mathbf{A}$, of size $(n \times n)$ is expressed as

$$|\lambda\mathbf{I} - \mathbf{A}| = \lambda^n + \alpha_1\lambda^{n-1} + \cdots + \alpha_{n-1}\lambda + \alpha_n = 0 \tag{B.24}$$

where α_k are *characteristic coefficients*, then

$$\mathbf{A}^n + \alpha_1\mathbf{A}^{n-1} + \cdots + \alpha_{n-1}\mathbf{A} + \alpha_n\mathbf{I} = \mathbf{0} \tag{B.25}$$

where $\mathbf{A}^k$ denotes $\mathbf{A}$ *multiplied* by $\mathbf{A}$, k times (or, $\mathbf{A}$ raised to the *power k*), and $\mathbf{0}$ denotes a matrix with *all* zero elements of the same size as $\mathbf{A}$.

Hermitian: the *transpose* of the *complex conjugate* of a matrix, $\mathbf{A}$, is called the *hermitian* of $\mathbf{A}$, and denoted by $\mathbf{A}^H$. A matrix, $\mathbf{U}$, whose hermitian equals its inverse (i.e. $\mathbf{U}^H = \mathbf{U}^{-1}$) is called a *unitary matrix*.

Differentiation of a matrix by a scalar: the *derivative* of a matrix, $\mathbf{A}$, with respect to a scalar, c, is defined as the matrix, $d\mathbf{A}/dc$, each element of which is the *derivative* of the corresponding element, da_{ij}/dc, of the matrix, $\mathbf{A}$.

Integration of a matrix: the *integral* of a matrix, $\mathbf{A}$, is defined as the matrix whose elements are *integrals* of the corresponding elements of $\mathbf{A}$.

Differentiation of a scalar by a vector: the derivative of a scalar by a vector is defined as the vector, each element of which is the *derivative* of the scalar with the *corresponding* element of the vector. For example, if a scalar z, is related to *two vectors*, $\mathbf{x}$ and $\mathbf{y}$ such that $z = \mathbf{x}^T\mathbf{y} = \mathbf{y}^T\mathbf{x}$, then the *partial derivative* of z with respect to the vector, $\mathbf{x}$, is $\partial z/\partial\mathbf{x} = \mathbf{y}$, and the partial derivative of z with respect to the vector $\mathbf{y}$ is $\partial z/\partial\mathbf{y} = \mathbf{x}$.

Quadratic form: the scalar, z, is called the *quadratic form* of the vector, $\mathbf{x}$, if it can be expressed as

$$z = \mathbf{x}^T\mathbf{Q}\mathbf{x} \tag{B.26}$$

where $\mathbf{Q}$ is a *square* matrix. It can be shown that the derivative of z with respect to $\mathbf{x}$ is $\partial z/\partial\mathbf{x} = 2\mathbf{Q}\mathbf{x}$.

Bilinear form: the scalar, z, is called the *bilinear form* of the vectors, $\mathbf{x}$ and $\mathbf{y}$, if it can be expressed as

$$z = \mathbf{x}^T\mathbf{Q}\mathbf{y} \tag{B.27}$$

where $\mathbf{Q}$ is a matrix. It can be shown that the derivative of z with respect to $\mathbf{x}$ is $\partial z/\partial \mathbf{x} = \mathbf{Q}\mathbf{y}$ and the derivative of z with respect to $\mathbf{y}$ is $\partial z/\partial \mathbf{y} = \mathbf{Q}^{\mathrm{T}}\mathbf{x}$.

Positive definiteness: if the quadratic form $\mathbf{x}^{\mathrm{T}}\mathbf{Q}\mathbf{x} > 0$ for all $\mathbf{x} \neq \mathbf{0}$, then the quadratic form is said to be *positive definite*. If the quadratic form $\mathbf{x}^{\mathrm{T}}\mathbf{Q}\mathbf{x} \geq 0$ for all $\mathbf{x} \neq \mathbf{0}$, then the quadratic form is said to be *positive semi-definite*.

References

1. Gantmacher, F.R. *Theory of Matrices*. Chelsea, New York, 1959.
2. Strang, G. *Linear Algebra and Its Applications*. Academic Press, New York, 1976.
3. Kreyszig, E. *Advanced Engineering Mathematics*. Wiley, New York, 1972.

Appendix C

Mass, Stiffness, and Control Influence Matrices of the Flexible Spacecraft

For the rotating, flexible spacecraft introduced in Example 6.2, the mass matrix, **M**, the stiffness matrix, **K**, and the control influence matrix, **d**, are as follows:

```
M =
 Columns 1 through 6
  2.8196e+001   1.6917e-001   8.5926e-003   2.6583e-001   8.5926e-003   1.7362e+000
  1.6917e-001   5.3857e-002   0             9.3214e-003  -2.9921e-003   0
  8.5926e-003   0             2.4550e-003   2.9921e-003  -9.2063e-004   0
  2.6583e-001   9.3214e-003   2.9921e-003   5.3857e-002   0             9.3214e-003
  8.5926e-003  -2.9921e-003  -9.2063e-004   0             2.4550e-003   2.9921e-003
  1.7362e+000   0             0             9.3214e-003   2.9921e-003   3.4081e-001
 -3.2381e-002   0             0            -2.9921e-003  -9.2063e-004  -5.0635e-003
  1.6917e-001   0             0             0             0             0
  8.5926e-003   0             0             0             0             0
  2.6583e-001   0             0             0             0             0
  8.5926e-003   0             0             0             0             0
  1.7362e+000   0             0             0             0             0
 -3.2381e-002   0             0             0             0             0


 Columns 7 through 12
 -3.2381e-002   1.6917e-001   8.5926e-003   2.6583e-001   8.5926e-003   1.7362e+000
  0             0             0             0             0             0
  0             0             0             0             0             0
 -2.9921e-003   0             0             0             0             0
 -9.2063e-004   0             0             0             0             0
 -5.0635e-003   0             0             0             0             0
  4.8275e-003   0             0             0             0             0
  0             5.3857e-002   0             9.3214e-003  -2.9921e-003   0
  0             0             2.4550e-003   2.9921e-003  -9.2063e-004   0
  0             9.3214e-003   2.9921e-003   5.3857e-002   0             9.3214e-003
  0            -2.9921e-003  -9.2063e-004   0             2.4550e-003   2.9921e-003
  0             0             0             9.3214e-003   2.9921e-003   3.4081e-001
  0             0             0            -2.9921e-003  -9.2063e-004  -5.0635e-003
```

```
Column 13
-3.2381e-002
      0
      0
      0
      0
      0
      0
      0
-2.9921e-003
-9.2063e-004
-5.0635e-003
 4.8275e-003
```

K = 1.0e+006*

Columns 1 through 12

0	0	0	0	0	0	0	0	0	0	0	0
0	3.4805	0	-1.7402	1.1602	0	0	0	0	0	0	0
0	0	2.0625	-1.1602	0.5156	0	0	0	0	0	0	0
0	-1.7402	-1.1602	3.4805	0	-1.7402	1.1602	0	0	0	0	0
0	1.1602	0.5156	0	2.0625	-1.1602	0.5156	0	0	0	0	0
0	0	0	-1.7402	-1.1602	1.7402	-1.1602	0	0	0	0	0
0	0	0	1.1602	0.5156	-1.1602	1.0313	0	0	0	0	0
0	0	0	0	0	0	0	3.4805	0	-1.7402	1.1602	
0	0	0	0	0	0	0	0	2.0625	-1.1602	0.5156	
0	0	0	0	0	0	0	-1.7402	-1.1602	3.4805	0	
0	0	0	0	0	0	0	1.1602	0.5156	0	2.0625	
0	0	0	0	0	0	0	0	0	-1.7402	-1.1602	
0	0	0	0	0	0	0	0	0	1.1602	0.5156	

Column 12	Column 13
0	0
0	0
0	0
0	0
0	0
0	0
0	0
0	0
0	0
-1.7402	1.1602
-1.1602	0.5156
1.7402	-1.1602
-1.1602	1.0313

```
d =   1   2   2
      0   0   0
      0   0   0
      0   0   0
      0   0   0
      0   0   0
      0   2   0
      0   0   0
      0   0   0
      0   0   0
      0   0   0
      0   0   0
      0   0   2
```

Answers to Selected Exercises

Chapter 2

2.4. The system is stable with $\omega_n = 377.27$ rad/s and $\zeta = 1.26 \times 10^{-6}$. Maximum overshoot $= 0.0069$ m, steady state deviation $= 0.007$ m, settling time $= 6100$ seconds.

2.5. The system is stable with $\omega_n = 5634.7$ rad/s and $\zeta = 1.409 \times 10^{-6}$. Maximum overshoot $= 3.13 \times 10^{-5}$ m, steady state deviation $= 3.15 \times 10^{-5}$ m, settling time $= 460$ seconds.

2.6. (a) $G(i\omega) = 1/(-m\omega^2 + ic\omega + k)$.

    (b) $\omega_n = (k/m)^{1/2}$; $\zeta = c/[2(km)^{1/2}]$.

    (c) $s_{1,2} = \pm i(k/m)^{1/2}$.

2.8. (a) Maximum overshoot $= 0$; settling time $= 0.8$ second; steady state output $= 0.2$.

    (b) Maximum overshoot $= -80\%$; settling time $= 5.33$ seconds; steady state output $= 0.5$.

    (c) Maximum overshoot $= 464\%$; settling time $= 18.5$ seconds; steady state output $= 1.0$.

2.9. Maximum overshoot $= 213.3$ units; steady state output $= 0$.

2.10. $|\delta_{max}| = 250$ m.

2.14. (a) $K_I = 1$, $K_D = 2.114$, $K_P = 2.394$;

    (b) maximum overshoot $= 21.6\%$; settling time $= 6.33$ seconds; steady state error $= 0$;

    (c) $0 \le K_I \le 5.8$;

    (d) Gain margin $= \infty$, phase margin $= 129.15°$, gain-crossover frequency $= 1.9$ rad/s.

2.18. (a) $G(s) = K(s+a)/[s^2(s+1)]$, $H(s) = K_1 + K_2 s + K_3 as(s+1)/(s+a)$.

    (b) Type $= 2$.

    (c) Plant is unstable.

    (d) $H(s) = [K_3 as^2 + s(K_1 + K_2 + K_3 a) + K_1 a]/(s+a)$.

(e) For $K_1 = 500$, $\omega_n = 1.83$ rad/s, $\zeta = 0.753$.

(f) Maximum overshoot $= 0$; settling time $= 1.5 \times 10^{-3}$ second; steady state error $= 0$.

(g) Gain margin $= \infty$; phase margin $= 178.04°$; gain crossover frequency $= 102.47$ rad/s.

2.22. (a) Gain margin $= \infty$; phase margin $= 3.63°$; gain crossover frequency $= 31.67$ rad/s.

(b) Phase is maximum at $\omega = 400$ rad/s; change in gain $= 40$ dB.

(c) Plant: max. overshoot $= 21\%$, settling time $= 3.8$ sec., steady state error $= -199$. Closed-loop system: max. overshoot $= 32.5\%$, settling time $= 0.25$ sec., steady state error $= 0.005$.

(d) Gain margin $= \infty$, phase margin $= 99.27°$, gain crossover frequency $= 43.88$ rad/s.

2.25. (a) Gain margin $= 28.09$ dB, phase margin $= 12.64°$.

(b) $\omega_o = 0.0775$ rad/s, $\alpha = 0.1$.

(c) Plant: maximum overshoot $= 91\%$, settling time $= 2000$ seconds. Closed-loop system: maximum overshoot $= 42\%$, settling time $= 70$ seconds.

2.28. $K = 0.7$, maximum overshoot $= 1.8$ m, settling time $= 12.25$ seconds. $|z(t)|_{max} = 29.9$ m/s^2.

Plant: gain margin $= \infty$, phase margin $= 20.98°$, gain crossover freq. $= 377.3$ rad/s. Closed-loop: gain margin $= \infty$, phase margin $= \infty$.

Chapter 3

3.1. (a) $\mathbf{A} = \begin{bmatrix} 0 & 1 & 0 \\ 0 & 0 & 1 \\ 2/17 & -10/17 & 0 \end{bmatrix}$; $\mathbf{B} = \begin{bmatrix} 0 \\ 0 \\ 1/17 \end{bmatrix}$; $\mathbf{C} = [0 \ 2 \ 5]$; $\mathbf{D} = 0$.

3.2. (a) $\mathbf{A} = \begin{bmatrix} -4 & -3 & 1 & -5 \\ 1 & 0 & 0 & 0 \\ 0 & 1 & 0 & 0 \\ 0 & 0 & 1 & 0 \end{bmatrix}$; $\mathbf{B} = \begin{bmatrix} 1 \\ 0 \\ 0 \\ 0 \end{bmatrix}$; $\mathbf{C} = [0 \ 1 \ -3 \ 1]$; $\mathbf{D} = 0$.

3.4. $x_1^{(1)}(t) = x_2(t)$; $x_3(t)x_2^{(1)}(t) = -2x_2(t)x_3^{(1)}(t) + a_{Mc}(t)$; $\theta_M(t) = x_1(t)$.

3.8. (a) $\mathbf{A} = \begin{bmatrix} -0.5 & 1.0724 & 0 & 0 \\ -1.0724 & -0.5 & 0 & 0 \\ 0 & 0 & -0.0025 & 0.0774 \\ 0 & 0 & -0.0774 & -0.0025 \end{bmatrix}$; $\mathbf{B} = \begin{bmatrix} -1.6771 \\ -0.6065 \\ 6.8075 \\ 6.3307 \end{bmatrix}$

$\mathbf{C} = \begin{bmatrix} 0.0122 & -0.0073 & 0.0132 & -0.0116 \\ 0.2922 & -0.7779 & -0.0001 & -0.0001 \\ -0.0074 & -0.6646 & -0.0353 & -0.0276 \end{bmatrix}$; $\mathbf{D} = \begin{bmatrix} 0 \\ 0 \\ 0 \end{bmatrix}$

Chapter 4

4.1. (a) $e^{\mathbf{A}t} = \begin{bmatrix} e^{-2t} & (3e^{-2t} - 2.5607e^{-0.5858t} - 0.4393e^{-3.4142t}) & 0.75(2e^{-2t} - e^{-0.5858t} - e^{-3.4142t}) \\ 0 & (1.2071e^{-0.5858t} - 0.2071e^{-3.4142t}) & 0.3535(e^{-0.5858t} - e^{-3.4142t}) \\ 0 & 0.707(-e^{-0.5858t} + e^{-3.4142t}) & (-0.2071e^{-0.5858t} + 1.2071e^{-3.4142t}) \end{bmatrix}$

(b) $\lambda_1 = -2$, $\lambda_2 = -0.5858$, $\lambda_3 = -3.4142$. Stable.

(c) $\mathbf{X}(t) = \begin{bmatrix} 1.45e^{-2t} - 0.1189e^{-0.5858t} - 0.3311e^{-3.4142t} \\ 0.056e^{-0.5858t} - 0.156e^{-3.4142t} \\ -0.0329e^{-0.5858t} + 0.5329e^{-3.4142t} \end{bmatrix}$

Chapter 5

5.1. (a) Controllable, (b) Uncontrollable.

5.4. (a) Yes. (b) No. No.

5.6. (a) Yes. (b) No. (c) No.

5.8. $\mathbf{K} = \begin{bmatrix} -0.3814 & -0.1654 & 0.3212 & 0.0615 & -0.6328 & -0.4675 \\ 2.1142 & 0.3183 & -0.3035 & -0.0720 & 1.0523 & 0.5360 \end{bmatrix}$
Maximum overshoots: $y_1(t)$ (13.6789 m/s^2); $y_2(t)$ (-0.1312 rad/s).
Settling time: 4 seconds.

5.9. $\mathbf{K} = \begin{bmatrix} -0.0033 & 0.0008 & 0.0003 & -0.0011 \\ 0.5702 & -0.6445 & 9.0813 & 0.2692 \end{bmatrix}$

5.12. (a) Unobservable. (b) Observable. (c) Unobservable.

5.15. (a) Observable. (b) Unobservable. (c) Unobservable.

5.16. (a) Unobservable. (b) Observable. (c) Observable. (d) Unobservable.

(e) $\mathbf{L} = [\,-15.28; \quad -30.393; \quad 6.9472; \quad 5.4\,]^T$.

5.18. $\mathbf{L}^{\mathbf{T}} = \begin{bmatrix} 0.0000 & -1.0685 & 4.2603 & -1.9726 \\ 0.0000 & -2.8493 & -1.9726 & -0.2603 \end{bmatrix}$

Chapter 6

6.2. $\mathbf{Q} = \begin{bmatrix} 10^{-4} & 0 & 0 & 0 \\ 0 & 10^{-4} & 0 & 0 \\ 0 & 0 & 0.01 & 0 \\ 0 & 0 & 0 & 0.01 \end{bmatrix}$; $\mathbf{R} = \mathbf{I}$

$\mathbf{K} = \begin{bmatrix} 0.0072 & 0.0796 & -0.0254 & -0.0967 \\ 0.0000 & -0.0073 & 0.0967 & -0.0254 \end{bmatrix}$

6.3. Settling time $= 0.1$ second. Maximum overshoot $= -0.0117$ rad/s.
Largest magnitude of $\delta_A(t) = 0.3$ rad. Largest magnitude of $\delta_R(t) = 0.03$ rad.

6.4. $\mathbf{K} = \begin{bmatrix} -0.0077 & -0.0280 & -0.0916 & 0.0269 & -0.0344 & 0.0261 \\ -0.0037 & -0.0597 & 0.1593 & 0.6495 & -1.0641 & 1.0249 \end{bmatrix}$

Settling time: 0.33 second. Maximum overshoots: -9×10^{-4} units $(y_1(t))$; -0.09 units $(y_2(t))$.

Largest magnitude of $u_1(t) = 7.73 \times 10^{-4}$ units. Largest magnitude of $u_2(t) = 5.27 \times 10^{-4}$ units.

6.5. $\mathbf{Q} = \begin{bmatrix} 5 & 0 \\ 0 & 10 \end{bmatrix}$; $\quad \mathbf{R} = \mathbf{I}$

$\mathbf{K} = \begin{bmatrix} 0.3665 & -4.9893 & -4.1433 & 5.7678 & -5.2177 & -3.0179 \\ 3.5047 & 19.3521 & -12.8400 & 11.7067 & 14.2192 & 39.5569 \end{bmatrix}$

6.8. $\mathbf{K_{o1}} = \begin{bmatrix} 0.5983 & 0.0656 & -0.3026 & 0.9993 \\ 0.0626 & -0.7147 & -0.3068 & 0.0363 \end{bmatrix}$; $\quad \mathbf{K_d} = \mathbf{0}$.

6.11. $\mathbf{K_{o1}} = \begin{bmatrix} -3.1623 & -15.1463 & -2.8724 & 7.7314 & 0 & 0 & 0 & 0 \\ 0 & 0 & 0 & 0 & -3.1623 & -15.0165 & -3.1197 & 7.5203 \end{bmatrix}$;

$\mathbf{K_d} = \mathbf{0}$.

6.12. $\mathbf{Q} = 10^{-5}\,\mathbf{I}$, $\mathbf{R} = \mathbf{I}$, $\mathbf{V} = 1000\mathbf{I}$. Maximum magnitude of $u_1(t)$: 0.92 units. Maximum magnitude of $u_2(t)$: 0.24 units.

6.14. Maximum overshoot of $\theta^{(1)}(t)$: -0.03 rad/s. Final value of $\theta^{(1)}(t) = -0.003$ rad/s.

Chapter 7

7.1. $x_m = 0.5164$, $R_x(0) = 0.0843$, $x_{ms} = 0.3502$.

7.2. $\omega_o = 7$ rad/s.

7.8. $s_1 = -16.0070$, $s_2 = -9.8148$, $s_3 = -2.5413$, $s_4 = -0.9606$, $s_{5,6} = -0.9336 \pm 6.3617i$.

7.12. $\mathbf{F} = \mathbf{B}$, $\mathbf{Z} = \mathbf{CC^T}$, $\Psi = \mathbf{0}$, $\mathbf{V} = 10^6 \mathbf{B^T B}$.

$\mathbf{L^T} = \begin{bmatrix} 4234.9 & -13\,360 & 10\,714 & -6993.3 & -15\,013 & 20\,143 \\ 5370.1 & -16\,994 & 13\,631 & -8889.8 & -19\,117 & 25\,652 \end{bmatrix}$

Largest magnitude of $u_1(t) = 0.053$ units.
Largest magnitude of $u_2(t) = 1.877$ units.

Chapter 8

8.1. (a) $F(z) = ze^{aT}/(ze^{aT} - 1)$.

8.2. (a) $f(kT) = -0.5025i(0.995i)^k + 0.5025i(-0.995i)^k$.

8.3. (a) $G(z) = z(z - 0.6375)/[(z - 1)(z - 0.8187)]$.

8.4. (a) $G(z) = [z(e^{-T} + 2T - 1) + 1 - e^{-T} - 2Te^{-T}]/[z^2 + 2z(T - 1) + 1 + 2Te^{-T}]$.

8.6. (a) Unstable for all $T > 0$.

8.9. (a) $y(\infty) = \infty$.

8.12. The closed-loop system is stable for $-0.06 < K < 0$. With $K = -0.002$, the closed-loop step response settles in about 1 second, with a maximum overshoot of about 6 per cent.

8.19. $\mathbf{A_d} = \begin{bmatrix} 0 & 1 & 0 \\ 0 & -1 & 1 \\ 1 & 0 & -1 \end{bmatrix}; \quad \mathbf{B_d} = \begin{bmatrix} 0 & 0 \\ 1 & 1 \\ 1 & 0 \end{bmatrix}$

$\mathbf{C_d} = \begin{bmatrix} 1 & 0 & 0 \\ 0 & 0 & 1 \end{bmatrix}; \quad \mathbf{D_d} = \begin{bmatrix} 0 & 0 \\ 0 & 0 \end{bmatrix}$

8.25. $\mathbf{K} = \begin{bmatrix} -0.0904 & 0.0226 & -0.2418 & 0.1948 \\ 0.0491 & 0.4488 & -2.5087 & 0.0211 \end{bmatrix}$

8.30. $\mathbf{K} = \begin{bmatrix} 0.9833 & 0.0899 & 0.0248 & 0.0067 & -0.0800 & -0.0403 \\ 0.1610 & -0.0469 & 0.0109 & -0.0120 & 0.1442 & 0.0762 \end{bmatrix}$

Chapter 9

9.2. (a) $\mu(\mathbf{G}) = 10.24$, (b) $\mu(\mathbf{G}) = 4.65$, (c) $\mu(\mathbf{G}) = 102.8$, (d) $\mu(\mathbf{G}) = 10.8$.

9.3. Maximum value of $\mu(\mathbf{P_{22}}(i\omega))$ is approximately 12.5 occurring at about $\omega = 100$ rad/s.

Index

Terms in italic refer to MATLAB commands